RHEOLOGICAL ASPECTS OF SOIL MECHANICS

Rheological Aspects of Soil Mechanics

L. Šuklje

Professor of Soil Mechanics,
University of Ljubljana, Yugoslavia

WILEY-INTERSCIENCE
A division of John Wiley & Sons Ltd
London New York Sydney Toronto

Library of Congress catalog card number 71–81751

SBN 471 83550 1

Made in Great Britain at the Pitman Press, Bath

Preface

In classical soil mechanics the stress states in soil are supposed to be governed by the equilibrium conditions combined either with a linear stress-strain relationship or with the Coulomb-Mohr failure criterion. Displacements of soil due to loading by structures are commonly computed according to the first of the above combinations, and physical parameters appearing in the computation of displacements are deduced from confined drained and unconfined undrained tests. The stability of slopes, retaining structures and foundations is analysed according to the limit equilibrium by using strength parameters as obtained from direct shear or triaxial tests. Terzaghi's one-dimensional consolidation theory is used when predicting the development of displacements; it is based on linear stress-strain relationship and on Darcy's seepage law.

Within the last three decades many attempts have been made to improve the analyses. The studies of stresses and displacements in the elastic half-space have been enlarged and applied to layered and anisotropic soils. The testing techniques have been perfected, pore-pressure effects further investigated and non-linear, anisotropic, viscous stress-strain relations established. The theory of consolidation has been generalized for three-dimensional seepage, for partial saturation and for complex stress-strain-time relationships. Specified solutions of the consolidation process have been made and simplified procedures proposed for the improved practical settlement computation. Various methods of numerical integration of the fundamental equations of the theory of plasticity have been elaborated and applied to stability, bearing capacity and earth pressure problems. Kinematic conditions and viscous effects on earth pressures have been investigated and long-term strength decreases analysed. Many further studies of soil properties, practical geotechnical problems, observations in nature and case records have been published.

In this book the fundamentals of classical soil mechanics are reproduced, and new concepts are presented and discussed. Rheological aspects governed the choice of the matters as well as the discussion of theories and methods in which the viscous soil properties do not appear.

Existing studies concerning the viscous properties of soils can be divided into two groups. In the first one certain rheological schemes are defined in an intuitive way, and the corresponding relationships between stresses, strains, seepage paths and time are deduced and compared with experimental observations. This comparison controls the applicability of the assumed rheological models. In the second group, the stress-strain-time relations are derived from experimentally established consolidation curves and presented in a way allowing the computation of the consolidation at various stress states. Both groups of studies have been presented in this book and their applicability or the limits of their applicability have been analysed and discussed.

The use of electronic computers largely facilitates the application of theories, based on complex rheological relationships, to various boundary conditions. Nevertheless there are still limitations in applying the theories to the solution of practical geotechnical problems. The difficulties in programming the solutions governed by non-linear relationships containing several parameters, as well as by complicated boundary conditions, are not the only reason for simplifying the problems. The determination of numerous physical parameters represents a cumbersome and delicate task even for a homogeneous soil. The appearance of several sets of such parameters in solving the displacements of heterogeneous and anisotropic soil bodies represents experimental and computational difficulties which cannot be accepted in engineering design. Thus, simplifying assumptions have to be made in defining the deformability of soils and in setting the boundary conditions. The ingenuity of Terzaghi's classical concept of soil mechanics was in isolating the main parameters influencing soil behaviour. When improving the classical concepts by considering new experimental and analytical investigations and computation means, special care must be taken in finding out the essential factors affecting the instantaneous and long-term deformability of large layers.

The variety of problems requires different simplifying assumptions in different cases. The purpose of presenting various theoretical concepts and analyses of the rheological behaviour of soils in this book is to clarify the applicability of various methods and the influence of various rheological factors, and to facilitate the choice of appropriate simplifications. The presentation and discussion of the subjects have been based on various typical rheological analyses. The review of existing studies has been completed by a list of further publications treating the soil as a rheological medium.

The author is deeply indebted to Dr. R. Radok, Professor of Applied Mathematics at the Flinders University of South Australia who encouraged him to write on this subject.

L. ŠUKLJE

Contents

Part A

Fundamentals given by the Mechanics of Continua

Part B

Experimental and Theoretical Fundamentals of One-Dimensional Consolidation at Zero Lateral Strain

Part C

Experimental and Theoretical Fundamentals of the Three-dimensional Consolidation Analysis

Part D

APPLICATION OF THE CONSOLIDATION ANALYSIS IN THE HALF-SPACE AND IN EARTH-CONSTRUCTIONS

Part E

LONG-TERM STABILITY OF SLOPES

21. Tensile and Bending Strength of Soils

Part F

Creep and Relaxation Effects on Earth Pressures

22. Kinematic Conditions of the Magnitude of the Earth Pressure Against a Retaining Structure

Part A

FUNDAMENTALS GIVEN BY THE MECHANICS OF CONTINUA

1

Introduction

In elementary physics, solid, liquid and gaseous states are distinguished. Soils, however, are not simple bodies which can be placed in either of these three groups. Generally they are composed of solid, liquid and gas. The solid part of soils is a porous medium. It is composed of numerous grains. The size and the form of the grains are governed by their mineralogical composition as well as by the mechanical effects to which they have been subjected. The grains of some minerals, especially those of quartz, limestone and feldspar, are developed rather uniformly in all three dimensions; their shape is either rounded or irregular and their size is obviously large enough to permit gravity to predominate over molecular surface forces during sedimentation. The clay minerals, on the other hand, being chiefly hydrous aluminosilicates, are characterized by scaly or tabular or rod-shaped particles and by small size. By a mechanism which is not fully understood, the surface of clayey microcrystals adsorbs a thin film of water molecules. It may be due to unsatisfied charges in the surface layers having a small binding action on water dipoles, which are thereby attracted and loosely held. The films of adsorbed water govern the physical properties of clays.

The coarser grains settle out under the prevailing influence of gravitational forces in a rather dense single-grained structure (Fig. 1.1 (*a*)).

The fine, scaly particles of the clay minerals settle out under the prevailing influence of molecular surface forces and bind into a honeycomb structure of greater porosity (Fig. 1.1 (*b*)). If the particles aggregate into flocks when settling to the bottom of the suspension, a sediment with a flocculated structure is formed.

The voids between single particles are filled with liquid (water and aqueous solutions) and with gas (air). If the voids are filled either with liquid (saturated soils) or with gas only (dry soils), the soil is called a two-phase system. If they are filled simultaneously with water and air (partly saturated soils), the soil is called a three-phase system.

The two-phase and the three-phase systems of soils resist the effects of external forces in a different manner from the simple solid continua. The factor governing the distribution of stresses in non-cohesive soils of single-grained structure and their resistance to the effects of external forces, is the friction between the contact surfaces of the grains. In cohesive soils, however, composed predominantly of clay minerals, the strength of the films of adsorbed water surrounding the grains accounts for the resistance of the soil to deformation.

In the mechanics of continua creep is defined as the slow development of deformation under a constant stress state. Creep occurs at a rate which either remains constant or varies with time. In soils every additional loading causes creep of the solid soil skeleton, although the intensity of creep and the time to eventual stabilization are quite different. The creep deformation can be stopped by releasing the applied stresses. This release of stresses is called relaxation.

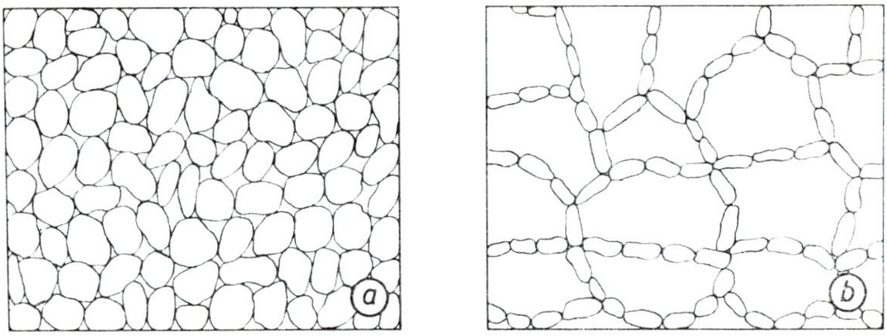

Fig. 1.1. Grain structures of soils: (a) single-grain structure; (b) honeycomb structure

The creep and the relaxation phenomena of soils are the main subject of this book.

At the contact surfaces the loads are transferred to the grains of the solid skeleton. Under the effects of stresses the grains become deformed and, due to stresses acting on the contact surfaces, move into new, more stable, positions. In relation to the final displacement of the soil surface, the volume deformations of the grains themselves are not important and can usually be neglected. In saturated two-phase systems the displacements of the grains into a denser state with smaller porosity require the squeezing out of the superfluous pore water and in partly saturated three-phase systems, the initial compression of the air bubbles as well as the subsequent squeezing out of both superfluous pore water and air. The friction between the solid skeleton and the pore water or air respectively restrains the squeezing out of the water and the air. This new factor affecting the speed of deformation cannot be treated separately from the effect of the creep of the solid soil skeleton.

In the phase of mobilization of friction, the grains of non-cohesive soils move more or less irregularly, although their movements are governed by the boundary conditions and by the actual state of stress. When mobile, the grains pack into either a denser or looser state. The statistical average values of such movements can hardly be considered to be analytical functions of position. For this reason the relationships deduced for continua can only be applied to deformations of non-cohesive soils with certain limitations. So long as the deformations are continuous, cohesive soils conform more closely to the deformation conditions of continua.

Despite these limitations the solutions obtained in mechanics of continua are largely used in strain and stress computations of soils. Thus the basic elements of the mechanics of continua will be given in the next chapter of this introductory part and some special applications will be presented in order to prepare the basis for the treatment of the problems dealt with in the other parts of the book.

2

Stress Tensor

2.1 STRESS STATE IN SPACE

2.1.1 General Stress Relationship

The stress tensor is defined by the matrix:

$$p_{rs} = \begin{bmatrix} p_{xx} & p_{xy} & p_{xz} \\ p_{yx} & p_{yy} & p_{yz} \\ p_{zx} & p_{zy} & p_{zz} \end{bmatrix} \tag{2.1}$$

Each stress component in this matrix has two subscripts. The first one indicates the direction of the normal of the stressed plane element, the second indicates the stress direction. The stress components with equal subscripts are normal stresses, the components with different subscripts are tangential (shear) stresses. An arbitrary subscript will be denoted by either α or β ($\alpha = x, y, z$; $\beta = x, y, z$). Tension stresses are considered positive.

Approximating the difference between the stress components acting in two parallel plane elements of an elementary orthogonal prism by

$$\frac{\partial p_{\alpha\beta}}{\partial \alpha}\, d\alpha$$

and considering also the volume forces and forces due to inertia, the equilibrium conditions for forces give three equations of the type:

$$\frac{\partial p_{xx}}{\partial x} + \frac{\partial p_{yx}}{\partial y} + \frac{\partial p_{zx}}{\partial z} + X = \rho x'' \qquad (x,y,z) \tag{2.2}$$

where X, Y, Z are the components of volume forces per unit of volume, ρ the mass per unit of volume and x'', y'', z'' the acceleration components. The equilibrium conditions for moments yield the solution:

$$p_{\alpha\beta} = p_{\beta\alpha} \tag{2.3}$$

Thus the tensor p_{rs} is symmetrical:

$$p_{rs} = \begin{bmatrix} p_{xx} & p_{xy} & p_{xz} \\ \cdot & p_{yy} & p_{yz} \\ \cdot & \cdot & p_{zz} \end{bmatrix} \tag{2.1-a}$$

Three boundary areas of the elementary tetrahedron as presented in Fig. 2.1 lie in the planes of the orthogonal coordinate system (x,y,z). The

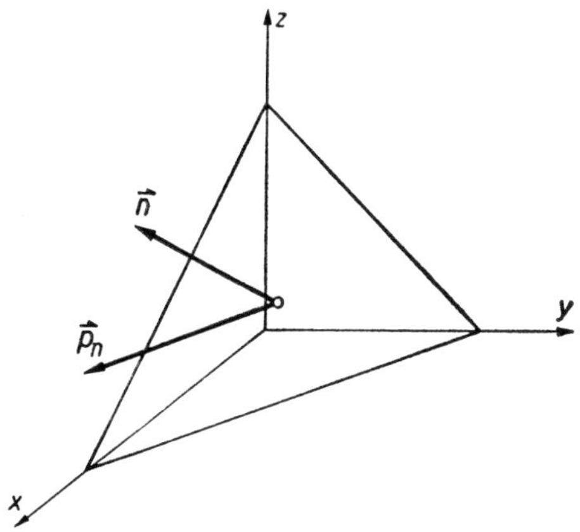

Fig. 2.1. Elementary tetrahedron whose oblique surface with the normal $\vec{n}$ is subjected to the stress vector $\vec{p}_n$

direction of the fourth elementary plane is given by the unit vector of the normal n with direction cosines (n,α). The volume forces of an infinitesimal element can be neglected when added to the surface forces. Thus, using the equilibrium conditions, the components of the stress vector p_n acting on the elementary plane of the normal n, are given by three equations of the following type:

$$p_{nx} = p_{xx}\,(n,x) + p_{yx}\,(n,y) + p_{zx}\,(n,z) \qquad (x,y,z) \tag{2.4}$$

Assuming the existence of directions $\vec{n}$ with zero tangential stresses, the components p_{nx} can also be expressed by the equation:

$$p_{nx} = p_n(n,x) \qquad (x,y,z) \tag{2.5}$$

where p_n denotes the magnitude of the stress vector $\vec{p}_n \equiv \vec{p}_{nn}$. Inserting (2.5) into (2.4), three homogeneous linear equations with direction cosines as unknown quantities are obtained. The condition that the determinant of these equations must be zero, leads to the equation:

$$p_n^3 - I_1 p_n^2 + I_2 p_n - I_3 = 0 \tag{2.6}$$

where:

$$I_1 = p_{xx} + p_{yy} + p_{zz} \tag{2.7}$$

$$I_2 = p_{xx}p_{yy} + p_{yy}p_{zz} + p_{zz}p_{xx} - p_{xy}^2 - p_{yz}^2 - p_{zx}^2 \tag{2.8}$$

$$I_3 = p_{xx}\,p_{yy}\,p_{zz} + 2p_{xy}\,p_{yz}\,p_{zx} - \\ - p_{xx}p_{yz}^2 - p_{yy}\,p_{zx}^2 - p_{zz}\,p_{xy}^2 \tag{2.9}$$

Equation (2.6) gives three real values of the so-called principal stresses p_{ii}, p_{jj}, p_{kk}. Using the additional condition

$$(n,x)^2 + (n,y)^2 + (n,z)^2 = 1 \tag{2.10}$$

the directions i,j,k can be determined. It can also be proved that they are orthogonal to each other. Further it can be verified that the principal stresses represent the extremum values of the normal stresses.

Thus the stress tensor can be presented also in the form:

$$p_{rs} = \begin{bmatrix} p_{ii} & 0 & 0 \\ 0 & p_{jj} & 0 \\ 0 & 0 & p_{kk} \end{bmatrix} \tag{2.11}$$

The principal stresses do not depend on the orientation of the system of coordinates (x,y,z) which govern the stresses appearing in the coefficients I_1, I_2 and I_3 of Equation (2.6). Thus these coefficients expressed by Equations (2.7), (2.8) and (2.9) are the invariants of the stress tensor. Their expressions in the terms of principal stresses are:

$$I_1 = p_{ii} + p_{jj} + p_{kk} \tag{2.12}$$

$$I_2 = p_{ii}\,p_{jj} + p_{jj}p_{kk} + p_{kk}\,p_{ii} \tag{2.13}$$

$$I_3 = p_{ii}\,p_{jj}\,p_{kk} \tag{2.14}$$

Taking the system of coordinates (x,y,z) in the principal stress directions (i,j,k), Equation (2.4) becomes:

$$p_{ni} = p_{ii}(n,i) \qquad (i,j,k) \tag{2.15}$$

Now, let us choose a new system of coordinates (ξ,η,ζ) so that the axis ζ has the direction of the normal $\vec{n}$, while the axes ξ and η lie in the plane with the normal $\vec{n}$, where they are arbitrarily orientated. The normal stress component $p_{\zeta\zeta}$ can then be expressed in the form

$$p_{\zeta\zeta} = p_{ni}(n,i) + p_{nj}(n,j) + p_{nk}(n,k)$$

Inserting the expressions (2.15), we have

$$p_{\zeta\zeta} = p_{ii}(n,i)^2 + p_{jj}(n,j)^2 + p_{kk}(n,k)^2 \tag{2.16}$$

The tangential stress component $p_{\zeta\xi}$ or $p_{\zeta\eta}$ can then be formed in a similar way:

$$p_{\zeta\xi} = p_{ni}(\xi,i) + p_{nj}(\xi,j) + p_{nk}(\xi,k) \qquad (\xi,\eta)$$

Inserting the expressions (2.15), we obtain:

$$p_{\zeta\xi} = p_{ii}(n,i)(\xi,i) + p_{jj}(n,j)(\xi,j) + p_{kk}(n,k)(\xi,k) \qquad (\xi,\eta) \tag{2.17}$$

2.1.2 Octahedral Stresses

The cosines of the angles between the normal to the area of an elementary octahedron (Fig. 2.2) and the principal stress axes have the value

$$(n,i) = (n,j) = (n,k) = \frac{1}{\sqrt{3}}. \tag{2.18}$$

For the normal stress belonging to the octahedral surface Equation (2.16) reduces to:

$$\sigma_{oct} = \frac{1}{3}(p_{ii} + p_{jj} + p_{kk}) = \frac{I_1}{3} \tag{2.19}$$

The components of the total stress vector obtained by Equation (2.4) give the total stress value p_{oct}:

$$p_{oct} = \sqrt{\left(\frac{p_{ii}^2 + p_{jj}^2 + p_{kk}^2}{3}\right)} \tag{2.20}$$

The tangential octahedral stress can then be found by the expression

$$\tau_{oct} = \sqrt{(p_{oct}^2 - \sigma_{oct}^2)}$$

giving

$$\tau_{oct} = \tfrac{1}{3}\sqrt{[(p_{ii} - p_{jj})^2 + (p_{jj} - p_{kk})^2 + (p_{kk} - p_{ii})^2]} =$$
$$= \sqrt{\left(\frac{2}{3}\right)}\sqrt{\left(\frac{I_1^2}{3} - I_2\right)} \tag{2.21}$$

Equations (2.19) and (2.21) show that the octahedral stresses are invariant quantities.

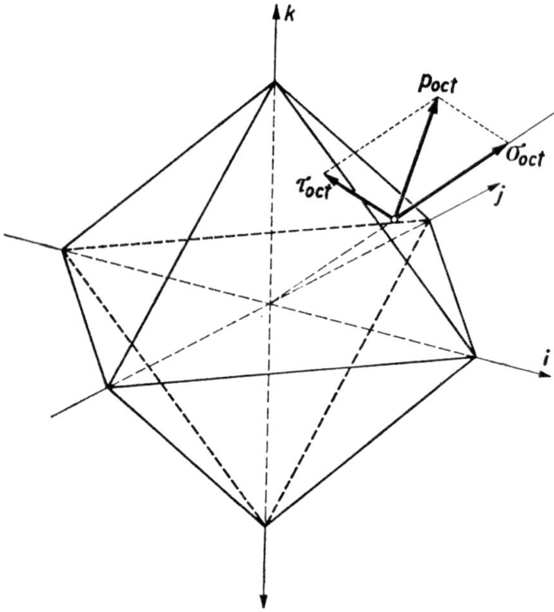

Fig. 2.2. Elementary octahedron with the vectors of octahedral stresses

In order to simplify the writing the following substitutions may be introduced:

$$p_{ii} \equiv \sigma_1, \qquad p_{jj} \equiv \sigma_2, \qquad p_{kk} \equiv \sigma_3 \tag{2.22}$$

So the octahedral normal and tangential stress can be expressed as follows:

$$\sigma_{\text{oct}} = \tfrac{1}{3} (\sigma_1 + \sigma_2 + \sigma_3) \tag{2.19-a}$$

$$\tau_{\text{oct}} = \tfrac{1}{3} \sqrt{[(\sigma_1 - \sigma_2)^2 + (\sigma_2 - \sigma_3)^2 + (\sigma_3 - \sigma_1)^2]} \tag{2.21-a}$$

Any stress tensor can be divided into a spherical (isotropic) component $p_{rs}{}^0$ and into a distortional (deviatoric) component $p_{rs}{}^d$:

$$p_{rs} = p_{rs}{}^0 + p_{rs}{}^d \tag{2.23}$$

The spherical component is

$$p_{rs}{}^0 = \sigma_{\text{oct}}\, \delta_{rs} \tag{2.24}$$

where δ_{rs} signifies Kronecker's delta:

$$\delta_{rs} = \begin{bmatrix} 1 & 0 & 0 \\ 0 & 1 & 0 \\ 0 & 0 & 1 \end{bmatrix} \tag{2.25}$$

The distortional component of the stress tensor is the following tensor:

$$p_{rs}{}^d = \begin{bmatrix} \sigma_1 - \sigma_{oct} & 0 & 0 \\ 0 & \sigma_2 - \sigma_{oct} & 0 \\ 0 & 0 & \sigma_3 - \sigma_{oct} \end{bmatrix} \tag{2.26}$$

It can be seen that the octahedral stresses of the spherical stress tensor are:

$$\sigma_{oct}{}^0 = \sigma_{oct}$$
$$\tau_{oct}{}^0 = 0 \tag{2.27}$$

and those of the distortional stress tensor are:

$$\sigma_{oct}{}^d = 0$$
$$\tau_{oct}{}^d = \tau_{oct} \tag{2.28}$$

In the orthogonal system of coordinates (σ_1, σ_2, σ_3) any stress state (stress tensor) can be presented by a point P (σ_1, σ_2, σ_3). The points on the octahedral normal (n), the direction cosines of which are

$$(n,i) = \cos \alpha_i = \frac{1}{\sqrt{3}}, \tag{2.29}$$

correspond to spherical stress states:

$$\sigma_1 = \sigma_2 = \sigma_3 = \frac{\sigma_{oct}}{3} \tag{2.30}$$

Straight lines in the coordinate system (σ_1, σ_2, σ_3) which are parallel to the straight line n, have the equation

$$\sigma_1 - C_1 = \sigma_2 - C_2 = \sigma_3 - C_3 \tag{2.31}$$

with C_1, C_2, C_3 as arbitrary constants. The points lying on such straight lines represent stress states with equal distortional components:

$$p_{rs}{}^d = \tfrac{1}{3} \begin{bmatrix} 2\,C_1 - C_2 - C_3 & 0 & 0 \\ 0 & 2\,C_2 - C_3 - C_1 & 0 \\ 0 & 0 & 2\,C_3 - C_1 - C_2 \end{bmatrix} \tag{2.32}$$

The plane Π passing through the origin of coordinates and perpendicular to the above defined line n has the equation

$$\sigma_1 + \sigma_2 + \sigma_3 = 0 \tag{2.33}$$

and all points on this plane represent pure distortional stress states with a zero spherical stress tensor.

Planes parallel to Π have the equation

$$\sigma_1 + \sigma_2 + \sigma_3 = C \tag{2.34}$$

where C is an arbitrary constant. All points on such a plane correspond to the states of stress with the same octahedral normal stress

$$\sigma_{\text{oct}} = \frac{C}{3} \tag{2.35}$$

Now let us examine Fig. 2.3. If a line n' is placed parallel to line n through the point P representing a stress tensor p_{rs}, the point of intersection P' of n' with the plane Π represents $p_{rs}{}^{d}$, the distortional part of the

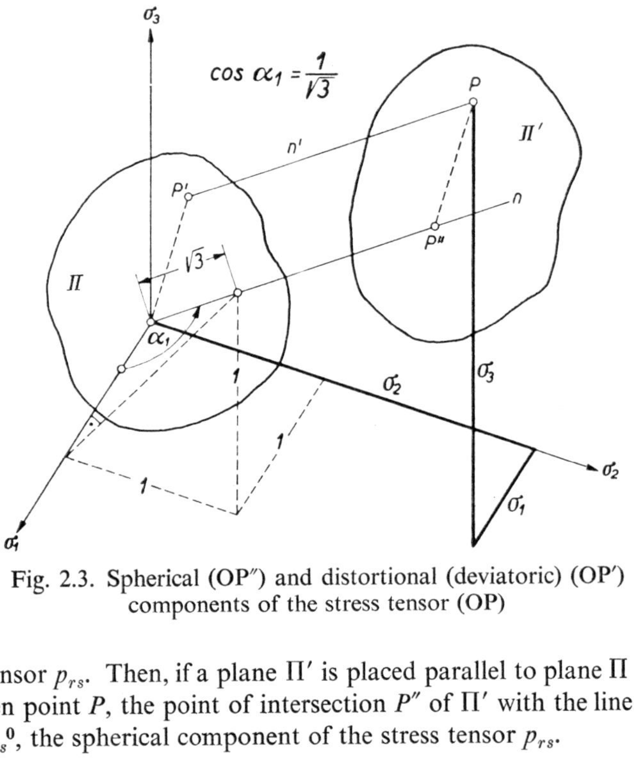

Fig. 2.3. Spherical (OP″) and distortional (deviatoric) (OP′) components of the stress tensor (OP)

stress tensor p_{rs}. Then, if a plane Π' is placed parallel to plane Π through the given point P, the point of intersection P'' of Π' with the line n represents $p_{rs}{}^{0}$, the spherical component of the stress tensor p_{rs}.

2.2 STATE OF PLANE STRESS

In the state of plane stress the stresses in one of the principal stress directions are zero ($p_{kk} = 0$). When analysing the stress relations the elementary tetrahedron can be replaced by an elementary prism with a cross-section as presented in Fig. 2.4.

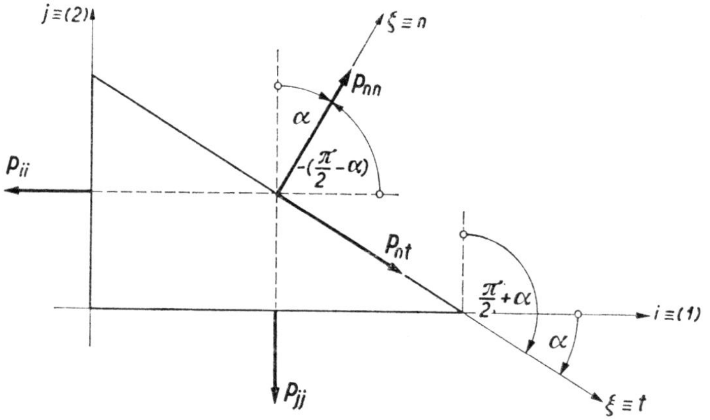

Fig. 2.4. State of plane stress on an elementary prism

Taking $\xi \equiv t$, $\zeta \equiv n$ and denoting (see Fig. 2.4) $(\xi, i) = (n,j) = \cos \alpha$, $(n,i) = -(\xi, j) = \sin \alpha$ Equations (2.16) and (2.17) reduce to:

$$p_{nn} = p_{ii} \sin^2 \alpha + p_{jj} \cos^2 \alpha \tag{2.36}$$

$$p_{nt} = (p_{ii} - p_{jj}) \sin \alpha \cos \alpha \tag{2.37}$$

Substituting

$$p_{nn} \equiv \sigma, \quad p_{nt} \equiv \tau, \quad p_{ii} \equiv \sigma_1, \quad p_{jj} \equiv \sigma_2 \tag{2.38}$$

Equations (2.36) and (2.37) can be written as follows:

$$\sigma = \frac{\sigma_1 + \sigma_2}{2} - \frac{\sigma_1 + \sigma_2}{2} \cos 2\alpha \tag{2.39}$$

$$\tau = \frac{\sigma_1 - \sigma_2}{2} \sin 2\alpha \tag{2.40}$$

These two equations are the basis of the construction of Mohr's circle of stress (Fig. 2.5). In the orthogonal plane system of coordinates (σ, τ) the tangential stresses are plotted versus normal stresses σ. The principal stresses σ_1 and σ_2 are represented by the points A and B on the σ-axis for the shear stresses vanish in the principal directions. The Mohr's circle passing through the points A and B has its centre on the axis σ at a distance $(\sigma_1 + \sigma_2)/2$ from the origin, the radius of the circle being $(\sigma_1 - \sigma_2)/2$.

The straight line parallel to the direction (1) of the principal stress σ_1 and passing through the point B intersects the stress circle at the point P, called the pole of the stress circle. The direction (2) passes through the points P and A, A being the end point of the stress vector $\vec{\sigma}_1$.

The stress vector $\vec{p}$, corresponding to an arbitrary direction t with the normal $\vec{n}$ (Figs. 2.4 and 2.5), can then be obtained in the following way. The direction t drawn through the pole P intersects the Mohr's circle at the point D. Its mirror image about the axis σ is the point E. The stress vector $\vec{p}$ is determined by the line connecting the origin 0 and the point E if it is assumed that the orientation of the positive (outside) normal $\vec{n}$ corresponding to the plane element in question, coincides with the σ-axis. (In Fig. 2.5 the inside of the plane element is shaded.) The stress vector p

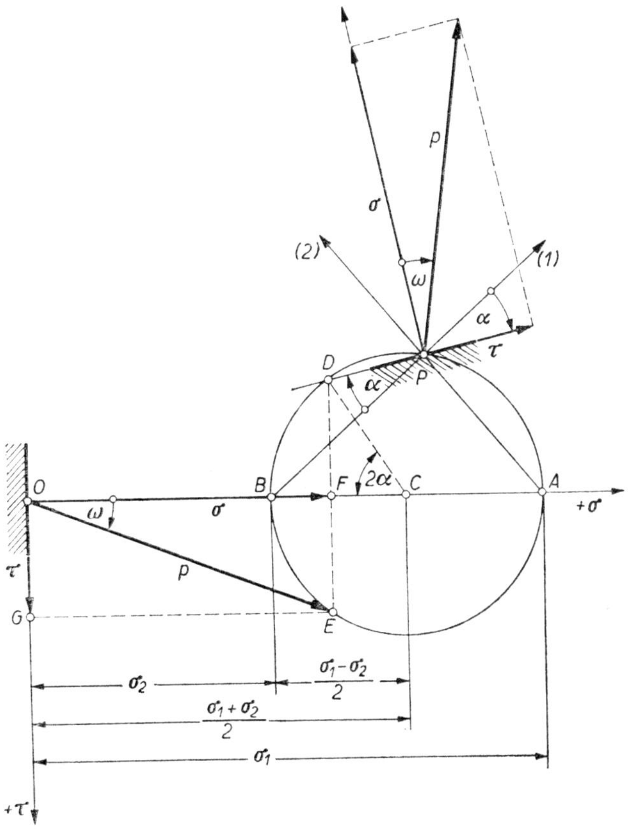

Fig. 2.5. Mohr's circle of stress

can be divided into normal stress and shear stress components in the directions of both axes of the system of coordinates. Figure 2.5 immediately proves the correctness of such a construction satisfying Equations (2.39) and (2.40).

In Fig. 2.5 the stress vector p with the components σ and τ is transposed into the pole P respecting the true orientation of the direction t of the elementary plane. The angle which the stress vector p makes with the normal vector n, is denoted by ω.

The construction explained above holds also in cases when either one or both principal stresses are negative (compressive stresses) and when the angle α is either greater than $\pi/2$ or negative.

Remark: In the preceding text the tension stresses have been taken as positive normal stresses—as is common in the general theory of elasticity. Now, in soil mechanics, the compression stresses are generally considered as positive quantities. In this case Equations (2.39) and (2.40) maintain their validity with unchanged signs if the positive tangential stresses in Fig. 2.4 are taken in the inverse direction also; that is in the opposite direction of the positive direction of the t axis. Neither does the construction of the Mohr's circle, as presented in Fig. 2.5, change, except that the right part of the axis τ at the origin 0 has to be shaded.

3

Strain Tensor

3.1 STATE OF STRAIN IN SPACE

3.1.1 Strain Relationships

Consider an elementary tetrahedron $ABCD$ (Fig. 3.1). Under the influence of applied stresses its joints undergo displacements the vectors of which are denoted by $\vec{u}$ (u_x, u_y, u_z).

Figure 3.2 shows the projection of the tetrahedron on the plane (x,y) in the positions before and after deformation. Using the symbols of Figs. 3.1 and 3.2 and assuming small differences between displacements of single joints, the linear strain in the direction x is defined by the equation

$$\varepsilon_{xx} = \frac{\mathrm{d}x' - \mathrm{d}x}{\mathrm{d}x} = \frac{u_{Bx} - u_{Ax}}{\mathrm{d}x} = \frac{\partial u_x}{\partial x} \tag{3.1}$$

Similar expressions define the linear strains in the directions y and z.

The tangential (shear) strain γ_{xy} in the plane (x,y) is defined as the change in the angle (originally a right angle) between the x and y directions of the sides of the elementary tetrahedron; thus, with the notations of Fig. 3.2, the expression for γ_{xy} is:

$$\gamma_{xy} = \vartheta_1 + \vartheta_2 \cong \tan\vartheta_1 + \tan\vartheta_2 = \frac{u_{By} - u_{Ay}}{\mathrm{d}x'} + \frac{u_{Cx} - u_{Ax}}{\mathrm{d}y'}$$

$$\gamma_{xy} = \frac{\partial u_y}{\partial x} + \frac{\partial u_x}{\partial y} \tag{3.2}$$

Analogously we can write the expressions for the shear strains γ_{yz} and γ_{zx} in the planes (y,z) and (z,x).

The strain tensor e_{rs} is defined by the symmetrical matrix:

$$e_{rs} = \begin{bmatrix} e_{xx} & e_{xy} & e_{xz} \\ \cdot & e_{yy} & e_{yz} \\ \cdot & \cdot & e_{zz} \end{bmatrix} \tag{3.3}$$

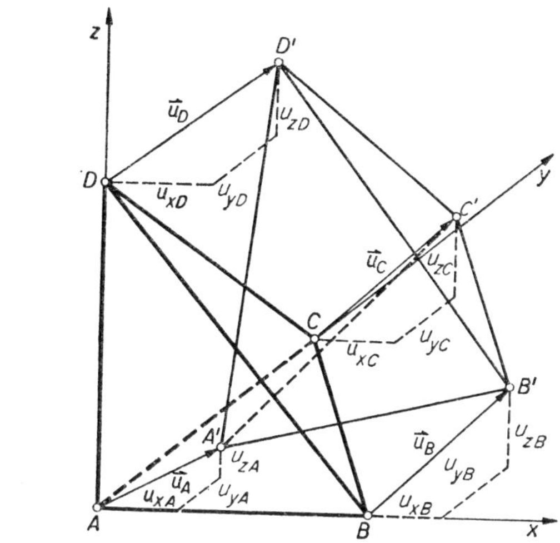

Fig. 3.1. Displacements of the joints of an elementary tetrahedron

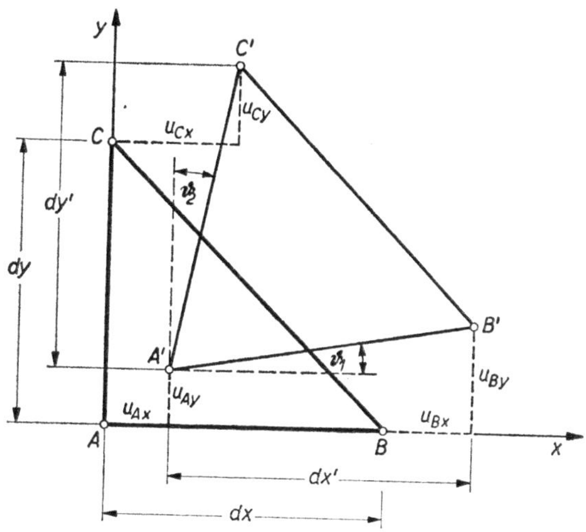

Fig. 3.2. Projection of the elementary tetrahedron in the (x,y) plane

the components of which are given by the expression:

$$e_{\alpha\beta} = \frac{1}{2}\left(\frac{\partial u_\alpha}{\partial \beta} + \frac{\partial u_\beta}{\partial \alpha}\right) \tag{3.4}$$

$$(\alpha = x, y, z; \;\; \beta = x, y, z)$$

Comparing expressions (3.1) and (3.2) we can see that the coefficients of the matrix (3.3) represent linear strains for equal subscripts, and half values of tangential strains for different subscripts:

$$e_{xx} = \varepsilon_{xx} = \frac{\partial u_x}{\partial x} \qquad (x,y,z) \tag{3.5}$$

$$e_{xy} = \frac{1}{2}\gamma_{xy} = \frac{1}{2}\left(\frac{\partial u_y}{\partial x} + \frac{\partial u_x}{\partial y}\right) \qquad (x,y,z) \tag{3.6}$$

As the strains are assumed to be very small, the direction cosines of the line elements after deformation can be considered equal to the direction cosines before deformation. Thus Equation (2.4) expressing the stress component p_n in terms of the stress components p, can analogously be applied for the strain tensor. Considering the deductions from Equation (2.4) the existence of three orthogonal principal directions (i,j,k) can then be concluded for the strain tensor e_{rs} as well. The shear strains $e_{\alpha\beta}$ $(\alpha \neq \beta)$ in the planes of principal directions are zero. The matrix of the strain tensor can then be expressed as follows:

$$e_{rs} = \begin{bmatrix} e_{ii} & 0 & 0 \\ 0 & e_{jj} & 0 \\ 0 & 0 & e_{kk} \end{bmatrix} \tag{3.7}$$

The invariants of the strain tensor are:

$$J_1 = e_{xx} + e_{yy} + e_{zz} = e_{ii} + e_{jj} + e_{kk} \tag{3.8}$$

$$J_2 = e_{xx}e_{yy} + e_{yy}e_{zz} + e_{zz}e_{xx} - (e_{xy}^2 + e_{yz}^2 + e_{zx}^2) =$$

$$= e_{ii}\,e_{jj} + e_{jj}\,e_{kk} + e_{kk}\,e_{ii} \tag{3.9}$$

$$J_3 = e_{xx}e_{yy}e_{zz} + e_{xy}e_{yz}e_{zx} - (e_{xx}\,e_{yz}^2 +$$

$$+ e_{yy}e_{zx}^2 + e_{zz}e_{xy}^2) = e_{ii}e_{jj}e_{kk} \tag{3.10}$$

Differentiating the displacement vectors $\vec{u}(u_x, u_y, u_z)$ by the time t we get the vectors of strain velocities $= \vec{v}(v_x, v_y, v_z)$ the components of which are:

$$v_\alpha = \frac{\partial u_\alpha}{\partial t} \qquad (\alpha = x, y, z) \tag{3.11}$$

The tensor of strain velocities is then defined by the matrix:

$$\dot{e}_{rs} = \begin{bmatrix} \dot{e}_{xx} & \dot{e}_{xy} & \dot{e}_{xz} \\ \cdot & \dot{e}_{yy} & \dot{e}_{yz} \\ \cdot & \cdot & \dot{e}_{zz} \end{bmatrix} \tag{3.12}$$

with the coefficients

$$\dot{e}_{\alpha\beta} = \frac{1}{2}\left(\frac{\partial v_\beta}{\partial \alpha} + \frac{\partial v_\alpha}{\partial \beta}\right) \tag{3.13}$$

The equations giving the strains in the elementary planes with the direction of the normal can be obtained by analogy with Equations (2.16) and (2.17).

$$e_{\zeta\zeta} = e_{ii}(\zeta,i)^2 + e_{jj}(\zeta,j)^2 + e_{kk}(\zeta,k)^2 \tag{3.14}$$

$$e_{\zeta\xi} = e_{ii}(\zeta,i)(\xi,i) + e_{jj}(\zeta,j)(\xi,j) + e_{kk}(\zeta,k)(\xi,k) \tag{3.15}$$

$$(\xi,\eta,\zeta)$$

Consider an elementary parallelepiped $dx\,dy\,dz$ the sides of which coincide with the principal strain directions (i,j,k). The volume strain e_v is defined by the quotient of the change of volume and the original volume of such a parallelepiped:

$$e_v = \frac{dx\,(1 + e_{ii})\,dy\,(1 + e_{jj})\,dz\,(1 + e_{kk}) - dx\,dy\,dz}{dx\,dy\,dz} \tag{3.16}$$

As small strains are assumed, the above equation reduces to

$$e_v = e_{ii} + e_{jj} + e_{kk} = J_1 \tag{3.17}$$

3.1.2 Octahedral Strains

Any strain tensor e_{rs} can be separated into a spherical (isotropic) component $e_{rs}{}^0$ and a distortional (deviatoric) component $e_{rs}{}^d$:

$$e_{rs} = e_{rs}{}^0 + e_{rs}{}^d. \tag{3.18}$$

$$e_{rs}{}^0 = e_{\text{oct}}\,\delta_{rs} = \frac{e_v}{3}\begin{bmatrix} 1 & 0 & 0 \\ 0 & 1 & 0 \\ 0 & 0 & 1 \end{bmatrix} \tag{3.19}$$

$$e_{rs}{}^d = \begin{bmatrix} e_{xx} - e_{\text{oct}} & e_{xy} & e_{xz} \\ \cdot & e_{yy} - e_{\text{oct}} & e_{yz} \\ \cdot & \cdot & e_{zz} - e_{\text{oct}} \end{bmatrix} \tag{3.20}$$

Substituting

$$e_{ii} \equiv e_1; \quad e_{jj} \equiv e_2; \quad e_{kk} \equiv e_3$$

and by analogy with the developments in section 2.1.2 the octahedral linear strain e_{oct} and the octahedral tangential (shear) strain γ_{oct} can be written as follows:

$$e_{oct} = \frac{e_1 + e_2 + e_3}{3} = \frac{e_v}{3} = \frac{J_1}{3} \tag{3.21}$$

$$\gamma_{oct} = \tfrac{2}{3}\sqrt{[(e_1 - e_2)^2 + (e_2 - e_3)^2 + (e_3 - e_1)^2]} = $$
$$= 2\sqrt{\left(\frac{2}{3}\right)}\sqrt{\left(\frac{J_1^2}{3} - J_2\right)} \tag{3.22}$$

Applying the above expressions to the spherical and distortional parts of the strain tensor, the following relations can be stated:

$$\left.\begin{aligned} e_{oct}^0 &= e_{oct} \\ \gamma_{oct}^0 &= 0 \end{aligned}\right\} \tag{3.23}$$

$$\left.\begin{aligned} e_{oct}^d &= 0 \\ \gamma_{oct}^d &= \gamma_{oct} \end{aligned}\right\} \tag{3.24}$$

3.2 STATE OF PLANE STRAIN

In the state of plane strain one of the principal strains vanishes ($e_{kk} \equiv e_3 = 0$). Figure 3.2 can then be interpreted as the cross-section of a prism perpendicular to its axis $k \equiv (3)$.

Based on the analogy with the stress tensor the linear strain $e_{\zeta\zeta} \equiv \varepsilon$ in an arbitrary direction (Fig. 3.3) of the (i,j) plane and the corresponding

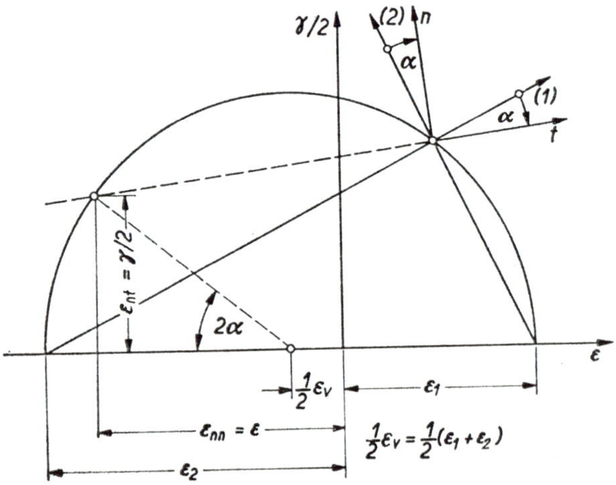

Fig. 3.3. Mohr's circle for the state of plane strain

half value of tangential strain $e_{\zeta\xi} = \gamma/2$ can be expressed in terms of principal strains by the equations:

$$\varepsilon = \frac{\varepsilon_1 + \varepsilon_2}{2} - \frac{\varepsilon_1 - \varepsilon_2}{2} \cos 2\alpha \tag{3.25}$$

$$\frac{\gamma}{2} = \frac{\varepsilon_1 - \varepsilon_2}{2} \sin 2\alpha \tag{3.26}$$

The above equations give rise to the construction of Mohr's circle for the state of plane strain (Fig. 3.3). The distance of the centre of the circle from the origin represents one-half of the volume strain $\varepsilon_v = \varepsilon_1 + \varepsilon_2$. Thus the circle with diameter $(\varepsilon_1 - \varepsilon_2)$ and with its centre at the origin represents the distortional component of the strain tensor, the volume strain of which must be zero.

3.3 FINITE DEFORMATIONS

The matrix (3.3) defining the strain tensor e_{rs} can be applied for finite strains as well. The relationships (3.4), (3.14), (3.15) and (3.16), however, do not hold any longer in the case of finite deformations since they have been based on the assumption of very small strains whose products can be neglected with respect to the strains themselves; furthermore small strains do not affect appreciably the directions of line elements. Herein the corresponding relationships for finite strains will be developed only where principal directions of strain maintain their orientation during deformation.

The finite linear strain $e_{\zeta\zeta}$ in an arbitrary direction ζ is defined as the unit elongation of the line element dl in this direction:

$$e_{\zeta\zeta} = \frac{dl' - dl}{dl} \tag{3.27}$$

$$dl' = (1 + e_{\zeta\zeta})\, dl \tag{3.27-a}$$

The components of the deformed line element in the principal directions (1), (2), (3) are:

$$\begin{aligned} dl_1' &= (1 + e_1)\, dl(\zeta, 1) \\ dl_2' &= (1 + e_2)\, dl(\zeta, 2) \\ dl_3' &= (1 + e_3)\, dl(\zeta, 3) \end{aligned} \tag{3.28}$$

The condition

$$dl'^2 = dl_1'^2 + dl_2'^2 + dl_3'^2$$

leads to the expression:

$$e_{\zeta\zeta} = \sqrt{[(1 + e_1)^2(\zeta,1)^2 + (1 + e_2)^2(\zeta,2)^2 + (1 + e_3)^2(\zeta,3)^2]} - 1 \quad (3.29)$$

The direction cosines of the line element dl' after deformation are

$$(\zeta',1) = \frac{1 + e_1}{1 + e_{\zeta\zeta}}(\zeta, 1); \quad (\zeta',2) = \frac{1 + e_2}{1 + e_{\zeta\zeta}}(\zeta,2);$$

$$(\zeta',3) = \frac{1 + e_3}{1 + e_{\zeta\zeta}}(\zeta,3) \quad (3.30)$$

Considering the well known relationship of analytical geometry

$$(\xi',\eta') = (\xi',1)(\eta',1) + (\xi',2)(\eta',2) + (\xi',3)(\eta',3) \quad (3.31)$$

and applying the relation (3.30) for the other two cartesian axes ξ',η', the cosine of the angle ϑ between the axes ξ',η' after deformation can be obtained. The change $\gamma_{\xi\eta}$ in the original right angle between ξ and η is then

$$\gamma_{\xi\eta} = \frac{\pi}{2} - \vartheta, \quad \sin \gamma_{\xi\eta} = \cos \vartheta = (\xi',\eta') \quad (3.32)$$

hence

$$e_{\xi\eta} = \frac{\gamma_{\xi\eta}}{2} = \frac{1}{2} \text{arc sin} \left\{ \frac{1}{(1 + e_{\xi\xi})(1 + e_{\eta\eta})} \cdot \right.$$

$$\left. \cdot [(1 + e_1)^2(\xi,1)(\eta,1) + (1 + e_2)^2(\xi,2)(\eta,2) + (1 + e_3)^2(\xi,3)(\eta,3)] \right\} \quad (3.33)$$

In Equation (3.16) defining the volume strain the products of linear strains can no longer be neglected. Thus the volume strain is

$$e_v = e_1 + e_2 + e_3 + e_1e_2 + e_2e_3 + e_3e_1 + e_1e_2e_3 \quad (3.34)$$

$$e_v = J_1 + J_2 + J_3 \quad (3.34\text{-a})$$

For infinitesimal strains, Equations (3.29), (3.33) and (3.34) reduce to the expressions (3.14), (3.15) and (3.17) respectively.

After Hencky the finite linear strains can be defined also by integrating the infinitesimal strain velocities with respect to time:

$$\bar{e}_{xx} = \int_{t_0}^{t} \dot{e}_{xx}\,\mathrm{d}t = \int_{t_0}^{t} \frac{\partial v_x}{\partial x}\,\mathrm{d}t = \int_{t_0}^{t} \frac{\partial}{\partial x}\left(\frac{\partial u_x}{\partial t}\right)\mathrm{d}t =$$

$$= \int_{t_0}^{t} \frac{\mathrm{d}u_x}{\mathrm{d}x} = \int_{dx}^{dx'} \frac{\mathrm{d}(\mathrm{d}x)}{\mathrm{d}x} = \ln\frac{\mathrm{d}x'}{\mathrm{d}x} = \ln\,(1 + e_{xx}) \qquad (3.35)$$

The corresponding definition of the finite volume strain is

$$\bar{e}_v = \ln\frac{\mathrm{d}V'}{\mathrm{d}V} = \ln\,(1 + e_{xx})(1 + e_{yy})(1 + e_{zz}) = \qquad (3.36)$$

$$= \bar{e}_{xx} + \bar{e}_{yy} + \bar{e}_{zz}$$

Thus the first invariant of the strain tensor again represents the volume strain. Nevertheless neither the transformation formulae (3.14) and (3.15) for infinitesimal strains nor Equations (3.29) and (3.33) for finite strains apply directly to logarithmic strains. The deductions of corresponding equations can be found in specialized literature dealing with plasticity theories (e.g. Hoffman and Sachs, 1953).

4

Rheological Equations

4.1 EQUATIONS OF RHEOLOGICAL ELEMENTS

4.1.1 Rheological Elements

In the foregoing chapters the stress tensor and the strain tensor were dealt with independently of each other. However rheology investigates and expresses the relationship between stresses and strains of various bodies. Thus, the general form of rheological equations is:

$$R(e_{rs}, \dot{e}_{rs}, p_{rs}) = 0 \tag{4.1}$$

The bodies whose rheological relations are expressed by a certain equation of the type (4.1) are commonly named after the scientist who first investigated them thoroughly. The range of rheological bodies extends from the rigid Euclidean body to the ideal Pascal liquid. Three fundamental rheological elements will be placed within this range first: the Hookean elastic body, the Newtonian viscous liquid and the Saint-Venant plastic body.

4.1.2 Euclidean Body

The rheological equation of the rigid Euclidean body is:

$$e_{rs} = 0 \tag{4.2}$$

All components of the strain tensor are zero. The spherical as well as the distortional strain tensors vanish. Applying Equation (4.2) to the strain velocities, their explicit form can be written:

$$\frac{\partial \dot{x}}{\partial x} = 0, \quad \frac{\partial \dot{x}}{\partial y} + \frac{\partial \dot{y}}{\partial x} = 0 \qquad (x,y,z) \tag{4.3}$$

The solution of the six equations (4.3) leads to the fundamental equations of the kinematics of rigid bodies.

4.1.3 Pascal's Liquid

Pascal's ideal liquid is incompressible and can only resist a uniform all-round pressure. Its rheological equations are:

$$e_v = 0 \tag{4.4}$$

$$p_{rs} = p_{rs}^{0} = -p\delta_{rs} \tag{4.5}$$

$$p_{rs}^{d} = 0 \tag{4.6}$$

Thus, the distortional component of the stress tensor vanishes.

4.1.4 Hookean Body

The linear relationships between the spherical components of the stress and strain tensor as well as those between their distortional components are characteristic of a Hookean body.

$$p_{rs}^{0} = 3Ke_{rs}^{0} \tag{4.7}$$

$$p_{rs}^{d} = 2Ge_{rs}^{d} \tag{4.8}$$

The constant of proportionality K is called the modulus of compressibility, and the constant G is called the modulus of distortion or shear modulus.

In the strength of materials the modulus of elasticity is defined by the equation

$$E = \frac{p_{zz}}{e_{zz}} \tag{4.9}$$

the stress p_{zz} belonging to a linear stress state:

$$p_{rs} = p_{zz}\begin{bmatrix} 0 & 0 & 0 \\ 0 & 0 & 0 \\ 0 & 0 & 1 \end{bmatrix} \tag{4.10}$$

Experimental evidence, e.g. tensile tests on steel bars or uniaxial unconfined compression tests on cylindrical samples of a pre-consolidated clay, shows that the strain tensor corresponding to the stress tensor p_{rs}, presented by Equation (4.10), can be expressed as follows:

$$e_{rs} = \frac{p_{zz}}{E}\begin{bmatrix} -\nu & 0 & 0 \\ 0 & -\nu & 0 \\ 0 & 0 & 1 \end{bmatrix} \tag{4.11}$$

The constant ν is called Poisson's ratio.

The spherical component of the stress tensor p_{rs} (Equation 4.10) is:

$$p_{rs}{}^{0} = \frac{p_{zz}}{3}\,\delta_{rs} \tag{4.12}$$

and the spherical component of the strain tensor (Equation 4.11) is:

$$e_{rs}{}^{0} = \frac{e_v}{3}\,\delta_{rs} = \frac{p_{zz}}{E}\cdot\frac{1-2\nu}{3}\,.\,\delta_{rs} \tag{4.13}$$

For the compressibility modulus K Equation (4.7) yields the expression:

$$K = \frac{p_{rs}{}^{0}}{3e_{rs}{}^{0}} = \frac{E}{3(1-2\nu)} \tag{4.14}$$

The distortional component of the stress tensor (4.11) is

$$p_{rs}{}^{d} = p_{rs} - p_{rs}{}^{0} = \frac{p_{zz}}{3}\begin{bmatrix} -1 & 0 & 0 \\ 0 & -1 & 0 \\ 0 & 0 & 2 \end{bmatrix} \tag{4.15}$$

while the distortional component of the strain tensor (4.11) is

$$e_{rs}{}^{d} = e_{rs} - e_{rs}{}^{0} = \frac{p_{zz}}{3E}\begin{bmatrix} -(1+\nu) & 0 & 0 \\ 0 & -(1+\nu) & 0 \\ 0 & 0 & 2(1+\nu) \end{bmatrix} \tag{4.16}$$

Inserting expressions (4.15) and (4.16) into Equation (4.8), the expression for the shear modulus G can be obtained:

$$G = \frac{p_{rs}{}^{d}}{2e_{rs}{}^{d}} = \frac{E}{2(1+\nu)} \tag{4.17}$$

The mechanical rheological model of the Hookean body is a completely elastic spring (Fig. 4.1).

Fig. 4.1. Hookean spring

Inserting expressions (4.7) and (4.8) into Equation (2.23) and considering the meaning of the strain tensor components as defined by the matrixes (3.19) and (3.20), the principal stresses in the directions $i = 1, 2, 3$ can be expressed in the form:

$$\sigma_i = 3K\frac{\varepsilon_v}{3} + 2G\left(\varepsilon_i - \frac{\varepsilon_v}{3}\right) \tag{4.18}$$

Substituting

$$K - \tfrac{2}{3}G = \lambda \tag{4.18-a}$$

$$G \equiv \mu \tag{4.18-b}$$

Equation (4.18) can be written as follows:

$$\sigma_i = \lambda\varepsilon_v + 2\mu\varepsilon_i \tag{4.18-c}$$

The coefficients λ and μ are called Lamé's coefficients.

4.1.5 Newton's Liquid

The Newtonian liquid is a compressible viscous body conforming to the following rheological equation:

$$p_{rs} = Ke_v\delta_{rs} + 2\eta^d\dot{e}_{rs}{}^d \tag{4.19}$$

While the spherical component of the stress tensor is the same as for a Hookean body, its distortional component is proportional to the distortional component of the tensor of strain velocities. The constant of proportionality η^d is called the coefficient of distortional viscosity; its dimension is [kp m^{-2} sec].

Pascal's liquid is a special case of Newton's liquid with the constants:

$$K \rightarrow \infty, \quad \eta^d = 0$$

The mechanical rheological model of the Newtonian liquid is a perforated piston moving in a cylinder containing a viscous liquid (Fig. 4.2), i.e. a dashpot.

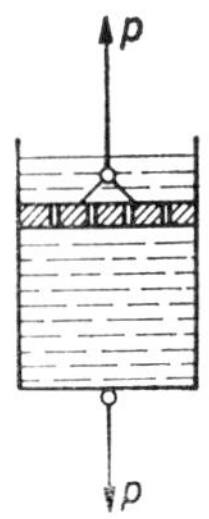

Fig. 4.2. Newton's dashpot

The viscous spherical consolidation of soil media can be expressed by a relationship analogous to the distortional component of Equation (4.19) for classical Newtonian liquid:

$$p_{rs}{}^{0} = 3\eta^{0}\dot{e}_{rs}{}^{0} \tag{4.20}$$

The factor of proportionality η^0 is the coefficient of spherical viscosity.

4.1.6 Saint-Venant's Body

The mechanical model of the Saint-Venant body is the frictional resistance between two rough plates pressed one against the other (Fig. 4.3 (*a*)). The

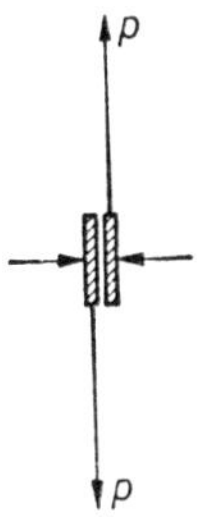

a

Fig. 4.3. (a) Saint-Venant's resistance

body is assumed to be incompressible:

$$e_v = 0, \quad e_{rs}{}^{0} = 0 \tag{4.21}$$

and cannot move so long as the distortional stresses do not exceed a certain resistance:

$$p_{rs}{}^{d} \leqslant \Theta_{rs} \rightarrow e_{rs}{}^{d} = 0 \tag{4.22}$$

$$p_{rs}{}^{d} > \Theta_{rs} \rightarrow p_{rs}{}^{d} = 2\lambda\dot{e}_{rs}{}^{d} \tag{4.23}$$

The difference between the distortional part of the expression (4.19) of the Newtonian body and the expression (4.23) of Saint-Venant's body lies in the fact that the coefficient of viscosity η^d represents a physical property of the body while the coefficient λ depends only on the invariants of the stress and strain tensors.

$$\lambda = \frac{p_{rs}{}^{d}}{2\dot{e}_{rs}{}^{d}} = \frac{\tau_{\text{oct}}}{\dot{\gamma}_{\text{oct}}} \tag{4.24}$$

Taking into account expressions (2.21) and (3.22), we obtain:

$$\lambda = \frac{1}{2}\sqrt{\left(\frac{I_1^2 - 3I_2}{\dot{J}_1^2 - 3\dot{J}_2}\right)} \tag{4.24-a}$$

Considering zero volume strains (Equation 4.21), Equation (4.23) can be written as follows:

$$p_{rs}{}^{d} = \frac{2\lambda}{\mathrm{d}t}\,\mathrm{d}e_{rs}{}^{d} = \frac{2\lambda}{\mathrm{d}t}\,\mathrm{d}e_{rs} \tag{4.25}$$

If applied to principal stress and strain directions, Equation (4.25) expresses the principle of the coaxiality of the tensor of distortional stress and of the tensor of strain increase, which is a fundamental principle of Saint-Venant's theory of plasticity.

4.1.7 Some further Rheological Elements

In the previous sections the spring, the dashpot and the slide block have been shown to represent the mechanical analogous elements for the Hookean, Newtonian and Saint-Venant's bodies respectively. Among other mechanical models the slide block of Képès, the stop and the regulator could be of interest in soil mechanics.

The slide block of Képès (Fig. 4.3 (*b*)) represents an element whose

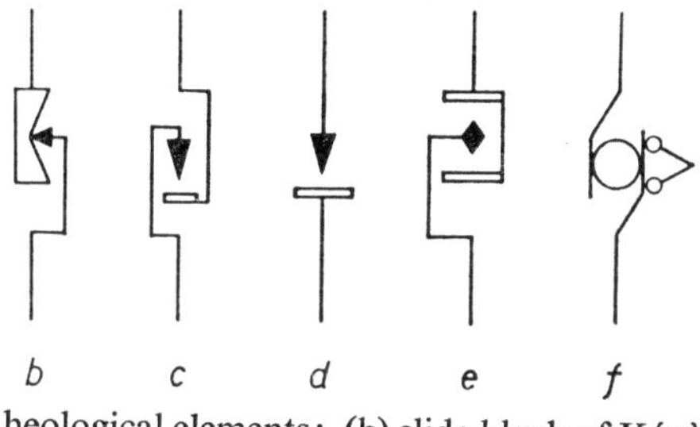

Fig. 4.3. (b–f) Rheological elements: (b) slide block of Képès; (c–e) stops; (f) regulator

stresses σ cannot exceed the limiting values which are proportional to the strain:

$$-K|\varepsilon| \leqslant \sigma \leqslant K|\varepsilon|$$

The stops (Figs. 4.3 (*c*), (*d*), (*e*)) confine the strain in one or in both directions. The model 4.3 (*e*) e.g. represents the requirement:

$$-\varepsilon_c \leqslant \varepsilon \leqslant \varepsilon_c$$

The regulator (Fig. 4.3 (*f*)) does not resist deformation until a certain limiting strain velocity $\dot{\varepsilon}_c$ has been reached. This limiting strain velocity cannot be exceeded even when the stress becomes very large:

$$-\dot{\varepsilon}_c \leqslant \dot{\varepsilon} \leqslant \dot{\varepsilon}_c$$

4.2 EQUATIONS OF RHEOLOGICAL BODIES CONSISTING OF TWO RHEOLOGICAL ELEMENTS

4.2.1 Method of Structural Models

By combining the rheological elements presented in the foregoing sections, various structural rheological models illustrating the stress-strain relationships of technical media can be obtained. The elementary models can be connected either (*a*) in parallel, or (*b*) in series. When connected in parallel, the distribution of the total stress to the constituent elements is governed by the condition of equal simultaneous strains of all elements. When connected in series, the whole stress is transferred to each element while the total strain equals the sum of the strains of the constituent elements.

The various structural models can be presented by structural equations using the following symbols:

N for the Newtonian dashpot
StV for Saint-Venant's body
H for the Hookean spring
|| for parallel connection
— for connection in series.

The most important rheological bodies consisting of two elements are:

Maxwell's body (M): M = H — N
Bingham's body (B): B = StV||N
Kelvin's body (K): K = H||N

Their rheological equations will be given in the following paragraphs.

4.2.2 Maxwell's Body

The structural model of the Maxwell body is shown in Fig. 4.4 (*a*) whilst Fig. 4.4 (*b*) represents the strain-time diagram corresponding to a sudden constant load application. Since the Hookean spring and the Newtonian dashpot are connected in series, the stresses in both elements are equal.

Maxwell's structural model can be applied either to distortional or to spherical tensor components. When applying it to the distortional stress—strain relationship, Equations (4.8) and (4.19) should be considered.

$$p_{rs}{}^{d} = 2G(e_{rs}{}^{d})_{\mathrm{H}} \ldots \mathrm{H} \tag{4.8-a}$$

$$p_{rs}{}^{d} = 2\eta^{d}(\dot{e}_{rs}{}^{d})_{\mathrm{N}} \ldots \mathrm{N} \tag{4.19-a}$$

Differentiating Equation (4.8) with respect to time, we obtain

$$\dot{p}_{rs}{}^{d} = 2G(\dot{e}_{rs}{}^{d})_{\mathrm{H}} \tag{4.26}$$

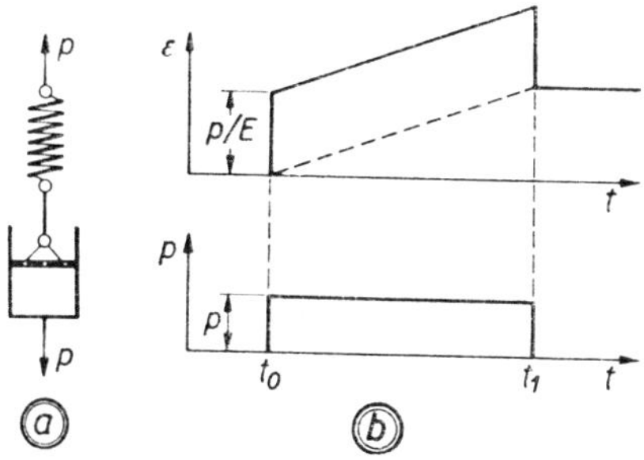

Fig. 4.4. Maxwell's body

The sum of the corresponding tensors of the Hookean $(\dot{e}_{rs}{}^{d})_{\mathrm{H}}$ and of the Newtonian $(\dot{e}_{rs}{}^{d})_{\mathrm{N}}$ elements represents the total strain velocity tensor:

$$\dot{e}_{rs}{}^{d} = (\dot{e}_{rs}{}^{d})_{\mathrm{H}} + (\dot{e}_{rs}{}^{d})_{\mathrm{N}}$$

Inserting expressions (4.26) and (4.19-a) the above equation becomes

$$\dot{e}_{rs}{}^{d} = \frac{\dot{p}_{rs}{}^{d}}{2G} + \frac{p_{rs}{}^{d}}{2\eta^{d}} \tag{4.27}$$

The solution of this differential equation can be obtained in the customary way. Firstly it may be written in the form:

$$\dot{p}_{rs}{}^{d} + \frac{G}{\eta^{d}} p_{rs}{}^{d} = 2G\dot{e}_{rs}{}^{d} \tag{4.27-a}$$

The corresponding homogeneous equation

$$\dot{y} + \frac{G}{\eta^{d}} y = 0 \tag{4.27-b}$$

has the solution:

$$y = y_0 \exp\left\{-\frac{G}{\eta^d}(t - t_0)\right\} \tag{4.27-c}$$

The solution of the inhomogeneous equation is then assumed to be:

$$p_{rs}{}^d = uy \tag{4.27-d}$$

Differentiating we obtain:

$$\dot{p}_{rs}{}^d = \dot{u}y + u\dot{y} \tag{4.27-e}$$

Inserting these results into Equation (4.27-a):

$$(\dot{u}y + u\dot{y}) + \frac{G}{\eta^d} uy = 2G\dot{e}_{rs}{}^d \tag{4.27-f}$$

rearranging:

$$\dot{u}y + u\left(\dot{y} + \frac{G}{\eta^d} y\right) = 2G\dot{e}_{rs}{}^d \tag{4.27-g}$$

and considering the homogeneous Equation (4.27-b) we get:

$$\mathrm{d}u = 2G\frac{\dot{e}_{rs}}{y}\,\mathrm{d}\tau \tag{4.27-h}$$

and after integration:

$$u = u_0 + \frac{2G}{y_0}\int_{t_0}^{t} \dot{e}_{rs}{}^d \exp\left\{\frac{G}{\eta^d}(\tau - t_0)\right\} \mathrm{d}\tau \tag{4.27-i}$$

Inserting (c) and (i) into (d) we obtain the final solution:

$$p_{rs}{}^d = \exp\left\{-\frac{G}{\eta^d}(t - t_0)\right\}\left[p_{rs0}{}^d + 2G\int_{t_0}^{t} \dot{e}_{rs}{}^d \exp\left\{\frac{G}{\eta^d}(\tau - t_0)\right\} \mathrm{d}\tau\right] \tag{4.28}$$

where τ denotes any time in the interval $t_0 \leqslant \tau \leqslant t$. With the creep velocity $\dot{e}_{rs}{}^d$ having a constant value: $\dot{e}_{rs}{}^d = c$, Equation (4.28) becomes:

$$p_{rs}{}^d = \exp\left[-\frac{G}{\eta^d}(t - t_0)\right] p_{rs0}{}^d + 2\eta^d c\left\{1 - \exp\left[-\frac{G}{\eta^d}(t - t_0)\right]\right\} \tag{4.29}$$

By stopping the creep, $c = 0$ and Equation (4.29) reduces to:

$$p_{rs}{}^d = \exp\left[-\frac{G}{\eta^d}(t - t_0)\right] p_{rs0}{}^d \tag{4.30}$$

Equation (4.30) represents the expression for the relaxation of the stress at constant strain. For $t \to \infty$ the stress tends to zero (Fig. 4.5).

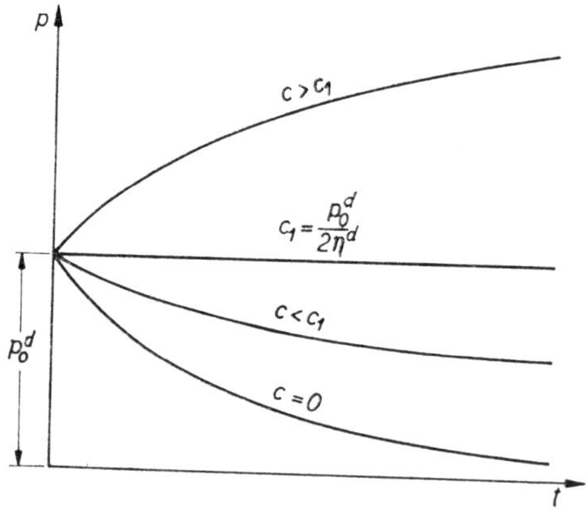

Fig. 4.5. Stress versus time plots of Maxwell's bodies at constant velocity

When c is equal to $c_1 = p_{rs0}{}^d/2\eta^d$, Equation (4.29) becomes:

$$p_{rs}{}^d = p_{rs0}{}^d \tag{4.31}$$

When $c < c_1$, the stress decreases during creep, and when $c > c_1$, the stress increases. The corresponding stress-time diagrams are shown in Fig. 4.5.

4.2.3 Bingham's Body

The structural model of the Bingham body is shown in Fig. 4.6 (*a*). The Saint-Venant body and the Newtonian dashpot are connected in parallel: $B = \mathrm{StV}\|\mathrm{N}$. The creep (flow) appears when the stress p_{rs} exceeds the

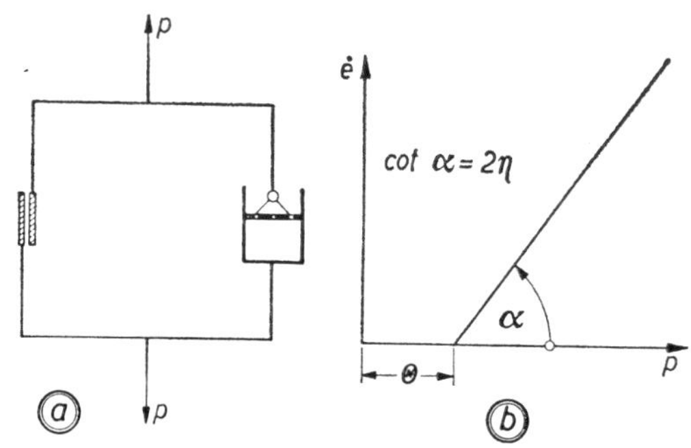

Fig. 4.6. Bingham's body

magnitude of the Saint-Venant body's resistance Θ_{rs} (Fig. 4.6 (*b*)). The further increase of stress is taken by the Newtonian element while the load on the Saint-Venant element remains constant.

The rheological equation of Bingham's body is:

$$p_{rs} = (p_{rs})_{\mathrm{StV}} + (p_{rs})_{\mathrm{N}} \tag{4.32}$$

Considering Equations (4.22), (4.23), (4.19) and (4.20), and assuming

$$p_{rs} > \Theta_{rs} \tag{4.33}$$

we obtain:

$$p_{rs} = \Theta_{rs} + 2\eta\dot{e}_{rs} \tag{4.34}$$

Thus:

$$\dot{e}_{rs} = \frac{p_{rs} - \Theta_{rs}}{2\eta} \tag{4.34-a}$$

This equation can be applied either to the distortional or to the spherical components of the strain-velocity and stress tensor. In the coordinate system $\dot{e}_{rs} \sim p_{rs}$ the relationship (4.34-a) is illustrated by a straight line which crosses the p_{rs}-axis at the distance Θ_{rs} from the origin and is inclined to this axis at the angle $\alpha = \text{arc cot } 2\eta$ (Fig. 4.6 (*b*)).

4.2.4 Kelvin's Body

The structural model of the Kelvin body is obtained by the parallel connection of the Hookean spring with the Newtonian dashpot: $K = \mathrm{H}\|\mathrm{N}$ (Fig. 4.7 (*a*)).

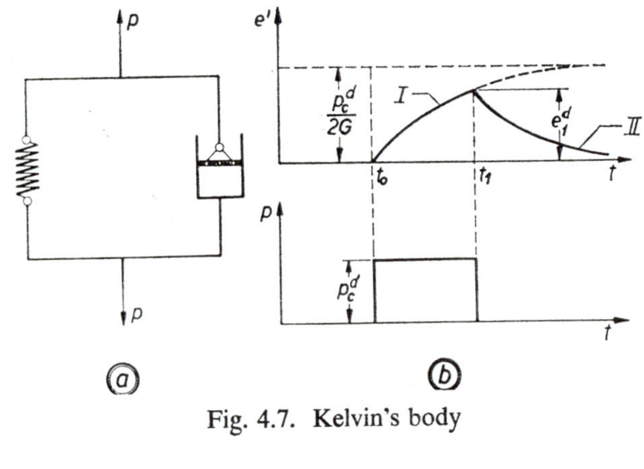

Fig. 4.7. Kelvin's body

The whole stress should be distributed to both elements according to the basic rheological equations of either their distortional [Equations 4.9 and 4.19] or spherical [Equations 4.8 and 4.20] stress and strain tensors.

Applying the Kelvin rheological model to the distortional stress and strain tensors, the following equation results:

$$p_{rs}{}^{d} = 2Ge_{rs}{}^{d} + 2\eta^{d}\dot{e}_{rs}{}^{d} \tag{4.35}$$

The solution of this differential equation—analogous to that of Equation (4.27)—yields the expression:

$$e_{rs}^{d} = \exp\left[-\frac{G}{\eta^{d}}(t-t_{0})\right]\left\{e_{rs0}{}^{d} + \frac{1}{2\eta^{d}}\int_{t_0}^{t} p_{rs}{}^{d}\exp\left[\frac{G}{\eta^{d}}(\tau-t_{0})\right]\mathrm{d}\tau\right\} \tag{4.36}$$

For the constant stress tensor

$$p_{rs}{}^{d} = p_{rsc}{}^{d}$$

Equation (4.36) becomes:

$$e_{rs}{}^{d} = \exp\left[-\frac{G}{\eta^{d}}(t-t_{0})\right]e_{rs0}{}^{d} + \frac{1}{2G}p_{rsc}{}^{d}\left\{1-\exp\left[-\frac{G}{\eta^{d}}(t-t_{0})\right]\right\} \tag{4.37}$$

In the case of zero initial strain $e_{rs0}{}^{d} = 0$, the strain increases as shown by the curve (*I*) in Fig. 4.7 (*b*). If at the time t_1 after the strain has reached the value $e_{rs1}{}^{d}$, the stress vanishes, the subsequent decrease of stress is governed by the equation:

$$e_{rs}{}^{d} = e_{rs1}{}^{d}\exp\left[-\frac{G}{\eta^{d}}(t-t_{1})\right] \tag{4.38}$$

This elastic after-effect, characteristic of Kelvin bodies, is illustrated by the curve (II) in Fig. 4.7 (*b*).

4.3 RHEOLOGICAL MODELS CONSISTING OF MORE THAN TWO ELEMENTS

By combining more than two rheological elements or connecting single elements and models consisting of two elements, a whole selection of structural models can be obtained. The rheological equations of bodies represented by such models can be obtained in an analogous way to those of two-element bodies, i.e. by superimposing the stresses of the bodies connected in parallel and by superimposing the strains of the bodies connected in series.

Two classical multiple rheological bodies are:

the Schwedoff body, the structural equation of which is

$$\text{Schw} = \text{H} - [\text{StV}\|(\text{H} - \text{N})] = \text{H} - [\text{StV}\|\text{M}]$$

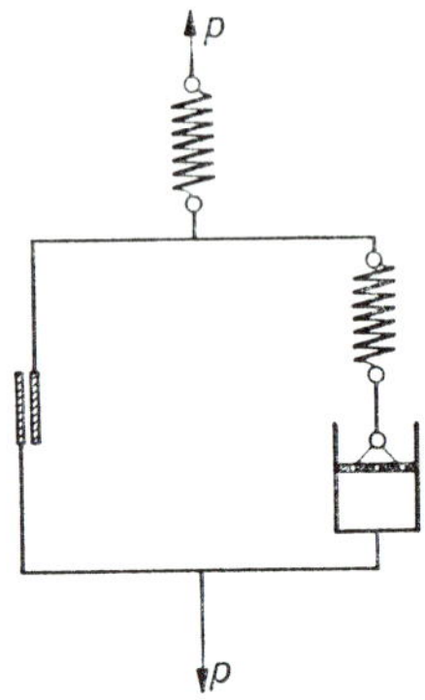

Fig. 4.8. Schwedoff's body

(Fig. 4.8) and the body of Schofield and Scott Blair with the following structural equation:

$$\text{Sch Sc B} = \text{N} - [\text{H}\|\text{N}] = \text{N} - \text{K}$$

(Fig. 4.9).

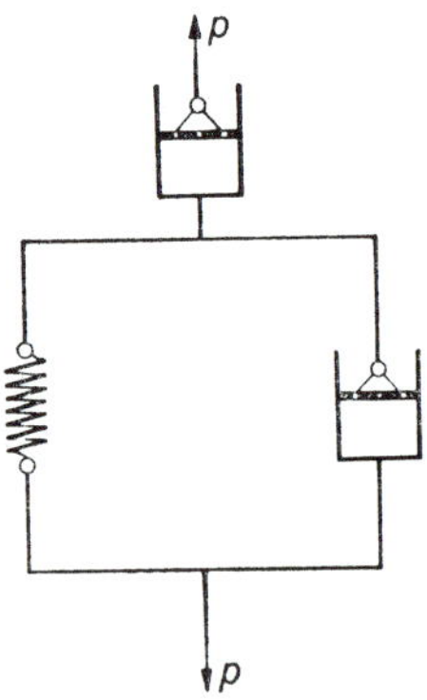

Fig. 4.9. Schofield and Scott Blair's body

Similar multiple rheological models will be dealt with when treating the different rheological consolidation theories.

4.4 LINEAR RHEOLOGICAL MODELS

4.4.1 Definitions and Examples

A rheological body is called Boltzmannean if the law of superposition can be applied to the relations between strains and stresses. The strain corresponding to the simultaneous action of several stresses can be obtained as the sum of strains due to the effects of single stresses.

Rheological models are called linear if there is a linear relationship between the stresses (σ) and strains (ε) and their successive derivatives with respect to time:

$$E_0\varepsilon + E_1\dot{\varepsilon} + E_2\ddot{\varepsilon} + \ldots = S_0\sigma + S_1\dot{\sigma} + S_2\ddot{\sigma} + \ldots \tag{4.39}$$

Any linear body is Boltzmannean but not every Boltzmannean body is linear. Nevertheless the term linear is commonly used as the synonym of the term Boltzmannean, i.e. it denotes the principle of superposition.

The Hookean spring and the Newtonian dashpot are linear elements while those described in sections 4.1.6 and 4.1.7 are not linear. Double or multiple models consisting of Hookean and Newtonian elements are also linear. Important models are the generalized Kelvin–Voigt model and the generalized Maxwell model. Their structural schemes are presented in Figs. 4.10 (*a*) and (*b*) respectively.

The effect of creep at constant stress σ_0 can be expressed in the form:

$$\varepsilon(t) = \sigma_0 f(t) \tag{4.40}$$

and the effect of relaxation at constant strain ε_0 by the relation:

$$\sigma(t) = \varepsilon_0 r(t) \tag{4.41}$$

$f(t)$ is called the creep function and $r(t)$ the relaxation function.

Considering Equations (4.9), (4.37) and (4.19) and assuming zero initial strain the creep function of the generalized Kelvin body applied to the distortional strain tensor can be written as follows:

$$f(t) = \frac{1}{2G_0} + \sum_{1}^{m} \frac{1}{2G_i}\left\{1 - \exp\left[-\frac{G_i}{\eta_i^d}(t - t_0)\right]\right\} \tag{4.42}$$

Expressing the initial stresses by corresponding strains:

$$p_{rs0}{}^d = 2Ge_{rs0}{}^d$$

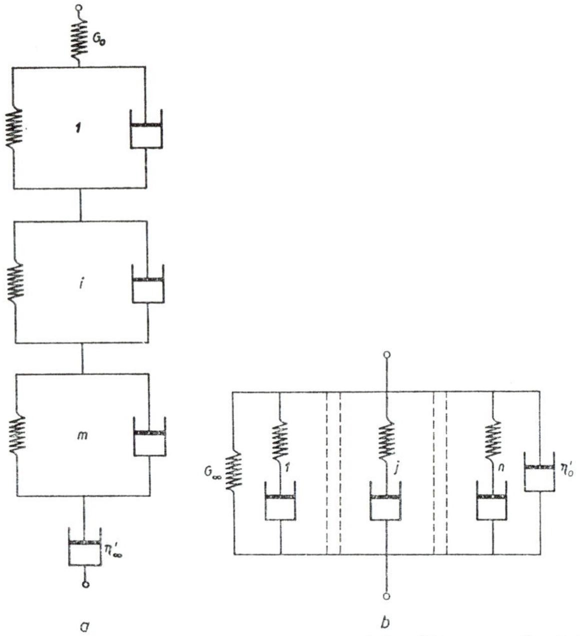

Fig. 4.10. (a) Generalized Kelvin-Voigt model; (b) generalized Maxwell model

and considering the relations (4.8), (4.30) and (4.19), the relaxation function of the generalized Maxwell body applied to the distortional stress tensor can be obtained in the form:

$$r(t) = 2G_\infty + \sum_{1}^{n} 2G_j \exp\left[-\frac{G_j}{\eta_j^d}(t - t_0)\right] + \eta_0\delta(t) \qquad (4.43)$$

$\delta(t)$ being Dirac's function which is defined in the following way. It is zero for all values of t except for $t = 0$, in which case it is infinite.

An important advantage of linear rheological models is that their differential or integral equations can be transformed to linear algebraic equations. In this way the problems of linear viscoelastic bodies can be treated in the same way as problems of elasticity.

4.4.2 Transformation of Linear Viscoelastic Models to Elastic Models

Consider a function $f(t)$ defined for $t \geqslant 0$. Its transformation after Carson is defined by the relation:

$$f^*(p) = p\int_0^\infty e^{-pt} f(t)\,\mathrm{d}t \qquad (4.44)$$

The function $f(t)$ is called the original and the function $f^*(p)$ the image. The relation between these functions may be expressed by the following symbol:

$$f(t) \supset f^*(p) \tag{4.45}$$

Referring to Mandel (1960) we shall make use of the following transformation:

$$\frac{d}{dt}\int_0^t f(u)\,g(t-u)\,du \supset f^*(p)\,g^*(p) \tag{4.46}$$

Now, $f(t)$ may denote a creep function:

$$\varepsilon(t) = \sigma f(t) \tag{4.46-a}$$

If the stress σ varies with time, the superposition of the effects of the initial stress σ_0 applied at time $t = 0$ and of successive stress increments $d\sigma = \sigma'(\tau)\,d\tau$ yields:

$$\varepsilon(t) = \sigma_0 f(t) + \int_0^t f(t-\tau)\sigma'(\tau)\,d\tau$$

Substituting

$$u = t - \tau$$

the above equation becomes:

$$\varepsilon(t) = \frac{d}{dt}\int_0^t \sigma(t-u)\,f(u)\,du \tag{4.47}$$

If the strain varies with time, the application of the principle of superposition to the relaxation function (4.41) yields an analogous expression:

$$\sigma(t) = \frac{d}{dt}\int_0^t \varepsilon(t-u)r(u)\,du \tag{4.48}$$

Using the transformation relation (4.46) the expressions (4.47) and (4.48) can be transformed as follows:

$$\varepsilon^*(p) = f^*(p)\sigma^*(p) \tag{4.49}$$

$$\sigma^*(p) = r^*(p)\varepsilon^*(p) \tag{4.50}$$

These relations have the form of Hooke's law, $r^*(p)$ representing the modulus of elasticity and $f^*(p)$ its inverse value:

$$f^*(p)\, r^*(p) = 1 \tag{4.51}$$

The relations (4.49) and (4.50) have been deduced for a linear stress state. They can be applied to any spherical or distortional stress tensor as well. By way of illustration the fundamental rheological relation (4.27) of the Maxwell body corresponding to distortional parts of stress and strain tensors and written in the form:

$$\frac{\mathrm{d}e_{rs}{}^{d}}{\mathrm{d}t} = \frac{1}{2G}\frac{\mathrm{d}p_{rs}{}^{d}}{\mathrm{d}t} + \frac{p_{rs}{}^{d}}{2\eta^{d}} \tag{4.27-a}$$

can be transformed into the expression:

$$pe_{rs}{}^{d*} = \left(\frac{p}{2G} + \frac{1}{2\eta^{d}}\right) p_{rs}{}^{d} \tag{4.52}$$

where the relation $p_{rs}{}^{d}(0) = 2Ge_{rs}{}^{d}(0)$ has been taken into account. Thus, the transformed creep function (image) is:

$$f^*(p) = \frac{1}{2G} + \frac{1}{2\eta^{d}p} \tag{4.53}$$

By using simple transformation computations or appropriate tables the corresponding original of the creep function can be found:

$$f(t) = \frac{1}{2G} + \frac{t}{2\eta^{d}} \tag{4.54}$$

The image of the relaxation function $r(p)$ is the inverse of the transformed creep function:

$$r^*(p) = \frac{1}{f^*(p)} = 2G\frac{p}{p + \dfrac{G}{\eta^{d}}} \tag{4.55}$$

By using transformation tables the original of the relaxation function can be obtained:

$$r(t) = G \exp\left(-\frac{G}{\eta^{d}}t\right) \tag{4.56}$$

The same result was obtained by solving the non-transformed differential Equation (4.27). [Cf. Equation (4.30) for $t_0 = 0$.]

The second example may be given for a Kelvin body whose fundamental rheological equation applied to the spherical part of the stress and strain tensors is:

$$p_{rs}{}^{0} = 3Ke_{rs}{}^{0} + 3\eta^{0}\dot{e}_{rs}{}^{0} \tag{4.35-a}$$

Its image is obtained as:

$$p_{rs}^{0*} = 3Ke_{rs}^{0*} + 3\eta^0 pe_{rs}^{0*} \tag{4.57}$$

Thus:

$$r^*(p) = 3K + 3\eta^0 p \tag{4.58}$$

and the corresponding original:

$$r(t) = 3K + 3\eta^0\delta(t) \tag{4.59}$$

where

$$\delta(t) = \text{Dirac's function.}$$

The transformed creep function is the inverse of the transformed relaxation function

$$f^*(p) = \frac{1}{r^*(p)} = \frac{1}{3K + 3\eta^0 p} \tag{4.60}$$

and the corresponding original creep function can be found by using transformation tables:

$$f(t) = \frac{1}{3K}\left[1 - \exp\left(-\frac{K}{\eta^0}t\right)\right] \tag{4.61}$$

which is in agreement with Equation (4.37).

If the total stress tensor $p_{rs}(t)$ is expressed by using the time-dependent Lamé's coefficients $\lambda(t)$ and $\mu(t)$:

$$p_{rs}(t) = \lambda(t)\varepsilon_v\delta_{rs} + 2\mu(t)e_{rs} \tag{4.62}$$

the transformation of Carson yields:

$$p_{rs}^* = \lambda^*(p)\varepsilon_v\delta_{rs} + 2\mu^*(p)e_{rs}^* \tag{4.63}$$

$$\varepsilon_v^* = \sum_{i=1}^{3} e_i$$

Lamé's coefficients $\lambda^*(p)$ and $\mu^*(p)$ of the transformed Maxwell body, for example, correspond to the following originals (cf. Equation 4.54):

$$\lambda^*(p) \subset \frac{1}{3\lambda} + \frac{t}{3\eta_\lambda} \tag{4.64}$$

$$\mu^*(p) \subset \frac{1}{2\mu} + \frac{t}{2\eta_\mu} \tag{4.65}$$

where:

$$\lambda = K - \tfrac{2}{3}G, \qquad \eta_\lambda = \eta^0 - \tfrac{2}{3}\eta^d$$

$$\mu \equiv G \quad , \qquad \eta_\mu \equiv \eta^d$$

5

Fundamental Applications of the Theory of Elasticity in Soil Mechanics

5.1 FUNDAMENTAL EQUATIONS OF THE MATHEMATICAL THEORY OF ELASTICITY

The mathematical theory of elasticity deals with stresses and strains of Hookean bodies, the fundamental rheological equations of which have been given in section 4.1.4. Applying Equations (2.23), (4.7), (4.8), (4.14) and (4.17) the stress tensor p_{rs} can be expressed by the following matrix:

$$p_{rs} = 2G \begin{bmatrix} \varepsilon_{xx} + \dfrac{\nu}{1-2\nu}\varepsilon_v & \varepsilon_{xy} & \varepsilon_{xz} \\ \cdot & \varepsilon_{yy} + \dfrac{\nu}{1-2\nu}\varepsilon_v & \varepsilon_{yz} \\ \cdot & \cdot & \varepsilon_{zz} + \dfrac{\nu}{1-2\nu}\varepsilon_v \end{bmatrix} \tag{5.1}$$

Inserting the components of this stress tensor into the equilibrium Equations (2.2) and assuming zero accelerations (x''), considering the relations (3.2) and using the Laplace operator

$$\Delta = \frac{\partial^2}{\partial x^2} + \frac{\partial^2}{\partial y^2} + \frac{\partial^2}{\partial z^2} \tag{5.2}$$

the following expression is obtained:

$$\Delta u_\alpha + \frac{1}{1-2\nu}\frac{\partial \varepsilon_v}{\partial \alpha} + \frac{A}{G} = 0 \tag{5.3}$$

$$(\alpha = x, y, z; \quad A = X, Y, Z)$$

Differentiating Equations (5.3) with respect to α, summarizing the obtained expressions for $\alpha = x, y, z$, and considering Equations (3.5) we obtain the relation:

$$\frac{2-2\nu}{1-2\nu}\Delta\varepsilon_v + \frac{1}{G}\left(\frac{\partial X}{\partial x} + \frac{\partial Y}{\partial y} + \frac{\partial Z}{\partial z}\right) = 0 \tag{5.4}$$

If the distribution of volume forces per unit of volume with the components X, Y, Z is uniform, then

$$\frac{\partial X}{\partial x} = \frac{\partial Y}{\partial y} = \frac{\partial Z}{\partial z} = 0 \tag{5.5}$$

and Equation (5.4) yields:

$$\Delta \varepsilon_v = 0 \tag{5.6}$$

This simplified form of Equation (5.4) holds for all homogeneous bodies including weightless bodies.

5.2 STRESSES AND STRAINS IN THE HALF-SPACE DUE TO A CONCENTRATED VERTICAL LOAD

5.2.1. Fundamental Equations for Stresses and Strains in Bodies of Revolution

The half-space is defined by the condition that each point of it shall lie below or above the plane $z = h$:

$$z \geqslant h \qquad \text{or} \qquad z \leqslant h$$

The equations for stresses and strains in the half-space can be deduced from the general relations of the state of stress in an arbitrary body of revolution. Its loading is assumed to consist of normal and tangential forces distributed symmetrically with respect to the axis of revolution and acting in the upper or lower surface of the body (Fig. 5.1).

Introducing cylindrical coordinates r, φ, z the components of the displacement vectors are denoted by ρ, t and w. Thus, the following strain components should be inserted into the matrix of the strain tensor (5.1):

$$\left.\begin{aligned} \varepsilon_{xx} &= \varepsilon_{rr} = \frac{\partial \rho}{\partial r} \\ \varepsilon_{yy} &= \varepsilon_{\varphi\varphi} = \frac{\rho}{r} \\ \varepsilon_{zz} &= \frac{\partial w}{\partial z} \end{aligned}\right\} \varepsilon_v = \frac{\partial \rho}{\partial r} + \frac{\rho}{r} + \frac{\partial w}{\partial z} \tag{5.7}$$

$$\varepsilon_{xz} = \varepsilon_{rz} = \frac{1}{2}\left(\frac{\partial w}{\partial r} + \frac{\partial \rho}{\partial z}\right) \tag{5.8}$$

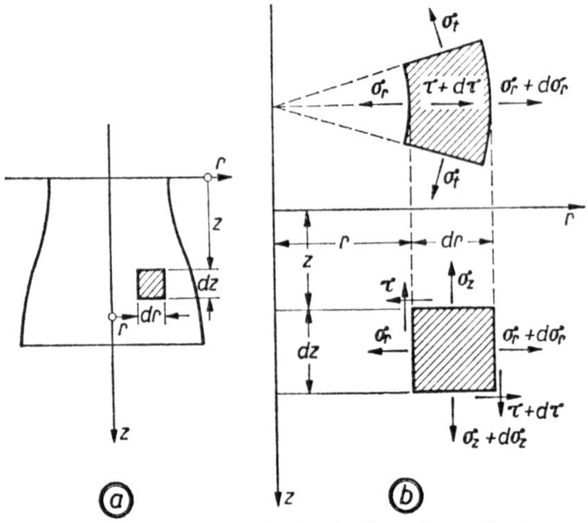

Fig. 5.1. Stresses in the bodies of revolution

The symmetry of the body and of the loading requires:

$$\left.\begin{aligned}\varepsilon_{xy} &= \varepsilon_{r\varphi} = 0\\ \varepsilon_{yz} &= \varepsilon_{\varphi z} = 0\end{aligned}\right\} \tag{5.9}$$

Using the equilibrium conditions in the directions z and r for a volume element (dz, dr, $r\ d\varphi$) (Fig. 5.1), neglecting the volume forces, applying Taylor's series to first approximation and dividing by $dz\ dr\ d\varphi$, we get:

$$\frac{\partial \sigma_z}{\partial z} r + \tau + \frac{\partial \tau}{\partial r} r = 0 \tag{5.10}$$

$$\sigma_r + \frac{\partial \sigma_r}{\partial r} r - \sigma_t + \frac{\partial \tau}{\partial z} r = 0 \tag{5.11}$$

According to the notation of Fig. 5.1 the following symbols have been introduced in the above equations:

$$\left.\begin{aligned}&\left.\begin{aligned}p_{xx} &= p_{rr} = \sigma_r\\ p_{yy} &= p_{\varphi\varphi} = \sigma_t\\ p_{zz} &\phantom{= p_{rr}} = \sigma_z\end{aligned}\right\}\ \text{normal stresses}\\ &p_{xz} = p_{rz} = \tau \ldots \text{shear stresses}\end{aligned}\right\} \tag{5.12}$$

The stresses in Equations (5.10) and (5.11) have to be expressed by the components of the stress tensor (5.1) considering Equations (5.9) and (5.8). Applying further the Laplace operator Δ

$$\Delta = \frac{\partial^2}{\partial z^2} + \frac{\partial^2}{\partial r^2} + \frac{1}{r}\frac{\partial}{\partial r} \tag{5.13}$$

the following equations result:

$$\frac{1}{1-2\nu}\cdot\frac{\partial\varepsilon_v}{\partial z}+\Delta w=0 \tag{5.14}$$

$$\frac{1}{1-2\nu}\cdot\frac{\partial\varepsilon_v}{\partial r}+\Delta\rho-\frac{\rho}{r^2}=0 \tag{5.15}$$

These two fundamental equations for the solution of stresses and strains in bodies of revolution have to be considered together with the preceding potential equation of homogeneous bodies:

$$\Delta\varepsilon_v=0 \tag{5.6}$$

5.2.2 Boussinesq's Equations

Boussinesq (1885) has applied Equations (5.14), (5.15) and (5.6) to the elastic isotopic, homogeneous half-space, where the half-space has been regarded as a body of revolution whose envelope and lower surface are

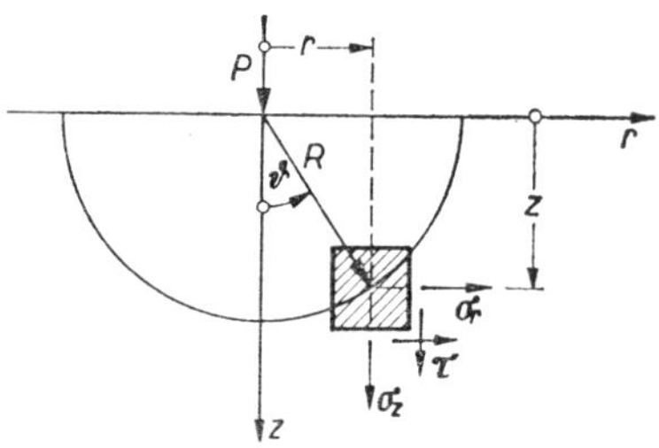

Fig. 5.2. Half-space subjected, on the upper surface, to a concentrated load

not loaded and tend towards infinity; the upper surface is loaded in the axis of the body of rotation by a concentrated load (Fig. 5.2). The solution of the problem should satisfy the following boundary conditions:

(1) At infinity ($r\rightarrow\infty$) the stresses vanish, and (2) the strains are zero; (3) the shear stresses are zero over all the boundary surface ($z=0$) of the half-space; (4) the normal stresses in this surface are zero also, except at the point of the load application; (5) the resultant of stresses acting in the surface of an arbitrary complete segment of the half-space, e.g. in the surface of a hemisphere as shown in Fig. 5.2, has the direction of the axis z and is in equilibrium with the force P.

From the above fundamental equations and boundary conditions Boussinesq obtained the following solutions:

(*a*) for strains:

$$\varepsilon_v = -\frac{P}{4\pi R^2}\cdot\frac{1}{G}\cdot 2(1-2\nu)\cos\vartheta \tag{5.16}$$

$$w = \frac{P}{4\pi R}\cdot\frac{1}{G}\{2(1-\nu)+\cos^2\vartheta\} \tag{5.17}$$

$$\rho = \frac{P}{4\pi R}\cdot\frac{1}{G}\left\{-\frac{(1-2\nu)\sin\vartheta}{1+\cos\vartheta}+\sin\vartheta\cos\vartheta\right\} \tag{5.18}$$

(*b*) for stresses:

$$\sigma_z = -\frac{3P}{2\pi R^2}\cos^3\vartheta \tag{5.19}$$

$$\sigma_r = \frac{P}{2\pi R^2}\left(-3\cos\vartheta\sin^2\vartheta+\frac{1-2\nu}{1+\cos\vartheta}\right) \tag{5.20}$$

$$\sigma_t = \frac{P}{2\pi R^2}(1-2\nu)\left(\cos\vartheta-\frac{1}{1+\cos\vartheta}\right) \tag{5.21}$$

$$\tau = -\frac{3P}{2\pi R^2}\cos^2\vartheta\sin\vartheta \tag{5.22}$$

5.3 VERTICAL STRESSES AND STRAINS DUE TO AN ARBITRARY SURFACE LOADING OF THE HALF-SPACE

5.3.1 Vertical Strains in the Axis of a Uniformly Loaded Circular Area

For applications in soil mechanics the integration of the Boussinesq equations for vertical stresses and strains in the axis of a uniformly loaded circular load is of special interest. The usefulness of these integrations for computing vertical displacements (settlements) of the half-space subjected to an arbitrary surface load will be shown in the following sections.

In order to obtain the vertical displacement w_0 of an arbitrary point M in the axis of a circular loaded area (Fig. 5.3), Equation (5.17) should be applied to a ring sectorial element of area $\mathrm{d}A$:

$$\mathrm{d}A = r\,\mathrm{d}\varphi\,\mathrm{d}r$$

The elementary concentrated load is then:

$$P = qr\,\mathrm{d}\varphi\,\mathrm{d}r \tag{5.23}$$

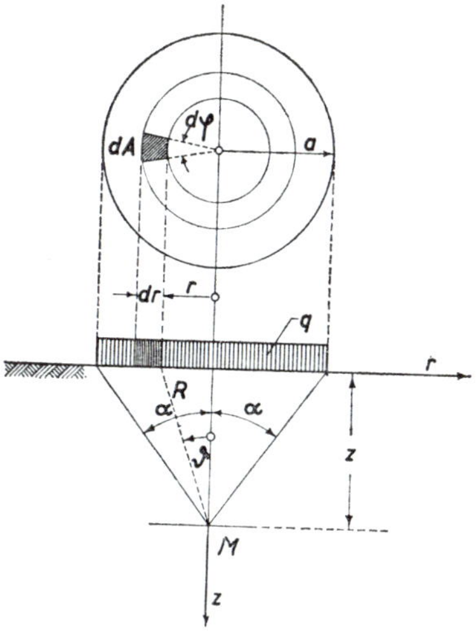

Fig. 5.3. Half-space subjected to a uniformly distributed circular load

The corresponding vertical displacement dw_0 in the depth z of the axis z is

$$dw_0 = \frac{qr\,d\varphi\,dr}{4\pi R} \cdot \frac{1}{G}\{2(1-\nu) + \cos^2\vartheta\} \tag{5.24}$$

From Fig. 5.3 it can be seen that

$$\left.\begin{aligned} r &= z\tan\vartheta, \qquad dr = z\frac{d\vartheta}{\cos^2\vartheta} \\ R &= \frac{z}{\cos\vartheta} \\ z &= a\cot\alpha \end{aligned}\right\} \tag{5.25}$$

Inserting the above expressions into Equation (5.24), integrating it between the limits $\varphi = 0$ and $\varphi = 2\pi$, $\vartheta = 0$ and $\vartheta = \alpha$, and considering the relation (4.17), we can find the following expression for the vertical displacement w of the point M lying at the depth z of the axis:

$$w_0 = \frac{2a}{E}\,\Phi_\nu(\alpha) \tag{5.26}$$

$$\Phi_\nu(\alpha) = \frac{1+\nu}{\sin\alpha}\{2(1-\nu) - (1-2\nu)\cos\alpha - \cos^2\alpha\} \tag{5.27}$$

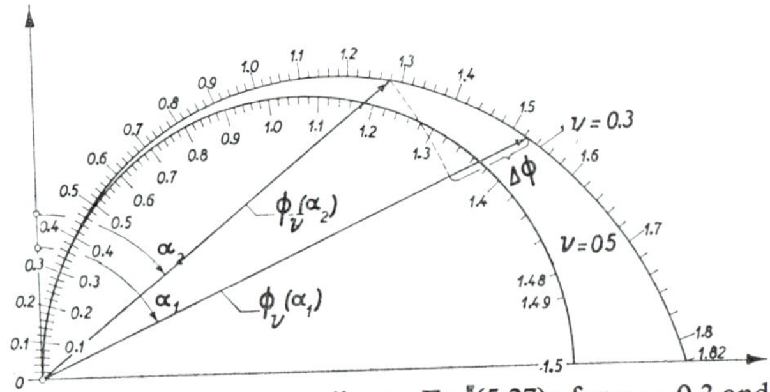

Fig. 5.4. $\phi_\nu(\alpha)$ charts corresponding to Eq.(5.27): for $\nu = 0.3$ and $\nu = 0.5$

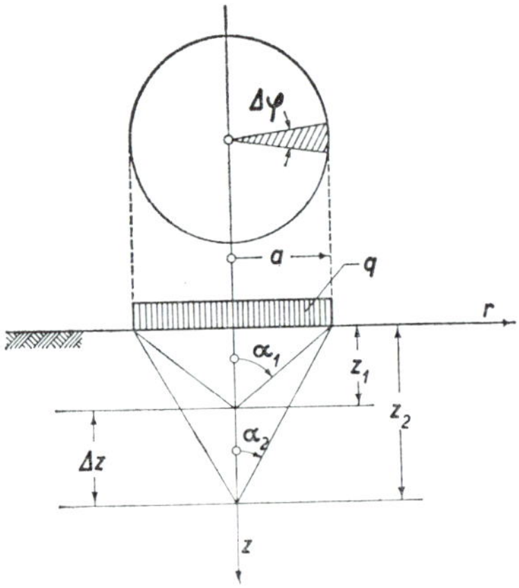

Fig. 5.5. Scheme for the computation of the vertical deformation of a horizontal layer $\Delta Z = z_2 - z_1$, subjected to a circular surface load

(Šuklje, 1960). The function $\Phi_\nu(\alpha)$ corresponding to Poisson's ratios $\nu = 0{\cdot}3$ and $\nu = 0{\cdot}5$, is presented in Fig. 5.4 in the system of polar coordinates (α, Φ_ν). In the case of $\nu = 0{\cdot}50$

$$\Phi_{0\cdot50}(\alpha) = 1{\cdot}5 \sin \alpha \tag{5.28}$$

the vector points of the function $\Phi_{0\cdot50}(\alpha)$ describe a circle of diameter 1·5.

The vertical deformation of a horizontal layer $\Delta z = z_2 - z_1$ (Fig. 5.5) in the axis of a uniformly distributed circular load can be obtained as the difference of the displacements w_{01} and w_{02} of the points of intersection of the axis z with the boundary surfaces of the layer:

$$\Delta w_0 = w_{01} - w_{02} = \frac{qa}{E}\{\Phi_\nu(\alpha_1) - \Phi_\nu(\alpha_2)\}$$

$$\Delta w_0 = \frac{qa}{E}\Delta\Phi \tag{5.29}$$

The determination of the coefficient $\Delta\Phi = \Phi_\nu(\alpha_1) - \Phi_\nu(\alpha_2)$ is shown in Fig. 5.4.

5.3.2 Vertical Deformation of a Horizontal Layer of the Half-Space Due to an Arbitrary Surface Load

A sector of the uniformly loaded circular surface of central angle $\Delta\varphi$ (Fig. 5.3) causes in the axis of the circular load a displacement Δw_s:

$$\Delta w_s = \Delta w_0 \frac{\Delta\varphi}{2\pi} \tag{5.30}$$

Expressing $\Delta\varphi = \dfrac{2\pi}{n}$ we obtain

$$\Delta w_s = \frac{\Delta w_0}{n} \tag{5.30*}$$

The settlement Δw of a layer of thickness Δz caused by an arbitrarily shaped uniformly loaded area in the vertical line passing through an arbitrary point A lying inside the load surface (Fig. 5.6) can be computed

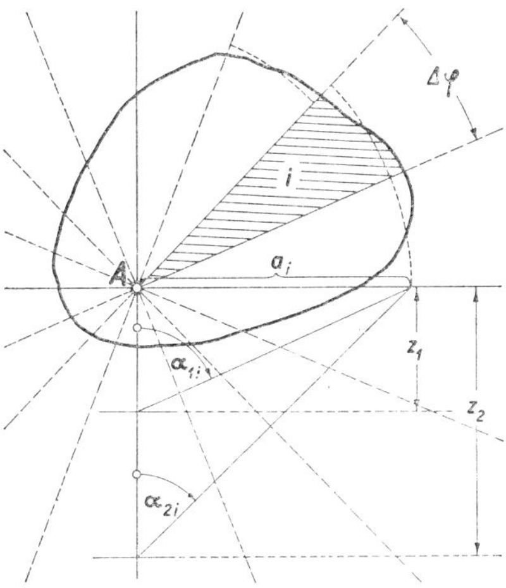

Fig. 5.6. Scheme for the computation of the vertical deformation of a horizontal layer $\Delta z = z_2 - z_1$ subjected to an arbitrarily shaped uniformly distributed surface load; deformation is computed along the vertical below the point A lying inside the loaded area

as the sum of the displacements Δw_s of single circular sectors equivalent to the parts of the load surface which are limited by the straight lines originating in the point A with angular intervals $\Delta\varphi = 2\pi/n$

$$\Delta w = \sum_{i=1}^{n} \Delta w_{si} \tag{5.31}$$

Taking into account expressions (5.30)* and (5.29), Equation (5.31) can be written in the form

$$\Delta w = \frac{1}{n} \sum_{i=1}^{n} \Delta w_{oi} = \frac{q}{nE} \sum_{i=1}^{n} a_i \Delta\Phi_i \tag{5.31-a}$$

In this equation a_i denotes the radius of the sectorial element i and $\Delta\Phi_i$ the coefficient $\Delta\Phi$ corresponding to the angles α_{1i} and α_{2i}. The radius a_i can be measured directly. In order to obtain the coefficient $\Delta\Phi_i$ as the difference of the values $\Phi_i(\alpha_1)$ and $\Phi_i(\alpha_2)$ the angles α_{1i} and α_{2i} given by the simple construction shown in Fig. 5.6 should be drawn in the diagram $\Phi = \Phi_v(\alpha)$ as illustrated in Fig. 5.4.

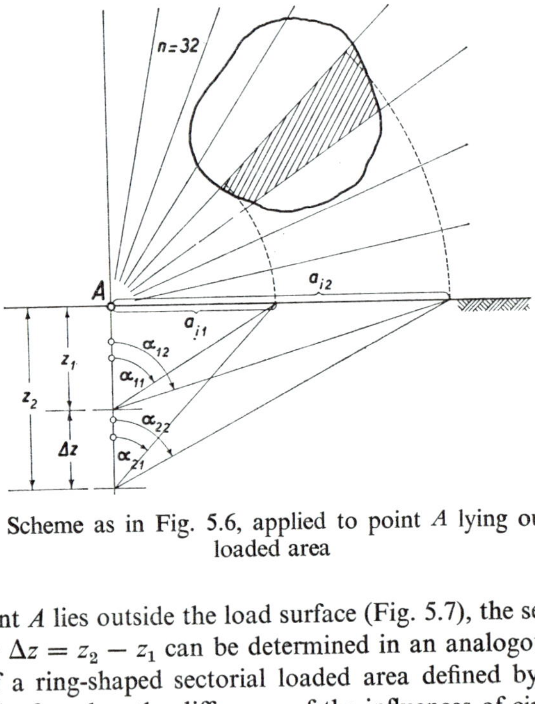

Fig. 5.7. Scheme as in Fig. 5.6, applied to point A lying outside the loaded area

If the point A lies outside the load surface (Fig. 5.7), the settlement Δw of the layer $\Delta z = z_2 - z_1$ can be determined in an analogous way. The influence of a ring-shaped sectorial loaded area defined by the radii a_2 and a_1 can be found as the difference of the influences of circular sectors the radii of which are a_2 and a_1.

Thus

$$\Delta w = \frac{q}{nE} \sum_{i=1}^{k} (a_{i2}\Delta\Phi_{i2} - a_{i1}\Delta\Phi_{i1}) \tag{5.32}$$

If the load intensity of the loaded area varies, the approximate settlement can be obtained by replacing the non-uniform load by a step load (Fig. 5.8). The settlements obtained for the loads should be superimposed:

$$\Delta w = \sum_{j=1}^{m} \Delta w_j \tag{5.33}$$

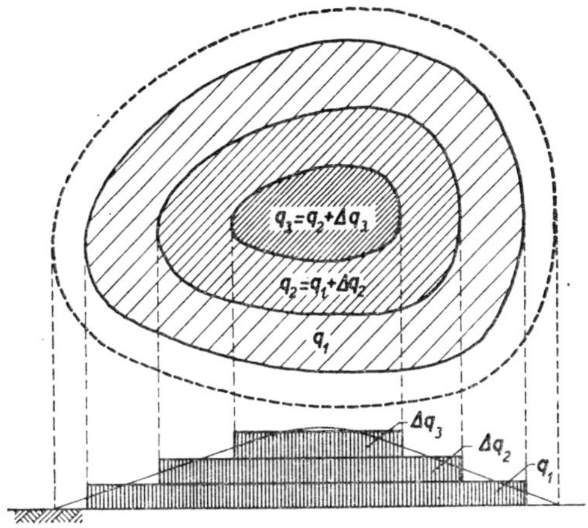

Fig. 5.8 Step load replacing a non-uniform load distribution

In the case of a uniform load the influence of each load q_j can be determined in the manner explained above.

5.3.3 Vertical Stresses for an Arbitrary Loading of the Half-Space

In the Boussinesq Equation (5.19) for the vertical stresses σ_z due to a concentrated load the force P can be replaced by the elementary load of a ring element according to Fig. 5.3 and Equation (5.23). Inserting the relations (5.25) and integrating throughout the loaded area, we obtain

$$\sigma_z = q(1 - \cos^3 \alpha) \tag{5.34}$$

Newmark (1942) first showed how Equation (5.34) can be applied to the computation of vertical pressures σ_z in an arbitrary point of the half-space subjected to an arbitrarily shaped load surface. For this purpose Equation (5.34) should be written in the form:

$$\tan \alpha = \frac{a}{z} = \sqrt{\left[\left(\frac{1}{1 - \dfrac{\sigma_z}{q}}\right)^{\frac{2}{3}} - 1\right]} \tag{5.35}$$

Then the quotients a/z have to be determined corresponding to progressive increases of the ratio σ_z/q by $\frac{1}{m}$, e.g. with $m = 10$ to progressive values of $\sigma_z/q = 0{\cdot}1, 0{\cdot}2, \ldots, 0{\cdot}9, 1{\cdot}0$.

Each loading ring between two successive circles contributes $\Delta\sigma_z = q/m$ to the pressure σ_z at the point lying at depth z below the centre of the circles. Dividing the surface of the half-space by radial straight lines originating at the centre of the circles, the contribution of each sectorial ring element to the pressure σ_z at the depth z below the centre becomes

$$\Delta\sigma_z = \frac{q}{mn} \tag{5.36}$$

n being the number of sectors: $\Delta\varphi = 2\pi/n$ (see Fig. 5.9).

The pressures due to an arbitrary load surface can then be determined in the following way. The load surface should be placed on the system of concentric circles and sectors defined above so that the point below which the stresses have to be computed coincides with the centre of the circles (Fig. 5.9). The scale of the loaded area must agree with the scale of the circles. Thus either the scale of the loaded area or the scale of the circles has to vary with the depth z for which the pressure σ_z is computed. If N is the number of sectorial ring elements covered by the load area (the

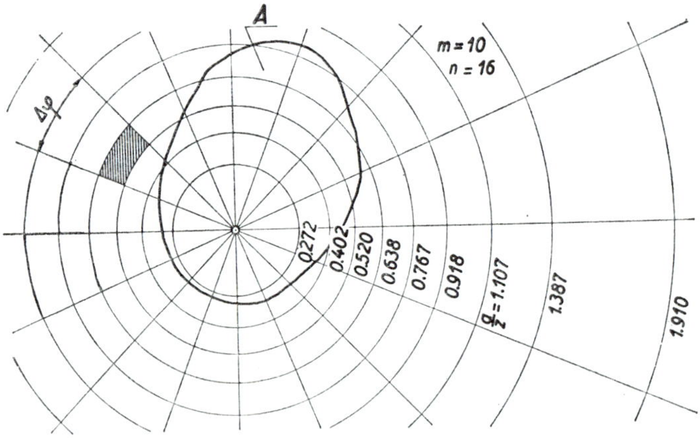

Fig. 5.9. Sectorial ring elements whose loading by q contributes $\frac{q}{mn}$ to the vertical stress in the depth z below the centre of circles (according to Newmark, 1942)

incompletely covered elements have to be taken with the percentage of the covering), the total pressure at depth z is:

$$\sigma_z = N\frac{q}{mn} \tag{5.37}$$

A non-uniform load can be replaced by a convenient step load (Fig. 5.8). The elements covered by the single step should be considered separately and the total pressure obtained as the sum of the single step influences.

The necessity of changing the scale of either the loaded area or circles is a shortcoming of the described procedure. It can be avoided by operating with whole sectorial elements which have already been applied usefully in the computation of displacements in section 5.3.2. The pressure σ_z can thus be obtained by the equation

$$\sigma_z = \frac{q}{n}\Sigma(1 - \cos^3\alpha) = \frac{q}{n}\Sigma\Psi(\alpha) \tag{5.38}$$

The function

$$\Psi(\alpha) = 1 - \cos^3\alpha \tag{5.39}$$

should be drawn in the system of polar coordinates on a sheet of transparent paper (Fig. 5.10). In the diagram $\Psi(\alpha)$ the depth z has to be marked

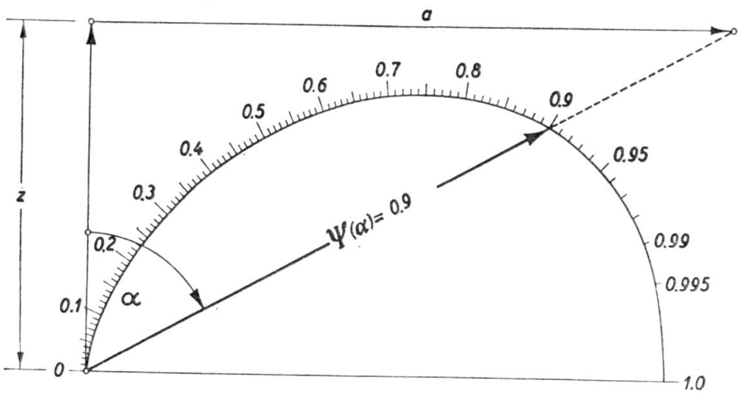

Fig. 5.10. Chart $\Psi(\alpha) = 1 - \cos^3\alpha$ (Equation 5.39)

off in the direction $\alpha = 0$ to the scale of the loaded area. Then the diagram should be placed on the plan of the load surface in such a way that its direction $\alpha = 0$ coincides with the normal of the sectorial element in question, and the end point of the co-ordinate z with the point A, below which the stresses have to be computed (Fig. 5.11). The coefficient $\Psi(\alpha)$ corresponding to the angle $\alpha = \arctan(a_i/z)$ can then be read off from the diagram in the way shown in Fig. 5.11. The product of the quotient q/n with the sum of all values $\Psi(\alpha)$ yields the pressure σ_z (Equation 5.38).

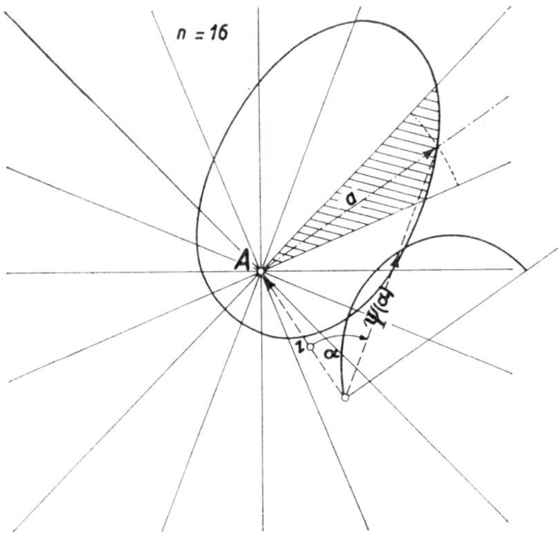

Fig. 5.11. Application of the $\Psi(\alpha)$ chart (Fig. 5.10) for the computation of vertical stresses σ_z in the depth z below the point A of a loaded area

The application of this method to the stresses along the verticals outside the loaded area, and to the stresses due to a non-uniform surface load distribution is analogous to the procedure explained for the displacement computation in such cases.

5.4 LAYERED HALF-SPACE

5.4.1 Application of the Solutions Given for Uniformly Loaded Circular Areas

5.4.1.1 *Stresses*

Using Burmister's method of numerical quadrature, Acum and Fox (1951) have presented exact solutions for the boundary stresses σ_{12} and σ_{23} (Fig. 5.12) in the axis of a circular, uniform load acting on the surface of a three-layer half-space. The solutions are given for a Poisson's ratio of $\nu = 0{\cdot}50$ and are based on the assumption of total friction along the contact surfaces. The vertical and radial stresses have been presented in eight tables as percentages of the uniform surface load q and as functions of the following parameters (Fig. 5.12):

$$k_1 = \frac{E_1}{E_2}, \quad k_2 = \frac{E_2}{E_3}, \quad H = \frac{h_1}{h_2}, \quad a_1 = \frac{a}{h_2} \tag{5.40}$$

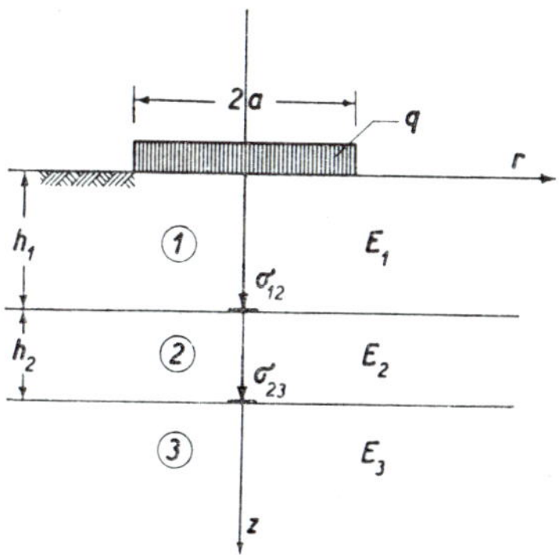

Fig. 5.12. Three-layer half-space

The tables contain results for the following combinations of the parameters k_1 and k_2: $k_1 = 5, 10, 50, 100$ and $k_2 = 5, 10, 20, 50, 100, 500$ and for the following eight combinations of the parameters H and a_1:

Table (Acum and Fox, 1951)	1	2	3	4	5	6	7	8
H	2	1	$\frac{1}{2}$	$\frac{1}{4}$	2	1	$\frac{1}{2}$	$\frac{1}{4}$
a_1	1	1	1	1	$\frac{1}{2}$	$\frac{1}{2}$	$\frac{1}{2}$	$\frac{1}{2}$

These tables enable the construction of the diagrams

$$\frac{\sigma_z}{q} = \Psi(H) \tag{5.41}$$

for certain ratios a_1 and for each combination of the above values of the parameters k_1 and k_2. The diagrams corresponding to the combination $k_1 = 10$, $k_2 = 5$ are shown in Fig. 5.13.

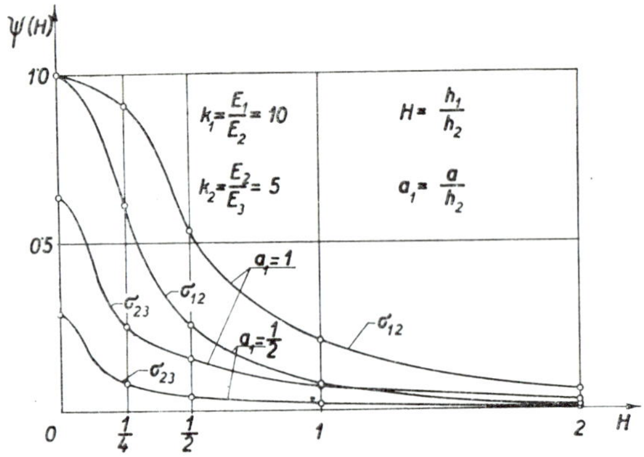

Fig. 5.13. $\frac{\sigma_{12}}{q}$ and $\frac{\sigma_{23}}{q}$ versus $H = \frac{h_1}{h_2}$ charts for $k_1 = 10$, $k_2 = 5$ and $a_1 = \frac{1}{2}$ and 1 alternatively; for notations see Fig. 5.2

The coefficients $\Psi(H)$ are analogous to the coefficients $\Psi(\alpha)$ according to Equation (5.38) for the homogeneous half-space. Thus replacing the coefficients $\Psi(\alpha)$ by the coefficients $\Psi(H)$, Equation (5.38) can be used for the layered half space as well:

$$\sigma_z = \frac{q}{n}\Psi(H) \tag{5.42}$$

Applying the solutions given by Fox (1948) the two-layer half-space can be treated in an analogous way. They correspond to the Poisson ratio $\nu = 0{\cdot}50$ and to two different assumptions concerning the friction along the contact surface of the layers: (*a*) frictionless contact, (*b*) slideless contact. In case (*b*) radial tension stresses result along the contact surface.

5.4.1.2 *Displacements*

Equation (5.26), applied for the angle $\alpha = \pi/2$, yields the settlement of the surface of the homogeneous half-space in the centre of the uniformly loaded circular area:

$$w_\infty = \frac{qa}{E} 2(1 + \nu)(1 - \nu) \tag{5.43}$$

Considering the relation (4.17) this equation can be written in the form:

$$w_\infty = \frac{qa}{G}(1 - \nu) \tag{5.44}$$

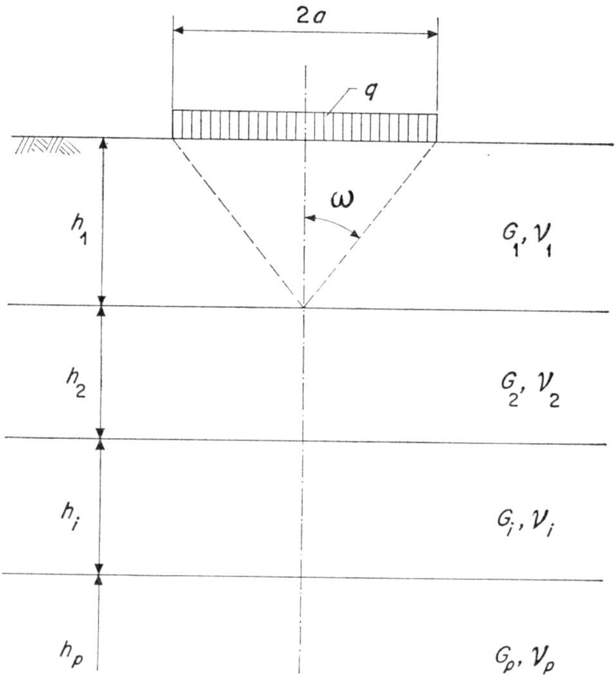

Fig. 5.14. Multi-layered isotropic half space subjected to a circular surface load

For a multi-layered isotropic half-space (Fig. 5.14) the above expression can be generalized as follows:

$$w_\infty = qa \frac{1 - \nu_1}{G_1} \Phi \tag{5.45}$$

where q and a are the unit surface load and the radius of the uniformly loaded circular area respectively; ν_1 and G_1 are the Poisson's ratio and the shear modulus of the first layer, and Φ is a function depending on the geometrical and elastic dimensionless parameters of the layers. In the homogeneous half space $\Phi = 1$. For a two-layer half-space the function Φ can be expressed as

$$\Phi(\omega, M, \nu_1, \nu_2) = 1 + G(\omega) \sin \omega \tag{5.46}$$

where $G(\omega)$ is given by Mandel's charts (Mandel, 1953, 1957, 1961) corresponding to different relations $M = G_1/G_2$ and to certain Poisson's ratios ν_1 and ν_2 ($\nu_1 = 0{\cdot}25$ and $\nu_2 = 0{\cdot}25$, $0{\cdot}50$ on available charts). For a three-layer half-space similar solutions are not yet available. Mandel (1953) has presented the solution in a general form and it should be applied to different ratios E_1/E_2, E_2/E_3, h_1/h_2, in the same way as Acum and Fox did for stresses. The results assuming zero friction along the contact sur-

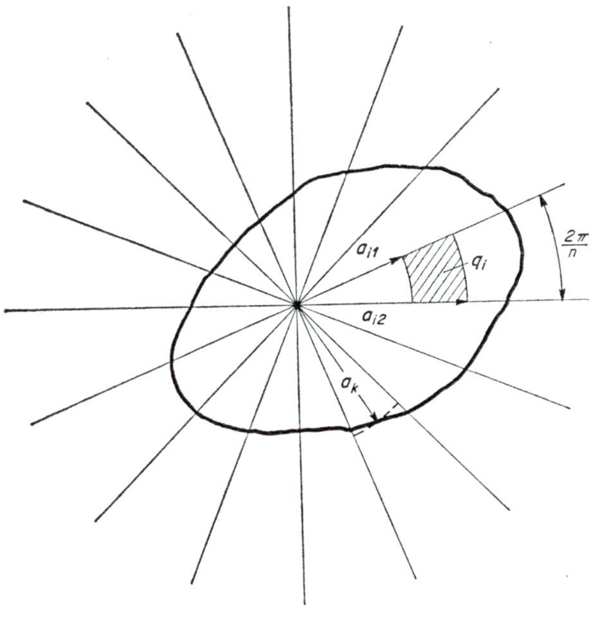

Fig. 5.15. Scheme for the application of Equation (5.46)

faces between the layers (Mandel) give results on the safe side as stated by Rappoport (1948). The pressures corresponding to friction between the layers are smaller and closer to the results obtained in the homogeneous half-space.

As for anisotropic soils, several solutions are reported by Florin (1959) to have been given by Lehnicky (1935), Stepanov (1950), Savin (1940) and Wolf (1935). For the homogeneous anisotropic half-space, Konning (1957) has presented solutions for the settlement of a concentrated load, as well as of a rigid circular plate. The integration for the centre of a uniformly loaded circular area would be desirable.

To obtain settlements of an arbitrary surface point due to an arbitrarily shaped and arbitrarily distributed surface load, expressions (5.32) and (5.45) yield the approximate solution in the form (Šuklje, 1960):

$$\rho_\infty = \frac{q_0}{n} \cdot \frac{1-\nu_1}{G_1} \sum_1^n \left\{ (a_{i1}\Phi_{i1} - a_{i2}\Phi_{i2}) \frac{q_i}{q_0} \right\} \tag{5.47}$$

(see Fig. 5.15). The loaded area must be divided into suitable ring elements for which a uniformly distributed load q_i can be assumed. The average surface pressure $q = Q/A$ can be introduced as the comparison pressure q_0.

Until charts for the multi-layered half-space with different geometrical and elastic parameters are available, the settlement of each layer can approximately be calculated as the difference between the displacements of the points at the two boundaries of the layer, as obtained in the homogeneous half-space with the elastic constants of the layer (Equation 5.32). As shown by Florin (1959), the differences in the stress distribution in the half-space are important only if the deformation moduli differ considerably.

5.4.2 Uniformly Loaded Rectangular Areas

Stresses and displacements in the upper layer of a two-layer rigid base soil system have been thoroughly investigated by Burmister (1956). Beneath the corner of a uniformly loaded (p) rectangular area ($a \,.\, b$, $b \leqslant a$) influence diagrams for vertical stresses (σ_z) and displacements (w) were prepared. Each influence diagram gives a family of influence curves for a constant value of the depth ratio, z/h, h being the thickness of the upper layer. Each curve is drawn for a constant value of the side ratio, b/a, for the basic argument, b/h, to define the influence values, $I_1 = \sigma_z/p$ and $I_2 = (w/b) \,.\, (E_1/p)$.

The influence diagrams were prepared for two values of Poisson's ratio: $\nu = 0{\cdot}2$ and $\nu = 0{\cdot}4$. Complete continuity of vertical and shear stresses and of vertical and horizontal displacements was considered to exist across the interface between the upper compressible layer and the immediate underlying, rigid base layer, with shear stresses being fully active at this interface.

As the principles of superposition of stresses and strains are valid, the influence charts can be applied for the computation of stresses and settlements in the vertical axis passing through any point A of the plain surface of the upper layer if the surface load can be divided into or composed of several positive or negative rectangular uniformly loaded areas whose sides have the directions of the x and y-axes and one corner of which coincides with the point A.

Where the application of the principles of superposition is not simple, as in the case of horizontal and shear stresses, circular influence diagrams have been prepared which readily permit graphic integrations for any shape of loaded area, located anywhere within the limits of the diagram, and for any distribution of foundation loading.

While Burmister's charts were obtained by integrating the basic solution of the two-layer rigid base soil system with a concentrated load, P, applied

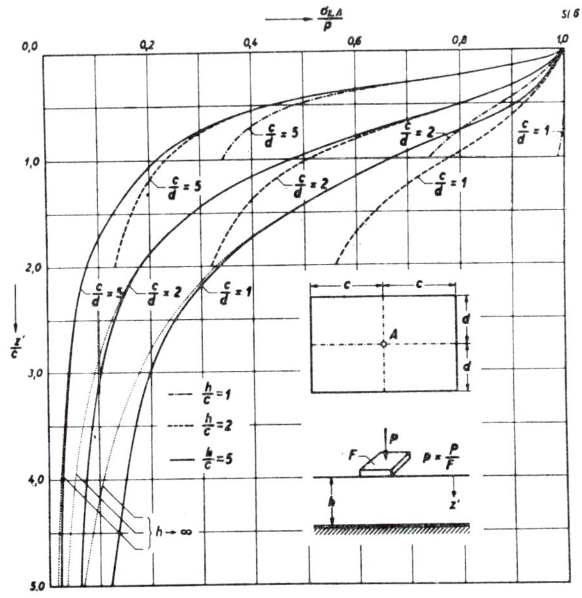

Fig. 5.16. Diagrams of vertical normal stresses below the centre of a uniformly loaded rectangular area for different thicknesses (h) of the compressible layer; after Sovinc (1961)

at the plain surface of a semi-infinite soil deposit, Sovinc (1961) developed similar solutions based on Woinowsky-Krieger equations for stresses and displacements in thick plates of limited dimensions. Sovinc's solutions are given for the Poisson's ratio $\nu = 0{\cdot}50$ assuming a frictionless contact between the elastic layer and the rigid base. Sovinc's diagrams for vertical stresses and settlements in the centre of a uniformly loaded rectangular area are reproduced in Figs. 5.16 and 5.17. In Fig. 5.16 Boussinesq's vertical stress distribution after Steinbrenner's integration (Steinbrenner, 1934) is presented by dotted lines $h \rightarrow \infty$.

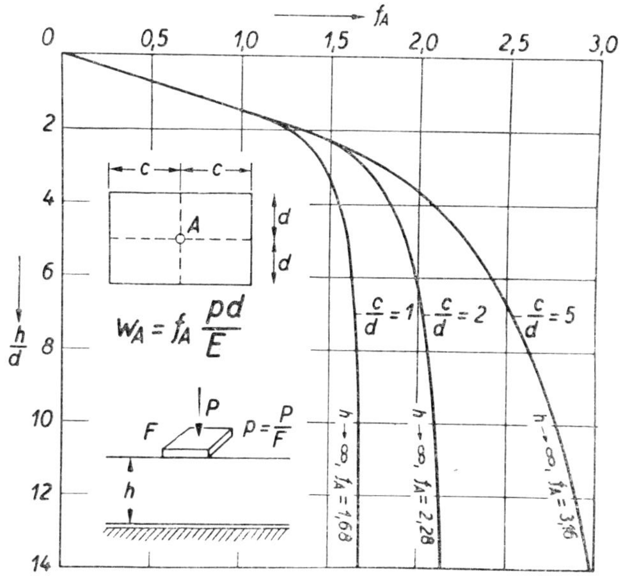

Fig. 5.17. Settlements of the centre of a uniformly loaded rectangular area, depending on the thickness (h) of the compressible layer. (Reproduced after Sovinc (1961))

6

Theories of Failure in Soil Mechanics

6.1 MOHR'S THEORY OF FAILURE

In Mohr's theory of failure the limiting stress states are determined by the function $\tau_f = f(\sigma')$. The curve presenting this function in the system of rectangular coordinates (σ,τ) is the envelope of all Mohr circles representing the limiting stress states. This envelope is characteristic of the material or, in soil mechanics, of the soil in a certain state of consistency. Such a failure envelope is shown in Fig. 6.1.

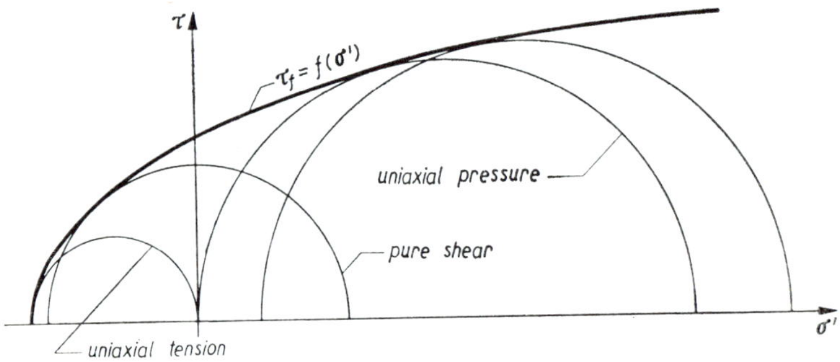

Fig. 6.1. Failure envelope of Mohr's stress circles

The failure envelope of normally consolidated cohesive soils is to good approximation a straight line:

$$\tau_f = c_0 + \tan \varphi' \, . \, \sigma' \tag{6.1}$$

τ_f is the shear strength and σ' the effective normal stress. The definition of the effective cohesion intercept c_0 and of the effective angle of shearing resistance φ' is shown in Fig. 6.2.

The envelopes of the limiting stress states for unloading or reloading pre-consolidated cohesive soils, the pre-consolidation pressure of which is denoted by σ_p', are usually curved lines forming a hysteresis loop

(Fig. 6.2). If in the first approximation this hysteresis is replaced by a straight line, the Coulomb law of shearing resistance can also be applied to preconsolidated or over-consolidated clays:

$$\tau_f = c' + \tan \varphi_r \, . \, \sigma' \tag{6.2}$$

According to experimental evidence the straight lines (6.2) corresponding to various pre-consolidation pressures σ_p' (Fig. 6.2) are approximately

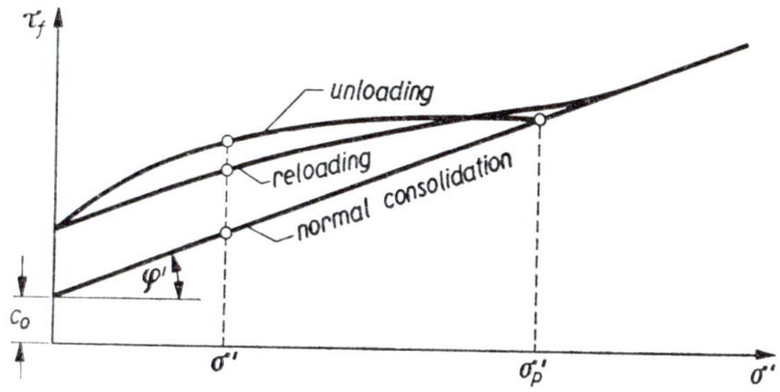

Fig. 6.2. Common formes of stress envelopes for normally consolidated and preconsolidated soils

parallel to each other. The friction angle φ_r can then be treated as a constant and the cohesion intercept c' can be expressed by the relation

$$c' = c_0 + \tan \varphi_c \, . \, \sigma_p' \tag{6.3}$$

Thus the shear strength law can be written for normally consolidated as well as for over-consolidated soils in the form

$$\tau_f = (c_0 + \tan \varphi_c \, . \, \sigma_p') + \tan \varphi_r \, . \, \sigma' \tag{6.4}$$

taking, in the case of normally consolidated soils, $\sigma_p' = \sigma'$. If the total pressures are denoted by σ and σ_p and the corresponding pore pressures by u and u_p respectively, the effective stresses are given by the relations

$$\sigma' = \sigma - u$$

$$\sigma_p' = \sigma_p - u_p$$

In Equations (6.1), (6.3) and (6.4) the cohesion intercept c_0 is small in value and can usually be neglected. According to Hvorslev (1937) Equation (6.4) can then be written in the form

$$\tau_f = \kappa \sigma_e + \tan \varphi_r \, . \, \sigma' \tag{6.5}$$

σ_e is the equivalent consolidation pressure corresponding, from the virgin curve $\sigma_e = f(e)$ of oedometer tests, to the void ratio e of the shear test sample occurring in the shear plane after shearing. Writing Equation (6.5) in the form

$$\frac{\tau_f}{\sigma_e} = \kappa + \tan\varphi_r \frac{\sigma'}{\sigma_e} \tag{6.5a}$$

the equation of a straight line in the system of coordinates

$$\left(\frac{\sigma'}{\sigma_e}, \frac{\tau_f}{\sigma_e}\right)$$

is obtained intersecting the axis of ordinates at the value κ and being inclined at the angle φ_r to the axis σ'/σ_e; κ and φ_r are constant characteristics of the soils for each state of consistency.

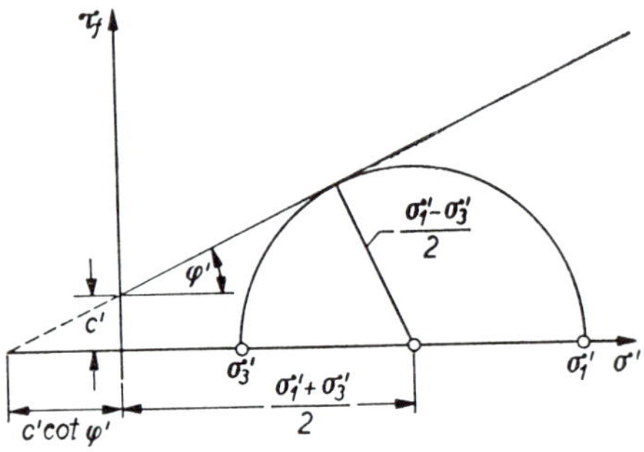

Fig. 6.3. Notation to Equation (6.6)

Using the notation of Fig. 6.3 the Coulomb law of shearing resistance (6.2) can be expressed, when interpreted according to Mohr's failure theory, as follows:

$$\frac{\sigma_1' - \sigma_3'}{2} = \left(c' \cot\varphi' + \frac{\sigma_1' + \sigma_3'}{2}\right) \sin\varphi'$$

$$\sigma_1' - \sigma_3' = 2c' \cos\varphi' + (\sigma_1' + \sigma_3') \sin\varphi' \tag{6.6}$$

In the case $c' = 0$, Equation (6.6) yields:

$$\frac{\sigma_3'}{\sigma_1'} = \frac{1 - \sin\varphi'}{1 + \sin\varphi'} = \tan^2\left(\frac{\pi}{4} - \frac{\varphi'}{2}\right) \tag{6.7}$$

In Mohr's theory failure is independent of the mean principal stress. In the two-parameter system the influence of the mean principal stress is partly taken into account when using the failure law:

$$\tau_{\text{oct}} = f(\sigma_{\text{oct}}) \tag{6.8}$$

6.2 FAILURE SURFACES IN THE COORDINATE SYSTEM OF PRINCIPAL STRESSES

6.2.1 Failure Conditions in Triaxial Tests Using Mohr's Failure Theory

Expressing the stress states in the coordinate system of principal stresses as shown in section 2.1, and considering the condition $\sigma_2' = \sigma_3'$, the Coulomb shear-strength law, expressed by Equation (6.6), is presented

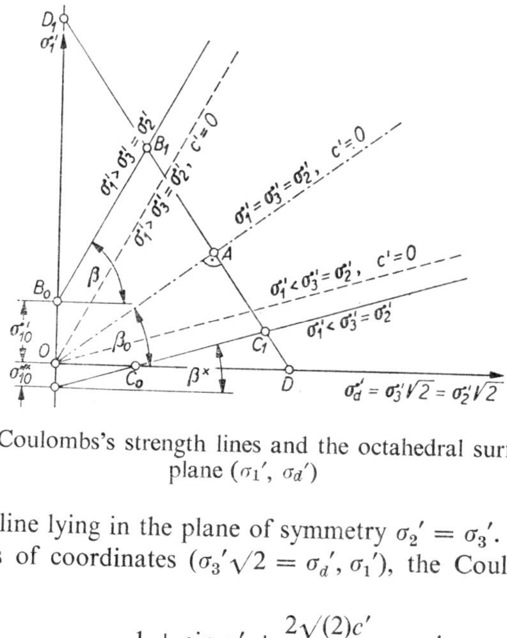

Fig. 6.4. Coulombs's strength lines and the octahedral surface in the plane (σ_1', σ_d')

by a straight line lying in the plane of symmetry $\sigma_2' = \sigma_3'$. In this plane with the axes of coordinates ($\sigma_3'\sqrt{2} = \sigma_d'$, σ_1'), the Coulomb law has the equation:

$$\frac{\sigma_1'}{\sigma_d'} = \frac{1 + \sin\varphi' + \dfrac{2\sqrt{(2)}c'}{\sigma_d'}\cos\varphi'}{(1 - \sin\varphi')\sqrt{2}} \tag{6.9}$$

The inclination coefficient of this straight line (Fig. 6.4) is

$$\tan\beta = \frac{\sqrt{2}}{2} \cdot \frac{1 + \sin\varphi'}{1 - \sin\varphi'} = \frac{\sqrt{2}}{2}\tan^2\left(\frac{\pi}{4} + \frac{\varphi'}{2}\right) \tag{6.10}$$

and its intercept on the axis σ_1' is

$$\sigma_{10}' = 2c' \frac{\cos \varphi'}{1 - \sin \varphi'} \tag{6.11}$$

Equations (6.9) to (6.11) are valid on condition

$$\sigma_1' \geqslant \sigma_3' = \sigma_2'. \tag{6.12}$$

When

$$\sigma_1' \leqslant \sigma_3' = \sigma_2' \tag{6.13}$$

the relation (6.9) should be replaced by the expression:

$$\frac{\sigma_1'}{\sigma_d'} = \frac{\sigma_1'}{\sqrt{(2)}\sigma_3'} = \frac{1 - \sin \varphi' - \dfrac{2\sqrt{(2)}c'}{\sigma_d'} \cos \varphi'}{(1 + \sin \varphi')\sqrt{2}} \tag{6.14}$$

In this case the Coulomb shear law is presented, in the coordinate system ($\sigma_d' = \sigma_3'\sqrt{2}, \sigma_1'$), by a straight line (Fig. 6.4) having the coefficient of inclination:

$$\tan \beta^\star = \frac{\sqrt{2}}{2} \cdot \frac{1 - \sin \varphi'}{1 + \sin \varphi'} = \frac{\sqrt{2}}{2} \tan^2 \left(\frac{\pi}{4} - \frac{\varphi'}{2}\right) \tag{6.15}$$

and intersecting the axis σ_1' at the intercept:

$$\sigma_{10}'^\star = -2c' \frac{\cos \varphi'}{1 + \sin \varphi'} \tag{6.16}$$

The octahedral normal, the equation of which is

$$\sigma_1' = \sigma_2' = \sigma_3' \tag{6.17}$$

has the direction coefficient (Fig. 6.4)

$$\tan \beta_0 = \frac{1}{\sqrt{2}} = \frac{\sqrt{2}}{2} \tag{6.18}$$

In Fig. 6.4 the octahedral surface with this normal is presented by the straight line $\overleftrightarrow{D_1 D}$. This surface will now be taken as the plane of projection. Fig. 6.5 illustrates the projections of the axes σ_1, σ_2 and σ_3 onto this plane. In addition, the projections of the contours of the octahedral area, the normal of which is OA (Fig. 6.4), [that is of the triangle $(D_1\ D_2\ D_3)$], and also the points of intersection between this octahedral area and the straight lines $\overleftrightarrow{B_0\ B_1}$ and $\overleftrightarrow{C_0\ C_1}$ representing the Coulomb shear strength law in the cases $\sigma_1' > \sigma_2' = \sigma_3'$ and $\sigma_1' < \sigma_2' = \sigma_3'$ respectively,

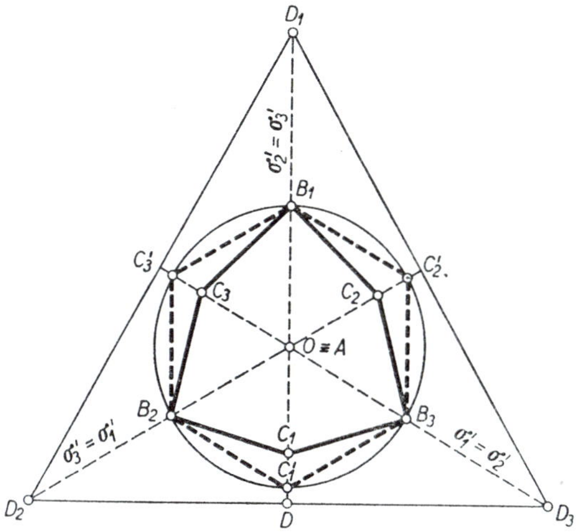

Fig. 6.5. Contours of failure surfaces on the octahedral plane: Coulomb's (thick full line), maximum shear stress (dashed line) and Mises's (circle) failure criterium

are included. The points of intersection of the straight lines representing the Coulomb straight envelopes of the Mohr circles are for the cases:

$\sigma_2' > \sigma_3' = \sigma_1'$	point B_2
$\sigma_2' < \sigma_3' = \sigma_1'$	point C_2
$\sigma_3' > \sigma_1' = \sigma_2'$	point B_3
$\sigma_3' < \sigma_1' = \sigma_2'$	point C_3

Connecting the points $B_1 - C_2 - B_3 - C_1 - B_2 - C_3 - B_1$, an irregular hexagon with axial symmetry is obtained. The lines connecting the joints of this hexagon to the corresponding intercepts on the coordinate axes B_{0i} or C_{0i} $(i = 1,2,3)$ (Fig. 6.4), define the edges of a truncated pyramid. These edges represent the failure states corresponding to the Mohr-Coulomb envelope for different possible combinations of stress states of common triaxial tests. Such failure states can be expressed by the following general equation (Hvorslev, 1960):

$$\{(\sigma_1' - \sigma_2')^2 - [2c' \cos \varphi' + (\sigma_1' + \sigma_2') \sin \varphi']^2\} . \{(\sigma_2' - \sigma_3')^2 - - [2c' \cos \varphi' + (\sigma_2' + \sigma_3') \sin \varphi']^2\} . \{(\sigma_3' - \sigma_1')^2 - - [2c' \cos \varphi' + (\sigma_3' + \sigma_1') \sin \varphi']^2\} = 0 \qquad (6.19)$$

Equation (6.19) contains three equations of the type (6.6), i.e. for the stress states of the type $\sigma_1' > \sigma_3' = \sigma_2'$ as well as three analogous equations for the stress condition of the type $\sigma_1' < \sigma_3' = \sigma_2'$.

The surfaces of this pyramid represent one of the possible failure relationships for the stress states, the mean principal stress of which is not equal to the greatest or smallest principal stress. This relationship applies also for such stress states in the case when the shear angle φ' remains constant.

6.2.2. Maximum Shear Stress as the Failure Criterion

In the case $\varphi' = 0$ the expression (6.19) reduces to

$$[(\sigma_1' - \sigma_2')^2 - 4c^2][(\sigma_2' - \sigma_3')^2 - 4c^2][(\sigma_3' - \sigma_1')^2 - 4c^2] = 0 \qquad (6.20)$$

From this equation six failure conditions can be deduced, e.g. two of them in the form

$$\pm(\sigma_1' - \sigma_2') = 2c \qquad (6.20a)$$

In the system of coordinates (σ_1, σ_2, σ_3), these six conditions are presented by the edges of a hexagonal prism, the axis being the octahedral normal and the projection on the octahedral plane the regular hexagon $B_1 - C_2' - B_3 - C_1' - B_2 - C_3'$ (Fig. 6.5). The edges of the prism lie in the planes of symmetry $\sigma_2' = \sigma_3'$, $\sigma_3' = \sigma_1'$, $\sigma_1' = \sigma_2'$. The surface of the prism—in the projection presented by the edges of the hexagon—is the geometrical position of the points representing stress states, the mean principal stress of which equals neither the greatest nor the smallest principal stress. This surface represents one of the possible failure criteria which takes account of the influence of the mean principal stress.

6.2.3 Specific Distortional Energy as the Failure Criterion

The specific deformation energy, i.e. the deformation energy per unit of volume, is given by the expression

$$W = \frac{1}{2E} \{(\sigma_1^2 + \sigma_2^2 + \sigma_3^2) - 2\nu(\sigma_2\sigma_3 + \sigma_3\sigma_1 + \sigma_1\sigma_2)\} \qquad (6.21)$$

Applying the above expression to the distortional component of the stress tensor:

$$\begin{bmatrix} \sigma_1 - \sigma_{oct} & 0 & 0 \\ 0 & \sigma_2 - \sigma_{oct} & 0 \\ 0 & 0 & \sigma_3 - \sigma_{oct} \end{bmatrix}$$

the specific distortional energy is obtained. After appropriate reduction the following expression can be obtained:

$$W_d = \frac{1}{12G}\{(\sigma_2 - \sigma_3)^2 + (\sigma_3 - \sigma_1)^2 + (\sigma_1 - \sigma_2)^2\} \tag{6.22}$$

Taking into account the relation (2.13) the above equation can also be written in the form:

$$W_d = \frac{3}{4G}\tau_{oct}^2 \tag{6.22a}$$

After Von Mises's hypothesis failure occurs when the specific distortional energy reaches a certain critical value. Considering Equation (6.22) the failure criterion can be expressed in the form

$$\tau_{oct}^2 = \text{const.} \tag{6.23}$$

As stated in section 2.1, the geometrical position of constant octahedral shear stresses τ_{oct} is the surface of a cylinder, the axis of which is the octahedral normal. The projection of this cylinder onto the octahedral area (Fig. 6.5) is a circle of radius $\overline{OB_1} = \overline{OB_2} = \overline{OB_3}$.

6.2.4 Failure Surfaces for Non-cohesive Soils

Concerning the role of the mean principal stress, an excellent review of the experimental evidence for non-cohesive as well as for cohesive soils was given by Hvorslev (1960). Here the principal statements only will be outlined.

For non-cohesive soils the experiences of Bishop and Eldin (1953) as well as those of Kirkpatrick (1957) lead to the conclusion that the magnitude of the shear angle φ' does not depend on whether the mean principal stress equals the maximum (points B in Fig. 6.6) or minimum (points C in Fig. 6.6) principal stress. Thus compression and extension triaxial tests yield the same value of the shear angle φ'.

On the other hand the theoretical considerations of Caquot (1934) as well as the experimental statements of Kirkpatrick prove that the angle φ'

increases somewhat for the intermediate values of the mean principal stress. Thus the cross-section of the corresponding failure surfaces with the octahedral plane (Fig. 6.6) should be a geometrical figure based on the outline of the hexagon representing the Mohr-Coulomb failure theory. In the Fig. 6.6 such a cross-section of the failure surface is presented by a dashed line.

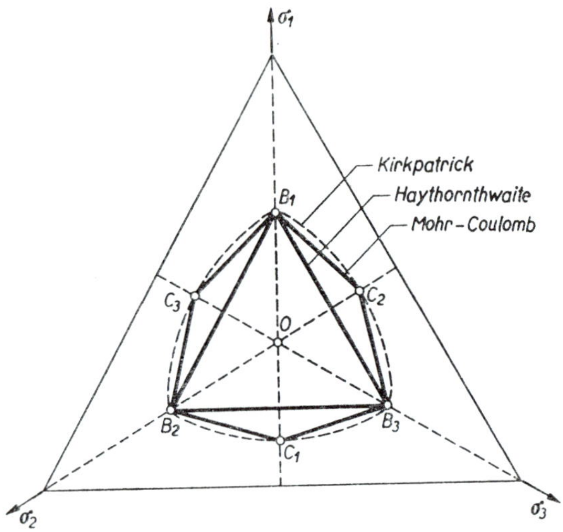

Fig. 6.6. Contours of failure surfaces on the octahedral plane according to Kirkpatrick (1957) and Haythornthwaite (1960)

On the contrary the test results obtained by Habib (1953), Peltier (1957) and Haythornthwaite (1960) lead to the conclusion that the shear angle φ' corresponding to extension tests (where the mean principal stress equals the greater principal stress) is smaller than the angle corresponding to compression tests. For this reason Haythornthwaite has proposed that the failure surfaces be presented as the edges of a triangular pyramid having the cross-section $B_1 - B_2 - B_3$ (Fig. 6.6) in the octahedral plane.

The incompatibility of the above mentioned conclusions requires further investigations.

6.2.5 Failure Surfaces for Cohesive Soils

For cohesive soils investigation has not yielded similar results either The extensive research work of Henkel (1958, 1959, 1960-a, 1960-b indicates, however, that the mean principal stress does not considerably affect the value of the shear angle φ' for states of stress represented by the points lying in the planes of symmetry of the system of co-ordinates (σ_1, σ_2, σ_3). Nevertheless the figure of failure lying in the octahedral area

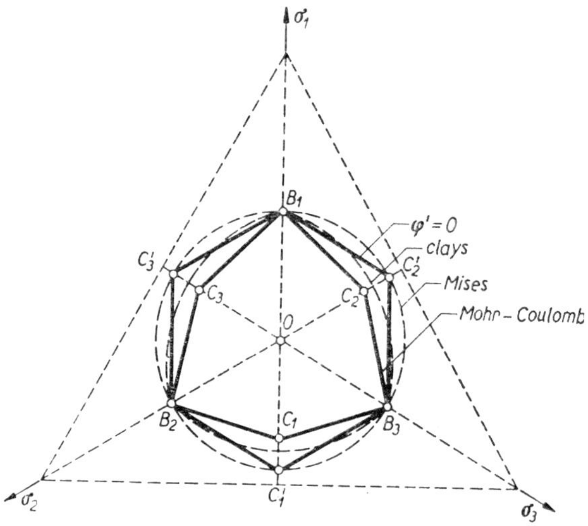

Fig. 6.7. Contours of failure surfaces for clays on the octahedral plane

is probably curved and has been drawn on the basis of the irregular hexagon which represents the Mohr-Coulomb theory (dashed line in Fig. 6.7). For the states corresponding to $\varphi' = 0$, Von Mises's failure cylinder could also serve as a good approximation of the surface of failure of over-consolidated cohesive soils.

7

Fundamental Applications of the Theories of Plasticity in Soil Mechanics

7.1 LIMITING STATES OF PLASTIC EQUILIBRIUM. KÖTTER'S DIFFERENTIAL EQUATIONS

For the limiting states of plastic equilibrium in soil mechanics the theories of plasticity have commonly been developed according to the assumption that plastic flow begins at stress states the Mohr circles of which touch the rupture envelope. Coulomb's straight line of shearing resistance

$$\tau_f = c' + \sigma' \tan \varphi' \tag{7.1}$$

is usually taken to be the envelope of the limiting stress circles. The definition of the shear characteristics, i.e. of the cohesion c' and of the shear angle φ', is shown in Fig. 7.1; σ' is the effective pressure.

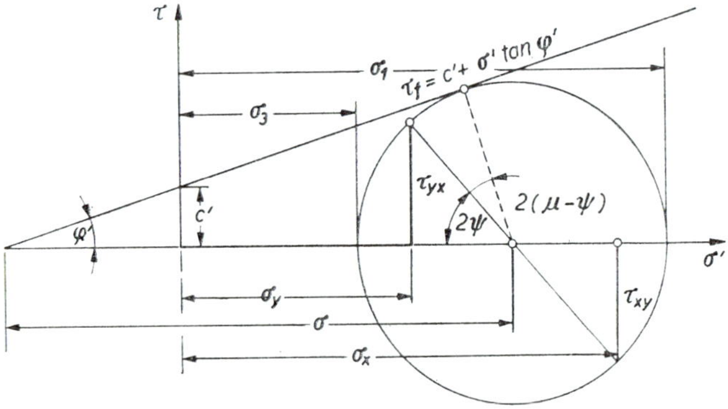

Fig. 7.1. Notation to Equations (7.1), (7.3) and (7.4)

In this way the soil is treated as a Saint-Venant body (see Section 4.1.6) applied to the Mohr-Coulomb hypothesis concerning failure or the beginning of plastic flow. Fundamental equations determining the states of stress can be obtained by introducing the expressions satisfying the condition of limiting equilibrium (7.1) into the equations of equilibrium (2.2), in which Equation (2.3) should also be taken into account. In this way two differential equations—first deduced by Kötter (1903)—can be obtained for two-dimensional states of stress. Since it was not possible to integrate these differential equations, they were not used for the solution of problems for a long period. Today, however, several methods are known for facilitating the iterative integration of Kötter's equations or for applying them to certain assumed rupture surfaces the forms of which are approximate, yet probable. Such methods will be presented shortly in the following Sections. Kötter's equations will be deduced in the forms applied to the special parameters of the methods treated.

Using such integration methods, the states of stress corresponding to limiting plastic equilibrium can be investigated if the stress-strain relationships as given for the Saint-Venant body by Equations (4.21) and (4.24) are neglected. Once the state of stress is solved these equations can be applied in order to obtain the displacements. In common cases, however, the solution is very complicated, even if the constant of proportionality λ is known.

7.2 SOKOLOVSKI'S METHOD

7.2.1 General Equations

Our investigations are limited to two-dimensional problems for which the equilibrium Equations (2.2) reduce to:

$$\left.\begin{aligned} \frac{\partial \sigma_x}{\partial x} + \frac{\partial \tau_{xy}}{\partial y} &= X \\ \frac{\partial \tau_{xy}}{\partial x} + \frac{\partial \sigma_y}{\partial y} &= Y \end{aligned}\right\} \tag{7.2}$$

If the x-axis is in the downward vertical direction (Fig. 7.2) and if the pure gravity potential field only is considered, the components X and Y of the volume forces per unit of volume are

$$X = \gamma, \qquad Y = 0$$

where γ is the volumetric weight.

From Fig. (7.1) the following equations can be deduced:

$$\left.\begin{matrix}\sigma_x\\ \sigma_y\end{matrix}\right\} = \sigma(1 \pm \sin\varphi\cos 2\psi) - c\cot\varphi \tag{7.3}$$

$$\tau_{xy} = \sigma\sin\varphi\sin 2\psi \tag{7.4}$$

The notation is given in Figs. 7.1 and 7.2, ψ being the angle between the direction x and the direction of the principal stress σ_1.

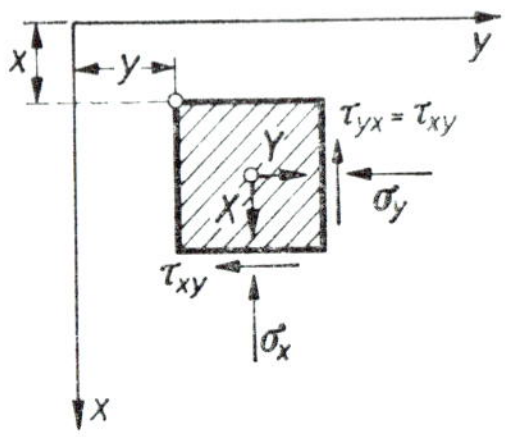

Fig. 7.2. Notation to Equations (7.2), (7.3) and (7.4)

Inserting Equations (7.3) and (7.4), which express the limiting equilibrium conditions, into the equilibrium Equation (7.2), the fundamental equations are obtained:

$$(1 + \sin\varphi\cos 2\psi)\frac{\partial\chi}{\partial x} + \sin\varphi\sin 2\psi\frac{\partial\chi}{\partial y} - \\ - \cos\varphi\left(\sin 2\psi\frac{\partial\psi}{\partial x} - \cos 2\psi\frac{\partial\psi}{\partial y}\right) = \frac{\cot\varphi}{2\sigma}X \tag{7.5}$$

$$\sin\varphi\sin 2\psi\frac{\partial\chi}{\partial x} + (1 - \sin\varphi\cos 2\psi)\frac{\partial\chi}{\partial y} + \\ + \cos\varphi\left(\cos 2\psi\frac{\partial\psi}{\partial x} + \sin 2\psi\frac{\partial\psi}{\partial y}\right) = \frac{\cot\varphi}{2\sigma}Y \tag{7.6}$$

The function χ is defined by the expression

$$\chi = \frac{\cot\varphi}{2}\ln\frac{\sigma}{\sigma_0}, \tag{7.7}$$

σ_0 being an arbitrary stress chosen in advance (Sokolovski, 1954).

Now, two new variables will be introduced: ξ and η

$$\left.\begin{matrix}\xi = \chi + \psi\\ \eta = \chi - \psi\end{matrix}\right\} \tag{7.8}$$

as well as the sign μ:

$$2\mu = \frac{\pi}{2} - \varphi \tag{7.9}$$

Multiplying Equation (7.5) firstly by $\sin(\psi - \mu)$ and secondly by $\sin(\psi + \mu)$, and Equation (7.6) firstly by $-\cos(\psi - \mu)$ and secondly by $-\cos(\psi + \mu)$, and adding the corresponding first and second pair of expressions obtained, we get:

$$\begin{aligned} \frac{\partial \eta}{\partial x} + \tan(\psi - \mu)\frac{\partial \eta}{\partial y} &= a \\ \frac{\partial \xi}{\partial x} + \tan(\psi + \mu)\frac{\partial \xi}{\partial y} &= b \end{aligned} \tag{7.10}$$

The meaning of the symbols a and b is:

$$\left.\begin{matrix} a \\ b \end{matrix}\right\} = \pm \frac{X\sin(\psi \pm \mu) - Y\cos(\psi \pm \mu)}{2\sigma \sin\varphi \cos(\psi \mp \mu)} \tag{7.11}$$

Inserting Equations (7.10) into the expressions for the total differentials:

$$\left.\begin{aligned} \mathrm{d}\xi &= \frac{\partial \xi}{\partial x}\mathrm{d}x + \frac{\partial \xi}{\partial y}\mathrm{d}y \\ \mathrm{d}\eta &= \frac{\partial \eta}{\partial x}\mathrm{d}x + \frac{\partial \eta}{\partial y}\mathrm{d}y \end{aligned}\right\} \tag{7.12}$$

and solving the expressions obtained with respect to the derivatives we obtain the equations:

$$\left.\begin{aligned} \frac{\partial \xi}{\partial x} &= \frac{b\,\mathrm{d}y - \tan(\psi + \mu)\,\mathrm{d}\xi}{\mathrm{d}y - \tan(\psi + \mu)\,\mathrm{d}x} \\ \frac{\partial \eta}{\partial x} &= \frac{a\,\mathrm{d}y - \tan(\psi - \mu)\,\mathrm{d}\eta}{\mathrm{d}y - \tan(\psi - \mu)\,\mathrm{d}x} \end{aligned}\right\} \tag{7.13}$$

as well as two analogous expressions for the derivatives $\partial\xi/\partial y$ and $\partial\eta/\partial y$.

If the numerator and the denominator of these expressions vanish simultaneously, the differential equations of the so-termed 'characteristics' result:

$$\frac{\mathrm{d}y}{\mathrm{d}x} = \tan(\psi - \mu), \quad \frac{\partial \eta}{\mathrm{d}x} = a \tag{7.14}$$

$$\frac{\mathrm{d}y}{\mathrm{d}x} = \tan(\psi + \mu), \quad \frac{\mathrm{d}\xi}{\mathrm{d}x} = 0 \tag{7.15}$$

Thus, two separate families of characteristics inclined to the x-axis at the angles $(\psi \mp \mu)$ exist. Now Fig. 7.1 shows that the angle between the rupture lines and the axis x has the same value. Consequently the characteristics are the rupture lines intersecting each other at the angle 2μ.

If in Equations (7.13) only the denominator vanishes, the differential equations of the lines of discontinuity result:

$$\text{either} \qquad \frac{dy}{dx} = \tan(\psi - \mu) \qquad \text{or} \qquad \frac{dy}{dx} = \tan(\psi + \mu). \tag{7.16}$$

The angle between these lines and the x-axis is also $(\psi \mp \mu)$; consequently the lines of discontinuity are either rupture lines or their envelopes.

In the lines of discontinuity the derivatives given by Equations (7.13) are indefinite and consequently the derivatives of stresses are indefinite as well; at the lines of discontinuity the stresses change in steps. The lines of discontinuity are boundary lines between elastic and plastic domains.

Taking into account the boundary conditions the solutions of Equations (7.14) and (7.15) may be reached by numerical progressive integration; the differentials should be replaced by finite differences.

7.2.2 Equations for Wedge-shaped Plastic Domains

Many soil mechanics stability problems can be solved by using the equations of limiting equilibrium valid for wedge-shaped plastic domains and limited by two boundary straight lines ϑ_0 and ϑ_1 (see Fig. 7.3).

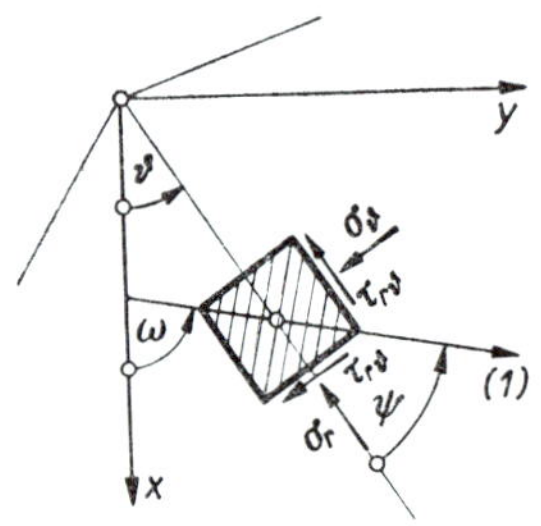

Fig. 7.3. Notation for stresses and stress directions in wedge-shaped plastic domains

Applying polar coordinates and considering $X = \gamma$, $Y = 0$, the equilibrium Equations (7.2) have the form:

$$\left.\begin{aligned} \frac{\partial \sigma_r}{\partial r} + \frac{1}{r}\frac{\partial \tau_{r\vartheta}}{\partial \vartheta} + \frac{\sigma_r - \sigma_\vartheta}{r} &= \gamma \cos\vartheta \\ \frac{\partial \tau_{r\vartheta}}{\partial r} + \frac{1}{r}\frac{\partial \sigma_\vartheta}{\partial \vartheta} + \frac{2\tau_{r\vartheta}}{r} &= -\gamma \sin\vartheta \end{aligned}\right\} \tag{7.17}$$

By analogy with Equations (7.3) and (7.4) and considering Fig. 7.1 we obtain:

$$\left.\begin{matrix}\sigma_r\\ \sigma_\vartheta\end{matrix}\right\} = \sigma(1 \pm \sin\varphi\cos 2\psi) - c\cot\varphi \tag{7.18}$$

$$\tau_{r\vartheta} = \sigma\sin\varphi\sin 2\psi \tag{7.19}$$

$$\sigma = \frac{\sigma_r + \sigma_\vartheta}{2} + c\cot\varphi \tag{7.20}$$

The notation is given by Fig. 7.3, ψ being the angle between the direction ϑ and the direction of the principal axis (1).

The following relationships are assumed:

$$\psi = \psi(\vartheta) \tag{7.21}$$

$$\omega = \omega(\vartheta) \tag{7.22}$$

$$\sigma = \gamma r s(\vartheta). \tag{7.23}$$

According to the assumption (7.23) the stresses along the rays ϑ are linearly proportional to the distances r. Considering the above assumptions the limiting equilibrium conditions (7.18) and (7.19) are to be inserted into the equilibrium Equations (7.17).

Hence

$$\sin\varphi\sin 2\psi\frac{ds}{d\vartheta} + 2s\sin\varphi\cos 2\psi\left(\frac{d\psi}{d\vartheta} + 1\right) + \\ + s(2 + \sin\varphi\cos 2\psi) = \cos\vartheta \tag{7.24}$$

$$(1 - \sin\varphi\cos 2\psi)\frac{ds}{d\vartheta} + 2s\sin\varphi\sin 2\psi\left(\frac{d\psi}{d\vartheta} + 1\right) + \\ + s\sin\varphi\sin 2\psi = -\sin\vartheta \tag{7.25}$$

Solving the above two equations with respect to the derivatives, we obtain:

$$\frac{d\psi}{d\vartheta} = \frac{\cos\vartheta - \sin\varphi\cos(2\psi + \vartheta) - s\cos^2\varphi}{2s\sin\varphi(\cos 2\psi - \sin\varphi)} - 1 \tag{7.26}$$

$$\frac{ds}{d\vartheta} = \frac{-\sin(2\psi + \vartheta) + s\sin 2\psi}{\cos 2\psi - \sin\varphi} \tag{7.27}$$

Knowing the boundary conditions either for ψ or s along two boundary rays ϑ, between which a plastic domain is assumed, the shape of the failure line as well as the stress coefficients s along it can be determined by progressive integration. If the values ψ_0 and ψ_n along the boundary lines

ϑ_0 and ϑ_n of a plastic field are known, different values s_0 can be assumed and integration can be started at ϑ_0. A certain, sufficiently small, but finite interval $\Delta\vartheta_1$ is chosen and by using Equations (7.26) and (7.27) the values $\Delta\psi_1$ and Δs_1 may be computed. They must correspond to the chosen interval $\Delta\vartheta_1$, to the given shear angle φ, to the initial values ϑ_0, ψ_0 as well as to the assumed coefficient s_0. The resulting new values:

$$\vartheta_1 = \vartheta_0 + \Delta\vartheta_1, \quad \psi_1 = \psi_0 + \Delta\psi_1 \quad \text{and} \quad s_1 = s_0 + \Delta s_1$$

are the starting values for an analogous computation of further differences $\Delta\psi_2$ and Δs_2 as well as of the corresponding values ψ_2 and s_2 for the ray $\vartheta_2 = \vartheta_1 + \Delta\vartheta_2$. Having arrived by this progressive computation at the straight line ϑ_n, the resulting ψ_n value must coincide with the boundary condition for ψ_n. The correct value of s_0 is the one for which the resulting and the given values of ψ_n are the same. The corresponding s_n value is obtained simultaneously.

7.3 BRINCH HANSEN'S METHOD

7.3.1 Fundamental Equations for Stresses and Forces along Rupture Lines

Starting with known boundary values Brinch Hansen (1953) applies Kötter's equations to obtain stresses along rupture surfaces. However he does not insist on exact, statically correct shapes for the rupture lines. He assumes a circular shape or, when investigating zone ruptures, a shape composed of circular and straight parts. In both cases the chosen rupture line shape must agree with the kinematic conditions of the structure in the state of rupture.

In Brinch Hansen's solution Kötter's equation is developed by using the equilibrium condition for an elementary area which is limited by rupture lines and pseudo-rupture lines. The axis of projection has the direction $\xi - \xi$ (Fig. 7.4) inclined towards the horizontal at an angle $(v + \varphi)$ and v represents the angle between the horizontal line and one of the boundary rupture lines (Fig. 7.4).

Considering, first, only stress states with constant stresses σ on the circumference of the element $ACDE$, the corresponding forces are evidently in equilibrium. Constant shear stresses τ having the directions shown in Fig. 7.4, which would be applied to the parallelogram element $BCDE$, would also be in independent equilibrium. Acting on the element $ACDE$ they give a resultant force $2\tau(\overline{AB})$.

Now the equilibrium conditions shall be stated for this force and other forces corresponding to differential increments of normal and shear stress.

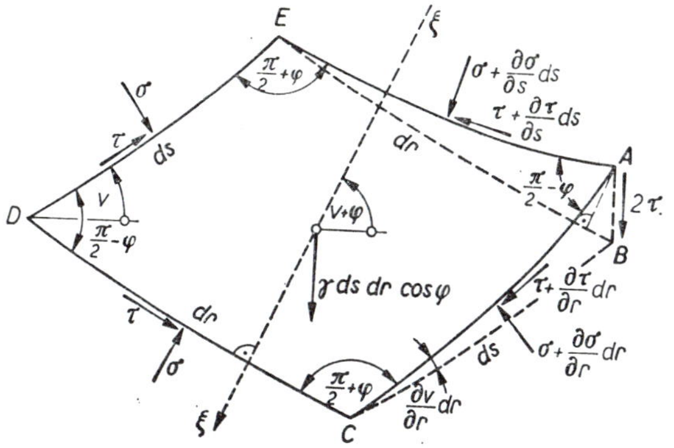

Fig. 7.4. Forces (stresses) affecting an elementary area limited by rupture lines and pseudo-rupture lines

The lengths of both sides of the elementary area *ACDE* are assumed to be ds and dr respectively. Thus we obtain:

$$\frac{\partial \sigma}{\partial s}\,\mathrm{d}s\,\mathrm{d}r - \frac{\partial \sigma}{\partial r}\,\mathrm{d}r\,\mathrm{d}s \sin\varphi + \frac{\partial \tau}{\partial r}\,\mathrm{d}r\,\mathrm{d}s \cos\varphi +$$

$$+ 2\tau \frac{\partial v}{\partial s}\,\mathrm{d}s\,\mathrm{d}r + \gamma\,\mathrm{d}s\,\mathrm{d}r \cos\varphi \sin(v + \varphi) = 0. \quad (7.28)$$

Inserting into Equation (7.28) Coulomb's yield condition:

$$\sigma = (\tau - c)\cot\varphi \quad (7.29)$$

Kötter's fundamental equation (Kötter, 1903) can be obtained in the form:

$$\frac{\partial \tau}{\partial s} + 2\tau \frac{\partial v}{\partial s}\tan\varphi + \gamma \sin\varphi \sin(v + \varphi) = 0 \quad (7.30)$$

The assumption that the rupture line is circular yields the condition

$$\frac{\partial s}{\partial v} = r \quad (7.31)$$

Under this condition Equation (7.30) reduces to

$$\frac{\partial \tau}{\partial v} + 2\tau \tan\varphi + \gamma r \sin\varphi \sin(v + \varphi) = 0 \quad (7.32)$$

Considering $r =$ const. the integration of the above differential equation yields:

$$\tau = C \exp(-2v \tan\varphi) + \gamma r \sin\varphi \cos\psi \cos(v + \varphi + \psi) \quad (7.33)$$

The angle ψ is defined by the equation

$$\tan \psi = 2 \tan \varphi \tag{7.34}$$

The integration constant c can be determined considering the boundary

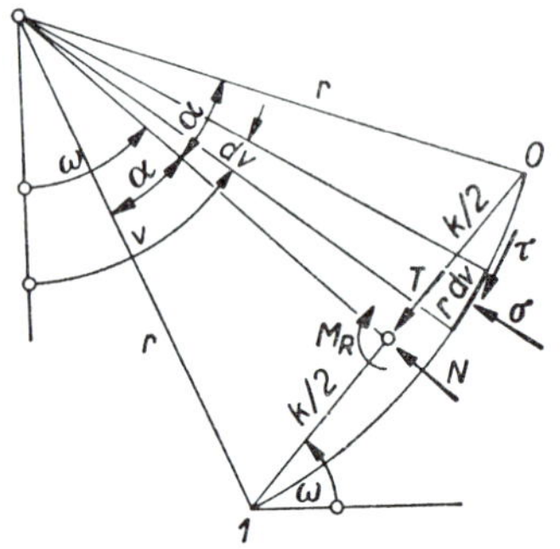

Fig. 7.5. Notation to the integration of Kötter's equations along a circular rupture line according to Brinch Hansen (1953)

conditions τ_0 and v_0 at the starting point O of the circular rupture line (Fig. 7.5).

$$C = \exp(2v_0 \tan \varphi)[\tau_0 - \gamma r \sin \varphi \cos \psi \cos (v_0 + \varphi + \psi)] \tag{7.35}$$

The circular rupture line is determined by the values (Fig. 7.5) ω, α and k. Fig. 7.5 shows the following relations:

$$v_0 = \omega + \alpha \tag{7.36}$$

$$v_1 = \omega - \alpha \tag{7.37}$$

$$r = \frac{k}{2 \sin \alpha} \tag{7.38}$$

Taking into account the relations (7.35)–(7.38), Equation (7.33) enables us to determine the tangential stress τ_1 at the point 1. The integration of stresses projected onto the chord 0–1 and onto its axes of symmetry, and the integration of moments with respect to the point of intersection of these two directions yield the components N, T and the moment M_R of the resultant R of the reaction forces acting on the arc $\widehat{01}$. The results can be presented in the form:

$$\tau_1 = \gamma k(\tau^x \sin \omega + \tau^y \cos \omega) + \tau_0 \tau^z \tag{7.39}$$

$$N = \gamma k^2(N^x \sin \omega + N^y \cos \omega) + \tau_0 k N^z - ck \cot \varphi \tag{7.40}$$

$$T = \gamma k^2(T^x \sin \omega + T^y \cos \omega) + \tau_0 k T^z \tag{7.41}$$

$$M_R = \gamma k^3(M^x \sin \omega + M^y \cos \omega) + \tau_0 k^2 M^z \tag{7.42}$$

The dimensionless coefficients τ^x, τ^y, τ^z; N^x, N^y, N^z; T^x, T^y, T^z and M^x, M^y, M^z are functions of the variable α and of the shear angle φ. The expressions defining these coefficients and the tables containing the values of these coefficients for different values of α and φ are given in Brinch Hansen's publications (1957 and 1960).

In order to obtain the proper failure line and corresponding stresses Brinch Hansen applies the so-called equilibrium method. The application of this method will be indicated in the following two sections for the case of line rupture as well as for the case of zone rupture in the backfill acting upon a retaining wall.

7.3.2 Line Rupture in the Backfill of a Retaining Wall

Let us consider a retaining wall under kinematic conditions such that the equations of limiting plastic equilibrium are satisfied only along a cylindrical surface dividing the upper and the lower elastic field, i.e. two domains with the states of stress whose Mohr circles are below the failure envelope. If the deformations inside the elastic fields are disregarded, the separating failure surface must be for kinematic reasons—in plane problems—of circular cylindrical shape. The projection of the surface of the cylinder in a vertical cross-section is a circular arc.

Consider a straight retaining wall 1–3 (Fig. 7.6) inclined anticlockwise

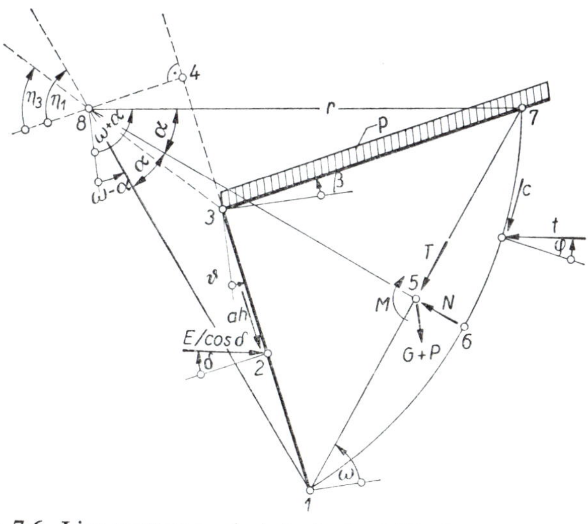

Fig. 7.6. Line rupture analysis in the backfill of a retaining wall according to Brinch Hansen (1953)

at an angle ϑ towards the vertical, and a straight ground surface forming an angle β with the horizontal in the anticlockwise-direction. The resultant G of the gravity force (γ) of the ground wedge (1–6–7–3) and the resultant P of the uniformly distributed vertical surcharge p (per unit area of the sloping surface) are considered to act at the centre (5) of the chord (1–7). These two forces as well as the moments M_G and M_P corresponding to such a reduction can be expressed in terms of the loading data γ and p, the geometrical data of the ground ϑ and β, and the parameters determining the circular rupture line (1–6–7) α and ω. The length k of the chord is not an independent parameter for it can be expressed by the length h of the retaining wall and by the angles ω, ϑ and β (see the triangle 1–7–3):

$$k = h \frac{\cos(\beta - \vartheta)}{\sin(\omega - \beta)} \tag{7.43}$$

The point of application of the resultant pressure on the retaining wall $E/\cos\delta$ is assumed to be at distance z_p from the foot of the wall 1 ($z_p = \overline{2\text{–}1}$, Fig. 7.6). The inclination δ of this resultant force towards the normal of the wall is assumed to be known. The adhesion mobilized along the wall is assumed to be known too.

Three equilibrium conditions should be considered with regard to the following forces and couples:

the active forces and moments of the ground wedge G, P, M_G, M_P;

the forces and the moment of reaction on the rupture line N, T and M_r; and

the forces of reaction on the ground area of the wall $E/\cos\delta$ and ah.

The unknown quantities appearing in these equilibrium equations are:

the parameters of the circular arc α and ω as well as the earth pressure E and its point of application determined by z_p.

As four unknown quantities have to be found, a further equation is needed. For this purpose the point 4 (Fig. 7.6) of the ground surface of the retaining wall whose displacement is a pure translation in the direction of the wall should either be known or assumed. If this point is known, then the centre of rotation of the ground wedge (1–6–7–3), i.e. the centre (8) of the circular rupture line, should lie somewhere in the normal passing through point 4.

The geometrical relations shown in Fig. 7.6 then yield the relationship:

$$\frac{z_p}{h} = \frac{\cos(\beta - \vartheta)\cos(\omega - \alpha - \vartheta)}{2\sin\alpha\sin(\omega - \beta)} \tag{7.44}$$

$$z_p = \overline{1\text{–}4} \text{ (Fig. 7.6)}$$

When establishing the equilibrium conditions one axis of projection has to coincide with the normal towards the direction of the force $E/\cos \delta$ and the other with the direction of the normal of the wall (1–3), while the point about which the moments act can be taken at the foot (point 1) of the wall. Using the notation of Fig. 7.6 the following equations result:

$$N \cos (\omega - \vartheta + \delta) - T \sin (\omega - \vartheta + \delta) - (G + P) \cos (\vartheta - \delta) - \\ - ah \cos \delta = 0 \quad (7.45)$$

$$E = N \sin (\omega - \vartheta) + T \cos (\omega - \vartheta) + (G + P) \sin \vartheta \quad (7.46)$$

$$Ez_p = N\frac{k}{2} - (G + P)\frac{k}{2} \cos \omega - M_R - M_G - M_P \quad (7.47)$$

By choosing and changing the angle α the parameters α and ω of the circular rupture line can then be found by trial and error from Equations (7.44) and (7.45). Equation (7.46) yields the earth pressure E while its point of application (z_p) can be found from Equation (7.47).

In the solution indicated above the shear stress τ_0 at the beginning of the rupture line (at the surface of the ground) which appears in expressions (7.40) to (7.42) determining N, T and M_r, is assumed to be known.

If the rupture line intersects the ground surface at the statically correct angle, i.e. in the direction of the shear characteristics corresponding to Rankine's theory (see Section 7.4), the shear stress τ_0 can be derived either from the equations of limiting plastic equilibrium or from the corresponding Mohr circle of stress. The most unfavourable circular rupture line, however, satisfying the equilibrium conditions does not usually meet the ground surface at the statically correct angle. Comparing the results obtained by the equilibrium method with the results obtained by the extreme method, in cases when this method gives theoretically exact solutions, Brinch Hansen (1953) concludes as follows. The value of the shear stress τ_0 should be found from the equilibrium condition for the forces in the elementary triangular prism shown in Fig. 7.7 projected on the direction normal to the pseudo-failure line which makes an angle $(\pi/2) + \varphi$ with the circular rupture line. The equilibrium condition yields:

$$\sigma_0 \sin (v_0 - \beta) + \sigma_0 \sin \varphi \cos (v_0 + \varphi - \beta) - \\ - \tau_0 \cos \varphi \cos (v_0 + \varphi - \beta) - p \cos \varphi \sin (v_0 + \varphi) = 0 \quad (7.48)$$

Considering Equation (7.1) (see Fig. 7.7) the following expression can be derived from the above condition:

$$\tau_0 = \frac{p \sin \varphi \sin (v_0 + \varphi) + c \cos \varphi \sin (v_0 + \varphi - \beta)}{\sin (v_0 - \beta)} \quad (7.49)$$

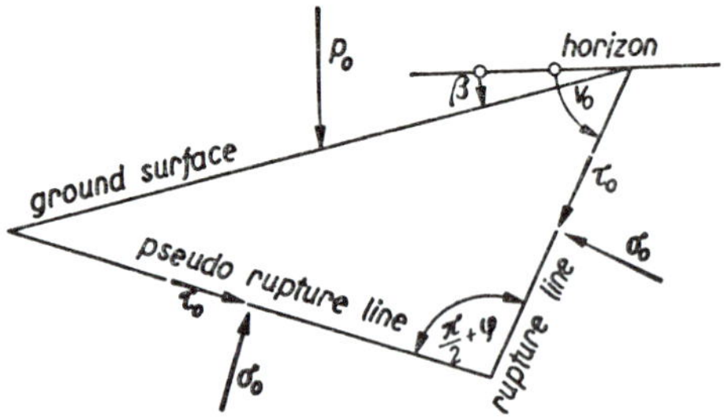

Fig. 7.7. Notation to the equilibrium condition defining the shear stress τ_0 at the external boundary of the failure line. (Reproduced after Brinch Hansen (1953))

The same value τ_0 results also in the case when the rupture line meets the surface at the statically correct angle $(v_0 - \beta)$. The value of this angle can be found from the second equilibrium condition referred to the normal of the rupture line as the axis of projection:

$$\tan\left(v_0 + \frac{\varphi}{2} - \frac{\beta}{2}\right) = \frac{c\cos\varphi\sin\beta \pm \sqrt{(c^2\cos^2\varphi + pc\sin 2\varphi\cos\beta + p^2(\sin^2\varphi - \sin^2\beta))}}{c\cos\varphi\cos\beta + p(\sin\varphi - \sin\beta)} \tag{7.50}$$

7.3.3 Zone Rupture in the Backfill of a Retaining Wall

If the conditions of limiting plastic equilibrium are satisfied either in the entire earth wedge above a boundary rupture surface or between two or several rupture surfaces—in the case when the plastic zone is surrounded by several elastic zones—the rupture is called a zone rupture. In such cases the kinematic conditions of the displacements of the retaining construction do not require a simple circular shape of the boundary and intermediate rupture lines. With reasonable approximation the rupture line can be considered to be composed of several circular and straight elements. Although Kötter's equations for such rupture lines are not accurately satisfied, approximately correct results can be obtained by using them on condition that the rupture lines composed of circular and straight elements satisfactorily approach the correct shape. If the rupture line contains at least two elements, i.e. a circular arc and a straight element, then it can always be shaped in such a way that it meets the ground surface and the wall at the 'statically correct' angles, i.e. so that the conditions of plastic equilibrium are fulfilled also at the starting and the finishing points of the rupture line.

For zone ruptures the earth pressure can in principle be determined at any point of the wall. Consequently the correct distribution of earth

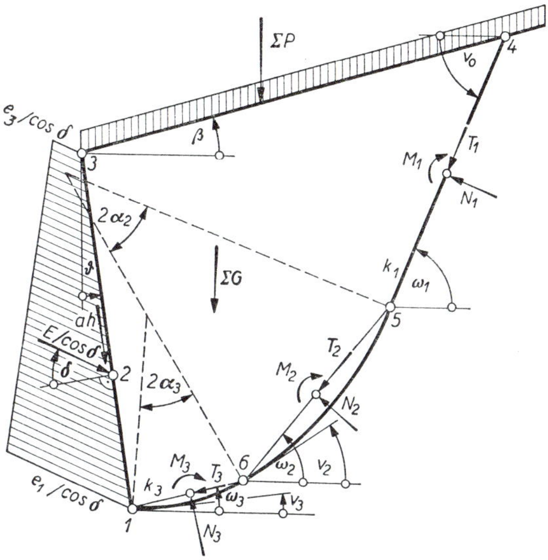

Fig. 7.8. Zone rupture analysis in the backfill of a retaining wall according to Brinch Hansen (1953)

pressure along the wall can be found. Yet, in order to simplify the computation, the linear distribution of pressure as shown in Fig. 7.8 is commonly assumed. This assumption yields the following relations:

$$E = \frac{h}{2}(e_3 + e_1) \tag{7.51}$$

$$Ez_p = \frac{h^2}{6}(2e_3 + e_1) \tag{7.52}$$

In order to obtain the parameters determining the composed rupture line (1–6–5–4) (Fig. 7.8) three equilibrium conditions should be applied and the following geometrical condition respected (Fig. 7.8):

$$\Sigma k \sin(\omega - \beta) = h \cos(\vartheta - \beta) \tag{7.53}$$

The central angles 2α can arbitrarily be chosen considering the condition:

$$\Sigma 2\alpha_i = \Sigma 2(\omega_i - v_i) \tag{7.54}$$

The angle v_0 at the starting point 4 can be computed using Equation (7.50); Equation (7.49) then permits the determination of the corresponding shear stress τ_0 in the straight element k_1 (Fig. 7.8).

The chords $k(k_1, k_2, k_3$ in Fig. 7.8) may be considered as independent unknown quantities that have to be determined by the equilibrium conditions. The corresponding angles $\omega(\omega_1 = v_0, \omega_2, \omega_3$ in Fig. 7.8) depend on the angles α and on the chords k.

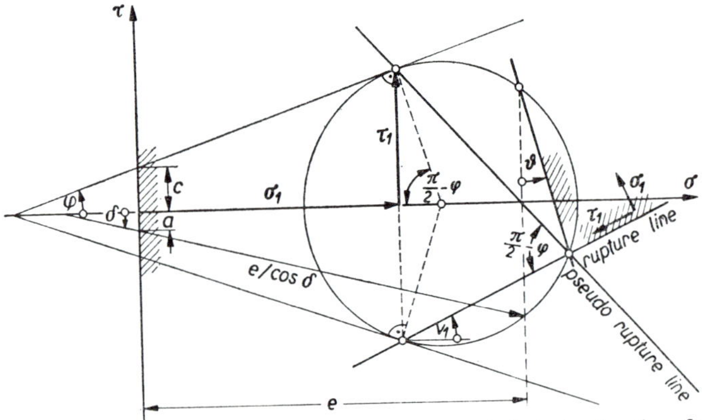

Fig. 7.9. Determination of earth pressures and their inclination from stresses at the end of rupture lines by using Mohr's stress circle

The shear stresses τ_1 at the finishing points of single failure-line elements can then be expressed by Equation (7.39) as functions of the angles α and of the chords k.

The relationship between the shear stress τ_1 and the angle v_1 at the wall on one side and the total earth pressure $e/\cos\delta$ on the wall, its inclination δ, and the adhesion a on the other side can be deduced from the Mohr circle for the limiting state of plastic equilibrium (Fig. 7.9). Knowing for example the shear stress τ_1 and the inclination v_1 and assuming the adhesion a to be given, the earth pressure $e/\cos\delta$ and its inclination δ can be determined by the construction shown in Fig. 7.9. Analytical expressions for the angle v_1 and the earth pressure e corresponding to this construction have been derived by Brinch Hansen (1960).

Therefore, if other quantities are expressed in the indicated way, the chords k remain the only unknown quantities in the equilibrium conditions. The equations should be solved by an iterative procedure.

Usually a two-part failure line may be chosen, composed of a circular arc and a straight line. Assuming the angle δ to be known, the chord k_1 and the length of the straight element k_2 can then be determined from the condition (7.53) and from the equilibrium condition applied to the projection on the normal to the direction of the force $E/\cos\delta$. The force E and the distance z_p can then be derived from Equations (7.51) and (7.52).

Detailed instructions and auxiliary expressions as well as tables facilitating the application of Brinch Hansen's equilibrium method are presented in the publications of the author of the method (Brinch Hansen, 1953, 1957, 1960).

7.4 RANKINE'S STATES OF PLASTIC EQUILIBRIUM

7.4.1 Conjugate Stresses in Infinite Plane Slopes

The straight line AB in Fig. 7.10 represents the surface of an infinite plane slope making an angle δ with the horizontal. The straight line b is parallel to the line $\overleftrightarrow{AB}$: i.e. $z = \text{const}$. Consider, between these two lines, a

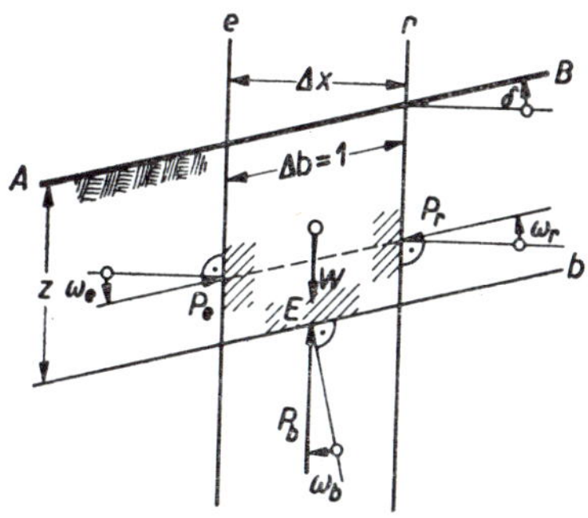

Fig. 7.10. Conjugate stresses in an infinite slope

prismatic element limited by two adjacent vertical cross-sections at a distance $\Delta b = 1$. The area of this element is

$$\Delta A = z \cos\delta\, \Delta b = z \cos\delta \tag{7.55}$$

The gravity force $W = \gamma z \cos\delta$ is supposed to be the only active force in such an element, the length of which, in the direction normal to the cross-section shown in Fig. 7.10, is equal to unity. This force and the reaction forces P_b at the base of the element and P_e, P_r on both lateral surfaces, make an equilibrium system:

$$\vec{W} + \vec{P_b} + \vec{P_e} + \vec{P_r} = 0 \tag{7.56}$$

Due to equal stress conditions on both lateral surfaces of the element, the forces $\vec{P}_e$ and $\vec{P}_r$ have the same magnitude, are parallel to each other, and the distance of the point of application from the soil surface $\overleftrightarrow{AB}$ should be equal for both of them. As the stress state below the surface $\overleftrightarrow{AB}$ is uniform for each plane z = const., the gravity force W and the reaction force P_b must pass through the centre E of the base Δb of the element, and the points of application of reaction forces P_e and P_r must be at the depth $(2/3)z$. The equilibrium conditions for moments with respect to the point E require the forces P_e and P_r to be in equilibrium with each other and their direction to be parallel to the surface $\overleftrightarrow{AB}$. Thus, the other two forces of the equilibrium system (7.56) (i.e. the forces $\vec{W}$ and $\vec{P}_b$), must be in independent equilibrium as well. Consequently, the reaction force $\vec{P}_b$ acting at the base of the element has an upward vertical direction and its magnitude is equal to the magnitude of the gravity force $\vec{W}$.

Stresses acting at different directions in elementary areas in such a way that the stress vector in one of the areas is parallel to the direction of the other area, and vice versa, are called conjugate stresses.

As shown in Fig. 7.10, the angles of deviation of the forces $\vec{P}_e$, $\vec{P}_r$ and $\vec{P}_b$ with respect to the normals of the corresponding areas, i.e. the angles $\omega_e(\omega_r)$ on the one hand and ω_b on the other, have equal values but different signs.

$$|\omega_e| = |\omega_r| = |\omega_b| = \delta \tag{7.57}$$

Thus the stresses at a depth z

$$p_e = p_r = \frac{2P_e}{z} \quad \text{and} \tag{7.58}$$

$$p_b = \frac{P_b}{\Delta \text{b}} = \frac{W}{\Delta \text{b}} = \gamma z \cos \delta \tag{7.59}$$

are two conjugate stresses with angles of deviation $(+\delta)$ or $(-\delta)$ with respect to the normals of the corresponding areas. In Fig. 7.11 the vectors of these two conjugate stresses are drawn with respect to the Mohr circle of the initial 'at rest' state of stress (the dotted circle in the figure). The stress p_b is greater than the stress p_e. The principal stress ratio is assumed to be known.

7.4.2 The Limiting 'Active' Stress State

The soil between the planes *e* and *r* (Fig. 7.10) may be assumed to become loosened in such a way that the shear strength becomes fully mobilized. Thus the limiting state of 'active' plastic equilibrium appears throughout the entire zone. At any point of this zone the stress state is represented by a Mohr circle touching the failure envelope. This envelope is supposed to be a straight line according to the Coulomb law of shearing resistance at zero cohesion:

$$\tau_f = \sigma' \tan \varphi \tag{7.60}$$

Neglecting the small lowering of the surface connected with loosening of the soil, the stresses p_b of the preceding stress state 'at rest' (the dotted circle in Fig. 7.11) cannot be considered to be affected by the loosening. Thus, the Mohr circle of the 'active' stress state of plastic equilibrium is governed by the following two conditions. Firstly, it must pass through the end-point of the stress vector $\vec{p}_b$ and secondly must touch the Coulomb straight failure envelope (Fig. 7.11). As the direction of the area stressed by $\vec{p}_b$ is known, the direction being parallel to the slope inclination, the pole *P* of the Mohr circle can be found by applying the construction which has been explained in Section 2.2 (Fig. 2.5). From this pole the directions of the failure surfaces, corresponding to stress vectors $\vec{OT_1}$ and $\vec{OT_2}$ (Fig. 7.11), can be drawn. Considering that the circumferential angle is equal to half the value of the central angle, the conclusion that the two systems of failure lines make the angles $(\pi/2 + \varphi)$ or $(\pi/2 - \varphi)$ with each other can immediately be drawn from Figure 7.11.

The determination of the principal stress directions is shown in Fig. 7.11 as well. The corresponding interpretation was given in Section 2.2. The directions of principal stresses halve the angles made by the failure lines. The circumferential angle above the arc $\widehat{T_2N_1}$ equals half the value of the circumferential angle above the arc $\widehat{T_2N_1T_1}$.

At the transition from the state of stress at rest into the limiting active state of stress, the pressure $\vec{p}_e$ acting on vertical areas decreases to the pressure $\vec{p}_a$ as shown in the Mohr circle construction. The ratio

$$\frac{p_a}{p_b} = K_a \tag{7.61}$$

is called the coefficient of active earth pressure related to the vertical direction. Similar coefficients of active earth pressure can be determined

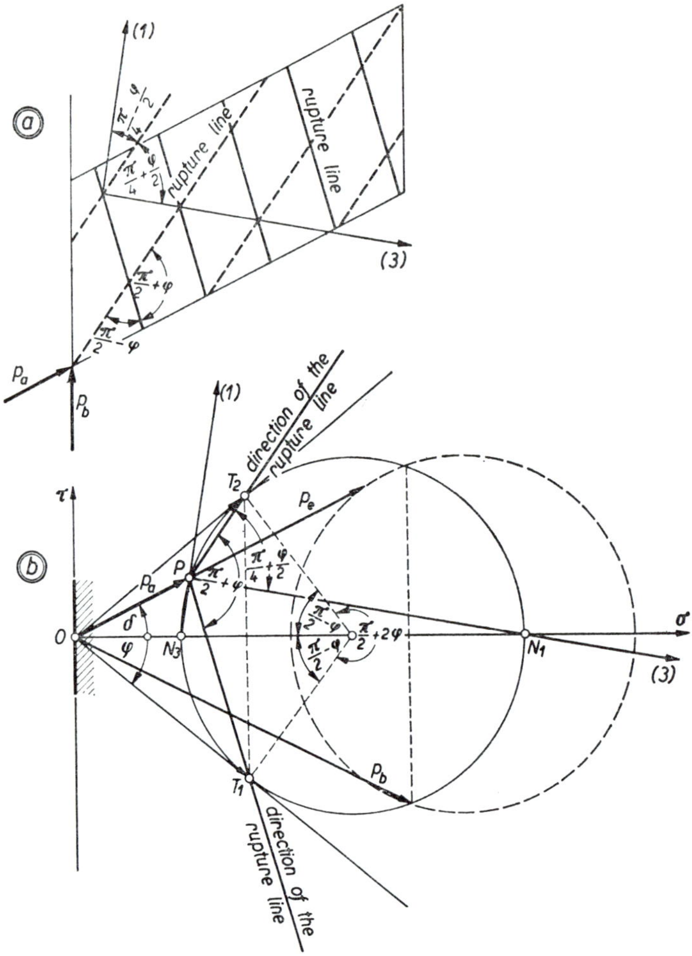

Fig. 7.11. Rankine's active stress state

for other directions also. Using the geometrical relations shown in Fig. 7.11 analytical expressions can be derived for such earth pressure coefficients.

7.4.3 The Limiting 'Passive' Stress State

Now the soil in the zone between the planes *e* and *r* (Fig. 7.10) may be supposed to be pressed in the direction parallel to the slope so that the

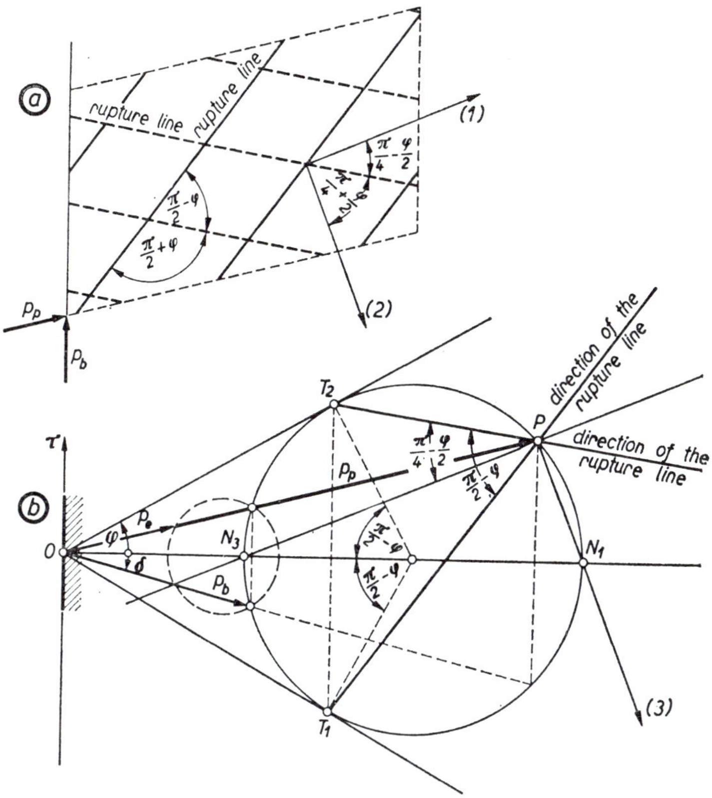

Fig. 7.12. Rankine's passive stress state

shear strength is fully mobilized throughout the entire zone. In this way the limiting state of 'passive' plastic equilibrium appears. The corresponding Mohr circle (Fig. 7.12) is governed by the conditions that it should pass through the end-point of the stress vector $\vec{p}_b$ and touch the Coulomb failure envelope ($\tau_f = \sigma' \tan \varphi$) on the side of the Mohr circle at rest opposite to the origin of the system of coordinates (σ', τ_f). (The influence of the small swelling of the soil surface owing to the transverse expansion of the soil and its slight effect on the pressure $\vec{p}_b$ has been neglected.) The construction of the Mohr circle is shown in Fig. 7.12. The direction of the area with the pressure $\vec{p}_b$ defines the pole P from which the directions of the rupture lines and principal stresses can then be drawn, as shown in Fig. 7.12. Again the principal stress directions coincide with the lines of symmetry of the angles ($\pi/2 - \varphi$) and ($\pi/2 + \varphi$) respectively made by the failure lines. (The circumferential angle corresponding to the arc $\widehat{T_2N_3}$ is equal to half the central angle $(\pi/2 - \varphi)$ and to half the double-angle above the arc $\widehat{T_2N_3T_1}$.)

When passing from the state of stress at rest into the limiting passive stress state, the stress $\vec{p}_e$ acting on vertical areas increases up to the value $\vec{p}_p$. The ratio

$$K_p = \frac{p_p}{p_b} \tag{7.62}$$

is called the coefficient of passive pressure related to the vertical direction. Similar coefficients of passive pressure can be determined for any other direction.

The analytical expressions for the directions of rupture surfaces and principal stresses for the magnitudes of passive pressures in either vertical or other directions and for the corresponding coefficients of the passive pressure can be derived from the geometrical relationship shown in the corresponding Mohr circles of limiting equilibrium. Neither these expressions nor other similar ones for the active state of stress have been reproduced in this book, yet, they can be found in appropriate literature (e.g. Kézdi, 1962).

7.5 THE METHOD OF CAQUOT AND KÉRISEL

7.5.1 Fundamental Differential Equations

Caquot and Kérisel (1949) applied the theory introduced by Boussinesq and completed by Résal. Boussinesq's fundamental differential equations will be presented in the form given by Caquot. Since it is not possible to integrate these equations, Boussinesq (1885) applied an approximate integration valid only for the computation of the active earth pressure in cases of positive and small angles ω and β (see Fig. 7.13). Résal (1903, 1910) extended these computations for some cases of negative angles ω and β.

By this method only wedge-shaped plastic domains are investigated. They are limited by the basic rupture line and by two rays β_1 and β_2 originating at the top point 1 (Fig. 7.13) of the wall. Along any ray β the stresses acting in areas having the direction of the ray are assumed to be parallel to each other and their magnitudes assumed to be proportional to the distance l of the elementary area from the point 1. Thus, the points of application of their resultants lie in the points of the lower third of the ray lengths.

Fig. 7.13 shows a triangular element (1–2–3) of such a wedge-shaped plastic zone. The length of the rays is assumed to be equal to unity $\overline{1\text{–}2} = \overline{1\text{–}3} = 1$ and the volume weight of the soil in the wedge $\gamma = 1$. Only the gravity forces $\mathrm{d}W$ are considered as active forces. Equilibrium conditions for the moments are established with respect to the centre of

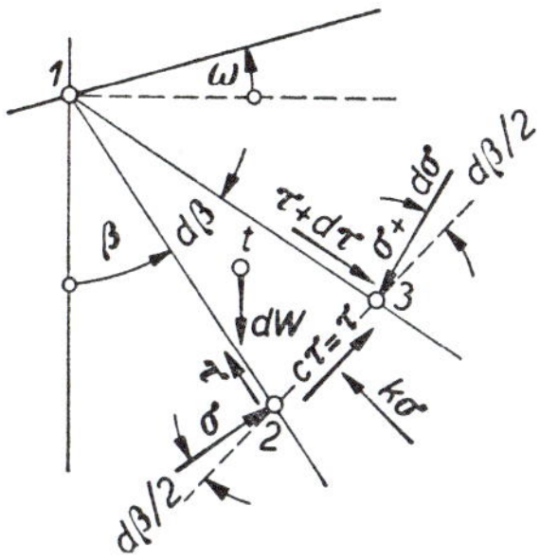

Fig. 7.13. Forces (stresses) affecting a triangular element of a wedge-shaped plastic zone

gravity '*t*' and for the top point '1' of the wall as well as for the projection of forces on the direction $\overleftrightarrow{1\text{–}3}$. Neglecting small members of second degree when summed with members of first degree, the first of the above mentioned equilibrium conditions yields

$$\tau c = \tau$$

after the notation in Fig. 7.13 while—after appropriate reduction—the other two equilibrium conditions lead to the following two differential equations:

$$\frac{\mathrm{d}\sigma}{\mathrm{d}\beta} = 3\tau - \sin\beta \tag{7.63}$$

$$\frac{\mathrm{d}\tau}{\mathrm{d}\beta} = m\sigma - \cos\beta \tag{7.64}$$

where

$$m = 2k - 1 \tag{7.65}$$

(for k see Fig. 7.13). Corresponding to the Coulomb envelope $\tau_f = \sigma' \tan\varphi$ (Fig. 7.14) the limiting stress state condition will be inserted into these equilibrium equations. The ratio of the normal stresses in two perpendicular directions can be derived from Fig. 7.14:

$$k = \frac{\overline{0\text{–}3'}}{\overline{0\text{–}2'}} = \frac{\sigma_0[1 + \sin\varphi\cos(\gamma - \alpha)]}{\sigma_0[1 - \sin\varphi\cos(\gamma - \alpha)]} \tag{7.66}$$

The application of the sine formula to triangle (012) yields

$$\sin \gamma = \frac{\sin \alpha}{\sin \varphi} \tag{7.67}$$

Inserting Equation (7.66), the value $m = 2k - 1$ yields:

$$m = \frac{1 + 3 \sin \varphi \cos (\gamma - \alpha)}{1 - \sin \varphi \cos (\gamma - \alpha)} \tag{7.68}$$

Considering further Equation (7.67) we obtain by appropriate transformation:

$$m = 1 + 4 \tan^2 \varphi \pm \frac{4}{\cos \varphi} \sqrt{(\tan^2 \varphi - \tan^2 \alpha)} \tag{7.69}$$

The above relationship represents the equation of an ellipse $m = m(\tan \alpha)$ expressing the condition of the limiting plastic state of stress which should

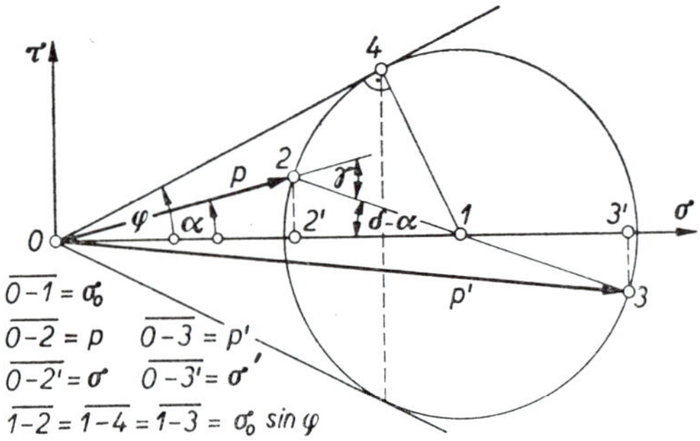

Fig. 7.14. Notation to Equation (7.66) according to Caquot and Kérisel (1949)

be considered when applying the equilibrium equations (7.63) and (7.64). The angle α (see Fig. 7.14) denotes the declination of the stress vector $\vec{p}$ (σ,τ) (Fig. 7.13) towards the normal of the ray β:

$$\tan \alpha = \frac{\tau}{\sigma} \tag{7.70}$$

The differential equations (7.63) and (7.64) connected with the condition (7.69) cannot be integrated. Caquot and Kérisel (1949), however, have developed two methods of progressive integration of these two equations. One of the two methods will be interpreted in the following paragraph. It can be applied particularly to the computation of the coefficients of passive earth pressure.

7.5.2 The Method of Progressive Integration of Boussinesq's Fundamental Equations

The ellipse $m = m(\tan \alpha)$ satisfying Equation (7.69) and corresponding to a certain shear angle φ, may be replaced by a polygon. This polygon must remain within the limits of two adjacent ellipses corresponding to the shear angles $\varphi - \varepsilon$ and $\varphi + \varepsilon$, ε being a small angle difference. An arbitrary side of this polygon lies along the straight line whose equation is

$$a \tan \alpha + bm + 1 = 0 \tag{7.71}$$

Hence

$$m = -\frac{1 + a \tan \alpha}{b} \tag{7.72}$$

This expression for m will be inserted into the Equation (7.64). The resulting Equation (7.73) and Equation (7.63) represent two differential equations which can be integrated:

$$\frac{d\sigma}{d\beta} = 3\tau - \sin \beta \tag{7.63}$$

$$\frac{d\tau}{d\beta} = -\frac{\sigma}{b} - \frac{a\tau}{b} - \cos \beta \tag{7.73}$$

Equation (7.63) has the following particular solution:

$$\tau_1 = c \sin \beta + d \cos \beta \tag{7.74}$$

$$\sigma_1 = 3d \sin \beta + (1 - 3c) \cos \beta \tag{7.75}$$

Inserting these two expressions into Equation (7.73), an appropriate transformation yields:

$$c = \frac{(3 - b)(b + 1)}{a^2 + (3 - b)^2} \tag{7.76}$$

$$d = \frac{-(b + 1)a}{a^2 + (3 - b)^2} \tag{7.77}$$

By omitting the last member, a complementary solution of the differential Equations (7.63) and (7.73) can be obtained

$$\frac{d\sigma}{d\beta} = 3\tau \tag{7.78}$$

$$\frac{d\tau}{d\beta} = -\frac{\sigma}{b} - \frac{a\tau}{b} \tag{7.79}$$

Differentiating Equation (7.78) with respect to β, taking into account Equation (7.79) and rearranging it, we obtain:

$$\frac{d^2\sigma}{d\beta^2} + \frac{a}{b}\frac{d\sigma}{d\beta} + \frac{3\sigma}{b} = 0 \tag{7.80}$$

The solution of this equation is

$$\sigma_2 = f \exp(\alpha_1\beta) + g \exp(\alpha_2\beta) \tag{7.81}$$

Equation (7.78) yields

$$\tau_2 = \frac{\alpha_1}{3} f \exp(\alpha_1\beta) + \frac{\alpha_2}{3} g \exp(\alpha_2\beta) \tag{7.82}$$

Expressions (7.81) and (7.82) have to be inserted into Equation (7.79). The comparison of the coefficients yields then

$$b\alpha^2 + a\alpha + 3 = 0 \tag{7.83}$$

Hence:

$$\alpha_{1,2} = \frac{-a \pm \sqrt{(a^2 - 12b)}}{2b} \tag{7.84}$$

The combination of the particular and complementary solutions gives the general solution:

$$\tau = c \sin\beta + d \cos\beta + \frac{\alpha_1}{3} f \exp(\alpha_1\beta) + \frac{\alpha_2}{3} g \exp(\alpha_2\beta) \tag{7.85}$$

$$\sigma = 3d \sin\beta + (1 - 3c)\cos\beta + f \exp(\alpha_1\beta) + g \exp(\alpha_2\beta) \tag{7.86}$$

Equation (7.64) may be written in the form

$$m\sigma = \frac{d\tau}{d\beta} + \cos\beta \tag{7.64'}$$

Inserting the derivative of the expression (7.85) we get:

$$m\sigma = -d \sin\beta + (1 + c)\cos\beta + \frac{\alpha_1{}^2}{3} f \exp(\alpha_1\beta) + \frac{\alpha_2{}^2}{3} g \exp(\alpha_2\beta) \tag{7.87}$$

The boundary values β_0, m_0, σ_0 and τ_0 can be obtained by Rankine's theory using either Mohr's circle as presented in Fig. 7.12 or the corresponding equations (see e.g. Caquot–Kérisel 1949, page 89). By applying these boundary values, the constants f and g can be found by Equations (7.85) and (7.86). Other constants c, d, α_1 and α_2 depend on the chosen values a and b, i.e. on the chosen first side of the polygon.

If the points A and B (Fig. 7.15) were not chosen very closely to m_0, the value m, computed for the same constants f and g and for σ_0 corresponding to Equation (7.87), would not be close enough to the boundary value m_0. Thus the last members in the above equations should be corrected by introducing a new unknown quantity in the boundary conditions. All three unknowns (f, h and γ_1) have to conform to these conditions. Equations (7.85) to (7.87) then become:

$$\tau = c \sin \beta + d \cos \beta + \frac{\alpha_1}{3} f \exp(\alpha_1 \beta) + \frac{\gamma_1}{3} h \exp(\gamma_1 \beta) \tag{7.88}$$

$$\sigma = 3d \sin \beta + (1 - 3c) \cos \beta + f \exp(\alpha_1 \beta) + h \exp(\gamma_1 \beta) \tag{7.89}$$

$$m\sigma = -d \sin \beta + (1 + c) \cos \beta + \frac{\alpha_1^2}{3} f \exp(\alpha_1 \beta) + \frac{\gamma_1^2}{3} h \exp(\gamma_1 \beta) \tag{7.90}$$

At the passive states of stress, the last member in Equations (7.85) respectively (7.88), and (7.86) respectively (7.89) is always very small when

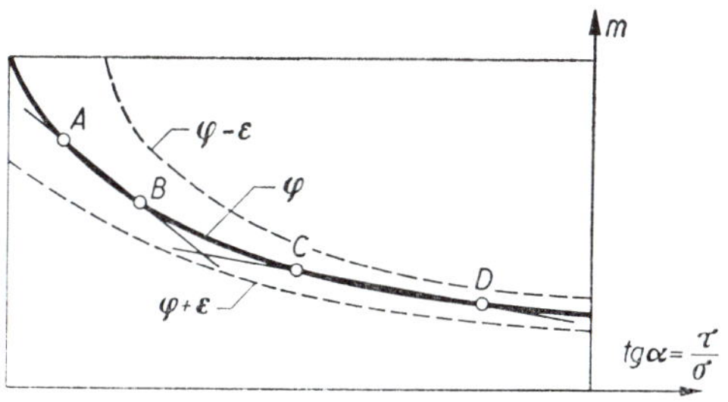

Fig. 7.15. Ellipse $m = m(\tan \alpha)$ (Equation (7.69))

compared to the last but one. This can be concluded by analysing the expressions α_1 and α_2 taking into account the negative values of a and b (see Fig. 7.15).

If once the constants f, h and γ_1 are determined by taking into account the boundary values β_0, σ_0, τ_0 and m_0, and by using the iterative trial and error method, the further values τ, σ and $m\sigma$ corresponding to various β values may be computed by the same equations. This procedure can be continued as long as the values $m = m\sigma/\sigma = f(\tan \alpha) = f(\tau/\sigma)$ remain close enough to the ellipse $m = m(\tan \alpha)$, i.e. in the interval of the adjacent ellipses corresponding to the angles $(\varphi - \varepsilon)$ and $(\varphi + \varepsilon)$. Afterwards a new angle β is chosen as the new origin with new boundary conditions so that the corresponding m value becomes closer to the ellipse $m = m(\tan \alpha)$, which had been constructed for the angle φ.

Applying this procedure, the given retaining wall, inclined at the angle β, will react with stresses whose vectors $\vec{p}(\sigma,\tau)$ form the resulting angle arc tan τ/σ with the normal of the wall. Usually, however, this angle is given in advance. If the wall displacements are large enough to permit full mobilization of friction between soil and wall, the value $\alpha_e = -\varphi$ is commonly assumed for the passive state of stress. After Caquot this boundary condition will be considered in the following way:

Equations (7.63) and (7.64) may be interpreted as a superposition of two states of stress: a weightless state of stress governed by the equations:

$$\frac{d\sigma}{d\beta} = 3\tau \tag{7.91}$$

$$\frac{d\tau}{d\beta} = m\sigma \tag{7.92}$$

and the Rankine state of stress in which only the gravity forces are taken into account.

The following notation will be introduced:

$$f = \tan \alpha = \frac{\tau}{\sigma} \tag{7.93}$$

Differentiating the above expression and inserting Equations (7.91) and (7.92) we obtain after approximate transformation:

$$d\beta = \frac{df}{m - 3f^2} \tag{7.94}$$

Inserting for m expression (7.69), the above expression becomes:

$$d\beta = \frac{df}{1 + 4\tan^2\varphi - 3f^2 \pm \dfrac{4}{\cos\varphi}\sqrt{(\tan^2\varphi - f^2)}} = \frac{df}{D} \tag{7.95}$$

Beginning at the wall with the value $f_0 = \tan(-\varphi)$ and using progressive numerical integration, the function $f = f(\beta)$ can be obtained. From Fig. 7.16 the equation of the envelope of the stress directions α can be derived in polar coordinates as follows:

$$\frac{dr}{r d\beta} = \tan \alpha = f$$

Hence:

$$r = \exp\left(\int f\, d\beta\right) = \exp \int \frac{f\, df}{D} \tag{7.96}$$

Using this equation, the values r corresponding to the angles β can be determined; the value $r = 1$ may be taken at the wall.

The computation procedure is then as follows:

The stresses obtained after Rankine's theory and making the angle $-\varphi = \arctan f_0$ with the normal of the wall, have to be subtracted from the corresponding stresses obtained for different angles β by Equations

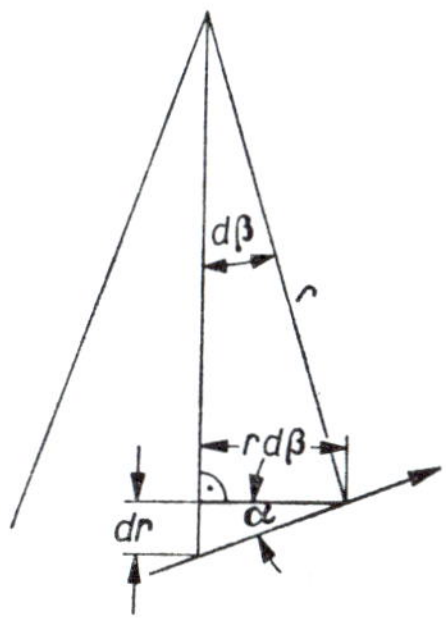

Fig. 7.16. Notation to Equation (7.96)

(7.88) to (7.90). Thus, the graph of the function $f = \tau/\sigma = f(\beta)$ can be determined. Further, the point of intersection between this graph and the curve $f = f(\beta)$, as obtained earlier by Equation (7.95), should be found yielding the angle β_i. Equation (7.96) then yields the value r_i^3 corresponding to β_i.

The difference between the earth pressure values corresponding, in the direction β_0, to the general solution after Equations (7.88) to (7.90) on the one hand and to the Rankine solution on the other, should be multiplied by r_0^3 and added to the Rankine value. This sum gives the resulting earth pressure on the wall acting at the angle $\alpha_0 = -\varphi$. (The Rankine pressure inclined at the angle $-\varphi$ at the wall, corresponds to a soil wedge whose surface inclination is parallel to the inclination of the pressure at the wall.)

Part B

EXPERIMENTAL AND THEORETICAL FUNDAMENTALS OF ONE-DIMENSIONAL CONSOLIDATION AT ZERO LATERAL STRAIN

8

One-Dimensional Consolidation without Seepage Resistance

8.1 GENERAL CONSIDERATIONS

The oedometer test introduced by K. Terzaghi (1923) is the classical test for the determination of compressibility and consolidation. In this test the cylindrical sample as a whole can move only in the direction of its vertical axis and only in this direction can the pore water and the pore air be squeezed out.

The radial displacements of the sample being zero, the study of creep phenomena by oedometer tests is limited to the stress states corresponding to this displacement condition. The design of the classical oedometer, however, does not permit the determination of the radial and the tangential pressures on the sides of the sample. For this reason the deformability is commonly expressed only by the ratio of the axial stress on the upper contact surface to the corresponding linear strain of the sample. This ratio is called the 'modulus of linear compressibility' (M_v) and its reciprocal the 'coefficient of compressibility' (m_v).

Some recent designs of the device (e.g. the 'plastometer' of Tan (1957)), are equipped with peripheral electrical strain gauges. The measurement of strains in the metal ring enables the determination of average radial stresses on the periphery of the sample. In order to measure the side friction some oedometers have special balance arrangements.

In triaxial tests the radial stresses on the periphery of the sample can be controlled and arbitrarily chosen. Thus, triaxial tests enable a better observation and a larger study of creep phenomena. These tests will be dealt with in Chapter 13. Since, however, the first studies of the influence of creep and of 'plastic resistance' (Taylor and Merchant, 1940; Taylor, 1942) on consolidation as well as the first rheological studies of the problem of complex consolidation were applied to oedometer tests, they will be treated prior to the other tests. Oedometer tests still provide the main experimental basis for practical computations of settlements and consolidation.

Creep phenomena are important especially for cohesive soils. Nevertheless they will first be studied on such oedometer samples in which the seepage resistance does not appear or in which it is so small that it does not influence the consolidation of the thin sample. Thus, the subject of our study in the next sections will be either saturated or partly saturated samples of permeable soils or dry granular systems of sandy, silty and clayey soils. In this way the effect of the seepage resistance will be eliminated in the initial treatment of the consolidation and our attention may be concentrated on pure creep. In soil mechanics the term 'creep' is used to express the increase of deformation with time at constant effective stress.

Under the conditions of oedometer test the axial linear strain ε_1 equals the volume strain ε_v:

$$\varepsilon_1 = \varepsilon_v = 3\varepsilon_{\text{oct}} \tag{8.1}$$

The spherical component of the strain tensor is given by the equation

$$\varepsilon_{rs}{}^0 = \frac{\varepsilon_1}{3}\,\delta_{rs} \tag{8.2}$$

where δ_{rs} is the delta of Kronecker defined by Equation (2.25), and the distortional component is

$$\varepsilon_{rs}{}^d = \frac{\varepsilon_1}{3}\begin{bmatrix} 2 & 0 & 0 \\ 0 & -1 & 0 \\ 0 & 0 & -1 \end{bmatrix} \tag{8.3}$$

As the ratios between the strains and stresses of normally consolidated samples are not linear, the definitions of the deformation constants of linear proportionality have to be limited to smaller stress intervals $\Delta\sigma$ and to corresponding strains $\Delta\varepsilon$. For the oedometer test on a homogeneous isotropic sample the expression of the compression modulus K defined by Equation (4.7) has the form:

$$K = \frac{\Delta\sigma_{\text{oct}}}{\Delta\varepsilon_v} = \frac{\Delta\sigma_{\text{oct}}}{\Delta\varepsilon_1} \tag{8.4}$$

and the shear modulus G can be expressed in the form:

$$G = \frac{\Delta\sigma_1 - \Delta\sigma_{\text{oct}}}{2\,.\,\frac{2}{3}\Delta\varepsilon_1}. \tag{8.5}$$

The modulus of linear compressibility being

$$M_v = \frac{\Delta\sigma_1}{\Delta\varepsilon_1} \tag{8.6}$$

the following relation holds:

$$M_v = K + \tfrac{4}{3} G \tag{8.7}$$

The compression modulus K and the distortional modulus G can, of course, be determined only in cases when the design of the oedometer permits the measurement of lateral stresses $\sigma_2 = \sigma_3$. Neglecting the lateral friction, the octahedral normal stress is

$$\sigma_{\text{oct}} = \tfrac{1}{3}(\sigma_1 + 2\sigma_3) \tag{8.8}$$

Inserting expressions (4.14) and (4.17) into Equation (8.7) for K and G, we get the relation:

$$M_v = \frac{E}{3(1 - 2\nu)} + \frac{4}{3} \cdot \frac{E}{2(1 + \nu)}$$

$$\frac{M_v}{E} = \frac{1 - \nu}{(1 - 2\nu)(1 + \nu)} \tag{8.9}$$

The ratios M_v/E and E/M_v dependent on Poisson's ratio ν, are shown in Fig. 8.1.

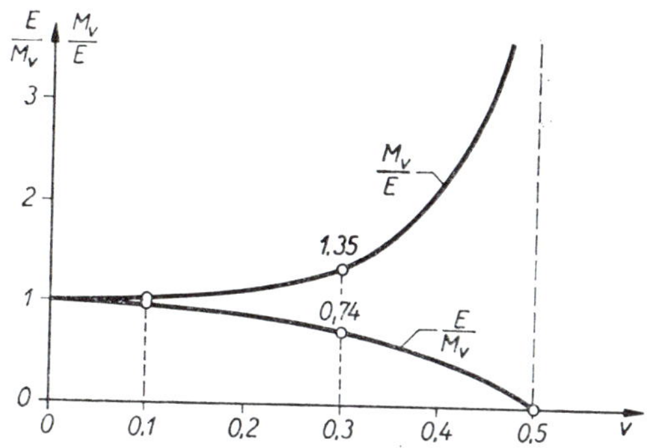

Fig. 8.1. Ratio between the modulus of linear compressibility M_v and the modulus of deformation E as a function of Poisson's ratio ν

Owing to creep the deformation constants change with time. Their development can be investigated by long term compression tests. Some examples will be shown in the next section.

8.2 CREEP PHENOMENA

The phenomenon of creep of the granular skeleton is not necessarily connected with the water surrounding the grains. In Fig. 8.2 (reproduced after Šuklje, 1953) the curves (*a*) present the development of the settlement of two oedometer samples of dry powder of a moor-clay (lacustral clay) containing about 37 per cent of clay particles (<0·002) and 62 per cent of silt (0·06/0·002). The mineralogical composition consisted predominantly of limestone (68 per cent). Before they were loaded, the samples had been compacted by shaking the oedometer. The initial load was as high as 0·274 kp/cm². The consolidation curves (*a*) correspond to

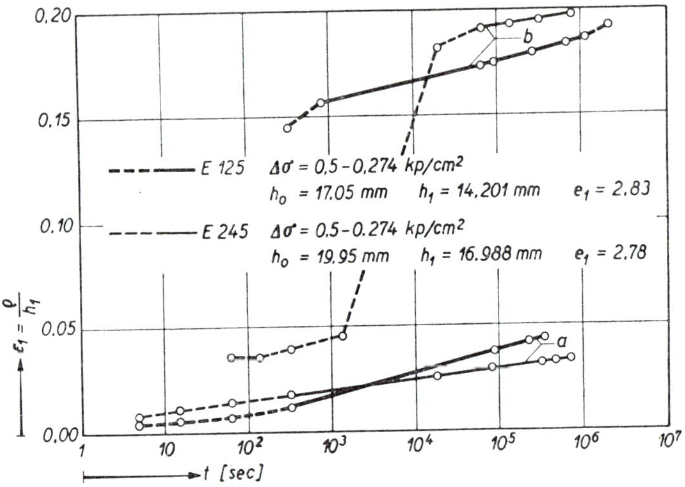

Fig. 8.2. Creep of two oedometer samples of the dry powder of a clay: (a) before, (b) after wetting

the loading step $\Delta\sigma = (0{\cdot}5 - 0{\cdot}274)$ kp/cm². Some minutes after the application of the additional load the strain curve had the form of a logarithmic line:

$$\varepsilon_1 = a + b \log_{10} t \tag{8.10}$$

which, after 120 hours of observation, still continued with the same dependence on time.

Afterwards both samples were flooded from below without changing the load. The infiltration of the water first caused a sudden important increase of strain (consolidation curves *b*). Then the creep continued with speeds decreasing in inverse proportion to time (logarithmic law). The logarithmic dependence on time, however, did not hold for the sample E 125. The retardation was slower than expected assuming an inverse proportionality to the square root of time.

Another example of the compressibility and creep of a dry grain skeleton is given in Fig. 8.3. By gently grinding a dried lump of soil in a mortar two equal samples of the powder of a lacustral chalk were prepared for testing. The samples contained about 21 per cent of clay particles (<0·002 mm), 77 per cent of silt (0·002/0·06) and 2 per cent of sand (0·06/0·2). The powder was filled freely into two oedometers of 10 cm diameter and

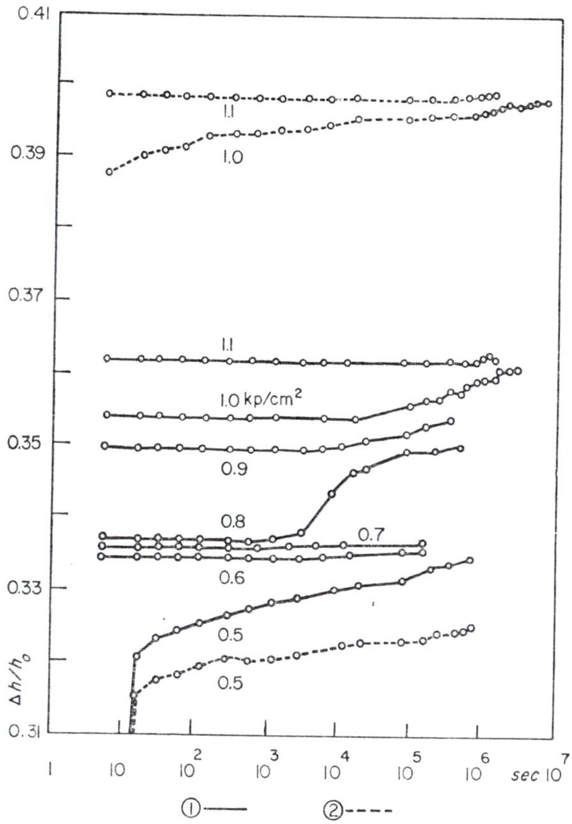

Fig. 8.3. Influence of load steps onto the consolidation of two oedometer samples of a dry grain skeleton (lacustral chalk)

of 4 cm height. Both samples* were first loaded to $\sigma = 0{\cdot}50$ kp/cm². Then the load on sample No. 2 (the dotted line 2) was increased to $\sigma = 1{\cdot}00$ and later to $1{\cdot}10$ kp/cm², while the load on sample No. 1 (the

* The samples were tested in the Soil Mechanics Laboratory of the University of Ljubljana (Yugoslavia). This note holds for all examples of laboratory tests presented in the text, if no special reference is given.

full line 1) was increased in steps to 0·60, 0·70, 0·80, 0·90, 1·00 and 1·10 kp/cm^2.

The consolidation curves presented in Fig. 8.3 prove that the settlement depends on the magnitude of the load steps and that the intensity of creep in successive steps is not uniform. If the loading is increased in large steps, the initially loose grain structure can more easily be destroyed and the strains are greater than if the loading is increased in several smaller steps to the same stress state. The increase of settlement and the intensity of creep in successive steps vary though the small loading intervals are of the same magnitude. In the case presented the settlement increased continuously. Intensive creep in a certain step can, however, be followed by moderate creep in the next one or two steps, the creep rate increasing again in the following load step.

Nevertheless, the rupture of the structure can be expected to occur at a certain stress state also in cases when the load increase is either continuous or increased in small steps. Hence the discontinuous passage in a denser structure, as obtained by continuous loading in a big step, can also occur under a slow load increase. This statement should be taken into account when predicting the consolidation of natural layers; because the dead weight loading of a construction is commonly continuous while the live load can be superimposed discontinuously.

Figure 8.4 presents the third example of the consolidation of dry grains in the oedometer. The soil tested was a dry fine sand (1 per cent $<$ 0·06; 50 per cent 0·06/0·2; 49 per cent 0·2/0·6). The initial porosity (e = 0·8896) was obtained by filling the sand freely into the oedometer.

The divergencies from the dashed logarithmic line of average values may be attributed to jerky movement of the grains when overcoming internal friction. They may also be partly due to overcoming the friction around the side of the sample and, possibly, on the surface of the piston. As the test was made in a room with a constant temperature of 10° C, the possibility of thermal effects being the cause of the irregular shape of the consolidation curve can be excluded.

Tests starting with a greater initial density show, naturally, less important creep effects.

Similar long term tests on wet sand were made by E. Nonveiller (1963). The tested sand was a fine sand (about 57 per cent 0·06/0·2) with about 3 per cent silty particles ($<$0·06), 30 per cent medium sand (0·2/0·6) and 10 per cent course sand (0·6/2·0). The sand contained 33 per cent calcite, 66 per cent quartz and 1 per cent other minerals. Four samples of this sand were tested in oedometers of 10 cm diameter and 3 cm high. Two samples were poured in the oedometer dry in a very loose state, loaded to

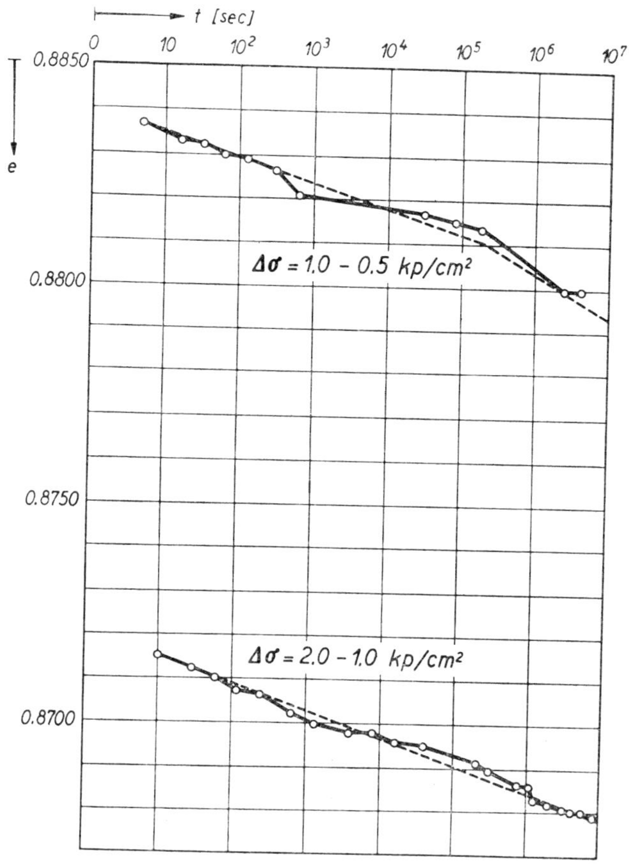

Fig. 8.4. Creep of a dry fine sand in oedometer test

0·25 kp/cm² and then saturated. The other two samples were compacted in the oedometer to a very dense state, loaded to 0·25 kp/cm² and saturated. The observed consolidation curves for load steps up to 1·50 kp/cm² are shown in Fig. 8.5 (*a*) (under 2 for loose and under 4 for dense samples). On the same figure (8.5 (*b*)) the vertical strains $\varepsilon = \Delta h/h$ for 1, 100 and 1,000 days are plotted against the logarithm of the vertical pressure p. The last values were obtained by extending the observed consolidation lines which can approximately be expressed by the equation:

$$\Delta h_t = \Delta h_1 + \alpha \log \frac{t}{t_1} \tag{8.11}$$

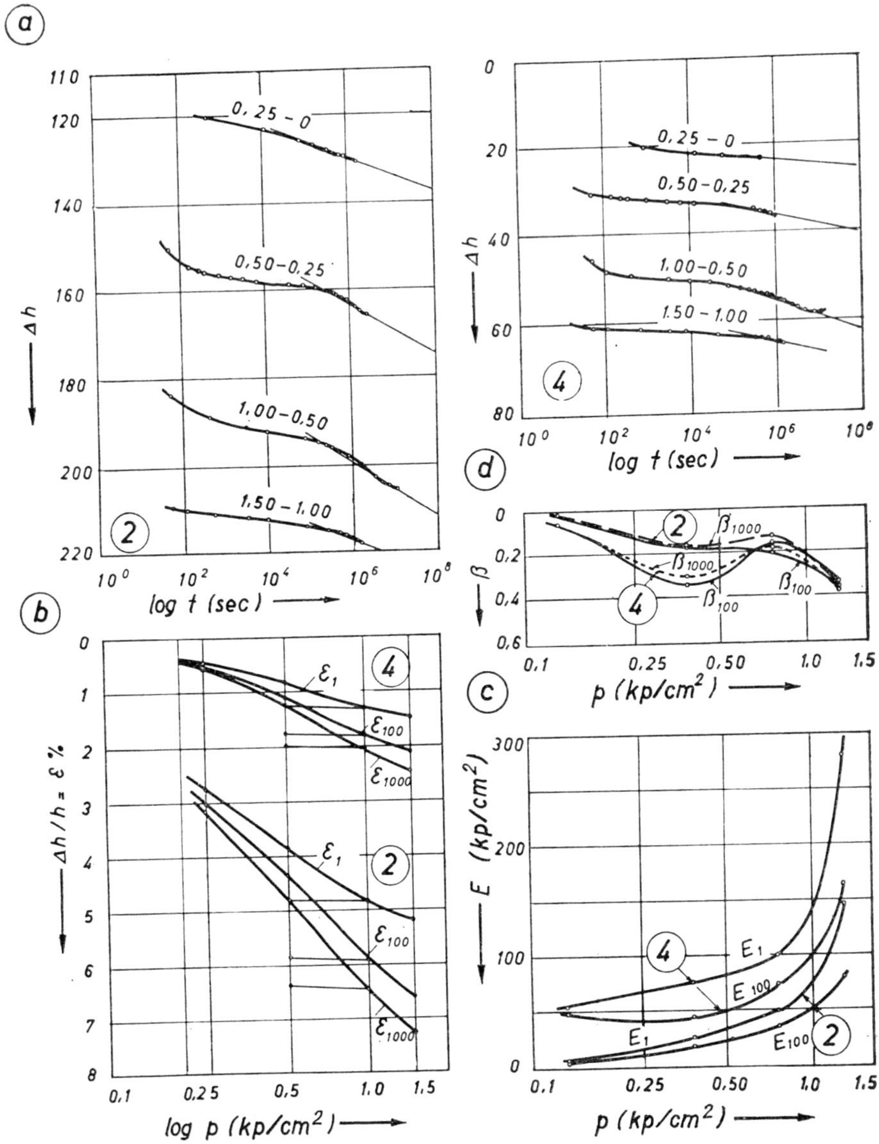

Fig. 8.5. Oedometer test results for a saturated fine sand (after Nonveiller, 1963): (a) consolidation curves, (b) vertical stress-strain plots for 1, 100 and 1,000 days test duration; (c) corresponding E – moduli; 2, loose sand, 4, dense sand; (d) corresponding coefficients β

Δh_t being the settlement at time t and Δh_1 the settlement at time t_1 (e.g. $t_1 = 1$ day) and α the settlement increment in a logarithmic cycle (i.e. the slope of the time—settlement curve on the semi-logarithmic plot).

The relationship (8.10) used by Buisman (1936) was first related to the secondary consolidation of saturated soils and this will be discussed in one of the following sections. The secondary phase of the consolidation of saturated soils occurs at very small pore pressures. Now some of the above examples prove that under the condition of no or of unimportant seepage resistance consolidation does not always develop with logarithmic dependence on time. In the examples presented the creep speed decreased at a slower rate. However, the creep of dense grain systems sometimes diminishes faster and if the period of observation is not very long and the measurements not very precise, the creep often seems to be completely terminated by the end of the test duration.

In every case the creep of the laboratory sample should be observed so long as an appropriate extrapolation can be made. A reliable extrapolation, however, will not always be rendered possible even in cases of long duration tests. Probable rheological relationships, based on suitable rheological models, facilitate the extrapolation though the same rheological scheme cannot be applied to every case.

8.3 ISOTACHES

Isotaches are graphs relating intergranular pressures σ' to void ratios e for certain constant consolidation speeds $\partial e/\partial t$ (Šuklje, 1957). In this definition it is assumed that the ratio between principal effective stresses is kept constant or if this is not possible that its change does not influence the consolidation speed.

Assuming the logarithmic law of creep decrease with time, the equation of the consolidation curve is:

$$e = e_1 - (\Delta e_1 + \alpha_e \log_{10} t) \tag{8.12}$$

In the system of coordinates t, e with a linear scale for void ratio e and a logarithmic scale for time t, such a consolidation curve is given by a straight line (thin line in Fig. 8.6). The consolidation speed is given by the derivative of expression (8.12) with respect to time:

$$v = \frac{\partial e}{\partial t} = -\frac{\alpha_e}{2{\cdot}3026\, t} \tag{8.13}$$

When the load is increased the void ratio decreases and the corresponding coefficient α_e in Equation (8.13) is a positive value.

Equation (8.13) facilitates the determination of the consolidation speed at any time for an arbitrarily shaped consolidation curve (thick line in Fig. 8.6). For this purpose the tangent line should be drawn at the corresponding point (t,e) of the consolidation curve and the logarithmic gradient α_e of this tangent then inserted into Equation (8.13).

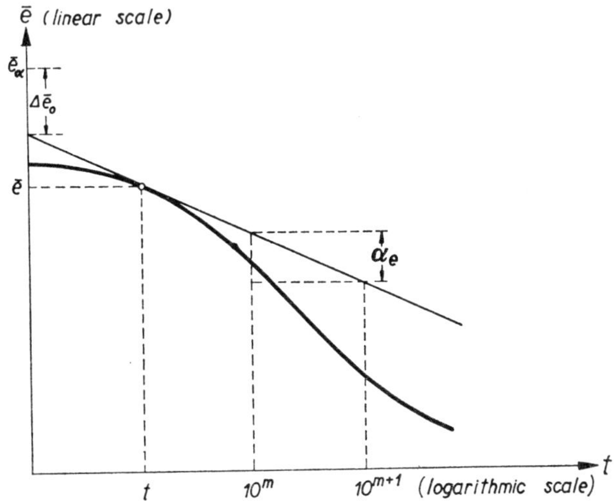

Fig. 8.6. Determination of the consolidation speed at time t of an arbitrarily shaped consolidation curve presented as a void ratio versus logarithm of time plot (see Equation (8.13))

The isotaches can be constructed in the following way. Graphs relating the consolidation speed v to the logarithm of time t are adapted to each consolidation curve $e = e(t)$, having previously extended the plot $e = e(t)$, by extrapolation beyond the time of laboratory observation. If the consolidation develops either at zero or negligible pore pressure excess, the effective pressure σ_1' remains approximately the same or unchanged throughout the consolidation curve $e = e(t)$ and throughout the corresponding graph $v = v(t)$. Hence for a given consolidation speed $v = \text{const}_v$, a void ratio e can be found for a certain effective pressure σ_1', corresponding to a certain consolidation curve $e = e(t)$ (see Fig. 8.7). By using several consolidation curves $e = e(t)$ and the corresponding graphs $v = v(t)$, several points (σ_1',e) of the isotache $v = \text{const}_v$ can be found.

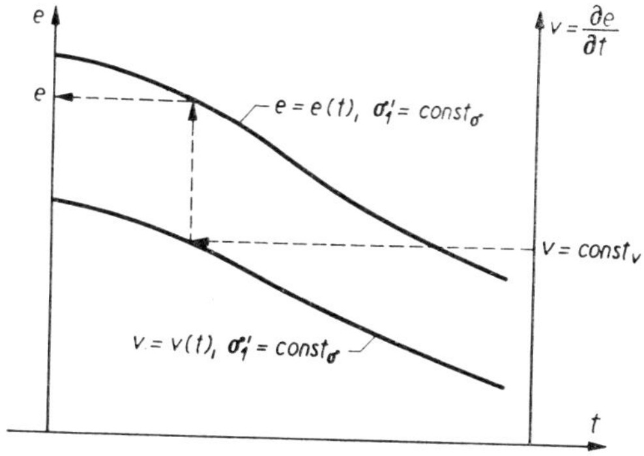

Fig. 8.7. Determination of single points of isotaches from $e = e(t)$ and $v = v(t)$ plots

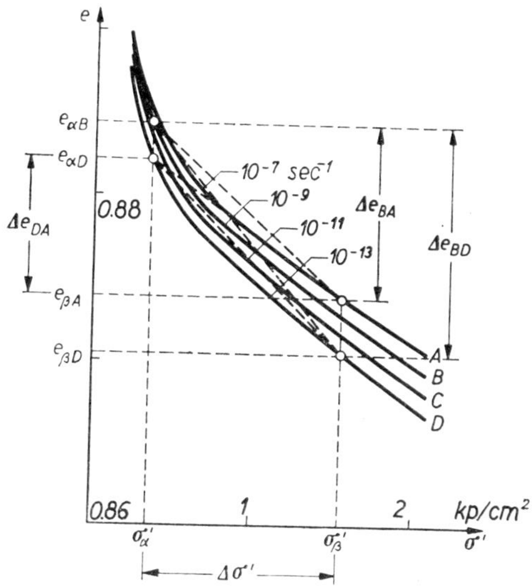

Fig. 8.8. Isotache set of a dry fine sand

Figure 8.8 represents the system of isotaches for a dry sample of sand the consolidation curves of which were shown in Fig. 8.4.

8.4 THE INFLUENCE OF CREEP ON THE MODULUS OF LINEAR COMPRESSION

In Equation (8.6) defining the modulus of linear compressibility M_v, the strain difference $\Delta\varepsilon_1$ corresponding to the stress increase $\Delta\sigma_1$, may be expressed by the initial void ratio e_α and its decrease Δe corresponding to the increase $\Delta\sigma_1'$ of the intergranular pressures.

$$\Delta\varepsilon_1 = \frac{-\Delta e}{1 + e_\alpha} \tag{8.14}$$

According to this equation the compression strains are denoted positive as is common in soil mechanics; during compression, the void ratios

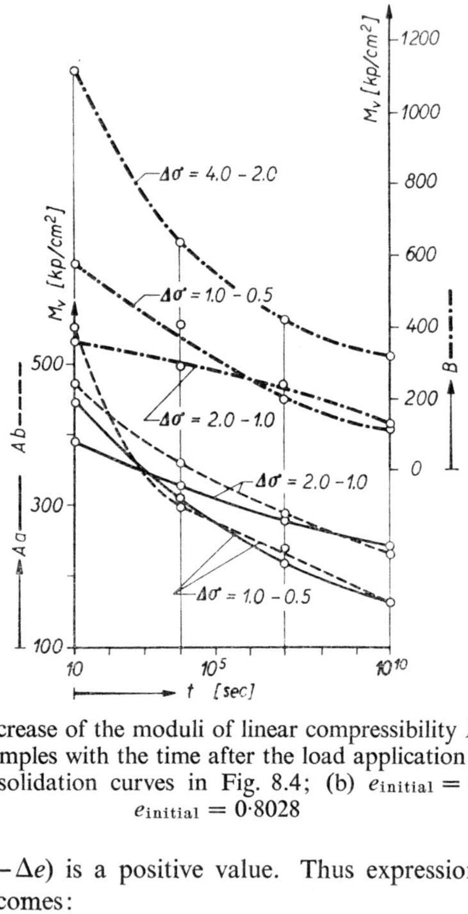

Fig. 8.9. Decrease of the moduli of linear compressibility M_v of three dry sandy samples with the time after the load application: (a) according to consolidation curves in Fig. 8.4; (b) $e_{\text{initial}} = 0{\cdot}8935$; (c) $e_{\text{initial}} = 0{\cdot}8028$

decrease and $(-\Delta e)$ is a positive value. Thus expression (8.6) for the modulus M_v becomes:

$$M_v = -\frac{\Delta\sigma'}{\Delta e}(1 + e_\alpha) \tag{8.15}$$

The isotache set, an example of which is presented in Fig. 8.8, proves that the initial void ratio e_α depends on the degree of preceding consolidation, while the void ratio difference $\Delta e = e_\beta - e_\alpha$ depends on the initial void ratio e_α and on the degree of consolidation at the new total pressure $\sigma_\beta = \sigma_\alpha + \Delta\sigma$.

In Fig. 8.9, the moduli of linear compression are plotted against the time elapsed after the application of the additional loading $\Delta\sigma = \sigma_\beta - \sigma_\alpha$. The graph *Aa* refers to the sample of dry sand, the consolidation curves and isotaches of which were presented in Fig. 8.4 and Fig. 8.8 respectively. The graph *Ab* shows the decrease in moduli of linear compression M_v with increasing creep settlements for another freely filled sand sample ($e_{in} = 0{\cdot}8935$), and graph *B* a similar relationship for a moderately pre-vibrated sample of the same dry sand ($e_{in} = 0{\cdot}8028$).

The moduli are related to the final void ratios of the preceding load step of the laboratory test. The graphs prove the importance of creep effects on the settlement of sandy soils of low density.

Similar graphs can be deduced from Nonveiller's consolidation curves shown in Fig. 8.5 (*a*). According to the logarithmic law of secondary compression (Equation 8.11) such graphs can be expressed by the relation

$$M_v = \frac{M_{v1}}{1 + \dfrac{\alpha}{\Delta h_1} \log \dfrac{t}{t_1}}$$

M_{v1} being the modulus of linear compression at time t_1; for α, Δh_1, t and t_1 see Equation (8.11). For $t_1 = 1$ day the moduli $M_{v1=1\text{day}}$ and $M_{vt=100\text{days}}$ are plotted against the vertical pressure σ in Fig. 8.5 (*c*) and the coefficients

$$\beta = \frac{\alpha}{\Delta h_1} = \frac{\alpha E_1}{\Delta\sigma h}$$

in Fig. 8.5 (*d*) where h is the height of the sample at the new load application.

8.5 INFLUENCE OF THE THICKNESS OF THE LAYER ON THE CONSOLIDATION SPEED

Creep of a dry grain skeleton occurs without any seepage resistance. There is no reason for the specific creep speed $v = \partial e/\partial t$ to be influenced by the layer thickness in such soils. Thus the relationships $e = e(t)$ as observed in oedometer tests can immediately be applied to the prediction of the consolidation of a thicker layer. In the case of uniform additional loading, similar to that applied in oedometer tests, the total settlement of a layer of initial thickness h_α is given either by the expression

$$\rho = \Delta\varepsilon_1\, h_\alpha = -\frac{\Delta e}{1 + e_\alpha}\, h_\alpha \tag{8.16}$$

or by the expression

$$\rho = \frac{\Delta\sigma\, h_\alpha}{M_v} \tag{8.16}$$

According to the decrease in void ratio e and to the decrease in the modulus of linear compression M_v respectively the settlement increases with time (see Fig. 8.9).

It is apparent that the logarithmic extrapolation of the consolidation curve cannot hold up to an arbitrary time t. Equation (8.12) would lead even to negative values of the void ratio e if a certain value of time t were exceeded. Within the probable economical duration of a technical construction, however, the void ratios e as obtained by Equation (8.12) generally remain within reasonable limits of positive values. For this reason the 'final' settlements, corresponding to a certain time at the end of the duration of the construction in question, can be computed by using appropriate 'final' moduli of compression. Similarly, the isotache set can be limited by a lower boundary specific consolidation speed $\partial e/\partial t$, which can still be relevant for practical computations. The consolidation speed $v = \partial e/\partial t = 10^{-13}\ \text{sec}^{-1} = 3{\cdot}156 \,.\, 10^{-6}\ \text{year}^{-1}$, for example, corresponds to the creep change of the void ratio $e = 0{\cdot}0003156$ in 100 years; i.e. assuming an initial void ratio of $e = 0{\cdot}80$ it corresponds to a specific deformation of 0·175 per cent in 100 years. Cases where further development of settlements would still be of interest are rare. The isotache $10^{-15}\ \text{sec}^{-1}$, however, can certainly be considered as the limiting isotache. Generally, the limiting isotache should be chosen by taking into account the thickness of the layer together with the character of the construction.

8.6 THE CONSTANTS OF THE FINAL SETTLEMENT

The limiting isotache, as defined in the previous section, represents the graph relating 'final' void ratios to effective pressures. The further change of the porosity is either small enough to be neglected or it occurs over a time when it is no longer relevant for the construction in question.

In cases when either the shape of the consolidation curve permits the determination of the asymptote $e_\omega = \text{const}$ of the curve or when complete consolidation is reached during the test, the limiting isotache corresponds to zero creep speed. In common practice, however, the void ratio as observed at the end of the single step consolidation of the oedometer test is assumed to determine the final strain corresponding to a given intergranular pressure; further development of creep is disregarded. This approximation cannot be accepted unless the creep effect, in comparison with the total settlement of the sample, is not important.

The graph relating the 'final' void ratios to effective pressures is called the 'curve of compression.' If the sample is consolidated from the

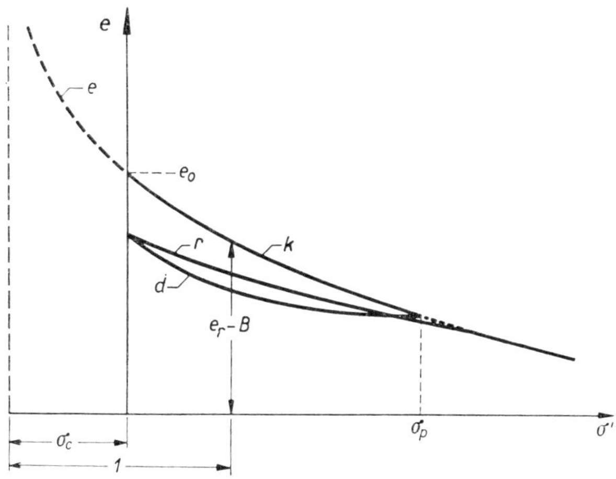

Fig. 8.10. Compression diagrams: *k*, virgin compression curve, *d*, decompression curve, *r*, recompression curve

liquid limit, the graph is called the 'virgin curve of compression.' The 'curve of decompression' $\sigma' = \sigma'(e)$ is related to the unloading of a sample which was pre-consolidated at a pressure $\sigma_p \geqslant \sigma'$. At the repeated loading in the interval $\sigma' \leqslant \sigma_p'$ the plot of 'final' void ratios against intergranular pressures represents the 'curve of recompression.'

After Terzaghi (1925) the virgin compression curve of clays (curve *k* in Fig. 8.10) agrees well with the equation:

$$e = -\frac{1}{C}\ln(\sigma' + \sigma_c') - B(\sigma' + \sigma_c) + e_r \tag{8.17}$$

Plotting the void ratio *e* against the logarithm of intergranular pressure σ' we can see, though, that Equation (8.17) can usually be satisfactorily replaced by the similar expression:

$$e = e_0 - (\Delta e_1 + C_c \log_{10} \sigma') \tag{8.18}$$

The meaning of the constants e_0 (the initial void ratio), Δe_1 and C_c (compression index) is shown in Fig. 8.11.

Differentiation of Equation (8.18) yields:

$$\frac{de}{d\sigma'} = -\frac{C_c}{2{\cdot}3026} \cdot \frac{1}{\sigma'} \tag{8.19}$$

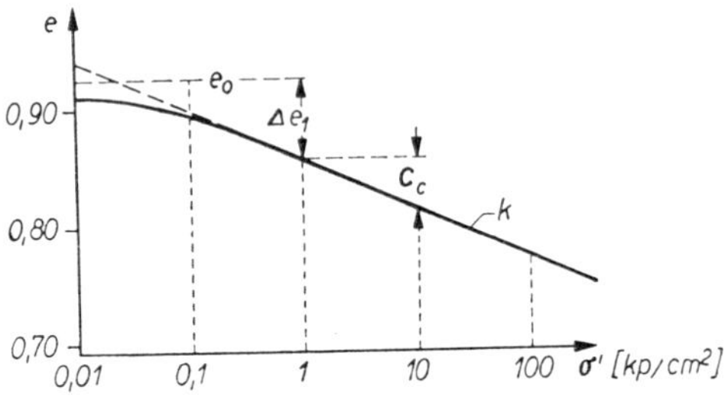

Fig. 8.11. Compression diagram as a semi-logarithmic plot

Applying Equation (8.15), by which the modulus of linear compression M_v was defined, to the differential increase $\Delta\sigma'$ of pressures, we get:

$$M_v = -\frac{d\sigma'}{de}(1 + e) \tag{8.20}$$

and inserting expression (8.19):

$$M_v = \frac{2{\cdot}3026}{C_c}(1 + e)\sigma' \tag{8.21}$$

Several authors use the terms C or K dependent on the compression index C_c as follows:

$$C = \frac{2{\cdot}3026}{C_c} \tag{8.22}$$

$$K = \frac{1}{C(1 + e)} = \frac{C_c}{2{\cdot}3026(1 + e)} \tag{8.23}$$

Fröhlich (1934) deals with constant coefficients K taking the value corresponding to the void ratio e_m in the middle of the pressure interval $\Delta\sigma$. Haefeli (1938) and Bendel (1944) also use constant coefficients K in the settlement analysis substituting in Equation (8.23) the approximation $e = e_\alpha = e_0 - \Delta e_1$ (Haefeli), or $e = e_0$ (Bendel).

The compression index C_c as well as the coefficients C and K are dimensionless values while the unit of the modulus of compression is kp/cm^2.

The reciprocal value of the modulus of linear compression M_v is called the coefficient of volume change m_v:

$$m_v = \frac{1}{M_v} = \frac{d\varepsilon}{d\sigma'} \tag{8.24}$$

If the effective pressure increases from σ_α to σ_β the void ratio decreases from e_α to e_β. Applying Equation (8.18), the void ratio decrease can be expressed as follows:

$$\Delta e = e_\beta - e_\alpha = - C_c \log_{10} \frac{\sigma_\beta'}{\sigma_\alpha'}$$

The corresponding strain $\Delta\varepsilon$ is

$$\Delta\varepsilon = - \frac{\Delta e}{1 + e_\alpha} = \frac{C_c}{1 + e_\alpha} \cdot \log_{10} \frac{\sigma_\beta'}{\sigma_\alpha'} \tag{8.25}$$

According to this equation the compression strains are positive values.

Equation (8.18) is less suitable for the curves of decompression (*d* in Fig. 8.10) and recompression (*r* in Fig. 8.10) which form a hysteresis loop. The recompression curve lying for the prevailing part above the decompression curve, intersects the latter at $\sigma' < \sigma_p$ and joins the virgin compression curve at $\sigma' > \sigma_p$, near to σ_p. The moduli of linear decompression and recompression respectively are essentially greater than the moduli of virgin compression corresponding to the same stress interval $\Delta\sigma'$.

9

Fundamentals of the One-Dimensional Consolidation of Saturated Soils

9.1 PRINCIPLE OF EFFECTIVE STRESS

9.1.1 Definition of Effective Stress in Homogeneous Stress Fields

Terzaghi's classical one-dimensional consolidation theory is based on the principle of effective stress. In this theory the effective stress represents that part of the total axial stress which governs the compression of the soil skeleton. The relationship between effective axial pressure and the volume change is assumed to be linear and viscous effects are neglected. At any time during consolidation the sum of the pressure of the liquid in the voids (u) and the effective axial pressure (σ_1') is equal to the total axial pressure (σ_1):

$$u + \sigma_1' = \sigma_1 \tag{9.1}$$

The pressure of the liquid in the voids is called the pore pressure.

The principle of effective stress can also be usefully applied when analysing the three-dimensional consolidation of soils having arbitrary visco-elastic properties. The effective stress tensor should then be defined as that part of the total stress tensor which governs the deformation speed of the soil skeleton. This definition applies for compressibility as well as for strength problems. In such a definition the amount of deformation is included; the latter is given by the integral of the consolidation speeds during the corresponding time interval. The influence of the effective stress tensor on the deformation speed depends on the rate of loading, porosity, structural anisotropy, stress and strain history of the soil. The strain history also involves the degree of mobilization of the attractive forces between the films of adsorbed water surrounding the grains, and the thixotropic effects.

Consider a finite volume element of the soil, having for example the form of a parallelepiped as shown (in normal projection) in Fig. 9.1. The stress states computed by mechanics of continua principles may be assumed to be homogeneous inside the observed element. The spherical component of the total stress tensor may be assumed to consist of two parts:

$$\sigma_{\text{oct}}\delta_{rs} = (\sigma_{\text{oct}}' + u)\delta_{rs} \tag{9.2}$$

u being equal to the pore pressure. Inside the voids the stress tensor $u\delta_{rs}$ represents the total actual stress state whilst inside the grains it represents that part of the total actual stress tensor which is transferred to the grains through the wetted surfaces and, by the same amount u, through the contact

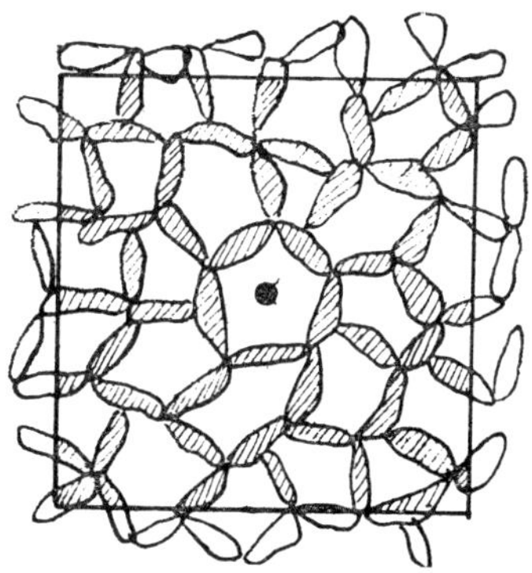

Fig. 9.1. Finite element of the grain skeleton

areas between the grains. Since the compressibility of the solid is assumed to be negligible in comparison with the deformability of the soil skeleton the stress tensor $u\delta_{rs}$ is considered to have no influence on the deformation of the soil. Only the remaining part of the total stress tensor governs the displacements of the grains in the soil skeleton. This remaining part is the effective stress tensor:

$$\begin{bmatrix} \sigma_1' & \cdot & \cdot \\ \cdot & \sigma_2' & \cdot \\ \cdot & \cdot & \sigma_3' \end{bmatrix} = (\sigma_{\text{oct}} - u)\delta_{rs} + \begin{bmatrix} \sigma_1 - \sigma_{\text{oct}} & \cdot & \cdot \\ \cdot & \sigma_2 - \sigma_{\text{oct}} & \cdot \\ \cdot & \cdot & \sigma_3 - \sigma_{\text{oct}} \end{bmatrix} \tag{9.3}$$

In isotropic soils, its spherical component

$$\sigma_{\text{oct}}'\delta_{rs} = (\sigma_{\text{oct}} - u)\delta_{rs} \tag{9.4}$$

governs the speed of the volume change, and its distortional part

$$\begin{bmatrix} \sigma_1 - \sigma_{oct} & \cdot & \cdot \\ \cdot & \sigma_2 - \sigma_{oct} & \cdot \\ \cdot & \cdot & \sigma_3 - \sigma_{oct} \end{bmatrix} =$$

$$= \begin{bmatrix} \sigma_1' - \sigma_{oct}' & \cdot & \cdot \\ \cdot & \sigma_2' - \sigma_{oct}' & \cdot \\ \cdot & \cdot & \sigma_3' - \sigma_{oct}' \end{bmatrix} \quad (9.5)$$

$$\sigma_i' = \sigma_i - u \qquad (i = 1,2,3) \quad (9.6)$$

the speed of distortional deformations. For the conditions of anisotropic deformability either stress tensor component influences either component of the strain velocity tensor.

It may be pointed out that both components of the total stress tensor analysed in the above way are related to the soil as a continuum and do not express either the real influence of the pore pressure or the total influence of the intergranular forces on the displacements of the soil skeleton. The influence of the pore pressure coincides, however, approximately with the effect of the spherical stress tensor $u\delta_{rs}$ and the influence of the intergranular pressure is approximately equal to the effects of the effective stress tensor provided that the contact areas between the grains are very small in comparison with the wetted surfaces of the grains.

Now, if the compressibility of the solid is taken into account, the preceding definition of the effective stress tensor is no longer exact. Denoting the modulus of (volume) compressibility of the solid by K_s and its reciprocal value, termed compressibility, by $C_s = 1/K_s$, the compression of the solid grains per unit volume of the soil is

$$\varepsilon_{vs} = (1 - n)C_s u \quad (9.7)$$

If in a soil of isotropic, linear and non-viscous deformability with no pore pressures the spherical part of the effective stress tensor $\sigma_{oct}'\delta_{rs}$ caused a volume strain ε_v, then in the same soil of the same porosity and the same stress history exposed to the total stress $\sigma_{oct} = \sigma'_{oct} + u$ the volume strain would be

$$\bar{\varepsilon}_v = \varepsilon_v + (1 - n)C_s u \quad (9.8)$$

Under zero pore pressure conditions the same volume strain would be produced by the effective stress $\bar{\sigma}_{oct}'$ given by the equation

$$\frac{\bar{\sigma}_{oct}'}{\sigma_{oct}'} = \frac{\varepsilon_v + (1 - n)C_s u}{\varepsilon_v} \quad (9.9)$$

Assuming for the soil skeleton a linear non-viscous stress-strain relationship

$$\sigma_{\text{oct}}' = K_c \varepsilon_v = \frac{\varepsilon_v}{C_c} \tag{9.10}$$

Equation (9.9) yields:

$$\bar{\sigma}_{\text{oct}}' = \sigma_{\text{oct}} - u\left[1 - (1 - n)\frac{C_s}{C_c}\right] \tag{9.11}$$

For soils of porosity $0{\cdot}2 \leqslant n \leqslant 0{\cdot}8$ and compressibility 10^{-3} cm²/kp $\leqslant C_c \leqslant 10^{-1}$ cm²/kp, the quantity $(1 - n)(C_s/C_c)$ varies between 2 and 0·005 per cent and has no practical importance. The compressibility of the solid was assumed to be $0{\cdot}25 \,.\, 10^{-5}$ cm²/kp. Thus the preceding effective stress definition in which the compressibility of the solid was disregarded, is fully justified.

The compressibility of the fluid (C_w) causes a decrease of the pore pressure and a corresponding increase of effective pressure. This effect is hardly noticeable in saturated soils, it is, however, significant in partly saturated soils.

9.1.2 Effective Stresses in Non-homogeneous Stress Fields

If the stress field is heterogeneous, i.e. if the total stress tensor and the pore pressures, or pore pressures only, vary from point to point, the definition of the effective stress tensor is no longer unambiguous. The excess pressure of the pore water surrounding a grain is not uniform and thus affects the movement of the grain. Different hydraulic potentials along single grains cause seepage forces and they should be considered as volume forces when establishing the equilibrium conditions for soil elements. The average pore pressure acting on the surfaces of the grains in a volume element of soil can be considered as a neutral pressure u in Equation (9.3) by which the effective stress tensor has been defined. As, however, the grains and the voids between them have finite dimensions the limit process cannot be applied and discontinuities appear at the boundary surfaces of adjacent elements.

If the pore pressure potential field only is heterogeneous, the discontinuities in the effective pressure at the boundary surfaces of single elements have to be attributed to the seepage resistance connected with shear stresses at the wetted surfaces of grains and with the variation of excess pore water pressures along them. If the total stress field is not homogeneous either, the differences in effective stress tensors at the boundaries of the observed elements have to be attributed partly to the change in total stresses and partly to the drainage process. In either case single grains do not remain in equilibrium under the influence of unequal pressures of the surrounding liquid and the corresponding equivalent pressures at the contact surfaces between the grains. They are, however, indifferent to the apparent pore pressure tensor corresponding to the average pore pressure or to the pore pressure in the centre of the element.

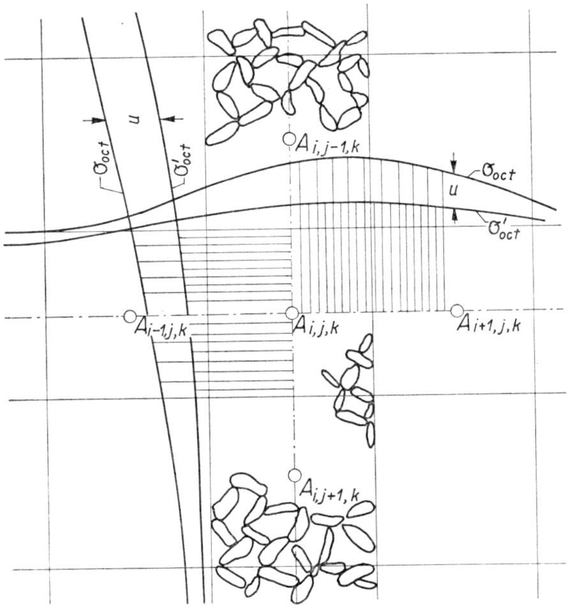

Fig. 9.2. Finite soil elements in non-homogeneous stress fields

Consequently the principle of effective stress cannot be applied to heterogeneous stress fields except as a working hypothesis expressing satisfactorily the role of real factors controlling the displacements of the soil skeleton. Provided that the seepage forces are considered in equilibrium conditions as volume forces, the approximate character of the principle of effective stresses is related mainly to the substitution of continuous diagrams of supplementary total spherical stresses by a steplike diagram and to a similar steplike transformation of the isochrones, i.e. of the diagrams of pore pressures corresponding, in a given time, to the observed direction passing through the observed point. In fine grained soils the steps can be considered to be very small though finite and in such soils the applications of the principle of effective stress seems to be fully justified in heterogeneous stress fields also.

9.1.3 Discussion of the Unique Concept of the Effective Stress Tensor Related to the Compression and Strength of Soils

The effective stress tensor defined as that part of the total stress tensor which governs the displacement of the soil skeleton also governs, in the author's opinion, the strength of the soil.

Let us consider that the initial stress state of a soil corresponds to a given total stress tensor $p_{rs,1} = p_{rs,1}{}^0 + p_{rs,1}{}^d$ and that the initial pore pressure is u_1. When changing the total stress tensor and the pore pressure, the soil skeleton will undergo certain displacements. Whatever the change in the stresses, the part $u\delta_{rs}$ of the total stress tensor never causes any rearrangement of the grains whilst the remaining part of the spherical component and the distortional component of the total stress tensor do. When Mohr circles representing the effective stress tensor as defined by Equation (9.3) approach the failure envelope, the distortional displacements begin increasing with constant and finally with increasing velocity and consequently failure occurs. Failure is due to the displacement velocity which results from the preceding rearrangements of the soil skeleton caused by either part of the effective stress tensor as defined by Equation (9.3). Consequently this effective stress tensor controls the strength of the soil.

If Mohr's failure theory is accepted, the term 'shear strength' is not clearly defined. In the author's opinion the separate analysis of the effective stress related to shear strength, particularly in homogeneous stress fields, is not satisfactorily justified. The reader will find the opposite point of view in thorough analyses of the effective stress principle in papers by Bishop and Eldin (1950), Bishop (1955), Skempton (1960) and Jennings (1960) as well as in the discussions concerning effective stress at the 'Pore Pressure and Suction in Soils' Conference (1960). Proceeding from Bishop's concept of the indifference of the grains' position to an all-around pressure u, the present author has analysed the effects of an increase of the total stress tensor appearing in the soil when treated as a continuum, considering the real transference of stresses in the soil as a two-phase system. Nevertheless it may be pointed out that as far as saturated soils are concerned, the separate analysis of the effective stress related to the shear strength of the soil leads to the same practical conclusion: to the unique value of effective stress given by Terzaghi's classical equation for either compression or strength problems.

9.2 GENERAL DIFFERENTIAL EQUATION OF THE ONE-DIMENSIONAL CONSOLIDATION OF SATURATED SOILS

9.2.1 Fundamental Conditions

The differential equation of the one-dimensional consolidation of soils was developed by K. Terzaghi (1923). The equation was based on the following assumptions: the voids of the soil are completely filled with water; both the water and the solid constituents of the soil are perfectly incompressible; Darcy's law is strictly valid; the coefficient of permeability k is a constant; the time lag of consolidation is due entirely to the low permeability of the soil; the soil is laterally confined; both the total and the effective normal stresses are the same for every point on any horizontal section through the soil and for every stage of the process of consolidation; the void ratio decrease is linearly proportional to the effective stress increase.

In its classical form Terzaghi's differential equation will be presented in the following section (9.3) as a special case of a more general form of the differential equation of consolidation allowing for a modified form of Darcy's law, for the variation of the coefficient of permeability with void ratio, for the viscous properties of soil and for non-linear stress-strain relationships. This generalized form of the consolidation equation will be developed by making use of Florin's appıoach (Florin, 1961).

The consolidation theory of saturated soils is based on Darcy's law expressing the linear relationship between the hydraulic gradient i and the filtration speed v:

$$v = ki \tag{9.12}$$

The constant of proportionality k is called the coefficient of permeability and its units are cm sec^{-1}. The filtration speed v is related to the total cross-section area of the soil. The average value of the true seepage speed v_e is approximately

$$v_e = \frac{v}{n} \tag{9.13}$$

n being the porosity.

Darcy's law in the form of Equation (9.12) is not applicable in all cases. As proved by Pavlovsky (1922) and Roza (1950) (see Florin 1959), in some dense clays the flow is extremely small (practically zero) until a certain hydraulic gradient i_0 is exceeded. For such clays Darcy's law may be generalized in the form (Florin, 1959):

$$\left.\begin{aligned} v &= k(i - i_0) \quad \text{when} \quad i \geqslant i_0 \\ v &= 0 \quad \text{when} \quad i < i_0 \end{aligned}\right\} \tag{9.14}$$

The hydraulic gradient is equal to the negative differential of the potential level, $\mathrm{d}H^*$, divided by the differential of the drainage path, $\mathrm{d}z$. Consolidation is governed only by that part of the hydraulic gradient which arises from the additional hydraulic potential field caused by loading. Denoting the corresponding excess pore pressure by u and the bulk density of the pore water by γ_w, the additional hydraulic (piezometric) height H is

$$H = \frac{u}{\gamma_w} \tag{9.15}$$

and the hydraulic gradient governing the consolidation process equals:

$$i = -\frac{\partial H}{\partial z} = -\frac{1}{\gamma_w}\frac{\partial u}{\partial z} \tag{9.16}$$

Two examples will be presented illustrating the difference between the excess pore pressure u and the total pore pressure $u^* = u + u_1$, as well as the difference between the hydraulic gradient i and the total hydraulic gradient i^*.

In Fig. 9.3 the plot $\sigma = \sigma(z)$ represents the vertical normal pressures along the axis (1)–(1) and the plot $u_t = u_t(z)$ the isochrone of the corresponding excess pore pressures at time t, both arising from the load q applied on the surface of the half-space. The preceding pore pressures u_1 were governed by the free water level (W.L.) in the pervious upper layer. At time t during consolidation the total pore pressure u^*, at depth z, exceeds the initial pore pressure u_1 by u:

$$u^* = u_1 + u$$

and at depth $z' = z + \mathrm{d}z$ the total pore pressure is:

$$u^{*\prime} = u_1' + u'$$

The corresponding hydraulic potential levels are determined by the respective heights H^* and $H^{*\prime}$ (Fig. 9.3):

$$H^* = z + \zeta + H$$

$$H^{*\prime} = z' + \zeta' + H'$$

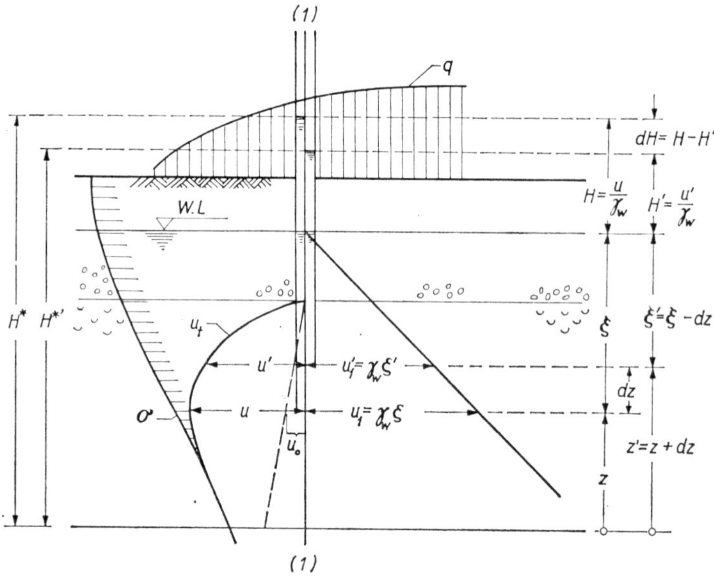

Fig. 9.3. Example illustrating the difference between the pore-pressure excess u and the total pore-pressure $u = u + u_1$; the total hydraulic gradient i^* coincides with the gradient i which governs the consolidation speed

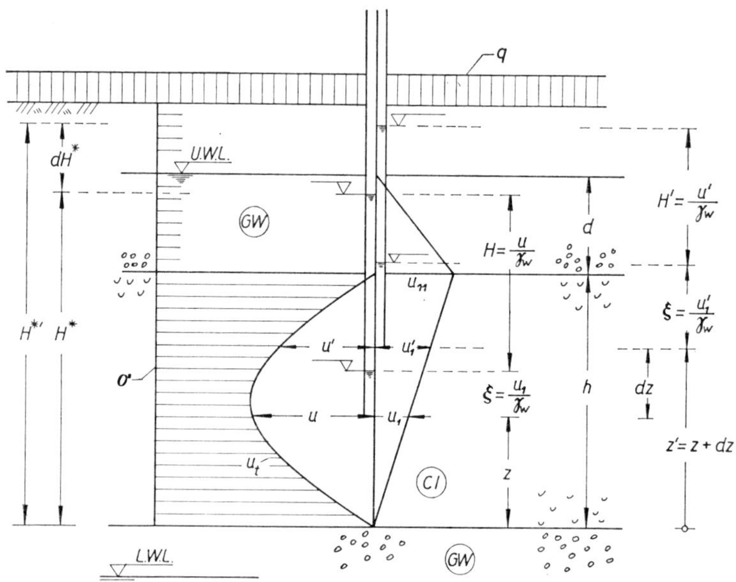

Fig. 9.4. Example illustrating that the total hydraulic gradients can differ from gradients controlling the consolidation speed

ζ and ζ' being the initial piezometric heights at the respective depths z and z':

$$\zeta = \frac{u_1}{\gamma_w}, \quad \zeta' = \frac{u_1'}{\gamma_w}$$

As $z + \zeta = z' + \zeta'$

$$\mathrm{d}H^* = H^{*\prime} - H^* = H' - H = \mathrm{d}H$$

and

$$i^* = -\frac{\partial H^*}{\partial z} = -\frac{\partial H}{\partial z} = i$$

As shown by the second example presented in Fig. 9.4, the total hydraulic gradient i^* must not equal the gradient i which governs the consolidation speed. The layer *CI* of thickness h, lying between two pervious layers *GW*, is assumed to contain seepage water with a stationary upper potential level U.W.L. at a height d above the upper boundary of the layer *CI*, and with a lower potential level at the lower boundary of the layer. Neglecting the seepage resistance of the pervious upper layer *GW*, the initial pore-pressure u_{11} at the upper boundary of the layer *CI* is $u_{11} = \gamma_w d$, and the pore-pressure in the position z above the lower boundary of the layer is

$$u_1 = \frac{u_{11}}{h} z = \gamma_w z \frac{d}{h}$$

Now, at a time $t = 0$ the surface of the upper layer *GW* is subjected to a uniform load q, and $u_t = u_t(z)$ is a possible isochrone of the excess pore-pressure at a time t during consolidation. The total pore-pressures u^* at time t and at depth z are (see Fig. 9.4):

$$u^* = u_1 + u = \gamma_w z \frac{d}{h} + u$$

while the total pore-pressures at depth $z' = z + dz$ are

$$u^{*\prime} = u_1' + u' = \gamma_w z' \frac{d}{h} + u$$

The corresponding potential levels are determined by the respective heights H^* and $H^{*\prime}$:

$$H^* = z + \zeta + H = z + \frac{u_1}{\gamma_w} + \frac{u}{\gamma_w}$$

$$H^{*\prime} = z' + \zeta' + H' = z' + \frac{u_1'}{\gamma_w} + \frac{u'}{\gamma_w}$$

and the total hydraulic gradient is:

$$i^* = -\frac{\partial H^*}{\partial z} = -\frac{(H^{*\prime} - H)}{\mathrm{d}z} = -\frac{u' - u}{\gamma_w \mathrm{d}z} - \left(\frac{z' - z}{\mathrm{d}z} + \frac{u_1' - u_1}{\gamma_w \mathrm{d}z}\right)$$

Considering $z' - z = \mathrm{d}z$ and inserting the above expressions for u_1' and u_1 we get:

$$i^* = -\frac{1}{\gamma_w}\frac{\partial u}{\partial z} - \left(1 + \frac{d}{h}\right) = i - \left(1 + \frac{d}{h}\right)$$

The gradient $i_1 = -\left(1 + \frac{d}{h}\right)$ controlling the preceding stationary seepage speed is constant. Thus only the part i of the total hydraulic gradient governs the consolidation speed.

If the initial hydraulic gradient is zero (Fig. 9.3), the limiting hydraulic gradient i_0 can be expressed in the form

$$i_0 = -\frac{\partial H_0}{\partial z} = -\frac{1}{\gamma_w}\frac{\partial u_0}{\partial z} \tag{9.17}$$

similar to Equation (9.16); H_0 and u_0 are the respective values of additional hydraulic heights and pressures.

Substituting

$$\bar{H} = H - H_0 \quad \text{and} \quad \bar{u} = u - u_0 \tag{9.18-a,b}$$

the generalized Darcy's law can be introduced in the consolidation equation in the form

$$v = -k\frac{\partial \bar{H}}{\partial z} \tag{9.19}$$

or

$$v = -\frac{k}{\gamma_w}\frac{\partial \bar{u}}{\partial z} \tag{9.20}$$

If stationary seepage corresponding to hydraulic gradients $i_1 > i_0$ preceded the consolidation process, the boundary condition

$$|\, i^* \,| = |\, i + i_1 \,| \leqslant i_0$$

should be considered in the consolidation analysis; gradients causing seepage in the upward direction have been noted as positive values.

If the directions of equipotential and seepage lines of the hydraulic field caused by additional loading do not coincide with the respective directions of the previous stationary hydraulic field, total resulting hydraulic gradients i^* should be considered in the consolidation analysis and corresponding boundary potentials taken into account (see Florin, 1961). Such cases will not be treated in the present book.

9.2.2 Deduction of the Differential Equation of Consolidation

Drainage is assumed to occur in a single direction, e.g. in the direction z (Fig. 9.5). If the drainage speed v is assumed to increase with the increasing ordinate z, the amount of pore water leaving a volume element $\mathrm{d}x\,\mathrm{d}y\,\mathrm{d}z$ in a time interval $\mathrm{d}t$, is

$$\frac{\partial v}{\partial z}\,\mathrm{d}z\,.\,\mathrm{d}x\,\mathrm{d}y\,.\,\mathrm{d}t \tag{9.21}$$

The corresponding volume decrease can be expressed in the form

$$-\frac{\partial n}{\partial t}\,\mathrm{d}x\,\mathrm{d}y\,\mathrm{d}z\,.\,\mathrm{d}t \tag{9.22}$$

if n is the porosity of the observed soil element at the observed time t. Equating the above expressions and dividing the equation by $\mathrm{d}x\,.\,\mathrm{d}y\,.\,\mathrm{d}z\,.\,\mathrm{d}t$ we get the continuity condition for the liquid phase of the soil:

$$\frac{\partial v}{\partial z}+\frac{\partial n}{\partial t}=0 \tag{9.23}$$

The movement of the water out of the element is accompanied by a movement of the grains in the opposite direction with a certain average

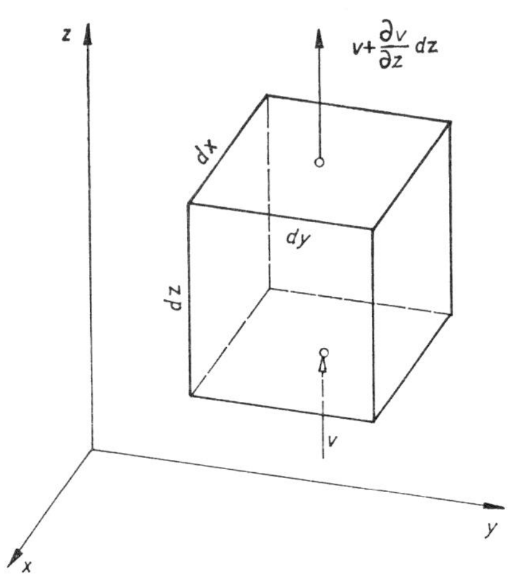

Fig. 9.5. Seepage through a soil element in one direction

real velocity $\mathring{v}_e$ and a corresponding apparent velocity $\mathring{v}$ related to the gross cross-sectional area of the soil. Their approximate relationship is:

$$\mathring{v}_e = \frac{\mathring{v}}{1 - n} \tag{9.24}$$

The continuity condition for the solid phase of the soil can be written by analogy with Equation (9.23):

$$\frac{\partial \mathring{v}}{\partial z} + \frac{\partial (1 - n)}{\partial t} = 0 \tag{9.25}$$

The addition of Equations (9.23) and (9.25) yields:

$$\frac{\partial v}{\partial z} + \frac{\partial \mathring{v}}{\partial z} = 0 \tag{9.26}$$

Substituting, in Equation (9.20), expression (9.13) for the seepage velocity v, the generalized Darcy's law becomes

$$v_e = -\frac{k}{\gamma_w n} \frac{\partial \bar{u}}{\partial z} \tag{9.27}$$

Now, as the grain skeleton also moves during consolidation, the real average velocity v_e of the liquid should be replaced by the relative velocity $v_e - \mathring{v}_e$ of the liquid with respect to the solid. Thus, Darcy's law reads:

$$v_e - \mathring{v}_e = -\frac{k}{\gamma_w n} \frac{\partial \bar{u}}{\partial z} \tag{9.28}$$

Substituting for v_e and $\mathring{v}_e$ expressions (9.13) and (9.24), Equation (9.28) becomes

$$\frac{v}{n} - \frac{\mathring{v}}{1 - n} = -\frac{k}{\gamma_w n} \frac{\partial \bar{u}}{\partial z} \tag{9.29}$$

or, considering the relation

$$\frac{n}{1 - n} = e \tag{9.30}$$

e being the void ratio,

$$v - e\mathring{v} = -\frac{k}{\gamma_w} \frac{\partial \bar{u}}{\partial z} \tag{9.31}$$

By differentiating the above equation with respect to z we get

$$\frac{\partial v}{\partial z} - e\frac{\partial \mathring{v}}{\partial z} - \mathring{v}\frac{\partial e}{\partial z} = -\frac{\partial}{\partial z}\left(\frac{k}{\gamma_w}\frac{\partial \bar{u}}{\partial z}\right) \tag{9.32}$$

Considering relation (9.26) we obtain

$$(1+e)\frac{\partial v}{\partial z} - \mathring{v}\frac{\partial e}{\partial z} = -\frac{\partial}{\partial z}\left(\frac{k}{\gamma_w}\frac{\partial \bar{u}}{\partial z}\right) \tag{9.33}$$

The second term on the left-hand side of the above equation is the product of two very small values and can be omitted when summed with the first term. Considering, further, Equation (9.23) for $\partial v/\partial z$, Equation (9.33) yields:

$$-\frac{\partial n}{\partial t}(1+e) = -\frac{\partial}{\partial z}\left(\frac{k}{\gamma_w}\frac{\partial \bar{u}}{\partial z}\right) \tag{9.34}$$

Substituting then

$$n = \frac{e}{1+e}, \quad \frac{\partial n}{\partial t} = \frac{1}{(1+e)^2}\frac{\partial e}{\partial t} \tag{9.35}$$

we get the fundamental form of the differential equation of one-dimensional consolidation:

$$\frac{\partial e}{\partial t} = \frac{1+e}{\gamma_w}\frac{\partial}{\partial z}\left(k\frac{\partial \bar{u}}{\partial z}\right) \tag{9.36}$$

It can be further developed in the following way:

$$\frac{\partial e}{\partial t} = \frac{\partial e}{\partial \sigma'}\frac{\partial \sigma'}{\partial t} \tag{9.37}$$

σ' being the effective stress in the drainage direction:

$$\sigma' = \sigma - u \tag{9.38}$$

Thus

$$\frac{\partial e}{\partial t} = \frac{\partial e}{\partial \sigma'}\left(\frac{\partial \sigma}{\partial t} - \frac{\partial u}{\partial t}\right) \tag{9.39}$$

Equating expressions (9.36) and (9.39) for $\partial e/\partial t$, considering Equations (9.17) and (9.18-b) and rearranging, we obtain:

$$\frac{\partial u}{\partial t} = \frac{\partial \sigma}{\partial t} - \frac{1+e}{\gamma_w \dfrac{\partial e}{\partial \sigma'}}\frac{\partial}{\partial z}\left[k\left(\frac{\partial u}{\partial z} - \gamma_w i_0\right)\right] \tag{9.40}$$

$\partial \sigma/\partial t$ denotes the increase of the total stress σ per unit of time during consolidation. The coefficient of permeability k and the limiting hydraulic gradient i_0 are allowed to vary with z.

9.2.3 General Form of the Void Ratio-Effective Stress Relationship

In Equations (9.36) or (9.40) the void ratio e depends on the actual effective stress σ' and on the stress history. The stress σ' may be assumed to vary with time τ:

$$\sigma' = \sigma'(\tau)$$

$$\sigma_{\tau+d\tau}' = \sigma_\tau' + \frac{\partial \sigma'}{\partial \tau} \mathrm{d}\tau \tag{9.41}$$

$(\partial\sigma'/\partial\tau)\,\mathrm{d}\tau$ being the effective stress increment in the time interval $\mathrm{d}\tau$. The soil may be supposed to have viscous properties and the creep function (see Equation 4.40) may be denoted by $f(t,\tau)$. If $\sigma'(\tau_1)$ is the effective stress at time τ_1 and $e(\tau_1)$ the void ratio at that time, then the void ratio at a later time t is given by the equation:

$$e(t) = e(\tau_1) - \sigma'(\tau_1)f(t,\tau_1) - \int_{\tau_1} \frac{\partial \sigma'}{\partial \tau} f(t,\tau)\,\mathrm{d}\tau \tag{9.42}$$

In the special case of non-viscous soil with a linear effective stress-strain relationship $f(t,\tau_1) = 0$, $f(t,\tau) = a$, and Equation (9.42) reduces to:

$$e(t) = e(\tau_1) - [\sigma'(t) - \sigma'(\tau_1)]a \tag{9.43}$$

yielding:

$$-\frac{e(t) - e(\tau_1)}{\sigma'(t) - \sigma'(\tau_1)} = \frac{-\Delta e}{\Delta \sigma'} = a \tag{9.43'}$$

The coefficient a is related to the coefficient of volume change m_v (see Equation 8.24) by the equation:

$$a = (1 + e)m_v \tag{9.44}$$

9.2.4 The Role of the Limiting Hydraulic Gradient in the Consolidation Analysis

In homogeneous soils the limiting hydraulic gradient i_0 can usually be taken as constant and in this case the differential equation of one-dimensional consolidation (9.36) reduces to

$$\frac{\partial e}{\partial t} = \frac{1 + e}{\gamma_w} \frac{\partial}{\partial z} \left[k \frac{\partial u}{\partial z} \right] \tag{9.36-a)*}$$

* By replacing u by $H\gamma_w$ this equation gets a more general form applied to arbitrary previous hydraulic fields (see Section 9.2.1, p. 130 and Florin, 1961, Equation 3.16). H^* denotes potential heights corresponding to the resulting hydraulic field.

Similarly Equation (9.40) becomes simplified as follows:

$$\frac{\partial u}{\partial t} = \frac{\partial \sigma}{\partial t} - \frac{1 + e}{\gamma_w \dfrac{\partial e}{\partial \sigma'}} \frac{\partial}{\partial z} \left[k \frac{\partial u}{\partial z} \right] \tag{9.40-a}$$

The limiting hydraulic gradient disappears in the differential equation, yet appreciably influences the boundary conditions. This influence will be illustrated by the example shown in Fig. 9.6 (see Florin, 1961, Fig. 57). A clay layer of thickness h lying on an impervious base is assumed to be subjected to a suddenly applied uniformly distributed load q.

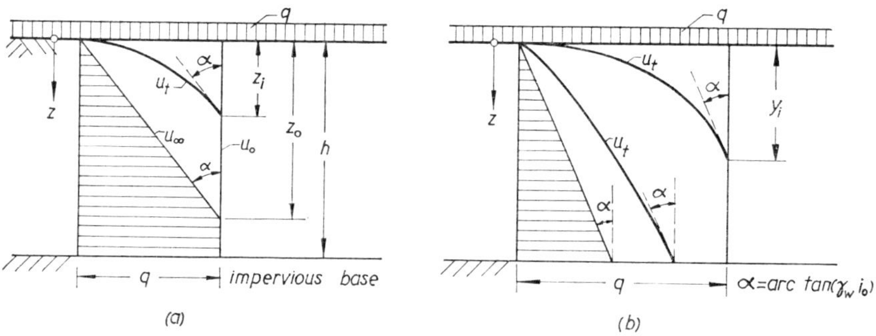

Fig. 9.6. Example illustrating the influence of the limiting hydraulic gradient onto the boundary conditions of the consolidation process

Two separate cases may be investigated. In case (*a*) (left hand side of Fig. 9.6) the end isochrone of excess pore-pressures u_∞

$$u_\infty = \gamma_w i_0 z$$

intersects the initial isochrone $u_0 = q$ at depth z_0

$$z_0 = \frac{q}{\gamma_w i_0}$$

When integrating the differential equation of consolidation (9.40-a), $\partial\sigma/\partial t = 0$ should be inserted and the following boundary conditions respected:

(1) $t = 0, \quad 0 \leqslant z \leqslant h \quad \rightarrow u = u_0 = q$

(2) $t > 0, \quad z_i \leqslant z \leqslant z_0 \rightarrow u = q$

(3) $t > 0, \quad z = 0 \qquad \rightarrow u = 0$

(4) $t > 0, \quad z = z_i \qquad \rightarrow u = q, \quad \dfrac{\partial u}{\partial z} = \gamma_w i_0$

In case (*b*) (right-hand side of the Fig. 9) the point of intersection of the end- and zero-isochrones lies below the impervious base of the layer:

$$z_0 = \frac{q}{\gamma_w i_0} < h$$

The solution should be performed in two steps. In the first step the same boundary conditions have to be considered as in case (*a*). This first step is limited by the time t_h at which the top of the isochrone u_t reaches the impervious base of the layer:

$$t = t_h, \quad z_i = h \rightarrow u = u(t_h, z)$$

In the subsequent period $t > t_h$ the following boundary conditions have to be considered:

$$t > t_h, \quad z = 0 \rightarrow u = 0$$

$$t > t_h, \quad z = h \rightarrow \frac{\partial u}{\partial z} = \gamma_w i_0$$

9.3 ONE-DIMENSIONAL CONSOLIDATION OF SATURATED SOILS WITH A LINEAR EFFECTIVE STRESS-STRAIN RELATIONSHIP

9.3.1 Terzaghi's Original Form of the Differential Equation of Consolidation

If all the special conditions that have been listed in the first section 9.2.1 are respected, the general differential equation of consolidation (9.40) reduces to Terzaghi's classical form:

$$\frac{\partial u}{\partial t} = \frac{\partial \sigma}{\partial t} + \frac{k(1+e)}{\gamma_w a} \frac{\partial^2 u}{\partial z^2} \tag{9.45}$$

In that instance the limiting hydraulic gradient was assumed to be zero and the linear effective stress-strain relationship accounted for by taking $\partial e / \partial \sigma' = -a$ (see Equation 9.43). Assuming a sudden load application without further total stress increase ($\partial \sigma / \partial t = 0$) and substituting

$$\frac{k(1+e)}{\gamma_w a} = \frac{k}{\gamma_w m_v} = c_v \tag{9.45-a}$$

Equation (9.45) becomes:

$$\frac{\partial u}{\partial t} = c_v \frac{\partial^2 u}{\partial z^2} \tag{9.46}$$

c_v is called 'the coefficient of consolidation.'

In Fig. 9.7 the line $\Delta\sigma = f(z)$ represents suddenly applied total additional vertical pressures in a layer of uniform thickness h with an impermeable base and a pervious cover. The curve u_t represents the pore pressure u at a given time t plotted against the depth z; these graphs of simultaneous pore-pressures are called isochrones. After Terzaghi's theory the initial or zero isochrone coincides with the line $\Delta\sigma = f(z) = u_0 = u_0(z)$. Thus

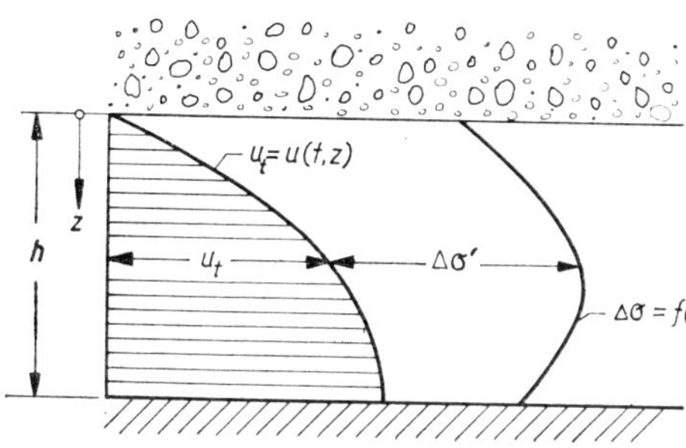

Fig. 9.7. Diagram of total additional vertical pressures $\Delta\sigma$ and the isochrone (u_t)

the additional effective pressures at time t are given by the distances $\Delta\sigma' = u_0 - u_t$ between the initial isochrone u_0 and the isochrone u_t appearing at time t.

Terzaghi (1925) designated Equation (9.46) as the 'thermo-dynamic equation for the time slope of the hydrodynamic stress fields' and showed that this equation is analogous to the differential equation for the non-stationary, one-dimensional flow of heat through isotropic bodies provided the symbols in the equation are assigned the folllowing physical meanings:

u = temperature, t = time, k = coefficient of heat conductivity, $\gamma_w m_v$ = heat capacity multiplied by unit weight, c_v = diffusivity.

The loss of water (consolidation) corresponds to the loss of heat (cooling) and the absorption of water (swelling) to an increase of the heat content of a solid body. Making use of this analogy the solutions obtained in connection with thermodynamic problems can be usefully applied in soil mechanics.

9.3.2 General Solution of Terzaghi's Differential Equation

The pore pressures varying with the depth z and with time t may be expressed in the form (Terzaghi-Fröhlich, 1936):

$$u(z,t) = A\varphi(t)\psi(z) \tag{9.47}$$

where A is a constant. The expression satisfies the differential Equation (9.46) for the following functions $\varphi(t)$ and $\psi(z)$:

$$u(z,t) = A \exp(-Kt) \sin \frac{\pi z}{2h} \tag{9.48}$$

$$K = \frac{\pi^2}{4} \cdot \frac{c_v}{h^2} \tag{9.49}$$

Consequently Equation (9.48) is the first particular solution of the differential Equation (9.46). Further particular solutions have the form:

$$u(z,t) = A_m \exp(-m^2 Kt) \sin\left(m \frac{\pi z}{2h}\right) \tag{9.50}$$

and their sum yields the general solution:

$$u(z,t) = \sum_{m=1}^{\infty} A_m \exp(-m^2 Kt) \sin\left(m \frac{\pi z}{2h}\right) \tag{9.51}$$

This general equation satisfies the boundary condition for $z = 0$ in any case:

$$u(z = 0,t) = 0 \tag{9.52}$$

The boundary condition for $t = 0$

$$u(z,t = 0) = \Delta\sigma(z) = f(z) \tag{9.53}$$

determines the factor A_m. Inserting $t = 0$ into Equation (9.51), we get:

$$f(z) = \sum_{m=1}^{\infty} A_m \sin \frac{m\pi z}{2h} \tag{9.54}$$

By using the appropriate transformation (see Terzaghi-Fröhlich, 1936) the above equation yields:

$$A_m = \frac{1}{h} \int_0^{2h} f(z) \sin \frac{m\pi z}{2h} \, dz \tag{9.55}$$

Applying these coefficients, the general solution (9.51) can be adapted to any given form of the initial isochrone $f(z)$. The settlement ρ_t of the layer of thickness h (Fig. 9.7) at time t can be obtained by integrating the differential settlements $d\rho_{dt}$ given by the equation:

$$d\rho_{dt} = d\varepsilon \, dz = m_v \, d\sigma' \, dz = -m_v \, du \, dz \tag{9.56}$$

Thus:

$$\rho_t = -m_v \int_{u=f(z)}^{u} \int_{z=0}^{h} du\, dz = m_v \int_{u}^{u=f(z)} \int_{z=0}^{h} du\, dz = m_v A_t \tag{9.57}$$

In Fig. 9.7 the double integral A_t is represented by the unshaded area between the initial isochrone $f(z)$ and the isochrone u_t. The final settlement:

$$\rho_\infty = m_v \int_{u=0}^{f(z)} \int_{z=0}^{h} du\, dz = m_v A_\infty \tag{9.58}$$

is then given by the total area between the initial isochrone $f(z) = \Delta\sigma$ and the final isochrone $u = 0$, multiplied by m_v. The degree of consolidation U is defined by the ratio

$$U = \frac{\rho_t}{\rho_\infty} \tag{9.59}$$

Inserting expressions (9.57) and (9.58), we get

$$U = \frac{A_t}{A_\infty}. \tag{9.60}$$

9.3.3 Hydrodynamic Consolidation at the Initial Isochrone u_0 = const

If the additional total pressure $\Delta\sigma$ in the drainage direction is constant throughout the layer thickness, the initial isochrone has the equation:

$$u_0 = u_0(z) = \Delta\sigma = \text{const} \tag{9.61}$$

In this case Equation (9.55) reduces to

$$A_m = \frac{\Delta\sigma}{h} \int_0^{2h} \sin \frac{m\pi z}{2h}\, dz \tag{9.62}$$

Inserting these coefficients A_m into Equation (9.51) and considering expression (9.49), we obtain:

$$u(z,t) = \frac{4\Delta\sigma}{\pi} \sum_{m=0}^{\infty} \frac{1}{2m+1} \sin\left[\frac{\pi}{2}(2m+1)\frac{z}{h}\right] \exp\left[-(2m+1)^2 \frac{\pi^2}{4} T_v\right] \tag{9.63}$$

By this equation the pore-pressures are related to the time factor T_v

$$T_v = \frac{c_v t}{h^2} \tag{9.64}$$

With $A_\infty = \Delta\sigma h$ Equation (9.60), expressing the degree of consolidation, becomes

$$U = 1 - \frac{1}{\Delta\sigma h}\int_0^h u(z,t)\,\mathrm{d}z$$

Inserting expression (9.63) for $u(z,t)$, and integrating it, we obtain:

$$U = 1 - \frac{8}{\pi^2}\sum_{m=0}^{\infty}\frac{1}{(2m+1)^2}\exp\left[-(2m+1)^2\frac{\pi^2}{4}T_v\right] \tag{9.65}$$

The series of this expression converges faster than the series in expression (9.63) for pore-pressures. In Fig. 9.8 the values U calculated by Equation

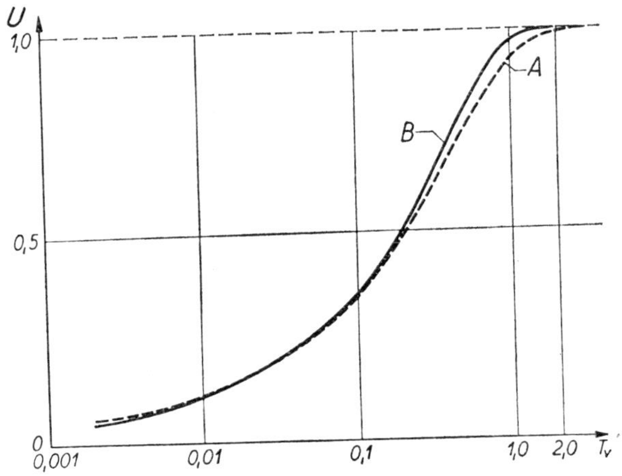

Fig. 9.8. Consolidation curves according to the Terzaghi theory at the initial isochrone u_0 = const; A, rigorous solution, B, assuming isochrones as parabolas of the second degree

(9.65) (see Terzaghi-Fröhlich, 1936) are plotted against the logarithms of the time factors T_v (curve A).

Inserting $z = h$ into Equation (9.63), the maximum values of the isochrones are obtained. At a certain time corresponding to the time factor $T_v = 0{\cdot}06$, the isochrones become displaced from the initial tangent isochrone $\Delta\sigma$ = const, and their maximum values become less than $\Delta\sigma : u_{h,t} < \Delta\sigma$. In Fig. 9.9 the resulting quantities $u_{h,t}/\Delta\sigma$ are plotted against the logarithms of the time factors $T_v > 0{\cdot}06$.

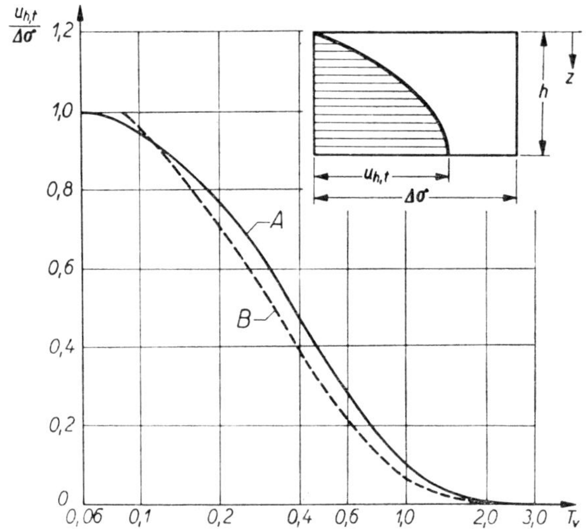

Fig. 9.9. Pore-pressure dissipation according to the Terzaghi theory at the initial isochrone u_0 = const; *A*, rigorous solution, *B*, assuming parabolic isochrones of the second degree

The diagrams $U = U(T_v)$ in Fig. 9.8 and $u_{h,t}/\Delta\sigma = g(T_v)$ in Fig. (9.9) can be compared, in the same figures, to analogous graphs corresponding to the assumption that the isochrones are parabolas of the second degree (curves *B*). Taking into account this assumption, the integration of the differential Equation (9.46) can be given in closed form. The isochrones depart from the initial isochrone $\Delta\sigma$ = const at the time factor

$$T_{vh} = \tfrac{1}{12} \tag{9.66}$$

In the domain $T_v \leqslant T_{vh}$ the following relations hold (see Fig. 9.10):

$$\frac{z_1}{h} = 2\sqrt{(3T_v)} \tag{9.67}$$

$$\frac{u_{z.t}}{\Delta\sigma} = 1 - \left(\frac{z_1 - z}{h}\right)^2 \tag{9.68}$$

while for $T_v > T_{vh}$ Equation (9.69) results:

$$\frac{u_{z,t}}{\Delta\sigma} = \exp\left(\frac{1}{4} - 3T_v\right)\left[1 - \left(\frac{h-z}{h}\right)^2\right] \tag{9.69}$$

(For the simple deduction of the above solutions, see Terzaghi and Fröhlich, 1936.)

The assumption that the isochrones are parabolas of the second degree requires a uniform consolidation process throughout the whole layer thickness, i.e. at any time there is to be a uniform pore water discharge throughout the layer. Now, as at a certain time of consolidation the inter-granular pressures vary with the depth z, the above assumption is not compatible with the supposition of a linear relationship between the strains and effective pressures. Nevertheless the comparison of the graphs

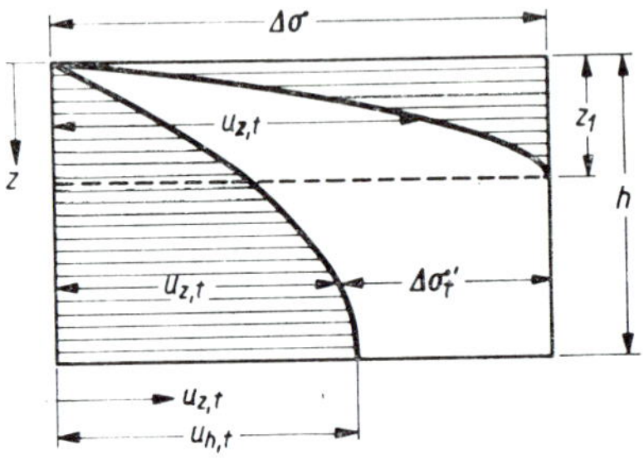

Fig. 9.10. Parabolic isochrones

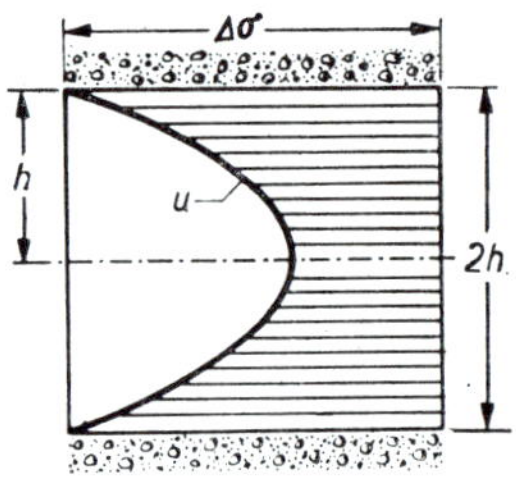

Fig. 9.11. Isochrones at bilateral drainage conditions

A and B in Figs. 9.8 and 9.9 shows that the degrees of consolidation as well as the maximum ordinates of the isochrones are not appreciably affected by this incompatibility of assumptions. Thus the parabolic shape of the isochrones can be accepted as a reasonable approximation.

In all the preceding considerations the base of the layer was assumed to be impervious; consequently only the upward seepage of the pore water had to be considered. If the lower contact of the layer is pervious too (Fig. 9.11), consolidation of the layer occurs by the pore water squeezing out in both vertical directions, the middle plane of the layer being the separating surface. In this case all the preceding deductions hold true if the half layer thickness is denoted by h (Fig. 9.11). Consolidation of the upper part of the layer occurs as in the case treated above of the layer with an impermeable base, and the lower part has to be regarded as the mirror image of the upper one.

9.4 THE PRIMARY AND SECONDARY CONSOLIDATION OF OEDOMETER TESTS

9.4.1 The Separation of the Primary and Secondary Consolidation Branches

In the Terzaghi consolidation theory only the seepage resistance is considered to retard the transition of the grain skeleton into a new equilibrium state that corresponds to the applied total stress state. The structural resistance is fully neglected. Its effect has been discussed in Chapter 8.1 when investigating the creep of soils whose consolidation is not affected by the seepage resistance. It can easily be understood that the structural resistance appears in saturated soils as well. In non-cohesive soils this resistance can be interpreted as the frictional resistance between single grains and as the interlocking resistance of the grain skeleton. In cohesive soils, the attraction forces acting between the bonds of adsorbed water surrounding the soil particles have to be overcome.

In the first, the so-called primary, phase of consolidation of oedometer tests, structural resistance often does not affect the character of consolidation which is governed by seepage resistance only. With decreasing pore pressures, however, structural resistance begins to prevail. When the pore pressure dissipates to immeasurably small values, the soil continues to creep being governed by its structural resistance only.

The secondary phase of consolidation was observed by Terzaghi as early as 1925 (Terzaghi, 1925). On the basis of the results of long term observation Buisman (1936) established a logarithmic relationship for secondary consolidation which was introduced when dealing with the consolidation without seepage resistance. Buisman expressed this relationship in the form

$$\frac{\rho}{h_\alpha \Delta\sigma} = \alpha_p + \alpha_s \log_{10} t \tag{9.70}$$

where ρ denotes the settlement, h_α the initial height of the layer or sample, $\Delta\sigma$ the pressure increment and t the time, α_ρ and α_s being the constants given by the graph relating the observed settlements to the logarithm of time. Buisman named the secondary phase of consolidation 'the secular phase,' thus indicating its long duration. Haefeli (1938) related the secondary consolidation at high pressures to the transition of soils into rocks (of sand into sandstone, of clay into shale). This transition, however, is not due to mechanical effects alone; it is, of course, also influenced by chemical and electro-chemical effects.

In cases when primary consolidation is mainly governed by Terzaghi's theory, secondary consolidation being controlled by the logarithmic relationship, the two branches of the consolidation curve can be separated by the logarithm of time fitting method as suggested by A. Casagrande. The construction is shown in Fig. 9.12 where the consolidation curve is

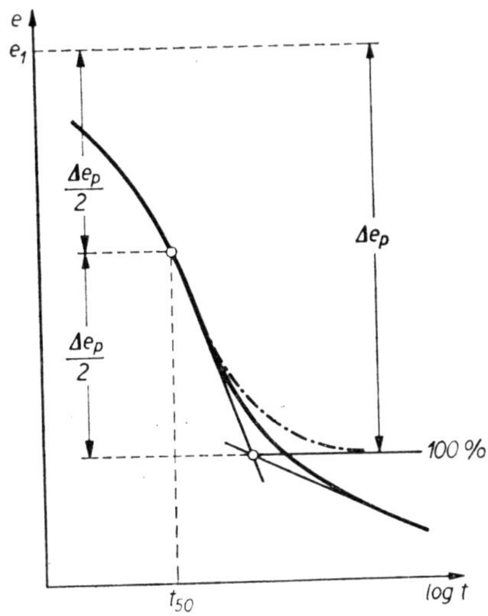

Fig. 9.12. The logarithm of time fitting method according to A. Casagrande

plotted against the logarithm of time scale. The point of 100 per cent primary compression can be determined by means of the intersection of the tangent at the point of inversion of the primary branch with the logarithmic straight line of secondary compression. The straight line parallel to the log t axis and passing through the point of intersection represents the asymptote of the primary consolidation curve. The corresponding coefficient of consolidation c_v can then be determined, for example, by using the relation

$$c_v = \frac{T_{50}}{t_{50}} h^2 \tag{9.71}$$

T_{50} being the time factor for $U = 0{\cdot}50$ (Fig. 9.8) and t_{50} the corresponding time, as given by the construction shown in Fig. 9.12.

In the square root of time plot Terzaghi's theoretical consolidation curve (Fig. 9.13 (*a*)) is a straight line up to about 60 per cent consolidation. The abscissa of the curve at 90 per cent consolidation is 1·15 times the abscissa of the extended straight line. First suggested by Taylor (1948) this characteristic of the theoretical curve is utilized to determine the point of 90 per cent consolidation on the laboratory time curve.

In the laboratory plot (Fig. 9.13 (*b*)) a straight line which is in close agreement with the observed points of the early portion of the compression

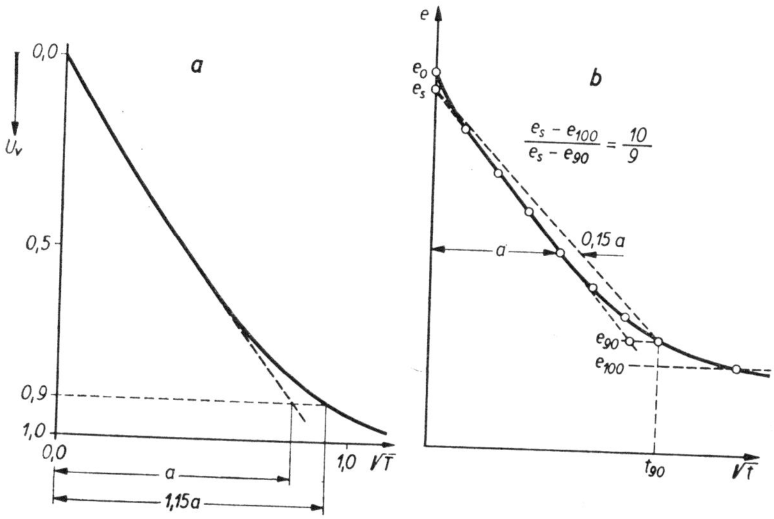

Fig. 9.13. The square root of time fitting method according to D.W. Taylor (1948): (a) theoretical curve; (b) laboratory curve

may be drawn. A second line, coinciding with the first line at zero time and having all abscissas 1·15 times as large as corresponding values on the first line, may next be drawn. The common point on these straight lines at zero time is generally a little below the point of zero compression, at least partly owing to the immediate compression of small amounts of gas in the pores (Taylor, 1948).

The coefficient of consolidation may be determined from Equation (9.64) in the form

$$c_v = \frac{T_{90}}{t_{90}} h^2 \tag{9.72}$$

T_{90} being the time factor for $U = 0{\cdot}90$ (Fig. 9.8): $T_{90} = 0{\cdot}848$ and t_{90} the corresponding time in the laboratory plot as given by the construction shown in Fig. 9.13 (*b*).

9.4.2 Various Shapes of Consolidation Curves

The primary branch of the oedometer consolidation curve does not always have the shape of Terzaghi's consolidation curve corresponding to a coefficient of consolidation as obtained by either the logarithm of time or the square root of time fitting method; and the secondary branch does not agree with Buisman's logarithmic law in all cases. Based on experimental evidence the following main types of both consolidation branches can be distinguished (Vidmar, 1956):

I. Primary consolidation branch:

1. having the shape of Terzaghi's curve corresponding to a coefficient of consolidation c_v determined by using the equation

$$c_v = \frac{k}{\gamma_w m_v}$$

   where k is the average value of the coefficients of permeability observed at the beginning and at the end of the load step and m_v the coefficient of compressibility obtained by either the logarithm of time or square root of time fitting method;
2. having the shape of Terzaghi's curve corresponding to a coefficient of consolidation c_v determined by either Casagrande's or Taylor's fitting method in cases when this coefficient is not compatible with the directly measured coefficient of permeability k, but corresponds to a smaller value of permeability;
3. S-shaped semi-logarithmic plot ($\log_{10} t - e$); although similar to Terzaghi's consolidation curve, the primary branch is rather more extended in the direction of increasing logarithm of time;
4. the primary and the secondary consolidation branches cannot be separated either by using Casagrande's or Taylor's fitting method; when plotted against the logarithm of time scale the consolidation curve already approaches a semi-logarithmic straight line in the primary phase; in its early stage it is concave towards the smaller void ratios and passes into the secondary branch without any distinguishing boundary.

II. Secondary consolidation branch:

1. having the shape of a straight line when plotted against the logarithm of time scale;
2. with a faster retardation of consolidation speed than that corresponding to the semi-logarithmic straight line, indicating a horizontal asymptote of final settlement;

3. with a slower retardation of consolidation speed than that corresponding to the semi-logarithmic straight line.

With any of the above types II/1–3 the following cases should be distinguished:

A—The secondary settlements extrapolated up to the end of the life-time t_e of the structure in question are sufficiently small to be neglected when compared with the primary settlements.
B—The secondary settlements related to the time t_e are several times smaller than the primary settlements.
C—The secondary settlements are approximately as important as the primary settlements.
D—The secondary character of the consolidation predominates.

All the distinctions cited above are related to the consolidation of oedometer tests.

The observed consolidation curves represent different combinations of the types of primary and secondary branches classified above. Some typical combinations are shown in Fig. 9.14 and are compared with the Terzaghi consolidation curves corresponding to the observed mean coefficient of permeability. The details are as follows:

(*a*) The combination I/1 + II/1 — B. The example (see Vidmar, 1955) represents the consolidation curve of a soil from the core of the Lokvarka dam (Croatia); w_L = 40 per cent, I_P = 11 per cent; grain-size distribution: 13 per cent < 0·002, 65 per cent 0·002/0·06,

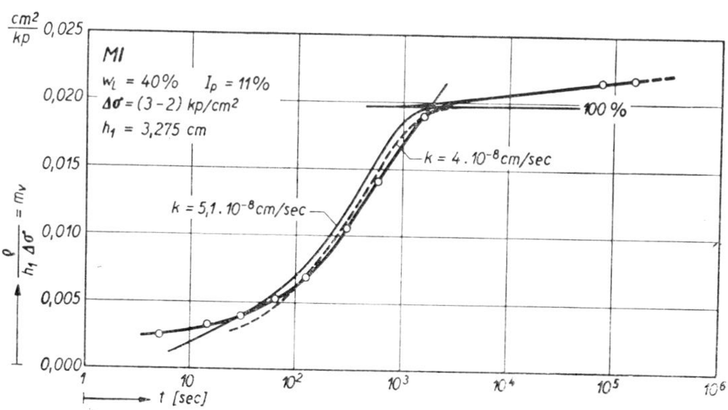

Fig. 9.14 (*a*). (See p. 149)

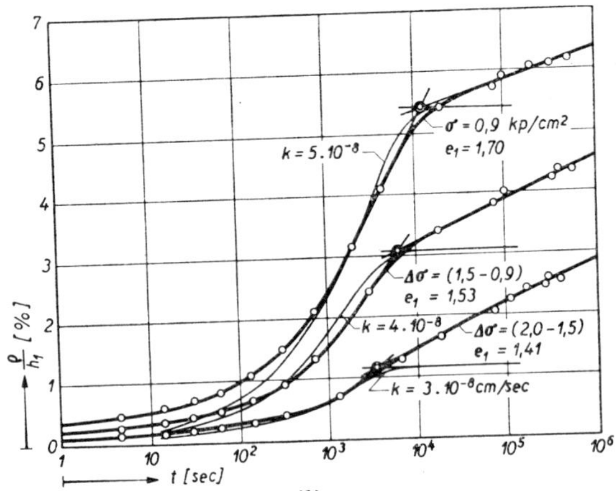
σ = 0,9 kp/cm²
e₁ = 1,70
k = 5.10⁻⁸
Δσ = (1,5 – 0,9)
e₁ = 1,53
k = 4.10⁻⁸
Δσ = (2,0 – 1,5)
e₁ = 1,41
k = 3.10⁻⁸ cm/sec
ρ/h₁ [%]
t [sec]

(*b*)

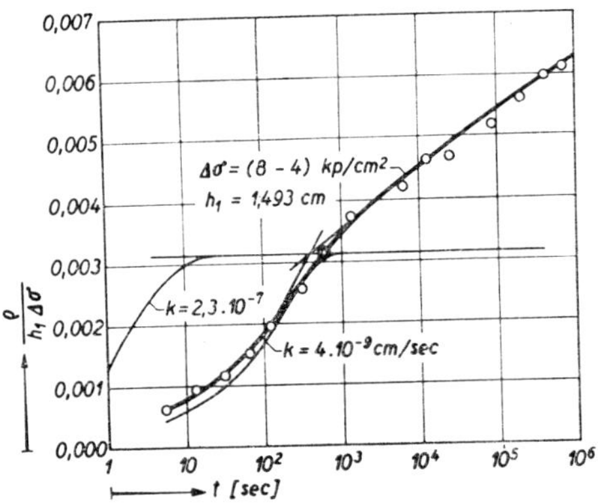
Δσ = (8 – 4) kp/cm²
h₁ = 1,493 cm
k = 2,3.10⁻⁷
k = 4.10⁻⁹ cm/sec
ρ/(h₁ Δσ)
t [sec]

(*c*)

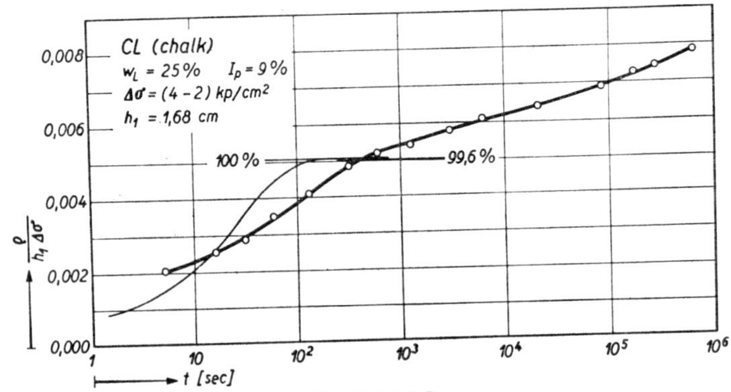
CL (chalk)
w_L = 25 %   I_p = 9 %
Δσ = (4 – 2) kp/cm²
h₁ = 1,68 cm
100 %
99,6 %
ρ/(h₁ Δσ)
t [sec]

Fig. 9.14 (*d*)

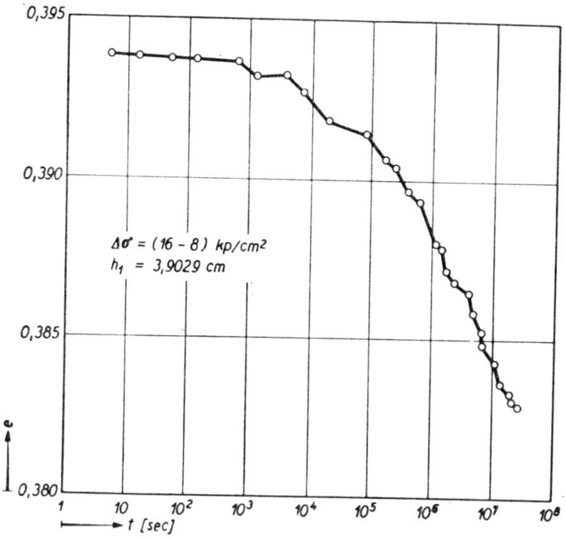

(*e*)

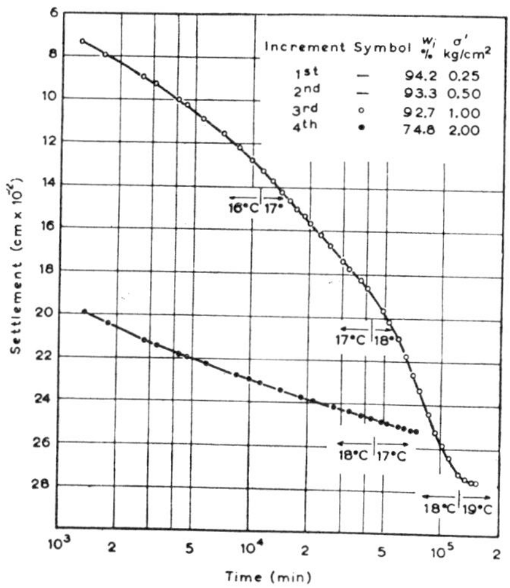

(*f*)

Fig. 9.14. Various shapes of the observed consolidation curves: (a) combination I/1 + II/1–B; (b) combination I/1 + II/1–C; (c) combination I/2 + II/1–C; (d) combination I/3 + II/3–C; (e) combination I/4 + II/1–D; (f) combination I/4 + II/2–D and I/4 + II/3–2–D

16 per cent 0·06/2 mm. The consolidation curve presented corresponds to the load increment $\Delta\sigma = (3 - 2)$ kp/cm^2. The average value of the coefficient of permeability observed directly was $k = 5{\cdot}1 \,.\, 10^{-8}$ cm/sec.

(*b*) The combination I/1 + II/1 – C. The curves represent the consolidation of three samples taken from the deposits of the Lake of Skadar (Yugoslavia) ($w_L = 66$ per cent, $I_P = 29$ per cent). The coefficients of permeability observed directly yield the Terzaghi curves which are drawn by thin lines and closely approach the observed curves of primary consolidation.

(*c*) The combination I/2 + II/1 – C. The graph is related to the load interval $\Delta\sigma = (8 - 4)$ kp/cm^2 of the consolidation tests on a lacustral chalk (from the Soča valley, Slovenia) sedimented in the oedometer.

(*d*) The combination I/3 + II/3 – C. This is shown by the consolidation curve of a remoulded sample of a lacustral chalk (from the Soča valley). The initial consolidation, faster than that corresponding to the hydrodynamic consolidation, is probably to be attributed to the compression of air bubbles due to incomplete saturation.

(*e*) The combination I/4 + II/1 – D. This represents an undisturbed sample of the lacustral chalk from the Soča valley; grain size distribution: 20 per cent $< 0{\cdot}002$, 77 per cent 0·002/0·06, 3 per cent 0·06/0·2; $\gamma_s = 2{\cdot}72$ p/cm^3, $w_L = 25{\cdot}3$ per cent, $I_P = 9{\cdot}8$ per cent, $I_c = 1{\cdot}13$ ($w_0 = 14{\cdot}2$ per cent); load increment: $\Delta\sigma = (16 - 8)$ kp/cm^2.

(*f*) The combination I/4 + II/2 – D and I/4 + II/3–2 D. The graph reproduced after *Lo* (1961) represents two settlement-time curves for undisturbed Lilla Edet Clay. In the upper curve corresponding to the load increment $\Delta\sigma = (1{\cdot}0 - 0{\cdot}5)$ kp/cm^2, the secondary compression is at first proportional to the logarithm of time. The sudden increase of the gradient in the semi-logarithmic plot appearing at a time of roughly $4{\cdot}5 \times 10^4$ min. has been attributed by *Lo* to a structural breakdown. The gradient of the plot begins to decrease again about the time 10^5 min. and the consolidation curve seems to tend towards a horizontal asymptote.

10

One-Dimensional Consolidation Analysis of Saturated Soils Based on Linear Rheological Models

10.1 LINEAR RHEOLOGICAL MODELS ASSUMING A LINEAR STRESS STATE

10.1.1 Soil as a Saturated Hookean Body in Series with a Saturated Kelvin Body

10.1.1.1 *Merchant's Fundamental Equation of Consolidation*

The disagreement between Terzaghi's consolidation theory and the behaviour of samples consolidated in the laboratory was first rationally investigated by Taylor and Merchant. In his thesis, completed under the supervision of D. W. Taylor, W. Merchant (1939) developed a new form of the differential equation of consolidation accounting for the observed secondary compression effects. In this theory, the void ratio decrease $\Delta e = \overline{GE}$ of an oedometer sample subjected to a total stress increase $\Delta\sigma = \sigma_b - \sigma_a$ (Fig. 10.1) is assumed to consist, at any time t during consolidation, of two parts, the primary and the secondary part:

$$\Delta e_t = \Delta e_{pt} + \Delta e_{st} \tag{10.1}$$

The 'primary compression' $\overline{GH} = \Delta e_{pt}$ is that part of Δe which would occur immediately under a sudden effective pressure increase $\Delta\sigma' = \sigma' - \sigma_a$. The relationship between the primary compression and the effective stress increase is assumed to be linear:

$$\frac{\Delta\varepsilon_{pt}}{\Delta\sigma'} = \frac{\dfrac{-\Delta e_{pt}}{1 + e_a}}{\Delta\sigma'} = m_{vp} \tag{10.2}$$

ε denoting the axial strain which, in oedometer tests, is equal to the volume strain, and m_v the coefficient of compressibility.

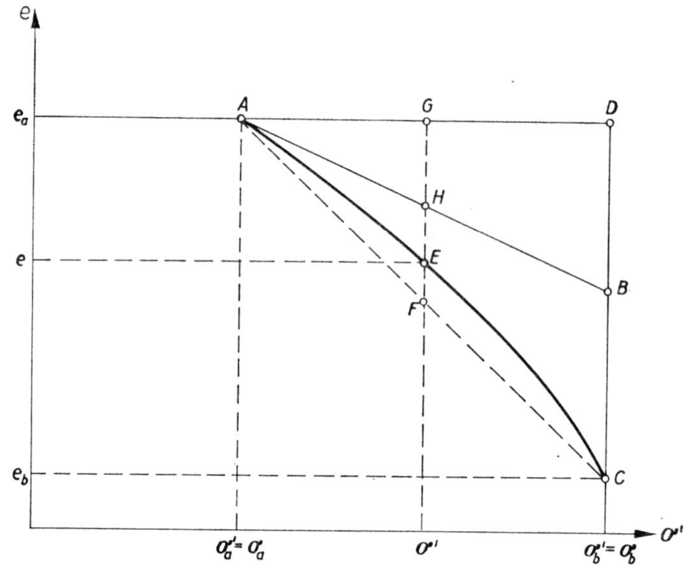

Fig. 10.1. Consolidation analysis according to Merchant and Taylor (1939)

The 'secondary compression' $\overline{HE} = \Delta e_{st}$ is that part of the total void ratio decrease Δe at time t which would develop in the time interval t under a constant effective stress $\sigma' = \sigma_a' + \Delta\sigma'$.

The secondary compression is assumed to tend to a final value $\Delta e_s = \overline{HF}$ given by the relationship:

$$\frac{\Delta\varepsilon_s}{\Delta\sigma'} = \frac{\dfrac{-\Delta e_s}{1 + e_a}}{\Delta\sigma'} = m_{vs} \tag{10.3}$$

The speed of occurrence of the secondary compression $\partial e_{st}/\partial t$ is supposed to be proportional to the undeveloped secondary compression $\Delta e_s' = \overline{EF}$ (Fig. 10.1):

$$\frac{\partial e_{st}}{\partial t} = -\mu|\Delta e_s'| \tag{10.4}$$

From Fig. 10.1 we can set off:

$$|\Delta e_s'| = \overline{EF} = \overline{HF} - \overline{HE} = |\Delta e_s| - |\Delta e_{st}| \tag{10.5}$$

Thus, considering Equation (10.3), Equation (10.4) becomes:

$$\frac{\partial e_{st}}{\partial t} = -\mu[(1 + e_a)m_{vs}\Delta\sigma' - (-\Delta e_{st})] \tag{10.4-a}$$

This equation can be rearranged as follows:

$$\Delta\sigma' = \frac{(-\Delta e_{st})}{(1 + e_a)m_{vs}} + \frac{\left(-\frac{\partial e_{st}}{\partial t}\right)}{\mu(1 + e_a)m_{vs}} \tag{10.6}$$

and written in terms of axial (volume) strains ε_{st}:

$$\Delta\sigma' = \frac{\varepsilon_{st}}{m_{vs}} + \frac{1}{\mu m_{vs}}\frac{\partial \varepsilon_{st}}{\partial t} \tag{10.7}$$

Comparing this equation with the fundamental Equation (4.35) of the Kelvin body it can be seen that in Merchant's theory the secondary compression corresponds to Kelvin's rheological scheme, m_{vs} standing

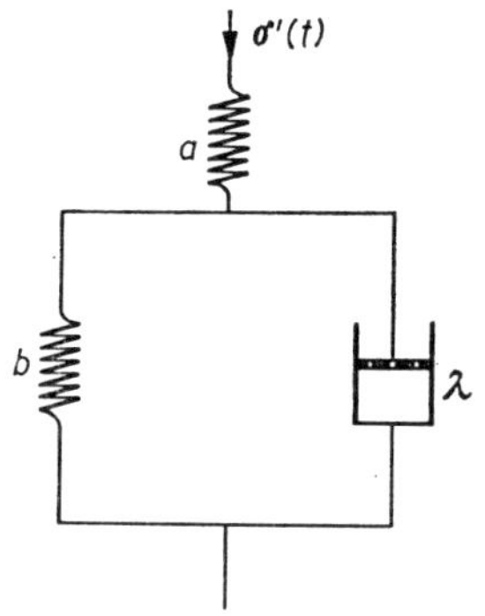

Fig. 10.2. Rheological model of Merchant's consolidation theory

for the coefficient of compressibility ($m_{vs} = 1/K$ where $K =$ the modulus of volume compressibility) and μm_{vs} for the reciprocal value of the coefficient of volume viscosity:

$$\frac{1}{\mu m_{vs}} = 3\eta^0 \tag{10.8}$$

On the other hand, as Equation (10.2) of the primary compression corresponds to Equation (4.8) of the Hookean body, the complete rheological model of Merchant's consolidation theory is the Hookean spring in series with the Kelvin element (Fig. 10.2) where:

$$a \equiv m_{vp}, \quad b \equiv m_{vs}, \quad \lambda \equiv \mu m_{vs}.$$

Equation (10.2) can be written in differential form:

$$\frac{\partial e_{pt}}{\partial \sigma'} = -m_{vp}(1 + e_a) \tag{10.2-a}$$

Thus:

$$\frac{\partial e_{pt}}{\partial t} = \frac{\partial e_{pt}}{\partial \sigma'}\frac{\partial \sigma'}{\partial t} = (1 + e_a)m_{vp}\frac{\partial u}{\partial t} \tag{10.9}$$

u standing for pore pressure. By differentiating Equation (10.1) we get

$$\frac{\partial e}{\partial t} = \frac{\partial e_{pt}}{\partial t} + \frac{\partial e_{st}}{\partial t} \tag{10.10}$$

Inserting Equations (10.9) and (10.4-a) the above equation yields:

$$\frac{\partial e}{\partial t} = (1 + e_a)m_{vp}\frac{\partial u}{\partial t} - \mu[(1 + e_a)m_{vs}(u_0 - u) - (-\Delta e_{st})] \tag{10.11}$$

By adding the following equation to the expression in brackets [ ]

$$(1 + e_a)m_{vp}(u_0 - u) + \Delta e_{pt} = 0 \tag{10.2-b}$$

we get:

$$\frac{\partial e}{\partial t} = (1 + e_a)m_{vp}\frac{\partial u}{\partial t} - \mu[(1 + e_a)(m_{vp} + m_{vs}) \,.\, (u_0 - u) + (e - e_a)] \tag{10.12}$$

This expression for $\partial e/\partial t$ may now be inserted into the differential equation of consolidation (9.36). In this case the coefficient of permeability k may be considered constant, the limiting hydraulic gradient i_0 zero and $1 + e_a$ written for $1 + e$. In this way we get the equation:

$$c_v\frac{\partial^2 u}{\partial z^2} = \frac{\partial u}{\partial t} - \frac{\mu}{m_{vp}}\left[m_v(u_0 - u) + \frac{e - e_a}{1 + e_a}\right] \tag{10.13}$$

$$m_v = m_{vp} + m_{vs} \tag{10.14}$$

$$c_v = \frac{k}{\gamma_w m_{vp}} \tag{10.15}$$

Taylor and Merchant (1940) solved the differential Equation (10.13) only for very small values of μt permitting the assumption:

$$\exp(-\mu t) = 1 \tag{10.16}$$

In 1961 Gibson and Lo (1961) presented the exact solution of the differential equation of consolidation based on the rheological model shown in Fig. 10.2. This solution will be reproduced in the following section. As shown by I. F. Christie (1964) the exact solution of Merchant's differential Equation (10.13) leads, as expected, to the identical result.

10.1.1.2 *Gibson and Lo's Solution*

Based on the rheological scheme consisting of a Hookean spring connected in series with a Kelvin element (Fig. 10.2) Gibson and Lo (1961) interpreted the process of consolidation in the following way.

When a time dependent effective stress $\sigma'(t)$ acts on an element of soil at a depth z, the spring (a) compresses instantaneously, but the settlement of the compound element $b \parallel \lambda$ is retarded because the dashpot λ is present. Since the transference of stress from pore water to soil skeleton is delayed due to the low permeability of the clay, the effective stress $\sigma'(t)$ increases gradually from zero to the full value of the applied stress. Hence, the compression of spring a is also gradual and is fully accomplished only when the applied stress has become fully effective.

Under the gradually increasing effective stress the Kelvin body commences to compress. Initially the full load is taken by the Newtonian dashpot and as compression proceeds it is progressively transferred to the spring b. This phenomenon of transference corresponds to the process of secondary consolidation which occurs under sustained effective stress. After a very appreciable time has elapsed, the full effective stress will be taken through springs a and b, the dashpot sustaining no load.

The total strain ε is the sum of the strain ε_H of the Hookean body and of the strain ε_K of the Kelvin element:

$$\varepsilon = \varepsilon_H + \varepsilon_K \tag{10.17}$$

The modulus of linear compression of the spring 'a' (Fig. 8.39) may be denoted by A, the modulus of the spring 'b' by B and the coefficient of viscosity of the dashpot 'λ' by η (kp sec cm^{-2}). Applying the relationships (4.7) for the Hookean body and (4.36) for the Kelvin element, and assuming zero strain ($e_{rs} = 0$) at time $t = t_0 = 0$, Equation (10.17) can be written as follows:

$$\varepsilon = \frac{\sigma'(t)}{A} + \frac{1}{\eta}\int_0^t \sigma'(\tau) \exp\left[\frac{B}{\eta}(\tau - t)\right] d\tau \tag{10.18}$$

In Gibson and Lo's theory the pore pressure dissipation is assumed to be governed by the fundamental differential equation of one-dimensional consolidation. Substituting

$$\frac{-\mathrm{d}e}{1+e} = \mathrm{d}\varepsilon \tag{10.19}$$

and taking $k = \text{const}$, $i_0 = 0$, Equation (9.37) can be presented in the form:

$$\frac{k}{\gamma_w}\frac{\partial^2 u}{\partial z^2} = -\frac{\partial \varepsilon}{\partial t} \tag{10.20}$$

(for z see Fig. 10.3). Considering further the relation

$$\mathrm{d}u = -\mathrm{d}\sigma' \tag{10.21}$$

and inserting, for ε, expression (10.18), Equation (10.20) becomes:

$$\frac{k}{\gamma_w}\frac{\partial^2\sigma'}{\partial z^2} = \frac{1}{A}\frac{\partial\sigma'}{\partial t} + \frac{\sigma'}{\eta} - \frac{B}{\eta^2}\int_0 \sigma'(z,\tau)\exp\left[\frac{B}{\eta}(\tau - t)\right]\mathrm{d}\tau \tag{10.22}$$

This equation, together with the boundary conditions (see Fig. 10.3)

$$\left.\begin{aligned} \frac{\partial\sigma'}{\partial z} &= 0, \quad z = h \\ \sigma' &= q(t), \quad z = 0 \end{aligned}\right\} 0 < t \leqslant \infty \tag{10.23}$$

completely defines the problem under consideration.

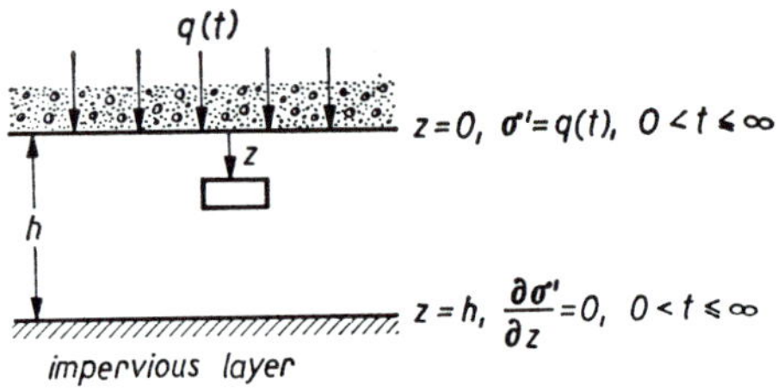

Fig. 10.3. Notation to Gibson and Lo's solution

Applying a Laplace transform to both sides of Equation (10.22), and using the convolution theorem for the last term on the right hand side, Gibson and Lo (1961) derived equations expressing settlements and pore water pressures for all types of monotonic loading $q(t)$.

For the simple case of a step load specified by

$$q(t) = 0 \qquad t < 0$$
$$q(t) = q_0 \qquad t > 0$$

Gibson and Lo obtained the following solutions:

(*a*) for the settlement $\rho(t)$ at time t:

$$\rho(t) = \left(\frac{1}{A} + \frac{1}{B}\right) q_0 h \left\{1 + \frac{8}{\pi^2} \sum_{n=0}^{\infty} \frac{1}{(2n+1)^2} \left[\frac{K - x_1}{x_1 - x_2} \exp(-x_2 t) - \frac{K - x_2}{x_1 - x_2} \exp(-x_1 t)\right]\right\} \tag{10.24}$$

(*b*) for the pore water pressures:

$$u(z,t) = \frac{4}{\pi}\left(1 + \frac{A}{B}\right) q_0 \sum_{n=0}^{\infty} \frac{1}{2n+1} \left\{\frac{x_1\left[\dfrac{B}{A+B} - \dfrac{x_2}{K_1}\right]}{x_1 - x_2} \exp(-x_1 t) - \frac{x_2\left[\dfrac{B}{A+B} - \dfrac{x_1}{K_1}\right]}{x_1 - x_2} \exp(-x_2 t)\right\} \sin\frac{(2n+1)\pi z}{2h} \tag{10.25}$$

In the above expressions the following substitutions have been made

$$\frac{x_1}{x_2} = \frac{(\alpha + K_1) \pm \sqrt{[(\alpha + K_1) - 4\beta K_1]}}{2} \tag{10.26}$$

$$K_1 = \frac{(2n+1)^2\pi^2\Theta}{4h^2} \tag{10.27}$$

$$\alpha = \frac{A + B}{\eta} \tag{10.28}$$

$$\beta = \frac{B}{\eta} \tag{10.29}$$

$$\Theta = \frac{kA}{\gamma_w} \tag{10.30}$$

$$K = \frac{B}{A + B} K_1 \tag{10.31}$$

For soil which is infinitely viscous ($\eta \to \infty$) or when there is no secondary compression ($B \to \infty$), Equations (10.24) and (10.25) reduce to the classical solution given by Terzaghi (Equations 9.65 and 9.63).

The solution for small values of time is

$$\rho(t) = \frac{2}{\sqrt{\pi}} \sqrt{(\Theta t)} \frac{q_0}{A} \left[1 + \frac{1}{3} \frac{A}{\eta} t + O\left(\frac{A^2 t^2}{\eta^2} \right) \right] \tag{10.32}$$

and for large values of time, the following expression has been formulated (Gibson and Lo, 1961)

$$\rho(t) = q_0 h \left[\frac{1}{A} + \frac{1}{B} \left\{ 1 - \exp\left(-\frac{B}{\eta} t \right) \right\} \right] \tag{10.33}$$

We observe that in the limit $t \to \infty$

$$\rho_\infty = \left(\frac{1}{A} + \frac{1}{B} \right) q_0 h \tag{10.34}$$

as expected.

From Equation (10.33), the parameters A, B and η, may be found in the following way:

Substituting:

$$\frac{1}{A} = a, \quad \frac{1}{B} = b \tag{10.35}$$

we write Equation (10.33) in the form:

$$\frac{\rho(t)}{h q_0} = \frac{\varepsilon(t)}{q_0} = (a + b) - b \exp\left(-\frac{t}{\eta b} \right) \tag{10.36}$$

Substituting $(a + b)$ from Equation (10.34):

$$a + b = \frac{\rho_\infty}{q_0 h} \tag{10.34-a}$$

Equation (10.36) becomes:

$$\frac{\varepsilon_\infty - \varepsilon(t)}{q_0} = b \exp\left(-\frac{t}{\eta b} \right) \tag{10.36-a}$$

Taking the logarithm of both sides we get

$$\log_{10} \frac{\varepsilon_\infty - \varepsilon(t)}{q_0} = \log_{10} b - 0{\cdot}434 \frac{t}{\eta b} \tag{10.37}$$

Hence a plot of $\log_{10}(\varepsilon_\infty - \varepsilon(t)/q_0)$ against t (Fig. 10.4) will yield the intercept $\log_{10} b$ and the slope $(-0{\cdot}434/(\eta b))$.

The value of 'a' may be determined either from the final settlement or from the relation (10.33) which should be applied for a suitable time t, e.g. for $t = t_a$ (see Fig. 10.4):

$$a = \frac{\varepsilon(t_a)}{q_0} - b\left[1 - \exp\left(-\frac{t_a}{\eta b}\right)\right] \qquad (10.38)$$

When applying the fitting method described above the final strain ε_∞ must be known. In order to avoid this requirement Equation (10.33)

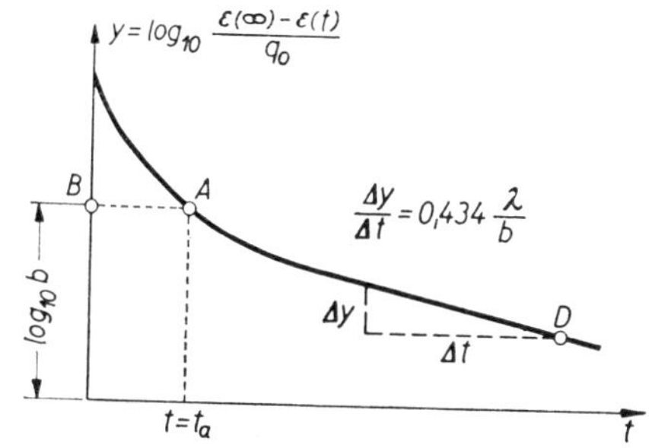

Fig. 10.4. Gibson and Lo's fitting method when the final strain is known

may be applied to t_1 and to $t_2 = t_1 + \Delta t_0$. By subtracting the corresponding expressions and by taking the logarithm of both sides we obtain the equation:

$$\log_{10}\frac{\varepsilon(t_2) - \varepsilon(t_1)}{q_0} = \log_{10}\left\{b\left[1 - \exp\left(-\frac{\Delta t_0}{\eta b}\right)\right]\right\} - 0{\cdot}434\frac{t_1}{\eta b} \qquad (10.39)$$

This equation should be applied to various times t_1, the interval $\Delta t_0 = t_2 - t_1$ being kept constant at any convenient value. A plot of

$$y = \log_{10}\frac{\varepsilon(t_2) - \varepsilon(t_1)}{q_0}$$

against t_1 (Fig. 10.5) yields the slope $(-\,0{\cdot}434/\eta b)$ and the coefficient of compressibility b of the spring (b) can then be computed from the intercept

$$y_0 = \log_{10}\left\{b\left[1 - \exp\left(-\frac{\Delta t_0}{\eta b}\right)\right]\right\}$$

The second and the third members of the right-hand side of Equation (10.32) are commonly (for small values of time) small enough to be neglected in the sum with unity. (The order of magnitude of the quotient

A/η is 10^{-6} sec^{-1}.) Thus, for the conditions of oedometer tests Equation (10.32) can be written in the simplified form:

$$\rho(t) = \frac{2aq_0}{\sqrt{(\pi)}} \sqrt{(\Theta)} \sqrt{t} \tag{10.40}$$

So, the plot of $\rho(t)$ against $\sqrt{t}$ yields the slope $(2aq_0/\sqrt{\pi})\sqrt{\Theta}$. The coefficient of compressibility a having being determined before, the coefficient of consolidation Θ can easily be found.

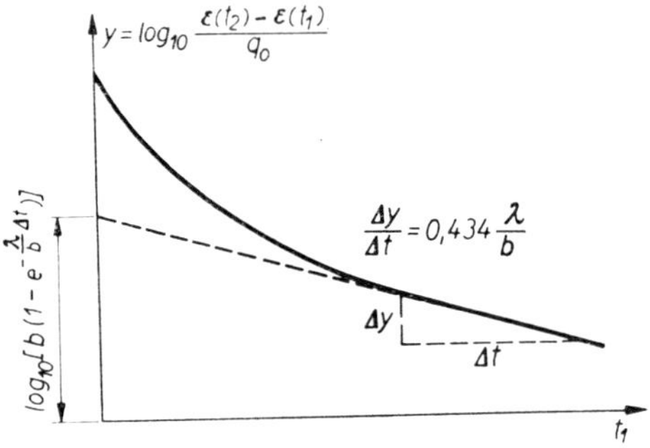

Fig. 10.5. Gibson and Lo's generalized fitting method

Knowing the parameters $A = 1/a$, $B = 1/b$, η and Θ the consolidation curve of a layer of arbitrary thickness h can be constructed by using Equation (10.24). For each load increment $\Delta\sigma = \sigma_\beta - \sigma_\alpha$, however, the essential parameters have to be found separately.

According to the basic rheological model consolidation always starts with zero additional intergranular pressures. Thus, Gibson and Lo's theory cannot be applied in cases when the effective pressures increase at the very beginning of the consolidation of the oedometer sample.

10.1.1.3 *Lo's Extension of the Theory*

According to Lo's examination of the secondary compression curves in a settlement versus logarithm of time plot (Lo, 1961 (*b*)) there are three types of secondary compression curves (Fig. 10.6). Curve Type I has a gentle curvature becoming concave upwards and the rate of secondary consolidation decreases with time. The curve becomes horizontal in a comparatively short time of the order of 20 days or so. Curves of Type II are characterized by the straight portion of the curve following the logarithmic law approximately for an appreciable range of time. The rate of compression decreases rapidly close to the ultimate stage of consolidation.

The Type III curve is concave downwards, continuous as in Type III (*b*) or abrupt as in Type III (*a*). The ultimate settlement is attained after an appreciable period (up to 100 days or more).

Type III appears in natural clayey deposits of loose structure whose particles are kept in position by some form of cementation partly of chemical and partly of mechanical nature. This loose structure is susceptible to breakdown when a critical stress (or more likely strain) leading to an accelerated compression is exceeded.

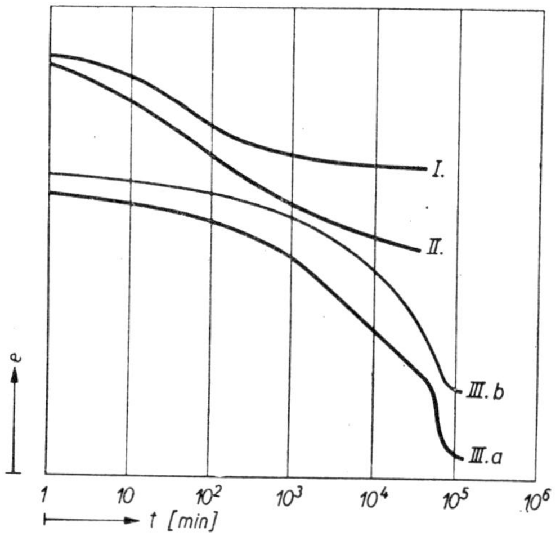

Fig. 10.6. Three types of secondary compression curves according to Lo (1961–a)

The examples presented by Gibson and Lo (1961) and Lo (1961 (b)) provide satisfactory agreement with their theory as far as the consolidation curves of oedometer tests of Types I and II are concerned. In order to describe the behaviour of clays showing Type III curve characteristics, Lo extended the theory by applying the generalized rheological model shown in Fig. 10.7. The model (Fig. 10.2) described previously is coupled in series to another similar Kelvin body through an element S which has the following mechanical properties. Below a certain critical value of either stress or strain, the element S is rigid and therefore sustains the stress $\sigma'(t)$ from the $b \parallel \lambda$ element without transmitting it to the $b_1 \parallel \lambda_1$ element. However, when this critical value is exceeded, the body S loses its rigidity and $\sigma'(t)$ also operates in the $b_1 \parallel \lambda_1$ element, i.e. for:

$$\left.\begin{aligned} 0 < t < t_k \qquad & \sigma_1'(t) = 0 \\ t \geqslant t_k \qquad & \sigma_1'(t) = \sigma'(t) \end{aligned}\right\} \tag{10.41}$$

The strain required to transform the property of S to bring the additional part $b_1 \parallel \lambda_1$ into action is assumed to be infinitesimal so that the element S does not contribute to the compression of the whole model.

For $t < t_k$ the equations given in the preceding sections remain unchanged. After the structural breakdown has occurred ($t \geqslant t_k$) the settlements (strains) for large values of time can be expressed by analogy

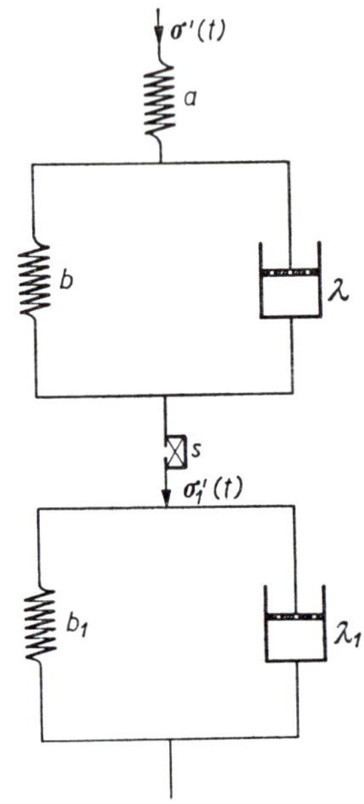

Fig. 10.7. Generalized rheological model according to Lo (1961–a)

with expression (10.24). Denoting the coefficient of compressibility of the spring, b_1, (see Fig. 10.7) by $b_1 = 1/B_1$ and the coefficient of viscosity of the dashpot, λ_1, by η_1 the equation:

$$\varepsilon(t) = q_0\left[a + b\left\{1 - \exp\left(-\frac{t}{\eta b}\right)\right\} + b_1\left\{1 - \exp\left(-\frac{t - t_k}{\eta_1 b_1}\right)\right\}\right] \tag{10.42}$$

applies to the combined model.

The time t_k at which breakdown occurs is given by the sudden change in the development of the observed consolidation curve.

Applying the fitting method described previously (Section 10.1.1.2) the parameters a, b and η may first be found from the part of the consolidation curve preceding breakdown ($t < t_k$). The strain $\varepsilon(t)$ given by expression

(10.42) and corresponding to an arbitrary time $t \geqslant t_k$ may be subtracted from the final strain $\varepsilon(\infty)$:

$$\varepsilon(\infty) = (a + b + b_1)q_0 \tag{10.43}$$

By taking the logarithm of the difference $\varepsilon(\infty) - \varepsilon(t)$ and rearranging the expression we obtain:

$$\log_{10}\left[\frac{\varepsilon(\infty) - \varepsilon(t)}{q_0} - b\exp\left(-\frac{t}{\eta b}\right)\right] = \log b_1 - 0{\cdot}434\frac{t - t_k}{\eta_1 b_1} \tag{10.44}$$

The plot of the left-hand side of this equation $y = \log_{10}$ [ ] against the logarithm of $(t - t_k)$ yields the intercept $y_0 = \log b_1$ and the slope $-0{\cdot}434/\eta_1 b_1$.

The critical time t_k varies, of course, with increasing layer thickness, the development of the critical effective pressure being dependent on the primary consolidation speed.

10.1.1.4 *Influence of the Layer Thickness on the Consolidation Process*

In order to prove the effect of the layer thickness in Gibson and Lo's solution, Equation (10.33) may be transformed into the following expression for the degree of consolidation U_v:

$$U_v = \frac{\rho_t}{(a + b)q_0 h} = 1 - m\exp(-T_\lambda) \tag{10.45}$$

where:

$$m = \frac{b}{a + b} = 1 - \frac{1}{M} \tag{10.46}$$

denotes the secondary consolidation ratio of the entire settlement of the oedometer sample, and:

$$T_\lambda = \frac{t}{\eta b} \tag{10.47}$$

the time factor of the secondary consolidation. Rewriting Equation (10.45) in the form:

$$1 - U_v = m\exp(-T_\lambda) \tag{10.45-a}$$

we have an expression for the undeveloped part of the consolidation. Now, as the time of the primary consolidation increases with the layer thickness, the above equation proves that the thicker the layer the smaller the subsequent secondary consolidation. For the examples presented by Gibson and Lo (1961) and Lo (1961 (b)) the order of magnitude of the coefficient $1/\eta b$ lies between 10^{-5} and 10^{-6} sec^{-1}. The corresponding secondary time factors T_λ for $t = 10^7$ sec for example, lie in the interval $10 \leqslant T \leqslant 100$ and the factors $\exp(-T_\lambda)$ in the interval

$$5 \cdot 10^{-5} \geqslant \frac{1 - U_v}{m} \geqslant 4 \cdot 10^{-44}.$$

Consequently, secondary compression practically does not occur at all in common cases when the primary consolidation of natural layers lasts longer than about half a year. The examples of the theoretical consolidation curves presented by Gibson and Lo (1961) and corresponding to the coefficient

$$M = \frac{1}{1 - m} = 1 + \frac{b}{a} = 1 \cdot 1$$

prove that the whole consolidation curve $U_v = f(T_v)$ has the character of primary consolidation, both approaching the Terzaghi theory and having a horizontal asymptote, provided that

$$N = \frac{h^2}{\eta b \Theta} \geqslant 1.$$

Since the order of the coefficient of consolidation for clays is commonly between 10^{-2} and 10^{-4} cm²/sec, and of the quantity $1/\eta b$ between 10^{-5} and 10^{-6} sec^{-1}, the condition $N \geqslant 1$ requires:

$$h \geqslant \sqrt{(\eta b \Theta)} = \text{from} \quad \sqrt{10} \quad \text{to} \quad 100 \text{ cm}.$$

This condition will be fulfilled in most cases of practical interest.

When the quantity $m = b/(a + b)$ and the coefficient of viscosity η are large enough, creep has more influence upon the shape of the primary consolidation and, as a result, the subsequent secondary consolidation can be rather important.

In Fig. 10.8 Gibson and Lo's theory has been applied to the observed oedometer consolidation curve $\Delta\sigma = (2 - 1)$ kp/cm² of an undisturbed sample of a lacustrine clay. Over a certain time interval the plot

$$\log_{10} \frac{\varepsilon(t_2) - \varepsilon(t_1)}{q_0} \text{ versus } t_1$$

approaches Equation (10.39) rather well. Naturally this equation does not hold for the logarithmic law of secondary compression of long duration. Consequently, according to Gibson and Lo's theory the final primary strains do not increase with the layer thickness in such a way as to be in agreement with the extrapolation of the observed logarithmic line of secondary compression. Comparison with Terzaghi's consolidation curves corresponding to a coefficient of consolidation $m_v = a + b$ proves to be in close agreement as expected according to the theory.

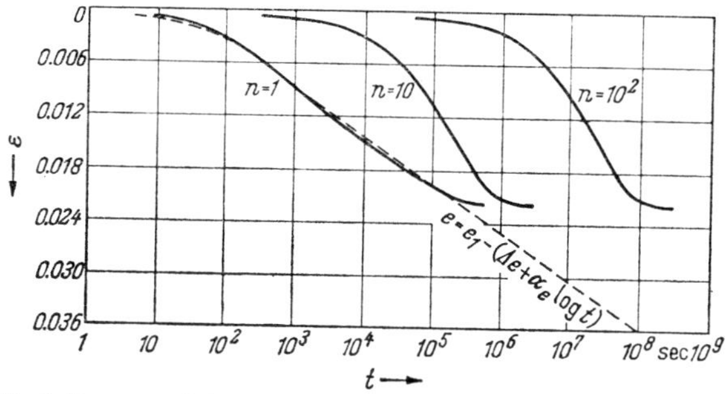

Fig. 10.8. Influence of the layer thickness onto the shape of the consolidation curve according to Gibson and Lo's solution. (Reproduced after Šuklje, 1964–b)

10.1.2 Soil as a Saturated Kelvin Body

10.1.2.1 *Taylor's Consolidation Theory B*

The approximate solution of Merchant's differential equation of consolidation (10.13) given by Taylor and Merchant (1940)* (see Section 10.1.1.1) was later called Theory *A* (Taylor, 1942) in order to distinguish it from a modified treatment of the consolidation process which was developed by Taylor (1942) and called Theory *B*.

In Theory *B* all secondary compression occurring subsequent to the primary compression is neglected and only the effect of secondary compression under the previous load increment is considered. The total effective pressure σ' is assumed to consist of three parts:

$$\sigma' = p_a + p_b + p_v \tag{10.48}$$

The meaning of single members of the above expression may be interpreted on the basis of the graph relating the void ratios e to the logarithm of effective pressures (Fig. 10.9). The semi-logarithmic straight line (a) represents the plot $e = e(\sigma')$ corresponding to the states at the end of the primary consolidation while the analogous straight line (b) corresponds to those void ratios in secondary consolidation branches which can be considered as the 'final' void ratios either with respect to the time elapsed after the load application (several decades or centuries) or with respect to

* As stated by Christie (1964) the degree of consolidation in this solution was incorrectly equated to $1 - (u_a/u_0)$, where u_a is the average excess pore-water pressure throughout the layer at time t. This substitution is not applicable to soils exhibiting viscous effects.

insignificant consolidation speed. Both straight lines are supposed to be parallel to each other.

The point 1 lying in the interval between the lines (*a*) and (*b*) may be assumed to represent the starting state. Thus, the primary consolidation has terminated while the secondary one is still going on. The total pressure may be increased by $\Delta\sigma$. The new total pressure is composed of the effective pressure σ' and the pore pressure u:

$$\sigma = \sigma' + u \tag{10.49}$$

The effective pressure changes with progressing consolidation. At the time t after the additional load application the initial void ratio e_1 has decreased to e_3; in the diagram (Fig. 10.9) the corresponding point

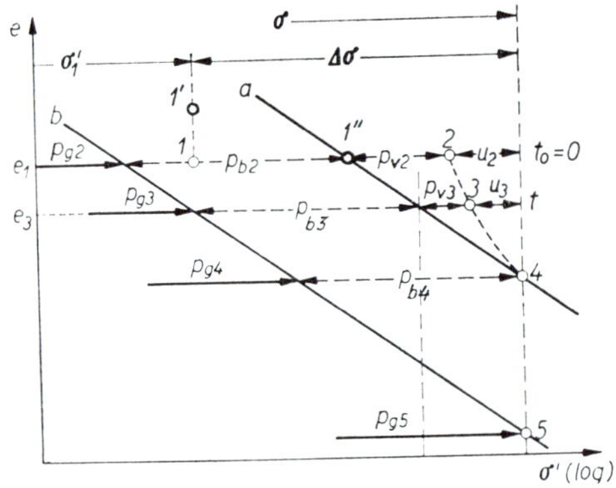

Fig. 10.9. Notation in Taylor's theory B

$(\sigma',e)_3$ is denoted by number (3); p_{g3}, p_{b3} and p_{v3} are the three parts of the effective pressure at time t explaining the Equation (10.48): p_g is the 'static resistance' depending on the actual void ratio e only and is independent on the degree of consolidation; it is given by the plot *b*. The remaining part of the effective pressure $p_s = p_b + p_v$ is called by Taylor 'plastic resistance'. After the theory *B* it is composed of the 'bond' p_b representing the plastic resistance existing at the void ratio $e = e_3$ at the end of primary compression, and of the viscous structural resistance p_v.

The basic assumptions of the theory are:

(*a*) In a sufficiently small stress interval the logarithmic line *a* (Fig. 10.9) is replaced by a straight line of slope:

$$\frac{-\mathrm{d}e}{\mathrm{d}(p_g + p_b)} = a \tag{10.50}$$

(*b*) The viscous structural resistance is proportional to the consolidation speed:

$$p_v = -\bar{\eta}\frac{\mathrm{d}e}{\mathrm{d}t} \tag{10.51}$$

In connection with Equation (10.48), the above assumptions allow the soil, in Taylor's Theory *B*, to be interpreted as a saturated Kelvin body.

The relations (10.50) and (10.51) should be considered in the fundamental differential equation of consolidation (9.36) in which constant permeability ($k = \text{const}$) and a zero limiting hydraulic gradient ($i_0 = 0$) may be assumed.

$$\frac{\partial e}{\partial t} = \frac{1+e}{\gamma_w}k\frac{\partial^2 u}{\partial z^2} \tag{10.52}$$

If the total stresses σ are constant throughout the layer thickness and do not vary during consolidation ($\mathrm{d}\sigma = 0$), the following relation can be deduced from Equations (10.48) and (10.49):

$$\mathrm{d}u = -\mathrm{d}\sigma' = -\mathrm{d}(p_g + p_b) - \mathrm{d}p_v \tag{10.53}$$

Thus Equation (10.52) can be written in the form:

$$\frac{\partial e}{\partial t} = -\frac{1+e}{\gamma_w}k\left\{\frac{\partial^2(p_g + p_b)}{\partial z^2} + \frac{\partial^2 p_v}{\partial z^2}\right\} \tag{10.54}$$

Considering relations (10.50) and (10.51) and making the approximation $1 + e = 1 + e_{\text{med}}$ Equation (10.54) becomes:

$$\frac{\partial e}{\partial t} - c_p\frac{\partial^2 e}{\partial z^2} - c_p a\bar{\eta}\frac{\partial^3 e}{\partial z^2 \partial t} = 0 \tag{10.55}$$

$$c_p = \frac{k(1 + e_{\text{med}})}{\gamma_w a} \tag{10.56}$$

Expression (10.55) is the fundamental differential equation of Taylor's Theory *B*.

Let us consider that consolidation starts at a stress-strain state represented, in Fig. 10.9, by point 1′. The initial compression 1′ – 1 may be attributed to incomplete saturation and the effective stress increase 1 – 1″ occurs immediately without decrease in void ratio. Thus, following the fundamental ideas of Theory *B*, the consolidation governed by Equation (10.55) is related to the transition from point 1″ to point 4 of the 'primary consolidation' plot *a*. With the notation and boundary conditions shown in Figs. 10.3 and 10.9 the integration of Equation (10.55) yields (Taylor, 1942, Equation 78):

$$e = e_4 + (e_1 - e_4)\left\{\exp\left(-\frac{t}{a\bar{\eta}}\right) - \frac{4}{\pi}\sum_{m=0}^{\infty}\frac{1}{2m+1}\sin\left[(2m+1)\frac{\pi}{2}\cdot\frac{z}{h}\right]\right.$$
$$\left.\left[\exp\left(-\frac{t}{a\bar{\eta}}\right) - \exp\left(-\frac{t}{\left(\frac{2h}{(2m+1)\pi}\right)^2\frac{1}{c_p} + a\bar{\eta}}\right)\right]\right\} \tag{10.57}$$

The average value of void ratio throughout the depth of the sample is

$$\bar{e} = \frac{1}{h}\int_0^h e\,\mathrm{d}z \tag{10.58}$$

and can be computed by substituting, in Equation (10.57), the factor

$$\sin\left[(2m+1)\frac{\pi}{2}\cdot\frac{z}{h}\right] \quad \text{by} \quad \frac{2}{(2m+1)\pi}.$$

The consolidation ratio at any point z is (see Fig. 10.9)

$$U_z = \frac{e_1 - e}{e_1 - e_4} \tag{10.59}$$

and the average consolidation ratio throughout the height of the sample is

$$U = \frac{e_1 - \bar{e}}{e_1 - e_4} \tag{10.60}$$

From Fig. 10.9 and Equations (10.50) and (10.51) the following relation can be derived:

$$\frac{e - e_4}{p_v + u} = a$$
$$u = \frac{e - e_4}{a} + \bar{\eta}\frac{\mathrm{d}e}{\mathrm{d}t} \tag{10.61}$$

Substituting e from (10.57) into (10.61) and putting

$$\frac{e - e_4}{a} = p_{v2} + u_2 = r_p'\,\Delta\sigma \tag{10.62}$$

(see Fig. 10.9) we obtain

$$u = \frac{4}{\pi} r_p' \Delta\sigma \sum_{m=0}^{\infty} \left\{ \frac{\sin\left[\frac{\pi}{2}(2m+1)\frac{z}{h}\right]}{(2m+1)\left[1 + \frac{\pi^2}{4}(2m+1)^2 J\right]} \exp\left(- \frac{T_k}{\left[\frac{2}{(2m+1)\pi}\right]^2 + J} \right) \right\} \tag{10.63}$$

where

$$T_k = \frac{c_p t}{h^2} \tag{10.64}$$

designates the so-called 'pore-phase time factor' and

$$J = \frac{a\bar{\eta}c_p}{h^2} = \frac{k\bar{\eta}(1 + e_{\text{med}})}{\gamma_w h^2} \tag{10.65}$$

the 'ratio of time factor'.

The quotient

$$U_z' = 1 - \frac{u}{r_p' \Delta p} \tag{10.66}$$

is termed the 'dissipation ratio'. In Fig. (10.10) some isochrones relating the dissipation ratio U_z' to the relative depth z/h (see Fig. 10.3) are presented corresponding to the ratio of the time factor $J = 0{\cdot}4$ and to various pore-phase time factors T_k. The example proves that a sudden increase of effective pressure immediately after the load application is characteristic of soils obeying Kelvin's rheological scheme.

10.1.2.2 *Florin's Solution*

When applying the fundamental form (9.37) of the differential equation of one-dimensional consolidation to soils exhibiting viscous properties, Florin (1953, 1961) made the approximation $1 + e = 1 + e_{\text{med}}$ and the assumptions that the coefficient of permeability k is constant and the limiting hydraulic gradient zero ($i_0 = 0$). The creep function (see Section 9.23) appearing in Equation (9.42) has been taken in the form:

$$f(t, \tau_1) = a_0 + a_1[1 - \exp\{-\gamma_1(t - \tau)\}] \tag{10.67}$$

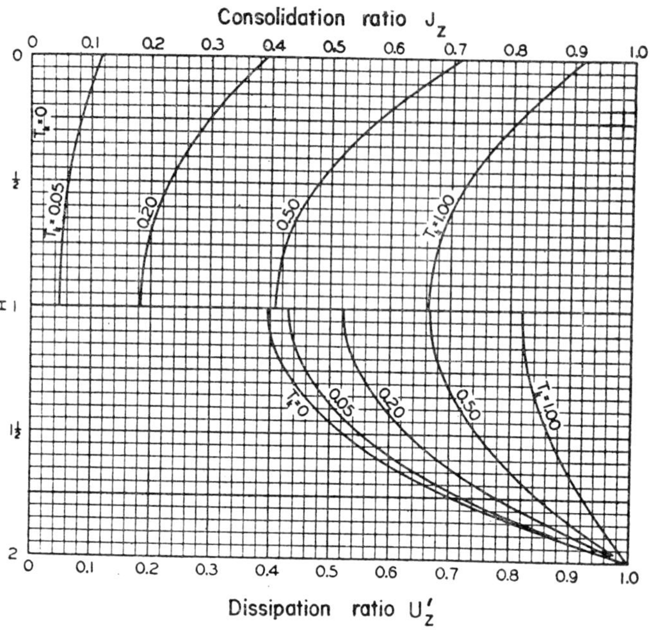

Fig. 10.10. Dissipation and consolidation ratios plotted versus the depth ratio according to Taylor's theory B; $J = 0{\cdot}4$. (Reproduced after Taylor, 1942)

It evidently corresponds to the rheological scheme treated in section 10.1.1: a Hookean spring of compressibility a_0 in series with a Kelvin element consisting of the parallel connection of a Hookean spring of compressibility $a_1 = 1/K$ and a Newtonian dashpot with strain velocity $K_1/\eta_1{}^0 = \gamma_1$.

In this way Equation (9.37) takes the form:

$$\frac{\partial}{\partial t}\Bigg[-\sigma'(\tau_1)\{a_0 + a_1[1 - \exp(-\gamma_1(t - \tau_1))]\} - \\ - \int_{\tau_1}^{t} \frac{\partial \sigma'}{\partial \tau}\{a_0 + a_1[1 - \exp(-\gamma_1(t - \tau))]\}\mathrm{d}\tau\Bigg] = \frac{1}{\gamma_w}(1 + e_{\mathrm{med}})\frac{\partial}{\partial z}\left(k\frac{\partial u}{\partial z}\right) \tag{10.68}$$

By integrating the expression

$$\int_{\tau_1}^{t} \frac{\partial \sigma'}{\partial \tau}\{\ \ \}\,\mathrm{d}\tau,$$

rearranging and substituting

$$\frac{\partial \sigma'}{\partial t} = -\frac{\partial u}{\partial t} \tag{10.69}$$

Florin has derived, from Equation (10.68), the following fundamental form of the differential equation applying for the rheological scheme explained above:

$$a_0 \frac{\partial^2 u}{\partial t^2} + \gamma_1(a_0 + a_1)\frac{\partial u}{\partial t} = \frac{1 + e_{\text{med}}}{\gamma_w} \frac{\partial}{\partial z}\left[k\left(\gamma_1 \frac{\partial u}{\partial z} + \frac{\partial^2 u}{\partial z \partial t}\right)\right] \tag{10.69'}$$

Taking $a_0 = 0$ and $k = \text{const}$, the above equation reduces to:

$$\frac{\partial u}{\partial t} = c\left(\gamma_1 \frac{\partial^2 u}{\partial z^2} + \frac{\partial^3 u}{\partial z^2 \partial t}\right) \tag{10.70}$$

$$c = \frac{(1 + e_{\text{med}})k}{\gamma_w \gamma_1 a_1} \tag{10.71}$$

Equation (10.70) governs the one-dimensional consolidation of soils the viscous properties of which correspond to the rheological scheme of the saturated Kelvin body. Thus its solution for the loading and boundary conditions presented in Fig. 10.3 ($q \equiv \Delta\sigma = \text{const}$) should give identical results to those obtained in Taylor's Theory *B*. Florin's solution (Florin, 1961, Equation (3.88)) corresponds exactly to Taylor's expression (10.63) if the following comparisons are taken into account:

Taylor (1942)	Florin (1961)
$2h$	h
$r_p'\Delta\sigma$	q
J	$\frac{4c}{h^2}$
$a\bar{\eta}c_p$	c
T_k	$\frac{4\gamma_1 c}{h^2}$
$\frac{1}{a\bar{\eta}}$	γ_1

10.2 LINEAR RHEOLOGICAL MODELS APPLYING FOR THE THREE-DIMENSIONAL STRESS-STATE

10.2.1 The Saturated Hookean Body for the Spherical Part and the Maxwell Body for the Distortional Part of the Stress-strain Relationship

10.2.1.1 *Fundamental Equations of Tan's Consolidation Theory*

Tan's theory of consolidation (Tan, 1954; 1957) was the first theory consciously related to a given rheological scheme. He presented this scheme in the way shown in Fig. 10.11. It must, however, be interpreted in such a

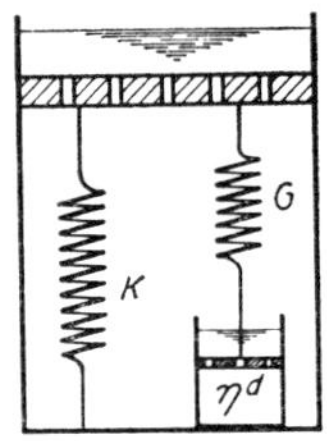

Fig. 10.11. Tan's rheological model

manner that the Hookean spring K is related to the spherical part and the Maxwell body $G - \eta^d$ to the distortional part of the stress-strain relationship.

$$p_{rs}{}^0 = 3Ke_{rs}{}^0 \tag{4.8}$$

$$\dot{e}_{rs}{}^d = \frac{\dot{p}_{rs}{}^d}{2G} + \frac{p_{rs}{}^d}{2\eta^d} \tag{4.27}$$

The last equation can be expressed in the form:

$$p_{rs}{}^d = \frac{\dfrac{\partial}{\partial t}}{\dfrac{1}{2\eta^d} + \dfrac{1}{2G}\dfrac{\partial}{\partial t}} e_{rs} \tag{10.72}$$

Thus Lamé's Equations (4.18-c) can be applied by substituting for Lamé's coefficients λ and μ the following differential operators (see Equations 4.18-a and 4.18-b):

$$\lambda = K - \frac{2}{3}\left\{\frac{\eta^d \dfrac{\partial}{\partial t}}{1 + \dfrac{\eta^d}{G}\dfrac{\partial}{\partial t}}\right\} \tag{10.73}$$

$$\mu = \frac{\eta^d \dfrac{\partial}{\partial t}}{1 + \dfrac{\eta^d}{G}\dfrac{\partial}{\partial t}} \tag{10.74}$$

In the case of one-dimensional consolidation at no lateral strain Lamé's Equation (4.18-c) may be applied to the effective stress $\sigma' \equiv \sigma_1'$ in the direction of the vertical axis $z \equiv (1)$:

$$\sigma' = \lambda\varepsilon_v + 2\mu\varepsilon_1 \tag{10.75}$$

Considering that at no lateral strain conditions the volume strain ε_v equals the axial strain ε_1, the above equation reduces to:

$$\sigma' = (\lambda + 2\mu)\varepsilon_1 \tag{10.76}$$

Denoting the displacements in the direction $z \equiv (1)$ by ρ and substituting

$$\varepsilon_1 = -\frac{\partial\rho}{\partial z} \tag{10.77}$$

Equation (10.76) becomes:

$$\sigma' = -(\lambda + 2\mu)\frac{\partial\rho}{\partial z} \tag{10.76-a}$$

The derivative of the above expression yields:

$$\frac{\partial\sigma'}{\partial z} = -(\lambda + 2\mu)\frac{\partial^2\rho}{\partial z^2} \tag{10.78}$$

Considering a sudden total pressure increase $\Delta\sigma = q = \text{const}$, the pore pressure decrease $\partial u/\partial z$ per unit of depth equals the corresponding effective pressure increase. Thus:

$$\frac{\partial u}{\partial z} = -\frac{\partial\sigma'}{\partial x} = (\lambda + 2\mu)\frac{\partial^2\rho}{\partial z^2} \tag{10.79}$$

Substituting λ and μ from (10.73) and (10.74), we obtain

$$\frac{\partial u}{\partial z} = \left[K + \frac{4}{3}\frac{\eta^d \frac{\partial}{\partial t}}{1 + \frac{\eta^d}{G}\frac{\partial}{\partial t}}\right]\frac{\partial^2 \rho}{\partial z^2} \tag{10.80}$$

The stress-strain relationship expressed by Equation (10.80) should now be considered when solving the general equation of one-dimensional consolidation (9.36). Assuming constant permeability (k = const) and zero limiting hydraulic gradient ($i_0 = 0$) it has the form

$$\frac{\partial e}{\partial t} = (1 + e)\frac{k}{\gamma_w}\frac{\partial^2 u}{\partial z^2} \tag{10.52}$$

Substituting

$$\frac{\partial \dfrac{e}{1+e}}{\partial t} = \frac{\partial n}{\partial t} = \frac{\partial \dfrac{\partial \rho}{\partial z}}{\partial t} = \frac{\partial^2 \rho}{\partial z \partial t}$$

Equation (10.52) becomes

$$\frac{k}{\gamma_w}\frac{\partial^2 u}{\partial z^2} = \frac{\partial^2 \rho}{\partial z \partial t} \tag{10.81}$$

10.2.1.2 *Tan's Solutions and their Discussion*

Tan's solutions of the differential Equations (10.80) and (10.81) are given in terms of the shear viscosity $\eta \equiv \eta^d$, of the 'time of relaxation' μ:

$$\mu = \frac{\eta}{G} \tag{10.82}$$

G being the shear modulus, and of the Poisson ratio ν:

$$\nu = \frac{3K - 2G}{2(3K + G)} \tag{10.83}$$

K denoting the modulus of (volume) compressibility. (Equation 10.83 has been derived from Equations 4.14 and 4.17.)

The resulting expression for the pore pressure u is (Tan, 1957 (*a*))

$$u = \frac{4q}{\pi}\sum_{m=0}^{\infty}\left\{\frac{(-1)^m}{2m+1}\cos\left[\frac{\pi}{2}(2m+1)\frac{h-z}{h}\right]\exp\frac{\omega g^2 t}{\mu}P(\omega,\lambda)\right\}_{\omega_1,\omega_2} \tag{10.84}$$

where:

$$\omega_{1,2} = \frac{-(1-\lambda) \pm \sqrt{[(1-\lambda)^2 + 4g^2\lambda]}}{2g^2} \tag{10.84-a}$$

and

$$\lambda = \frac{\pi^2}{2} \cdot \frac{1-\nu}{1-2\nu} \cdot \frac{k\eta}{\gamma_w h^2} \cdot (2m+1)^2 \tag{10.84-b}$$

$$P(\omega,\lambda) = \frac{(\omega+1)\lambda}{g^2\omega^2 + \lambda} \tag{10.84-c}$$

$$g^2 = \frac{1+\nu}{3(1-\nu)} \tag{10.85}$$

The corresponding settlements ρ are given by the equation:

$$\rho = \frac{hq}{K}\left\{1 - (1-g^2)\exp\left(-\frac{g^2t}{\mu}\right) - \frac{8}{\pi^2}\sum_{m=0}^{\infty}\left[\frac{1}{(2m+1)^2} \cdot \frac{\exp\frac{\omega g^2 t}{\mu}(1+g^2) - (1-g^2)\exp\left(-\frac{g^2t}{\mu}\right)}{\omega+1} P(\omega,\lambda)\right]_{\omega_1,\omega_2}\right\} \tag{10.86}$$

In expressions (10.84) and (10.86) the sums obtained for the values of ω_1 and ω_2 have to be added.

If even for $m = 0$, the coefficients λ are large enough to allow the approximation $\lambda - 1 \cong \lambda$, Equation (10.84) reduces to a simpler expression (Tan, 1957, Equation 5.1) which can be written in the form

$$u = \exp\left(\frac{Gg^2}{\eta}t\right) u_T \tag{10.87}$$

where u_T denotes the pore pressure obtained by Terzaghi's theory for the same time t and for the coefficient of consolidation c_v:

$$c_v = \frac{k}{\gamma_w} \cdot \frac{K}{g^2} = \frac{k}{\gamma_w} M_v \tag{10.88}$$

Considering expression (10.87) Tan concludes that the pore pressures resulting from his theory are greater than the pore pressures obtained by Terzaghi's theory. Now, this conclusion is limited by the condition $\lambda - 1 \cong \lambda$ and by the definition of the coefficient of consolidation that has to be used when calculating the pore pressure u_T.

Equation (10.87) proves that the initial pore pressures ($t = 0$) in Tan's theory are equal to those obtained by Terzaghi's theory, i.e. equal to $q = \Delta\sigma$. This conclusion is an expected consequence of the rheological scheme of Tan's theory. Thus, Tan's theory cannot be applied in cases when high plastic resistance causes the effective pressure to increase immediately after the load application.

Under the same condition of high values for the coefficient λ, $(\lambda - 1 \cong \lambda)$, Tan obtained a reduced expression (Tan, 1957, Equation 7.1) for the settlement. It can be transformed into the following equation giving the degree of consolidation U_v:

$$U_v = \frac{\rho_t}{\rho_\infty} = \frac{\varepsilon_t}{\varepsilon_\infty} = 1 - (1 - g^2)\exp\left(-\frac{Gg^2}{\eta}t\right) - $$
$$- g^2 \exp\left(\frac{Gg^2}{\eta}t\right)(1 - U_T) \qquad (10.89)$$

where the final strain ε_∞ is:

$$\varepsilon_\infty = \frac{\rho_\infty}{h} = \frac{q}{K} \qquad (10.90)$$

and U_T denotes the degree of consolidation obtained from Terzaghi's theory (expression 9.65) using the coefficient of consolidation c_v (see Equation 9.64) given by Equation (10.88).

For soils having no flow properties ($\eta \to \infty$) Equation (10.89) yields:

$$U_{v(\eta \to \infty)} = g^2 U_T \qquad (10.91)$$

or:

$$\varepsilon(t) = \varepsilon_\infty g^2 U_T \qquad (10.92)$$

Inserting Equations (10.90) and (10.85) and considering relations (10.88) and (8.9) the above expression may be written:

$$\varepsilon_t = \frac{q}{\frac{K}{g^2}} U_T = \frac{q}{M_v} U_T = \varepsilon_{\infty T}\, U_T \qquad (10.93)$$

where

$$\varepsilon_{\infty T} = \frac{q}{M_v} = m_v q \qquad (10.94)$$

$\varepsilon_{\infty T}$ is the final strain given by Terzaghi's theory. Thus, with $\eta \to \infty$ Tan's theory reduces to the classical Terzaghi theory. Comparing Equations

(10.90) and (10.94) it can be seen that, except in the case when $\nu = 0{\cdot}50$, a discontinuity in the end strain appears when limiting $\eta \to \infty$.

In Tan's interpretation the end value of primary consolidation of thin oedometer samples corresponds to the value $\varepsilon_{\infty T}$ after Equation (10.94) and the term

$$\frac{\varepsilon_\infty}{\varepsilon_{\infty T}} = g^2 = \frac{3(1-\nu)}{1+\nu} \tag{10.95}$$

represents then the ratio between the final settlement and the value at the end of primary compression.

Concerning the average value of the lateral pressure, Tan has derived for small values of time t the following formula:

$$\frac{\sigma_l}{q} = 1 - \frac{2(1-2\nu)}{1-\nu}\sqrt{\left(\frac{c_v t}{\pi h^2}\right)} \tag{10.96}$$

In Tan's oedometers (type II) the lateral pressures can be measured. Thus, the initial part of the plot $\sigma_l/q = f(\sqrt{(t)})$ allows the Poisson ratio ν to be ascertained if the coefficient of consolidation c_v is known. This coefficient (c_v) as well as the coefficient of compressibility m_v of primary consolidation (see Equation 10.94) can be obtained from the primary branch of the consolidation curve by using Taylor's fitting method. Finally, the relation (10.95) enables the value of the final settlement to be obtained as well as the corresponding value of the modulus of volume compressibility K (Equation 10.90).

In order to determine the coefficient of shear viscosity η Tan uses simple shear plastometers (Tan, 1964) in which the samples are subjected to homogeneous angular deformations. Plotting the rate of shear $\dot{\gamma} = \mathrm{d}\gamma/\mathrm{d}t$ against shear stress τ the coefficient of shear viscosity η is given by the gradient of the plot, $\eta = \mathrm{d}\tau/\mathrm{d}\dot{\gamma}$.

The consolidation curves of thick layers of soils obeying Tan's rheological scheme are expected to be close to those governed by Terzaghi's theory with a coefficient of compressibility equal to $m_v = 1/K$.

The expression (10.90) for final strains is valid only for a purely spherical state of stress. Therefore, Tan's theory involves a hydrostatic state of stress at the end of consolidation except in the limit case $\eta \to \infty$. The measurement of lateral pressures (Terzaghi, 1925; Bishop, 1958; Tan, 1961) does not confirm this conclusion to be correct. The acceptance of this conclusion of Tan's theory would lead to an important overestimate of the relaxation effects in earth pressure calculations (Šuklje, 1963 (*b*)).

10.2.2 The Saturated Kelvin Body for the Spherical Part and the Maxwell Body for the Distortional Part of the Stress-strain Relationship

Anagnosti's (1962, 1963) rheological scheme (Fig. 10.12) consists of a Kelvin body for the spherical part and a Maxwell body for the distortional part of the stress-strain relationship. The theory of consolidation based

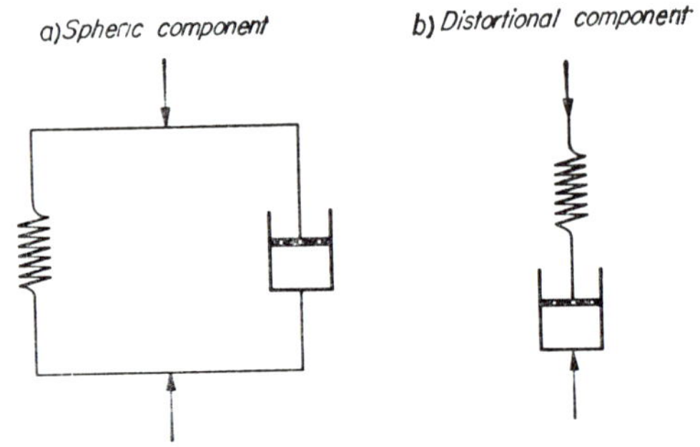

Fig. 10.12. Anagnosti's rheological model

on such a combined rheological scheme can be considered as a generalization of Tan's theory. When the coefficient of spherical viscosity η^0 becomes zero ($\eta^0 = 0$), Anagnosti's solutions are expected to conform to the results of Tan's theory.

By analogy with Equation (4.35), the fundamental relationship of spherical components of the stress and strain tensors may be written in the form:

$$p_{rs}{}^0 = 3\,K\,e_{rs}{}^0 + 3\,\eta^0\,\dot{e}_{rs}{}^0 \tag{10.97}$$

while the relationship of distortional components is the same as in Tan's theory:

$$\dot{e}_{rs}{}^d = \frac{\dot{p}_{rs}{}^d}{2G} + \frac{p_{rs}{}^d}{2\eta^d} \tag{4.27}$$

Expressing the above equations in the form

$$p_{rs}{}^0 = 3\left(K + \eta^0 \frac{\partial}{\partial t}\right) e_{rs}{}^0 \tag{10.97'}$$

$$p_{rs}{}^d = \frac{2\eta^d \dfrac{\partial t}{\partial}}{1 + \dfrac{\eta^d}{G}\dfrac{\partial}{\partial t}}\, e_{rs}{}^d \tag{10.72'}$$

use can be made of the transformation of linear viscoelastic models into elastic models (see Section 4.4.2). Lamé's coefficients λ and μ have to be replaced, in Equation (4.18-d), by the following operators (cf. Equations 4.18-b and 4.18-c):

$$\mu = \frac{\eta^d \frac{\partial}{\partial t}}{1 + \frac{\eta^d}{G}\frac{\partial}{\partial t}} \tag{10.74}$$

$$\lambda = \left(K + \eta^0 \frac{\partial}{\partial t}\right) - \frac{2}{3}\mu$$

yielding

$$\lambda = \frac{K + \left[\eta^0 + \left(\frac{K}{G} - \frac{2}{3}\right)\eta^d\right]\frac{\partial}{\partial t} + \frac{\eta^0\eta^d}{G}\frac{\partial^2}{\partial t^2}}{1 + \frac{\eta^d}{G}\frac{\partial}{\partial t}} \tag{10.98}$$

Thus the factor of proportionality $(\lambda + 2\mu)$ in Equation (10.76)

$$\sigma' = (\lambda + 2\mu)\varepsilon_1 \tag{10.76}$$

denotes

$$\lambda + 2\mu = \frac{K + \left[\left(\frac{K}{G} + \frac{4}{3}\right)\eta^d + \eta^0\right]\frac{\partial}{\partial t} + \frac{\eta^0\eta^d}{G}\frac{\partial^2}{\partial t^2}}{1 + \frac{\eta^d}{G}\frac{\partial}{\partial t}} \tag{10.99}$$

Now, in the differential equation of consolidation (10.52)

$$\frac{\partial e}{\partial t} = (1 + e)\frac{k}{\gamma_w}\frac{\partial^2 u}{\partial z^2} \tag{10.52}$$

(e = void ratio) the following substitutions may be inserted:

$$\frac{\partial \frac{e}{1 + e}}{\partial t} = -\frac{\partial \varepsilon_1}{\partial t} \tag{10.100}$$

$$\frac{\partial^2 u}{\partial z^2} = -\frac{\partial^2 \sigma'}{\partial z^2} = -(\lambda + 2\mu)\frac{\partial^2 \varepsilon_1}{\partial z^2} \tag{10.101}$$

In this way, Anagnosti applies the differential equation of consolidation in the form:

$$\frac{k}{\gamma_w}(\lambda + 2\mu)\frac{\partial^2 \varepsilon_1}{\partial z^2} + \frac{\partial \varepsilon_1}{\partial t} = 0 \tag{10.102}$$

where, for $(\lambda + 2\mu)$, the operator (10.99) has to be inserted.

For the suddenly applied uniform load q and boundary conditions presented in Fig. 10.3 Anagnosti gives, for the mean value of the linear strain ε_1 at time t, the approximate solution corresponding to the assumption

$$\frac{\partial^2 \varepsilon_1}{\partial t^2} = 0 \tag{10.103}$$

The solution is:

$$\varepsilon_t = \frac{q}{K}\left\{1 - \frac{8}{\pi^2}\sum_{m=0}^{\infty}\frac{1}{(2m+1)^2}\exp\frac{-Kt}{\eta^0 + \left(\frac{4}{3} + \frac{K}{G}\right)\eta^d + \lambda^2}\right\} \tag{10.104}$$

where

$$\lambda^2 = \frac{\gamma_w}{k}\left(\frac{2h}{\pi}\right)^2\frac{1}{(2m+1)^2} \tag{10.105}$$

For the mean value of the lateral pressure Anagnosti (1962) deduced the following expression:

$$\sigma_{b\,\mathrm{med}} = q\left\{1 + \frac{8}{\pi^2}\sum_{m=0}^{\infty}\frac{1}{(2m+1)^2}\cdot\frac{6\eta^d}{4\eta^d + 3(\eta^\circ + \lambda^2)}\cdot\right.$$
$$\left.\cdot\left[\exp\left(-\frac{G}{\eta^d}t\right) - \exp\frac{-Kt}{\eta^0 + \left(\frac{4}{3} + \frac{K}{G}\right)\eta^d + \lambda^2}\right]\right\} \tag{10.106}$$

As stated before (see Equation 4.30) the stress in a Maxwell body tends to zero if the strain is kept constant. Thus, in Anagnosti's scheme the distortional stresses vanish at the end of consolidation and a purely hydrostatic state of stress remains. Equations (10.104) and (10.106) agree with this conclusion. For $t \to \infty$ they give:

$$\varepsilon_\infty = \frac{q}{K}, \quad \sigma_{b\infty} = q \tag{10.107}$$

This result of Anagnosti's rheological scheme does not agree with the real soil properties.

10.3 GENERAL DISCUSSION OF THE APPLICABILITY OF LINEAR RHEOLOGICAL MODELS TO THE CONSOLIDATION PROCESS

The great advantage of linear rheological models is that they allow the application of the fundamental differential equations of the theory of elasticity by substituting for Lamé's coefficients the corresponding differential operators (see Section 4.4.1). In this way the effective stresses and strains can be found in the plane or in the space subjected to certain loads. For this purpose the parameters governing the spherical strains as well as those governing the deviatoric strains have to be known. In the preceding sections two solutions have been presented in which both parts of the stress and strain tensors are considered and which, consequently, permit the determination of the elastic and viscous parameters. Appropriate methods must be carried out in order to obtain the viscoelastic parameters from the consolidation curves and stress measurements.

In Tan's and Anagnosti's solutions the deviatoric stress-strain relationship has been represented by the rheological scheme of the Maxwell body. Anagnosti's solution in which a Kelvin body governs the spherical stress-strain relations was shown to be a generalization of the older theory of Tan who assumed the compression to be of elastic character only.

As proved by the analysis of the results obtained (see foregoing sections), the following disadvantages of the above rheological scheme for soils can be ascertained:

(1) If a confined specimen is exposed to supplementary loading and if it is assumed that a final deformation is reached, the distortional stresses tend to disappear and the soil approaches a hydrostatic state of stress. The observations do not prove that such a final stress state really occurs.

(2) In the half-space exposed to a limited surface load the end state of stress cannot be hydrostatic. Thus the distortional strains obeying the law of the Maxwell body, must continue to occur at a certain final speed. Such a conclusion does not correspond to the observations except in the case of overloaded soils at the upper limit of their bearing capacity. In safe foundation design, the bearing capacity is not allowed to be fully mobilized.

(3) In Tan's solution volume creep is not considered at all and in Anagnosti's solution Kelvin's rheological model does not express the secondary consolidation as increasing with the logarithm of time as is commonly observed. Although the logarithmic law of secondary compression cannot extend beyond a certain finite time limit, it has been proved to extend for

a number of years in some long duration tests (see e.g. Cox, 1936). Comparison of the consolidation curves of some clay samples sedimented in the laboratory with the natural void ratios of some deposits, whose approximate values of pre-consolidation pressure and geological age are known, proves that secondary compression can be of very long duration. For centuries and milleniums its speed has been decreasing probably in inverse proportion to the time elapsed after the load application or even slower.

Besides the shortcomings mentioned above which have to be attributed to the rheological schemes used, the resulting viscoelastic parameters can be applied only to soils that can be assumed to be homogeneous and isotropic and whose deformation characteristics do not depend on the stress path. The practical applicability of consolidation theories based on linear rheological models is to be limited to particular cases.

The deduction of both spherical and distortional viscoelastic parameters requires rather extensive experimental work and appropriate interpretation. The determination of viscoelastic coefficients governing one-dimensional consolidation, and related to axial stresses only, is simpler. These, however, are not sufficient to be used for the computation of stresses and strains in general plane and space problems. The application of consolidation theories based on such uniaxial stress-strain relationships has to be limited to loading cases and boundary conditions which are consistent with the assumptions. Nevertheless, the studies of consolidation related to linear stress states facilitate a rather simple and clear rational analysis of viscous effects on the consolidation process and prove to what extent linear rheological models can express the development of effective stresses and strains in soils.

The saturated Hookean body connected in series with a saturated Kelvin body was shown to be the general rheological scheme for the existing consolidation analysis of this type. In Taylor's Theory *B* and in the special case of Florin's solution, the compressibility of the Hookean body is zero and the soil is treated as a pure saturated Kelvin body. The advantage of these solutions is that they apply for saturated layers (samples) whose consolidation, after a sudden load application, begins with the presence of supplementary effective pressures in the drainage direction. This phenomenon appears in thin samples of soils of moderate permeability and pre-consolidated or subjected to high pressures (see the next Chapter).

When adding a saturated Hookean spring in series with a Kelvin body for saturated soils, the full pore pressure must appear at the very beginning of one-dimensional consolidation, just as in Terzaghi's theory. This classical theory can be treated as a special case of the general viscoelastic model; it corresponds to the condition of either zero or infinite viscosity of the Newtonian dashpot. For arbitrary intermediate values of the coefficient of viscosity ($0 < \eta < \infty$, η . . . kp sec cm^{-2}) the secondary consolidation was proved to be negligible for layers of greater thickness having usual soil properties (see Section 10.1.1.4). In this way the viscous phenomena of such soils would be important for the interpretation of the tests only; the consolidation of thick layers would be governed to a satisfactory approximation by the elastic parameter(s) and the coefficient of permeability.

Taylor (1942) when establishing the fundamental assumptions of theory *B* (see Section 10.1.2.1) was fully aware of the inability of his selected rheological scheme to express satisfactorily the long-term secondary effects. He interpreted the viscous effects—represented in Kelvin's model by the Newtonian dashpot—as the effects of secondary compression in the previous load increment; the secondary compression occurring subsequent to the primary compression was disregarded.

11

One-Dimensional Consolidation Analysis of Saturated Soils Considering Non-Linear Stress–Strain Relationships

11.1 ANALYSIS OF THE CONSOLIDATION PROCESS BY THE ISOTACHE METHOD

11.1.1 Some Introductory Observations

The isotaches corresponding to one-dimensional confined compression were defined and their construction described in Section 8.1.3 dealing with isoconsolidation under no seepage resistance. When constructing the isotaches corresponding to the consolidation curves of saturated soils, the pore pressure dissipation has to be taken into account. The intergranular pressures change during consolidation. Thus, the graphs relating the pore pressures or the complementary values of effective stresses ($\sigma' = \sigma - u$) with time must first be derived from each consolidation curve.

If the pore water drainage occurs in the upward direction only, the pore pressures can be measured at the lower surface of the oedometer sample by connecting the voids below the porous stone to a pore-pressure measuring device. Now, the development of the pore pressure is to some extent affected by the measuring system itself.* The main influential factors are the deformability of the tube connecting the oedometer to the manometer, the compressibility of the water in the space below the sample and in the tube, and especially the compressibility of the air left in the water. Thus, an additional quantity of water is needed to accommodate the material of the measuring device, the air and the water to the new pressure. During this inflow the pore pressure is reduced at the undrained boundary of the sample.

* Recently, the measurement of pore-pressures has been impoved by applying electronic transducers placed as near as possible to the measuring points. By the use of such transducers the below mentioned inaccuracy of the measurement can be reduced.

The opposite procedure is required when the accommodation is completed and the drainage occurs only in the upward direction. As the additional quantity of water accumulated in the measuring device has to be squeezed out through the soil sample, the consolidation process is slower and the pore pressure dissipation retarded. The lower the permeability and porosity of the soil the more important are the retardation effects.

The above interpretation of the disagreement between computed and measured pore pressures points to the conclusion that the computed results are expected to be closer to the actual pore pressures when the undrained boundary of the sample is completely confined or when the sample is drained from both boundary surfaces, the maximum pore pressure appearing in the middle surface of the sample.

The isotaches are plots of void ratios (e) against axial effective pressures (σ'), each isotache corresponding to a certain consolidation speed ($\partial e/\partial t$). They can easily be constructed from the secondary parts of oedometer consolidation curves. If their primary parts shall be used as well, the average void ratios ($\bar{e}$) occurring along the height of the sample should be taken and plotted against the average axial effective pressures ($\bar{\sigma}'$).

In order to determine the average axial pressures, the shape of the isochrone and not merely its maximum ordinate should be known. As in common consolidation tests the measurement of pore pressure inside the sample is not possible, the difficulties connected with the observation of initial pore pressures are not the only shortcomings of the direct method of determining the pore pressures during consolidation. The shape of the isochrone must also be assumed when applying the measured values of the pore pressure at the lower boundary of the sample. Now, if the shape of the isochrones must be assumed, then the boundary values can be more simply and more accurately determined if calculated from the data of the consolidation curve and the data of the graphs relating permeability coefficients to void ratios. These graphs can be plotted on the basis of direct observation of the permeability at the beginning or at the end of each load step.

The method of calculating the pore pressures corresponding to the assumed isochrone shape will be discussed in the following sections.

11.1.2 Mean Effective Pressures Corresponding to Parabolic Isochrones

The isochrones may be supposed to have the shape of parabolas of the nth degree:

$$u = u_0\left\{1 - \left(\frac{z}{h}\right)^n\right\} \tag{11.1}$$

The notation is explained by Fig. 11.1. Differentiating the above equation gives

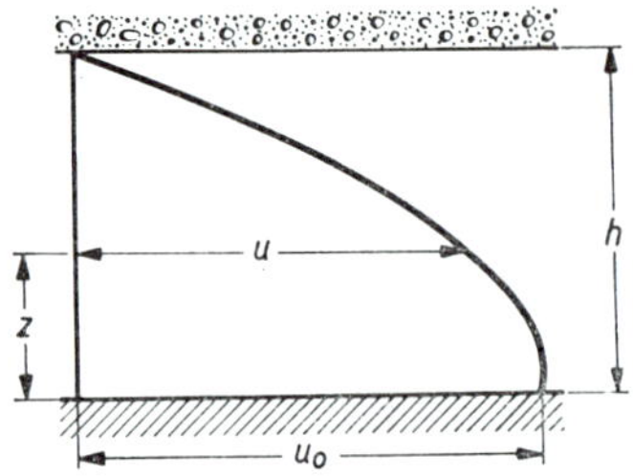

Fig. 11.1. Parabolic isochrone. Notation to Equation (11.1)

$$\frac{\partial u}{\partial z} = -\frac{nu_0}{h^n} z^{n-1} \tag{11.2}$$

Assuming the validity of Darcy's law the seepage speed at the depth z is:

$$v = ki = \frac{k}{\gamma_w}\left(-\frac{\partial u}{\partial z}\right) \tag{11.3}$$

$$v = \frac{k}{\gamma_w} \cdot \frac{nu_0}{h^n} z^{n-1} \tag{11.3'}$$

With $z = h$, i.e. at the pervious contact, the outlet seepage speed is

$$v_w = \frac{k}{\gamma_w} \cdot \frac{nu_0}{h} \tag{11.4}$$

The outlet seepage speed is equal to the negative value of the consolidation speed v relative to the whole thickness of the sample.

$$v_w = -v_h \tag{11.5}$$

$$v_h = \frac{\partial h}{\partial t} = h_s \frac{\partial \bar{e}}{\partial t} \tag{11.6}$$

h_s is the height of the voidless solid phase of the sample (Fig. 11.2) and

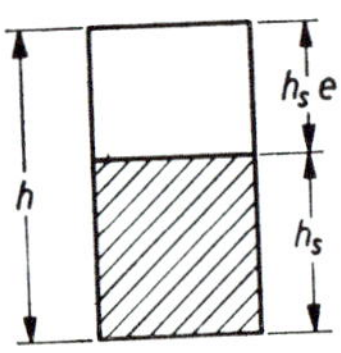

Fig. 11.2. Notation in Equation (11.6)

$\bar{e}$ is the average void ratio at time t. If the consolidation curve is plotted against the logarithm of time, expression (8.13) should be inserted into Equation (11.6):

$$v_h = -h_s \frac{\bar{\alpha}_e}{2{\cdot}3026t} \tag{11.6'}$$

$\bar{\alpha}_e$ is the logarithmic cycle gradient of the void ratio decrease relative to the tangent line to the consolidation curve at time t (see Fig. 8.6).

Inserting expressions (11.4) and (11.6′) into Equation (11.5) we obtain:

$$\frac{k}{\gamma_w} \cdot \frac{nu_0}{h} = h_s \frac{\bar{\alpha}_e}{2{\cdot}3026t} \tag{11.7}$$

Thus considering the relation

$$h = h_s(1 + \bar{e}) \tag{11.8}$$

the maximum pore pressure of the parabolic isochrone can be expressed either in the form:

$$u_0 = \frac{\gamma_w}{2{\cdot}3026n} h_s^2 \frac{\bar{\alpha}_e(1 + \bar{e})}{kt} \tag{11.9}$$

or as follows:

$$u_0 = \frac{\gamma_w}{2{\cdot}3026n} \cdot \frac{\bar{\alpha}_e}{1 + \bar{e}} \cdot \frac{h^2}{kt} \tag{11.9'}$$

If the isochrones are parabolas of the second degree, $n = 2$ and

$$u_0 = \frac{\gamma_w}{4{\cdot}6052} \cdot \frac{\bar{\alpha}_e}{1 + \bar{e}} \cdot \frac{h^2}{kt} \tag{11.10}$$

(Šuklje, 1955 and 1957).

If the consolidation curve $\bar{e} = \bar{e}(t)$ is plotted against the square root of time (Fig. 11.3), the tangent line at point $(t,\bar{e})$ has the equation:

$$\bar{e} = \bar{e}_\alpha - \Delta\bar{e}_0 - \bar{\beta}_e\sqrt{t} \tag{11.11}$$

Differentiating, we get:

$$\frac{\partial\bar{e}}{\partial t} = -\frac{\bar{\beta}_e}{2\sqrt{t}} \tag{11.12}$$

Inserting the above expression into Equation (11.6) the relation (11.5) yields:

$$u_0 = \frac{\gamma_w}{2n} \cdot \frac{\bar{\beta}_e}{1 + \bar{e}} \cdot \frac{h^2}{k\sqrt{t}} \tag{11.13}$$

Now, the mean effective pressure $\bar{\sigma}'$ is given by the equation:

$$\bar{\sigma}' = \sigma - \frac{1}{h}\int_{z=0}^{h} u \, dz \tag{11.14}$$

Inserting expression (11.1) for the pore pressure u produces

$$\bar{\sigma}' = \sigma - \frac{u_0}{h}\int_{0}^{h}\left\{1 - \left(\frac{z}{h}\right)^n\right\} dz \tag{11.14'}$$

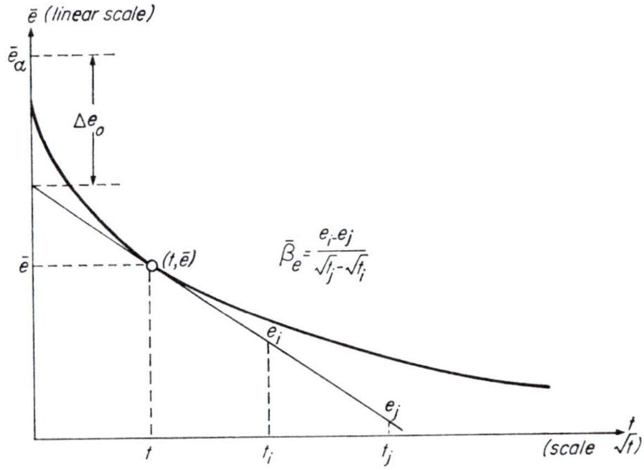

Fig. 11.3. Determination of the consolidation speed at time t of an arbitrarily shaped consolidation curve presented as a void ratio versus square root of time plot (see Equation 11.11)

and after integration:

$$\bar{\sigma}' = \sigma - \frac{n}{n+1} u_0 \tag{11.15}$$

where expression (11.9), (11.9′), (11.10) or (11.13) has to be inserted for u_0.

Equations (11.9) or (11.9′) and (11.13) corresponding to loading by steps $\Delta\sigma$, hold only for values

$$u_0 \leqslant \Delta\sigma \tag{11.16}$$

When Equations (11.9), (11.9′) or (11.10) yield $u_0 > \Delta\sigma$, semiparabolic isochrones (Fig. 11.4) reaching their maximum values $u_0 = \Delta\sigma$ already at a distance y_1 from the pervious contact should be used. By analogy with expression (11.4) the outlet seepage speed is then:

$$v_w = \frac{k}{\gamma_w} \cdot \frac{n\Delta\sigma}{y_1} \tag{11.17}$$

In this case, Equation (11.5) yields:

$$y_1 = \frac{Nn}{\gamma_w} \cdot \frac{1 + \bar{e}}{\bar{\vartheta}_e} \cdot \frac{kt^m}{h} \tag{11.18}$$

where the values of N, $\bar{\vartheta}_e$ and m are given by the equations

(*a*) $$N = 2{\cdot}3026,\ \bar{\vartheta}_e = \bar{\alpha}_e,\ m = 1 \tag{11.19}$$

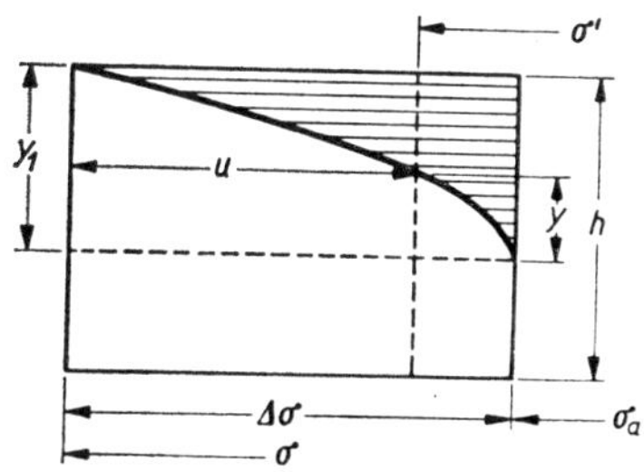

Fig. 11.4. Half-parabolic isochrones at $u_0 = \Delta\sigma$

for plots with the logarithm of time scale, and by the equations

(*b*) $$N = 2,\ \bar{\vartheta}_e = \bar{\beta}_e,\ m = \tfrac{1}{2} \tag{11.20}$$

for plots with the square root of time scale. In both cases the value $n = 2$ must be inserted for parabolic isochrones of the second degree. The average additional effective pressures $\Delta\bar{\sigma}'$ are then given by the equation

$$\Delta\bar{\sigma}' = \frac{\Delta\sigma}{n + 1} \cdot \frac{y_1}{h} \tag{11.21}$$

Inserting expression (11.18) for y_1, we obtain:

$$\Delta\bar{\sigma}' = \frac{(\Delta\sigma)^2}{n + 1} \cdot \frac{Nn}{\gamma_w} \cdot \frac{1 + \bar{e}}{\bar{\vartheta}_e} \cdot \frac{kt^m}{h^2} \tag{11.22}$$

All equations derived in this section hold also for continuous loading if the isochrones are also assumed to be parabolic. In Equation (11.15) the actual total pressure σ is to be considered; the total pressure increment appearing in Equation (11.22) represents the difference between the actual total pressure and the total pressure at the preceding equilibrium state.

11.1.3 Mean Effective Pressures Assuming an Isochrone Shape according to the Terzaghi Consolidation Theory

The isochrones according to the Terzaghi consolidation theory were expressed by the series (9.63). In Fig. 8.14 their maximum ordinates are plotted against the time factor T_v.

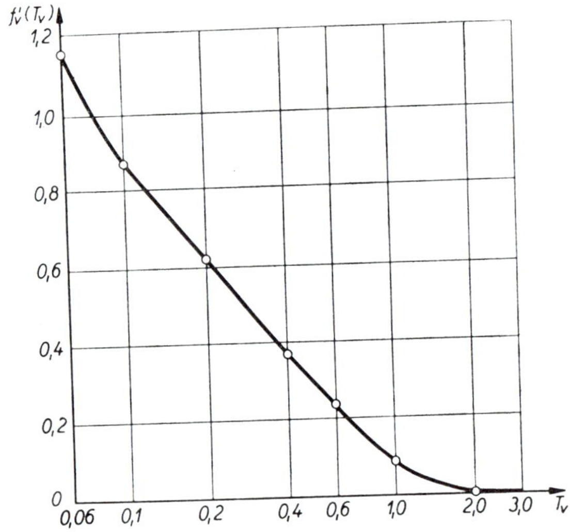

Fig. 11.5. Function $f_v(T_v)$ according to Equation (11.24)

Differentiating Equation (9.63) with respect to z and inserting the resulting values $(\partial u_{z,t}/\partial z)_{z=0}$ into Equation (11.3) the outlet drainage speed becomes:

$$v_w = \frac{k}{\gamma_w} \cdot \frac{2\Delta\sigma}{h} \cdot f_v(T_v) \tag{11.23}$$

$$f_v(T_v) = \sum_{m=0}^{\infty} \exp\left\{-(2m+1)^2 \frac{\pi^2}{4} T_v\right\} \tag{11.24}$$

In Fig. 11.5 the function $f_v(T_v)$ is plotted against the logarithm of the time factor T_v.

Inserting expressions (11.23) and (11.6) into Equation (11.5) yields:

$$f_v(T_v) = \frac{\gamma_w}{2N\Delta\sigma} \cdot \frac{\bar{\alpha}_e'}{1 + \bar{e}} \cdot \frac{h^2}{kt^m} \tag{11.25}$$

where the values (11.19) or (11.20) should be inserted if the consolidation curves are plotted against the logarithm of time or square root of time. In Fig. 11.5 the time factor T_v corresponding to the value $f_v(T_v)$ resulting from Equation (11.25) should be found; the values $\bar{\alpha}_e$ and $\bar{e}$ for any time t that have to be inserted are given by the consolidation curve $\bar{e} = \bar{e}(t)$, and the coefficient of permeability by the corresponding $k = k(\bar{e})$ plot while the height h is given by the expression (11.8).

From diagram A of Fig. 9.5 the degree of consolidation U_v may then be found corresponding to the time factor T_v obtained above. Finally the mean effective pressure should be determined by applying the relation:

$$\bar{\sigma}' = \sigma - (1 - U_v)\Delta\sigma \tag{11.26}$$

In the classical Terzaghi consolidation theory a linear relationship between the increase of effective pressures $\bar{\sigma}'$ and strains ε as well as the constancy of the permeability coefficient k are assumed. These assumptions are not compatible with a general consolidation process where the coefficients of permeability change during consolidation and where the assumed linear relationship $\Delta\bar{\sigma}' = M_v\Delta\varepsilon$ can be related, at best, to the primary consolidation phase. Consequently a step-like plot $\bar{\sigma}' = \bar{\sigma}'(\varepsilon)$, as shown in Fig. 11.6, corresponds to the application of

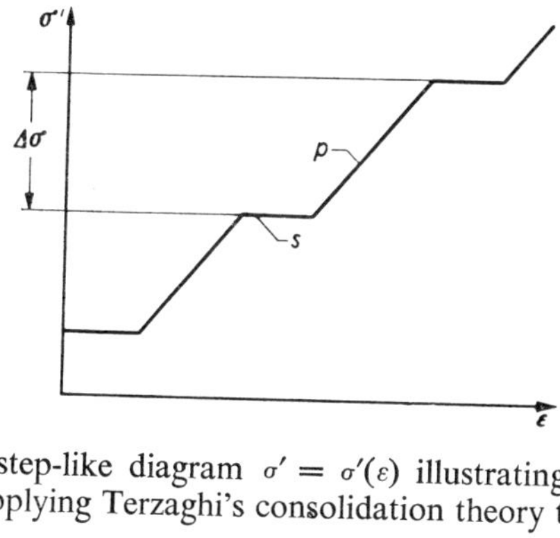

Fig. 11.6. The step-like diagram $\sigma' = \sigma'(\varepsilon)$ illustrating the assumptions made when applying Terzaghi's consolidation theory to viscous soils

Terzaghi's equations when deriving the pore pressures from the observed consolidation curves. Thus the application of the method treated above should be limited to common cases of zero initial additional effective pressure. Even in these cases the resulting pore pressure dissipation will not be very close to the classical one if in Equation (11.25) the coefficient of permeability k is related to the decreasing void ratio $\bar{e}$.

The method discussed in this section cannot be applied to continuous loading.

11.1.4 Comparison of Various Computed and Measured Pore Pressures

In Figs. 11.7, 11.8 and 11.9 three examples of pore pressure dissipation are given. In these the results from the methods treated above are compared to the measured values.

Fig. 11.7 (*a*) represents the consolidation curve $\bar{e} = \bar{e}(t)$ and the graph $k = k(\bar{e})$ relating the coefficient of permeability to the average void ratio of a Tertiary clay which was sedimented in the oedometer itself. The characteristics of the clay are: $w_L = 29$ per cent, $I_p = 10$; 21 per cent $< 0{\cdot}002$, 76 per cent 0·002/0·06, 3 per cent 0·06/2 mm; and the curves are for the load increment: $\Delta\sigma = (4 - 2)$ kp/cm². Using the plots $\bar{e} = \bar{e}(t)$ and $k = k(t)$ the pore pressure dissipation curves $u_0 = u_0(t)$ were constructed corresponding (*a*) to parabolic isochrones of second degree (curve *a*) and (*b*) to isochrones having the shape resulting from Terzaghi's theory. In the same diagram (Fig. 11.7 (*b*)) they are compared (*c*) with the pore pressures measured directly and (*d*) with Terzaghi's pore pressure dissipation curves corresponding to the average value of the coefficient of permeability, as measured in the interval $\Delta\sigma$, and to the coefficient of volume change as obtained by Casagrande's fitting method.

The comparison shows that the assumption of parabolic isochrones of second degree leads to a somewhat faster pore pressure dissipation than the assumption of a linear relationship between the effective pressures and strains during primary compression. This statement agrees with the comparison of the approximate and rigorous solutions of the differential equation of the Terzaghi theory, if the approximate solution is based on the assumption of parabolic isochrones of second degree (Terzaghi and Fröhlich, 1936).

The reason for the differences between curves (*b*) and (*d*) has already been indicated at the end of the preceding section. Slower initial and faster final pore pressure dissipation, as obtained by the Terzaghi theory (curve *d*), is to be attributed to the fact that the true coefficient of permeability exceeds the average value during the first half of the primary consolidation and does not reach it during the second half.

The inaccuracy of the measured pore pressure values in the first phase of

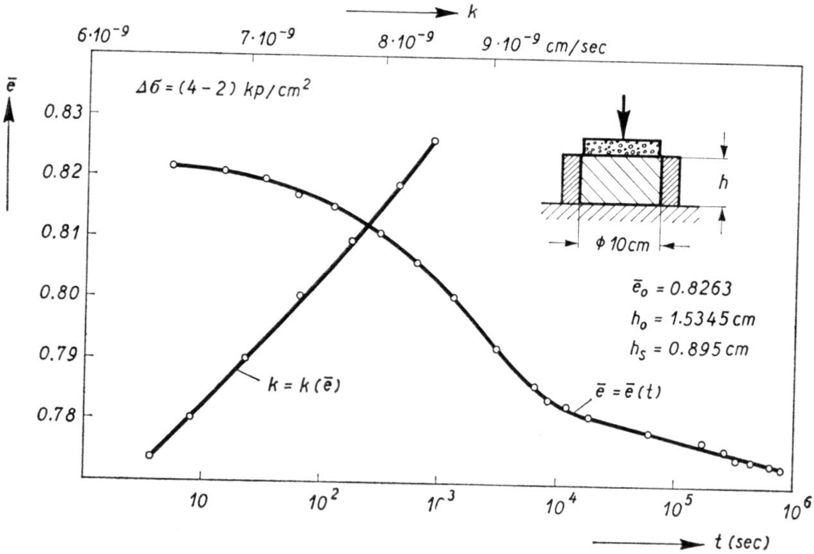

(*a*)

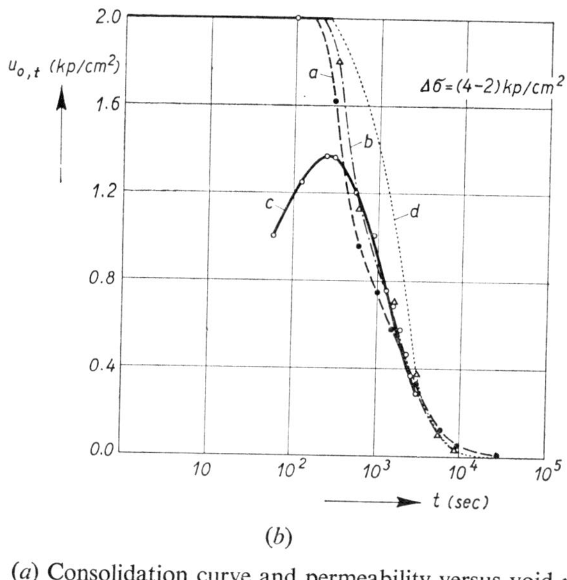

(*b*)

Fig. 11.7. (*a*) Consolidation curve and permeability versus void ratio plot of a clayey oedometer sample subjected to a suddenly applied axial load $\Delta\sigma$; (*b*) Comparison of corresponding measured (line c) and computed maximum pore-pressure dissipation (lines a, b and d): line a and b, deduced from the above $\bar{e} = \bar{e}(t)$ and $k = k(\bar{e})$ plots by assuming parabolic isochrones of second degree (line a) and isochrones having the shape of Terzaghi's isochrones (line b); line d, according to Terzaghi's theory at an average k-value consolidation has already been discussed in Section 11.1.1. The accommodation of the measuring device to the new pressure concerns: (*a*) the compressibility of the water in the measuring device, (*b*) the deformability of the device itself, and (*c*) the compression of air bubbles. In the case presented in Fig. 11.7, the accommodation being completed, the measured pore pressure dissipation approached, at its closest, curve (*a*) corresponding to parabolic isochrones with $n = 2$. Due to the influence of the decompression of the water, measuring device and air bubbles, the measured values are expected to exceed the predicted pore pressures in the second phase of consolidation.

Fig. 11.8 represents the correlation between the measured pore pressures, those obtained by Equation (11.10) (parabolic isochrones with $n = 2$), and those corresponding to Taylor's theory *B*. The comparison is for a Boston Blue clay sample whose consolidation curve was presented by Taylor (1942, Fig. 50 (*b*)).

In Fig. 11.9 the pore pressure dissipation curves obtained by applying Equation (11.10) are compared with the measured pore pressures (lines connecting small circles). Part (*a*) shows some load steps of a normally consolidated lacustrine clay of high plasticity and part (*b*) the consolidation curves $\Delta\sigma = (4 - 2)$ and $(8 - 4)$ kp/cm^2 of the lacustrine chalk presented in Fig. 8.31 (*a*). In example (*a*) full pore pressures equal to the total load increase appear at the very beginning of consolidation. In example (*b*) however, the high initial plastic resistance allows the immediate increase of the intergranular pressures.

Generally, the primary consolidation is expected to be closer to the Terzaghi theory the smaller the permeability, the smaller the pre-consolidation and the thicker the sample (layer). On the contrary, after a sudden load increment, the initial effective pressure increase is the higher the thinner the sample and the smaller the permeability; moreover it depends on the degree of preconsolidation and the amount of the total pressure affecting the density of the soil.

11.1.5 Construction of the Isotaches

In order to construct the isotache set, the mean axial effective pressures have to be plotted against time: $\bar{\sigma}' = \bar{\sigma}'(t)$. This plot should be drawn above the same time scale (logarithm of time or square root of time) as the consolidation curve $\bar{e} = \bar{e}(t)$ and the graph between the consolidation velocities $\partial\bar{e}/\partial t$ and the time $t: \partial\bar{e}/\partial t = \omega(t)$ (Fig. 11.10). By applying the three graphs the corresponding pairs of values $\bar{e}$ and $\bar{\sigma}'$ can be adapted to any chosen constant velocity $\partial\bar{e}/\partial t = \text{const}$ (see Fig. 11.10).

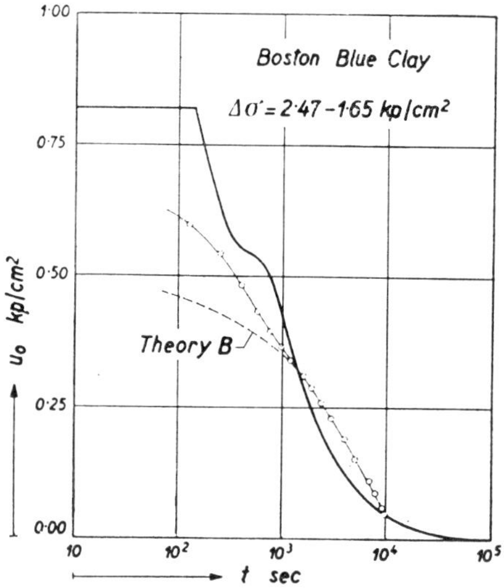

Fig. 11.8. Correlation between measured pore pressures (thin full line), those obtained from Equation (11.10) with $n = 2$ (full thick line) and those corresponding to Taylor's theory B (dashed line): Boston Blue Clay (Taylor, 1942). Reproduced after Šuklje, 1957

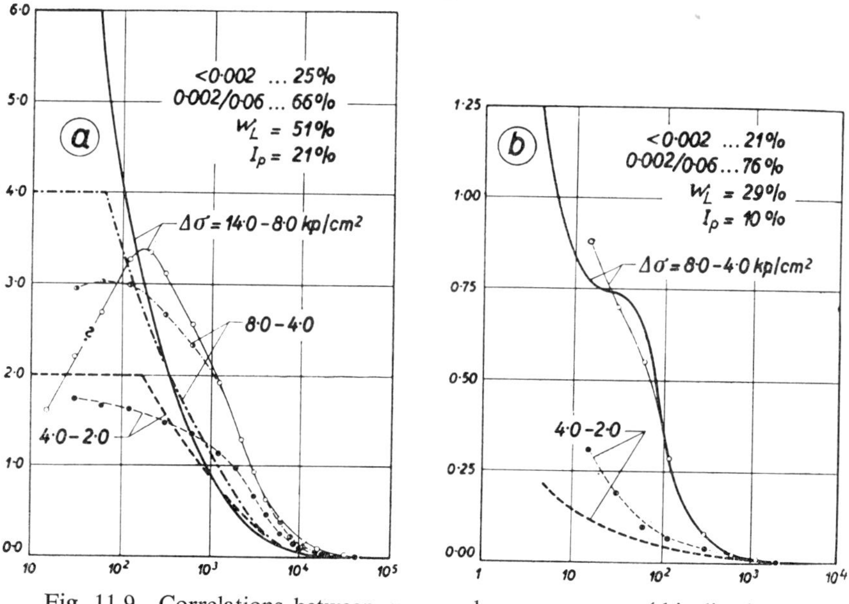

Fig. 11.9. Correlations between measured pore pressures (thin lines) and those obtained from Equation (11.10) with $n = 2$ (thick lines): for normally consolidated samples of (*a*) a lacustrine clay, (*b*) a lacustrine chalk (consolidation curves in Fig. 11.11). Reproduced after Šuklje, 1957

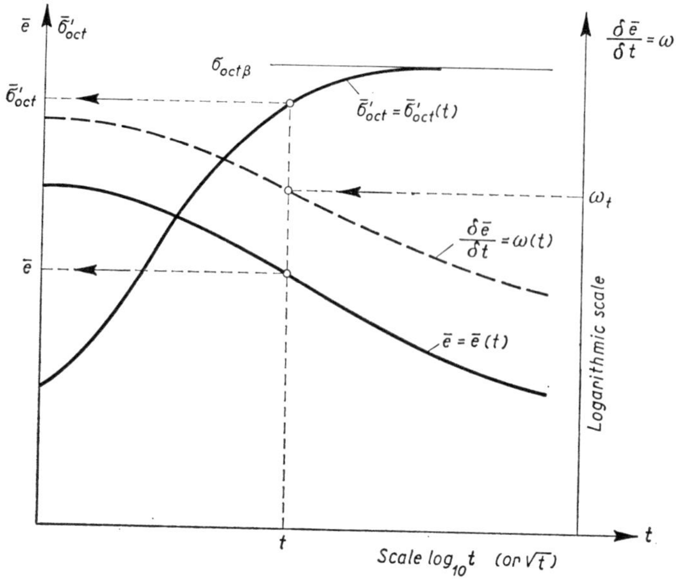

Fig. 11.10. Determination of single points of isotaches from $\bar{e} = \bar{e}(t)$, $\frac{\partial \bar{e}}{\partial t} = \omega(t)$ and $\bar{\sigma}' = \bar{\sigma}'(t)$ plots. Notification σ_{oct} applies to the three-dimensional consolidation analysis (Sections 14.4.2 and 16.4). For the one-dimensional consolidation at no lateral strain the sign σ_{oct} should be replaced by $\sigma \equiv \sigma_1$

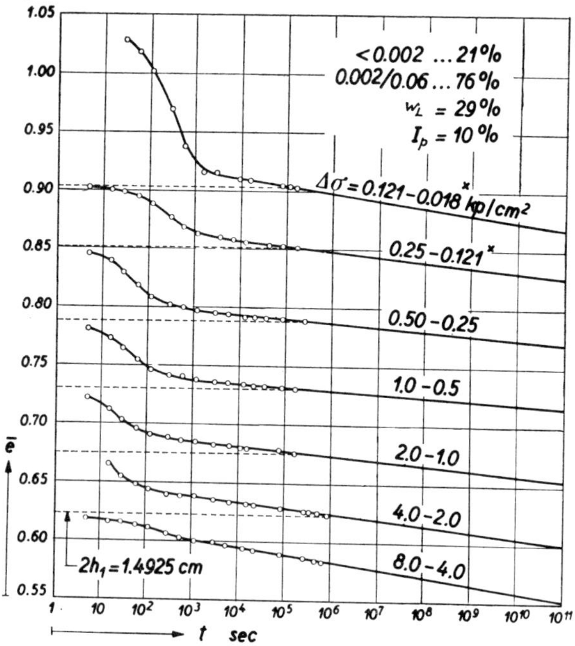

Fig. 11.11. (*a*)

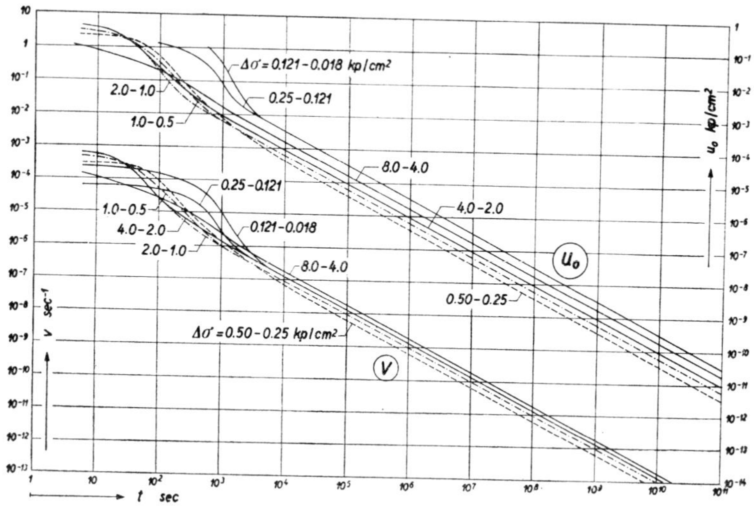

Fig. 11.11. (*b*)

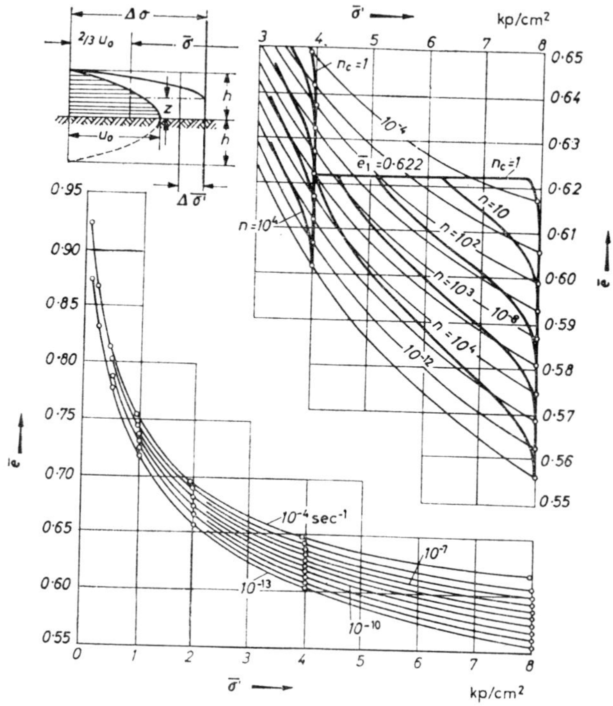

Fig. 11.11. (*c*)

Fig. 11.11. A normally consolidated lacustrine chalk: (*a*) consolidation curves; (*b*) pore-pressure dissipation (u_0) and consolidation speed versus time plots (v); (*c*) corresponding isotaches $\bar{e} = [\bar{e}(\bar{\sigma}')]_{\delta\bar{e}/\delta t=\text{const}}$ and compression curves $\bar{e} = [\bar{e}(\bar{\sigma}')]_{\Delta\sigma,h}$ for a suddenly applied load increment $\Delta\sigma$ and various layer thicknesses $h = nh_1$, $2h_1 = 1{\cdot}4925$ cm. Reproduced after Šuklje, 1957

Other points ($\bar{\sigma}'$,$\bar{e}$) of the same isotache $\partial\bar{e}/\partial t$ = const can be found in similar diagrams corresponding to other load steps $\Delta\sigma = \sigma_\beta - \sigma_\alpha$.

Two examples of the construction of isotache sets are presented in Figs. 11.11 and 11.12.

Fig. 11.11 (*a*) presents the consolidation curves $\bar{e} = \bar{e}(t)$ of a lacustrine chalk (from the Soča valley)—the particles were sedimented in the oedometer itself. The graphs $\partial\bar{e}/\partial t = \bar{v}_e(t)$ and $u_0 = u_0(t)$ corresponding to the

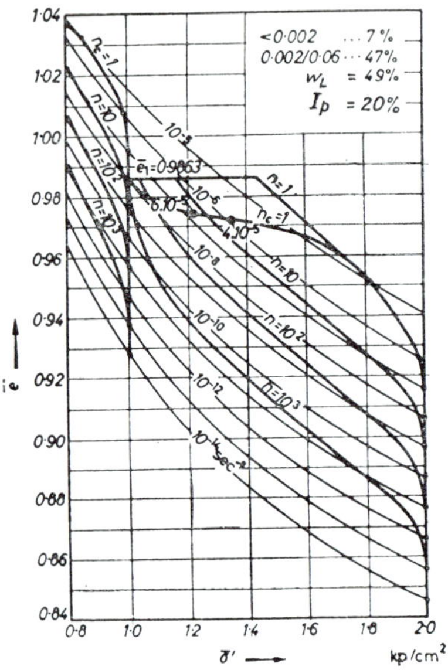

Fig. 11.12. A normally consolidated clay: isotaches, and compression curves for a suddenly applied load $\Delta\sigma = (2{\cdot}0 - 1{\cdot}0)kp/\text{cm}^2$ and various layer thicknesses $h = nh_1$, $2h_1 = 3{\cdot}585$ cm. Reproduced after Šuklje, 1957

same load intervals are plotted in Fig. 11.11 (*b*), and the resulting isotaches in Fig. 11.11 (*c*). The logarithmic lines of secondary compression were extended beyond the time of observation and parabolic isochrones of second degree were assumed. For the load increment $\Delta\sigma = (8 - 4)$ kp/cm² various graphs $\bar{\sigma}' = \bar{\sigma}'(\bar{e})$ corresponding to various layer thicknesses and to the same initial void ratio $\bar{e}$ are drawn in an enlarged detail of the isotache set (Fig. 11.11 (*c*); see next section).

In Fig. 11.12 the isotache set derived from the consolidation curves of an undisturbed sample of lower permeability (sediment of the Lake of Skadar) is shown. The grain size distribution and the plasticity characteristics are given in the figure. The isotaches presented are related to the effective pressures in the interval between 0·8 and 2·0 kp/cm² and the curves for various layer thickness ratios n are also drawn in. The figure shows that in the oedometer test the initial consolidation speeds after a sudden load increase do not always agree with a continuous isotache set and that they have to be neglected when constructing the isotaches. This disagreement with the fundamental hypothesis of the isotache method may be explained by thixotropic effects appearing at a sudden load application. As the transfer of the immediate surcharge on the pore water requires a certain compression of the water, the whole additional stress is initially taken up by the films of adsorbed water at the solid contacts. Thus, the structure becomes unstable, the pore pressures can increase beyond the values of the additional total stresses and the consolidation speed increases until the stabilizing thixotropic effects regain control.

11.1.6 The Construction of the Consolidation Curve of a Layer of Arbitrary Thickness Using the Isotache Set

Using the isotache set the consolidation curve $\bar{e} = \bar{e}(t)$ of a layer of an arbitrary initial thickness $h_{\alpha n}$ can be constructed by respecting the following two conditions:

(*a*) By Equations (11.15), (11.22) and (11.26) respectively, the mean effective pressures $\bar{\sigma}'$ can be determined for any position $(\bar{\sigma}', \bar{e})$ and inclination $(\bar{\alpha}_e')$ of the consolidation curve, taking into account the relationship $k = k(\bar{e})$ and the height h_{sn} of the solid. In the isotache set the point given by the void ratio $\bar{e}$ of the consolidation curve and the resulting effective stress should correspond to the same consolidation speed as obtained by Equations (8.13) or (11.12) for the same values $\bar{e}$, t and $\bar{\alpha}_e$ or $\bar{\beta}_e$.

(*b*) The consolidation curve must be continuous.

Considering the above conditions the consolidation curve may be gradually constructed by trial and error. For small values of time it is recommended that the consolidation curve be plotted to the square root of time scale. First, in order to facilitate the construction, Terzaghi's consolidation curve should be drawn by the iterative method. This should correspond to a drainage length h and be in agreement with Casagrande's or Taylor's fitting method. It should also join the extrapolated line of secondary compression of the oedometer sample.

In Fig. 11.13 the method described is carried out for the load increment $\Delta\sigma = (8 - 4)$ kp/cm² applied to layers of lacustrine chalk of various thicknesses the oedometer consolidation curves and isotache set of which have been presented in Fig. 11.11 (*a*) and (*c*). In Fig. 11.14 another example of the application of the method is given for the load increment $\Delta\sigma = (2 - 1)$ kp/cm² applied to layers of the lacustrine clay, the isotaches of which have been shown in Fig. 11.12. The consolidation curves

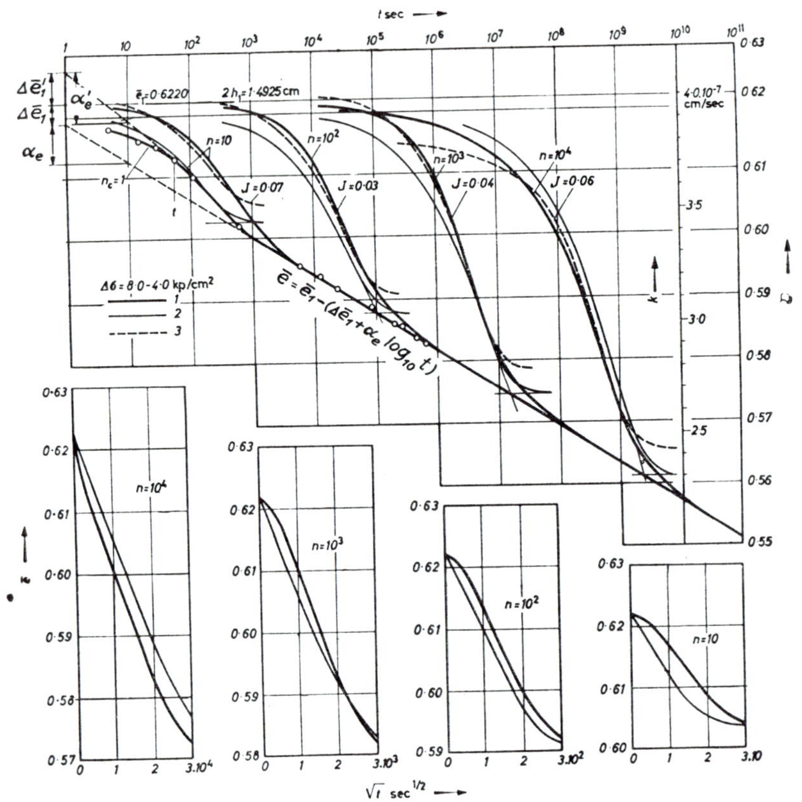

Fig. 11.13. Consolidation curves of the lacustrine chalk layers whose isotache set has been presented in Fig. 11.11–*c*, for various layer thicknesses $h = nh_1$; 1 . . . corresponding to the isotache set and to parabolic isochrones of second degree, 2 . . . according to Terzaghi's theory and logarithmic law of secondary consolidation, 3 . . . according to Taylor's theory B at appropriate time factors *J*. Reproduced after Šuklje, 1957

constructed by the isotache method are compared, in the same figures, with the corresponding Terzaghi consolidation curves (thin lines) reaching to the extended secondary consolidation line of the oedometer test. In Fig. 11.13 the comparison with the consolidation curves according to Taylor's theory *B* is also given. The corresponding ratios of the time factor J are marked in the diagram. They are very small for thick layers as expected by Taylor, but with increasing layer thickness their decrease is not continuous.

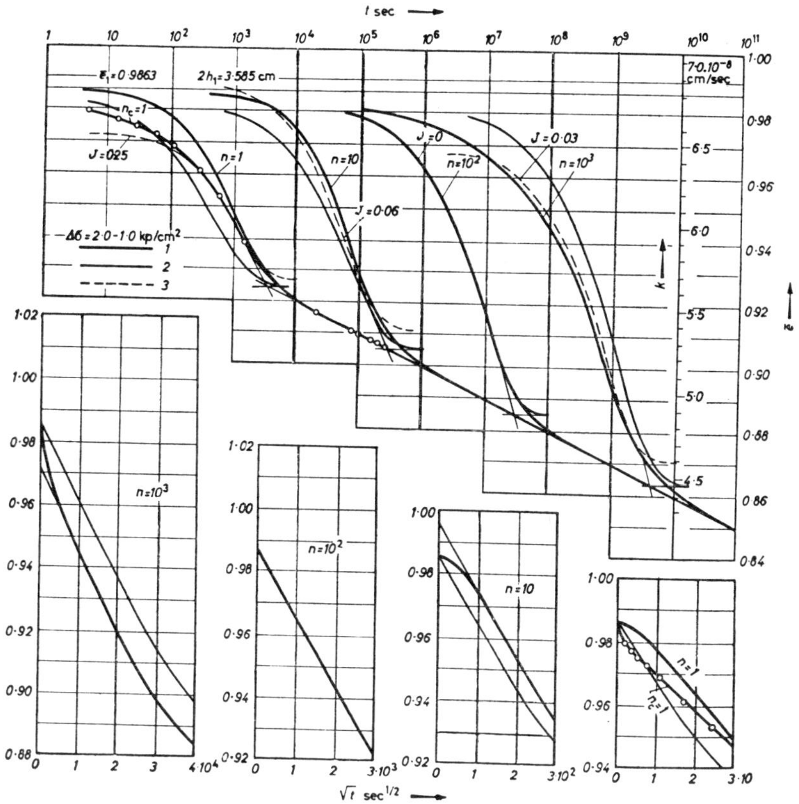

Fig. 11.14. Consolidation curves of the lacustrine clay layers whose isotache set has been presented in Fig. 11.12. For notation see Fig. 11.13. Reproduced after Šuklje, 1957

11.1.7 Discussion of the Isotache Method

When defining the isotaches in order to analyse the one-dimensional compression of oedometer tests, the consolidation speed has been assumed to be independent of the change in the effective principal stress ratio occurring during consolidation. The velocity of compression has been supposed to be a function only of the void ratio and of the main principal effective pressure. As a result of the experimental evidence, however, it depends also on the rate of load increase and on the character of the preceding loading.

The effect of a sudden load application increasing the initial consolidation speed of samples whose structural resistance is small was discussed and interpreted in the preceding section.

Moreover the consolidation speeds of pre-consolidated samples are greater than corresponding to the isotache set resulting from the consolidation curves observed by the first loading of soils consolidated from the liquid limit. In pre-consolidated soils the structural resistance of the soil skeleton to the transition to the porosity attained at the previous load is evidently smaller than at first loading from the same void ratio. Thus the isotache set of a pre-consolidated soil in the range below the pre-consolidation pressure is expected to be different from the set corresponding to the normal consolidation of the same soil.

In the extreme case of pure hydrodynamic consolidation following Terzaghi's theory all isotaches condense to a single straight line corresponding to any consolidation speed imposed by the drainage gradients of the isochrones. For any actual intergranular stress state represented by a point $(\bar{\sigma}', \bar{e})$ on the line $\bar{\sigma}' = \bar{\sigma}'(\bar{e})$ complete equilibrium is reached. As structural resistance does not exist in this case, the transition of grains to new equilibrium positions occurs with a speed of effective pressure increase as given by the hydrodynamic law. Thus, this special case is also included in the general consolidation scheme of the isotache method.

Owing to our excluding the role of lateral pressures the isotaches as defined and applied in the preceding sections cannot represent completely the behaviour of soils under the conditions of confined compression tests. Nevertheless the accepted definition has proved to be a useful working hypothesis.

The isotache set facilitates a clear interpretation of the influence exerted by former geological pre-consolidation on the speed of compression due to loading by a structure. Assume, for example, an actual overburden effective pressure of 4 kp/cm^2 applied, in a vertical direction, to the lacustrine chalk, whose consolidation isotaches were presented in Fig. 11.11 (*c*). The layers were not pre-consolidated and the actual consolidation state may be represented by a void ratio of $e = 0{\cdot}62$. If a surcharge of 1 kp/cm^2 is applied the character and the speed of consolidation will naturally depend on the thickness of the layer. In any case, however, the initial speed cannot be smaller than $5{\cdot}10^{-10}$ sec^{-1} or greater than $2{\cdot}10^{-7}$ sec^{-1}. Now, if the previous consolidation at $\sigma' = 4$ kp/cm^2 advanced in such a way that a void ratio of 0·60 was reached before the load increase of 1 kp/cm², then the initial consolidation speed should be within the interval from $7 \cdot 10^{-14}$ to $4 \cdot 10^{-11}$ sec^{-1} (see the isotache set). Consequently, over a certain period of, for example, 100 years, the previous consolidation will affect the settlement of the layer by reducing not only its magnitude but also its consolidation speed.

If the same soil was previously subjected to a geological loading exceeding the assumed one (for example, to the weight of gravel deposits which were subsequently eroded away), the soil will be denser because the prevailing plastic deformations remain after unloading. If, for example, a pre-consolidation vertical pressure of 8 kp/cm² decreasing the void ratio to 0·59 (see Fig. 11.11 (*c*)) is assumed and if the reversible part of the deformation at unloading to 4 kp/cm² is disregarded, the initial consolidation speed at reloading from 4 to 5 kp/cm² would be less than $5 \cdot 10^{-13}$ sec^{-1}. The speed would decrease during further consolidation. Now, taking account of some void ratio increase during reloading and particularly taking account of the change in the isotache set as interpreted before, the true consolidation speed is expected to be greater. However, the consolidation cannot develop faster than would be expected according to Terzaghi's theory where the coefficient of volume change is defined by the straight line connecting the points of the actual and of the preceding porosity and stress [see the diagram $\bar{\sigma}' = \bar{\sigma}'(\bar{e})$]. After reaching the preceding void ratio, further consolidation is governed again by the primary isotache set.

Similar conclusions can be drawn from the sets of secondary compression lines

$$e = [e(\sigma')]_{t=\text{const.}}$$

as introduced by Taylor (1942) and applied for the computation of the 'delayed consolidation' by Bjerrum (1967). The author believes that it is an advantage of the isotache sets to allow the consolidation analysis by using the same principle in the primary as well as in the secondary phase.

Fig. 11.15 represents the isotache set deduced from the compression lines $e = [e(\sigma')]_{t=\text{const.}}$ of a normally consolidated marine clay (see Bjerrum, 1967, Fig. 14). This isotache set was used in order to obtain the lines $e = e(t)$ of an 8·80 m thick layer of the plastic marine clay subjected to an average increase of vertical effective stresses from 9·1 to 10·6 t/m² (corresponding to the soil and loading data of the case record—Scheitlies gate—presented in Fig. 26 of Bjerrum's Rankine Lecture). The following alternative assumptions have been considered: initial average void ratio: $\bar{e}_0 = 1 \cdot 38$, $\bar{e}_0 = 1 \cdot 345$; average coefficient of permeability:

$$\bar{k} = 10^{-7} \text{ cm sec}^{-1}, \qquad \bar{k} = 10^{-8} \text{ cm sec}^{-1}$$

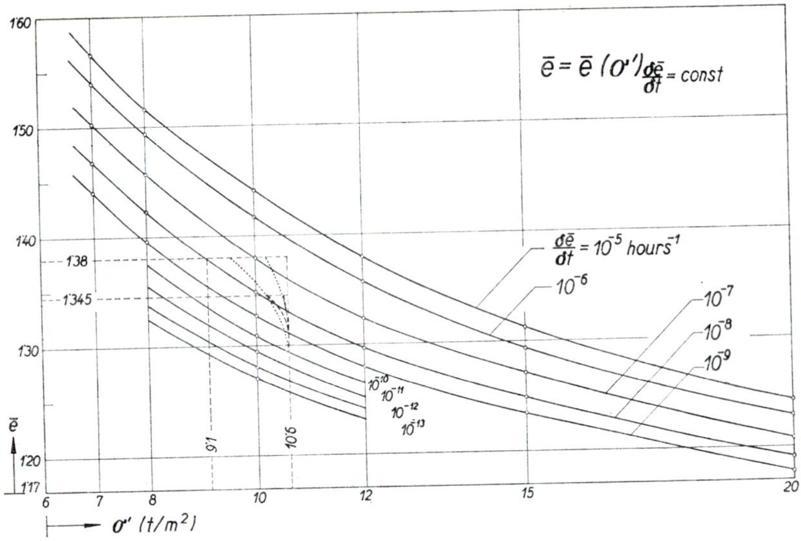

Fig. 11.15. Isotache set of a normally consolidated marine clay

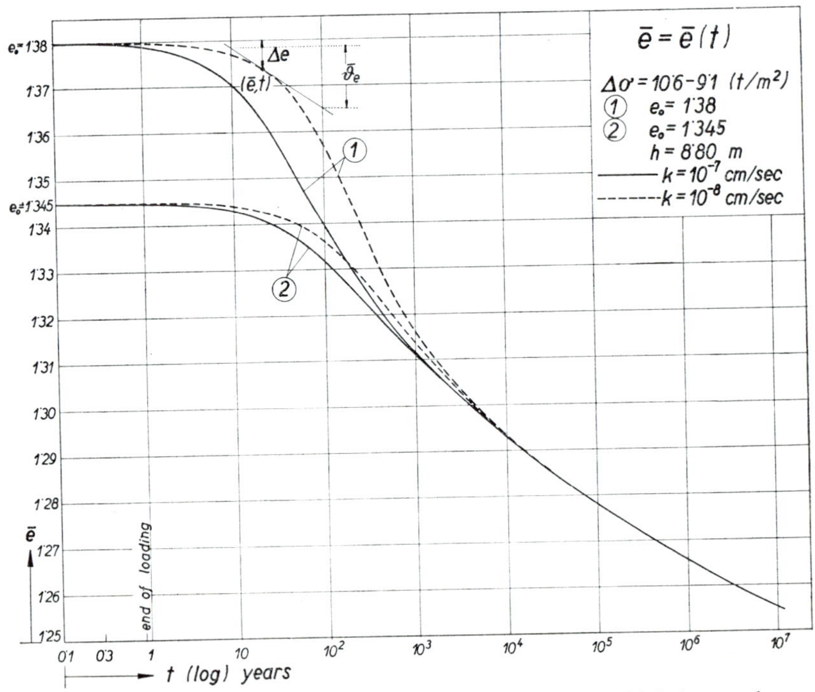

Fig. 11.16. Consolidation lines $\bar{e} = \bar{e}(t)$ of an 8·80 m thick layer whose isotache set is presented in Fig. 11.15; drainage has been assumed to occur only in the upward direction

The resulting consolidation lines $\bar{e} = \bar{e}(t)$ are shown in Fig. 11.16 and pore-pressure dissipation lines $\bar{u} = \bar{u}(t)$ in Fig. 11.17.

Isochrones in the form of parabolas of second degree were assumed; $\bar{u}$ denotes the average pore-pressure of the isochrone. The results prove

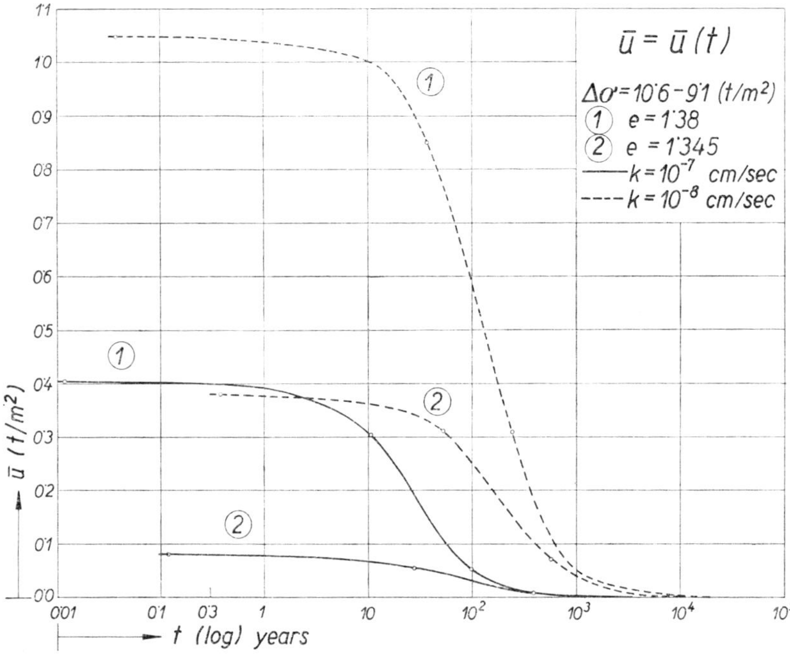

Fig. 11.17. Pore-pressure dissipation lines corresponding to the consolidation lines $\bar{e} = \bar{e}(t)$ presented in Fig. 11.16

the remarkable influence of the initial void ratio and of the coefficient of permeability onto the pore-pressure dissipation and the development of the consolidation.

11.1.8 Relaxation Effects at Linear Consolidation. Swelling Pressures

In Taylor's Consolidation Theory *B* the bond p_b (see Fig. 10.9) represents the release in effective stress needed to terminate the compression at the end of primary consolidation.

A more general illustration of relaxation effects is given by isotache sets. Consider, for example, the isotaches presented in Fig. 11.11 (*c*). In the position $\bar{\sigma}' = 8$ kp/cm², $\bar{e} = 0{\cdot}59$ the soil creeps with speed $\partial\bar{e}/\partial t = 3{\cdot}5 \cdot 10^{-8}$ sec^{-1}. In order to reduce the creep velocity to $\partial\bar{e}/\partial t \leqslant 10^{-13}$ sec^{-1} which practically means a complete end to further deformation, the effective pressure $\bar{\sigma}'$ must be reduced to $\bar{\sigma}' \leqslant 4{\cdot}7$ kp/cm². In laboratory test conditions, however, it would hardly be possible to adapt the pressure relaxation to creep speeds less than 10^{-10} sec^{-1} and the apparent relaxation effect corresponding to the complete cessation of movement would seem to be smaller.

The isotache set can also be used to give information on stress relaxation effects corresponding to an arbitrary reduction of the creep speed. Nevertheless, it may be pointed out again (see Section 11.1.6) that pre-consolidation affects the isotache set and that the isotaches corresponding to virgin

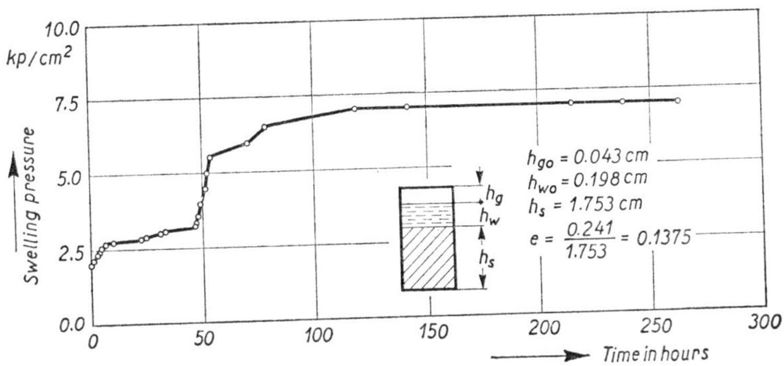

Fig. 11.18. Swelling pressure versus time plots of a partly saturated clayey sample at constant void ratio $e = 0{\cdot}1375$

consolidation curves do not represent well the behaviour of pre-consolidated soils.

Thus, the isotache set obtained from the consolidation curves of adequately pre-consolidated samples should be applied when analysing relaxation effects.

Pre-consolidated (or over-consolidated) saturated or partly saturated clays swell when exposed to moisture or when flooded. The swelling is partly due to elastic reversible deformations of the soil skeleton appearing as the capillary forces decrease when the capillary water menisci vanish. The main part of the swelling displacements, however, is to be attributed to the passage (suction) of water from the domain of positive pressures into the voids having negative pore pressures.

If the swelling is prevented the stresses increase. The pressure that reduces the speed of the creep or terminates it completely is called the swelling pressure.

Figure 11.18 presents the gradual increase of swelling pressure of a partly saturated sample of a Tertiary marine clay exposed to the effect of a humid atmosphere only. The void ratio of the sample was $e = 0{\cdot}1375$ and the initial degree of saturation $S_r = 0{\cdot}82$. During a period of 120 hours the swelling pressure increased until it reached the constant value of 7 kp/cm^2.

In laboratory tests even higher swelling pressures can be observed and their effects demonstrated in retaining constructions. The greater the possibility of displacement of the construction the smaller are the swelling pressures.

11.2 CONSOLIDATION ANALYSIS BASED ON THE LOGARITHMIC CREEP FUNCTION

11.2.1 The Derivation of Brinch Hansen's Model Law of Simultaneous Primary and Secondary Consolidation

Brinch Hansen's (1961) model law for simultaneous primary and secondary consolidation applies for the case of one-dimensional consolidation in which an extensive, homogeneous clay layer of thickness $2h$ (drained at both sides) or h (drained at one side only) is subjected to a vertical stress increase $\Delta\sigma \equiv p$ throughout the layer. In the initial phase of consolidation Brinch Hansen's law approaches the relation

$$\bar{\sigma} = \beta\sqrt{t} \tag{11.27}$$

which is a good approximation of pure primary consolidation in the domain $T_v \leqslant 0{\cdot}4$ (see Section 9.4.1), while for greater values of time t the law covers the following equation of pure secondary consolidation:

$$\varepsilon_\infty = \frac{p}{K_s} \log \frac{t + t_s}{t_s} \tag{11.28}$$

In the above equations $\bar{\sigma}$ denotes the mean value of the additional effective pressures, β a constant of proportionality, t_s and K_s the constants of the secondary consolidation and ε_∞ the strain in the secondary phase.

The secondary consolidation is assumed to increase with the logarithm of time and to develop simultaneously with the primary consolidation.

The increment $d\bar{\sigma}_0$ of the effective pressure at time t_0 causes a compression strain $d\varepsilon$ at any further time t:

$$d\varepsilon = \frac{d\bar{\sigma}_0}{K_s} \cdot \log \frac{t_s + t - t_0}{t_s} \tag{11.29}$$

The modulus K_s and the time t_s are characteristic constants of the clay at the given state. Integration of the above equation yields the total strain at time t:

$$\varepsilon = \frac{1}{K_s} \int_{t_0=0}^{t} \log \frac{t_s + t - t_0}{t_s} \cdot \frac{d\bar{\sigma}_0}{dt_0} \, dt_0 \tag{11.30}$$

Inserting the relation (11.27) for $\bar{\sigma}_0$ and neglecting t_s (a small value commonly less than 1 sec) when added to t, we obtain, after integration and appropriate transformation, the expression:

$$\varepsilon \cong \frac{\bar{\sigma}}{K_s} \log \frac{4t}{e^2 t_s}, \quad e = 2{\cdot}71828 \tag{11.31}$$

If the consolidation follows the Terzaghi theory, the degree of consolidation U_v, smaller than 0·50, can be expressed with sufficient accuracy by the equation:

$$U_v = \sqrt{\left(\frac{4}{\pi} T_v\right)} = \sqrt{\frac{t}{t_v}} \tag{11.32}$$

where

$$t_v = \frac{\pi \gamma_w h^2}{4kM_v} \tag{11.33}$$

k denotes the coefficient of permeability and M_v the modulus of linear compressibility (tangent modulus). The degree of consolidation U_v is defined by the expression:

$$U_v = \frac{\varepsilon}{\varepsilon_c} = \frac{\bar{\sigma}}{p} \tag{11.34}$$

ε denoting the compression strain at time t and ε_c the final compression. Thus, Equation (11.32) can be written in the form:

$$\varepsilon = \varepsilon_c \sqrt{\frac{t}{t_v}} \tag{11.35}$$

Differentiating with respect to time (t) we get:

$$\frac{\mathrm{d}\varepsilon}{\mathrm{d}t} = \frac{\varepsilon_c}{2\sqrt{(tt_v)}} \tag{11.36}$$

Respecting the definition:

$$\varepsilon_c = \frac{p}{M_v} \tag{11.37}$$

and inserting

$$\sqrt{\frac{t_v}{t}} = \frac{p}{\bar{\sigma}}$$

(see Equations 11.32 and 11.34), Equation (11.36) becomes:

$$\frac{\mathrm{d}\varepsilon}{\mathrm{d}t} = \frac{p^2}{2M_v t_v \bar{\sigma}} \tag{11.38}$$

Inserting, further, $M_v t_v$ from Equation (11.33) we get:

$$\frac{\mathrm{d}\varepsilon}{\mathrm{d}t} = \frac{p^2 c_s}{2K_s h^2 \bar{\sigma}} \tag{11.39}$$

in which the symbol c_s denotes:

$$c_s = \frac{4kK_s}{\pi\gamma_w} \tag{11.40}$$

Now, the expression (11.31) may be differentiated with respect to t and then equated to expression (11.39). The resulting differential equation:

$$\frac{\mathrm{d}\bar{\sigma}}{\mathrm{d}t} \cdot \frac{1}{K_s} \log \frac{4t}{e^2 t_s} + \frac{\bar{\sigma}}{K_s} (\log e) \frac{1}{t} = \frac{1}{\bar{\sigma}} \cdot \frac{p^2 c_s}{2K_s h^2} \tag{11.41}$$

has the solution:

$$\bar{\sigma} = p\sqrt{\left(\frac{c_s t}{h^2}\right)} \cdot \frac{\sqrt{\left(\log \frac{4t}{e^3 t_s}\right)}}{\log \frac{4t}{e^2 t_s}} \tag{11.42}$$

By inserting the above expression into Equation (11.31) we obtain the expression for the initial phase of consolidation:

$$\varepsilon_0 \cong \frac{p}{K_s}\sqrt{\left(\frac{c_s t}{h^2} \log \frac{t + 5t_s}{5t_s}\right)} \tag{11.43}$$

In this expression, in order to avoid negative values of $\log(t/5t_s)$ at $t < 5t_s$, a very small value $5t_s$ has been added to the time t of the logarithmic factor.

Now, expression (11.43) governing the first part of the primary consolidation (moderate values of time t), as well as expression (11.28) governing the secondary consolidation (large values of t) are combined with satisfactory accuracy in the following general expression (Brinch Hansen, 1961):

$$\varepsilon = \frac{p}{K_s} : \sqrt[6]{\left\{\frac{h^6}{c_s^3} : \left[t \log \frac{t + 5t_s}{5t_s}\right]^3 + 1 : \left[\log \frac{t + t_s}{t_s}\right]^6\right\}} \qquad (11.44)$$

The parameters K_s, c_s and t_s appearing in this equation can be determined in the following way:

The first part of the consolidation curve may be plotted against the square root of time ($\sqrt{t}$) and the second part against the logarithm of

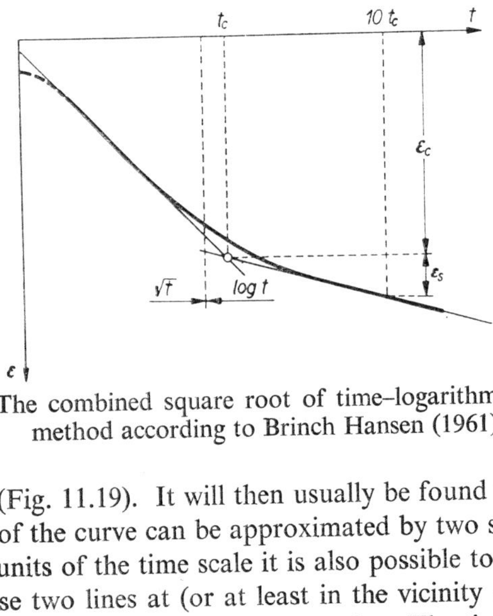

Fig. 11.19. The combined square root of time–logarithm of time fitting method according to Brinch Hansen (1961)

time ($\log_{10} t$) (Fig. 11.19). It will then usually be found that the first and the last parts of the curve can be approximated by two straight lines. By changing the units of the time scale it is also possible to obtain the intersection of these two lines at (or at least in the vicinity of) the boundary ordinate between the $\sqrt{t}$ and the $\log_{10} t$ scales. The time curve will then furnish us with the three observed quantities t_c, ε_c and ε_s, as defined in Fig. 11.19.

By using these quantities, the parameter t_s can be determined to first approximation by the equation

$$\log \frac{t_c}{t_s} \cong \frac{\varepsilon_c}{\varepsilon_s} - \left[\frac{1{\cdot}1\ \varepsilon_s}{\varepsilon_c}\right]^2 - 0{\cdot}13 \qquad (11.45)$$

while a more accurate result can be obtained from the expression:

$$\frac{\varepsilon_c}{\varepsilon_s} = \frac{A + 0{\cdot}434}{A + 0{\cdot}297} B \tag{11.46}$$

$$A = \log \frac{t_c + 50t_s}{50t_s}$$

$$B = \log \frac{t_c}{t_s}$$

The parameter c_s can then be found by the equation:

$$c_s \cong \frac{A}{t_c}\left[\frac{Bh_0}{A + 0{\cdot}297}\right]^2 \tag{11.47}$$

and the parameter K_s by the equation:

$$K_s = \frac{p}{\varepsilon_s} \tag{11.48}$$

Inserting the values c_s and K_s obtained above Equation (11.40) yields the coefficient of permeability k. (For the derivation of expressions 11.45 to 11.48 see Brinch Hansen, 1961.)

11.2.2 Comparison with the Classical Solution

By equating expressions (11.32) and (11.34) for the degree of consolidation U_v and by inserting expression (11.33) for t_v we obtain:

$$\varepsilon = \varepsilon_c \sqrt{\left(\frac{4kM_v}{\pi\gamma_w h^2} t\right)} \tag{11.49}$$

Following the classical theory this approximate expression for the first part of the primary consolidation can now be compared with the corresponding expression (11.43) of Brinch Hansen's theory. Equating the two expressions and inserting Equation (11.48) we get:

$$M_v = \frac{p\varepsilon_s}{\varepsilon_c^{\,2}} \log \frac{t + 5t_s}{5t_s} \tag{11.50}$$

Thus the modulus of linear compression applying for creep effects depends not only on the strain ε_c corresponding to the end of primary consolidation of the oedometer sample, but also on the logarithmic decade increment ε_s of the secondary consolidation and on the time t.

From the oedometer consolidation curve given in Fig. 11.14 (an undisturbed clay sample, $n_c = 1$) the consolidation curves corresponding to various layer thicknesses have been derived by using Brinch Hansen's method. The resulting consolidation curves (see Šuklje, 1961, Fig. 28) are comparable with the curves obtained by the isotache method and are close to the Terzaghi curves extending into the extrapolated logarithmic lines of secondary compression.

Brinch Hansen's model law requires zero additional effective pressures $\bar{\sigma}$ at the very beginning of consolidation when $t = 0$ (see Equation 11.27). Thus, it cannot be applied when the plastic resistance gives rise to a faster pore pressure dissipation of the sample (e.g. in cases presented in Figs. 11.11 and 11.13).

11.3 SOIL AS A NON-LINEAR KELVIN BODY FOR THE SPHERICAL AND THE DISTORTIONAL STRESS–STRAIN RELATIONSHIPS

Stroganov (1963) treats the soil as a non-linear viscoelastic medium. The relation between spherical and distortional components of the stress and strain tensors is based on the rheological model of a generalized Kelvin body in which the modulus of compression K as well as the shear modulus G are considered to be non-linear functions of the state of stress. The fundamental equations are:

(*a*) for the distortional part of the stress and strain tensor:

$$\tau_{\mathrm{oct}} = G\gamma_{\mathrm{oct}} + \eta^{d}\dot{\gamma}_{\mathrm{oct}} \tag{11.51}$$

$$G = \frac{\bar{G}\mu}{\mu + \bar{G}\gamma_{\mathrm{oct}}}(H + \sigma_{\mathrm{oct}}) \tag{11.52}$$

(*b*) for the spherical part:

$$\sigma_{\mathrm{oct}} = K\varepsilon_{v} + \eta\dot{\varepsilon} \tag{11.53}$$

$$K = \frac{\alpha(\varepsilon_0 + \varepsilon_v)^{\beta} - \sigma_0}{\varepsilon_v} \tag{11.54}$$

In the above equations the common symbols τ_{oct}, σ_{oct} and γ_{oct} have been used for shear and normal octahedral stress and for shear octahedral strain respectively, and the following further notation has been employed:

μ = $\tan\psi$ = coefficient of friction on octahedral area
H, σ_0 = constant hydrostatic pressures representing the effect of the cohesion
$\bar{G}H$ = 'initial shear modulus'

ε_v = volume strain corresponding to the octahedral pressure σ_{oct}
ε_0 = volume strain corresponding to the pressure σ_0
η, η^d = coefficients of the spherical and distortional viscosity respectively
$\dot{\varepsilon}$ = velocity of volume strain

Taking into account the effect of dilatation, i.e. of the volume strain due to the distortional stress increment, the total volume strain may be expressed in the form:

$$\varepsilon = \varepsilon_v - \nu\gamma_{oct} \qquad (11.55)$$

Based on the above fundamental relations and superimposing the spherical and distortional components of the stress tensor, Stroganov has derived the following expressions which are presented in explicit form:

$$\sigma_x = \left[1 + \frac{2\bar{G}\tan\psi}{\tan\psi + \bar{G}\gamma_{oct}}\left(\varepsilon_x + \frac{\varepsilon}{3}\right)\right]\{H + \alpha[(\varepsilon_0 + \varepsilon) + \nu\gamma_{oct}]^\beta - \\ - \sigma_0 + \eta\dot{\varepsilon}\} + 2\eta^d\left(\dot{\varepsilon}_x - \frac{\dot{\varepsilon}}{3}\right) - H \qquad (11.56)$$

and similarly for σ_y, σ_z;

$$\tau_{xy} = \frac{\bar{G}\tan\psi}{\tan\psi + \bar{G}\gamma_{oct}}\{H + \alpha[(\varepsilon_0 + \varepsilon) + \nu\gamma_{oct}]^\beta - \\ - \sigma_0 + \eta\dot{\varepsilon}\}\gamma_{xy} + \eta^d\dot{\gamma}_{xy} \qquad (11.57)$$

and similarly for τ_{xy}, τ_{yz}.

Stroganov has applied the above general equations to one-dimensional deformation. Among the resulting expressions the equations determining the coefficients of the initial and final lateral pressures of the confined compression tests may be cited:

$$K_0 = \frac{\sigma_3}{\sigma_1} = \frac{1 - \frac{2}{3}\bar{G}H\dfrac{\varepsilon_0}{\sigma_0\left(1 + \frac{2}{\sqrt{3}}\nu\right)}}{1 + \frac{4}{3}\bar{G}H\dfrac{\varepsilon_0}{\sigma_0\left(1 + \frac{2}{\sqrt{3}}\nu\right)}} \qquad (11.58)$$

$$K_\infty = \frac{\sigma_3}{\sigma_1} = \frac{1 - \dfrac{\sqrt{3}}{3}\tan\psi}{1 + \dfrac{2\sqrt{3}}{3}\tan\psi} \qquad (11.59)$$

Thus, for non-cohesive soils ($H = 0$) the Stroganov theory requires the initial coefficient of lateral pressure to be equal to unity (hydrostatic state of stress) whilst the final value of this coefficient depends only on the coefficient of friction for cohesive as well as for non-cohesive soils.

Combining the above stress–strain relationship with the differential equation of one-dimensional consolidation (Equation 10.20) Stroganov (1963) also investigated the development of consolidation of oedometer samples of saturated soils. For the initial distribution of pore pressures

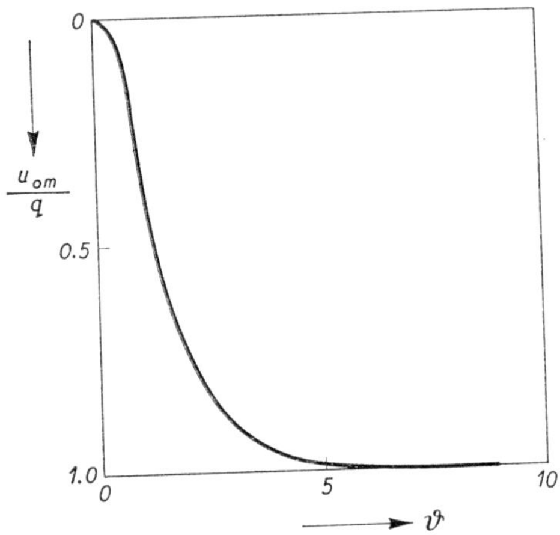

Fig. 11.20. The maximum ordinate u_{om} of the zero isochrone as a function of the coefficient $\vartheta = h\sqrt{\frac{\gamma_w}{k(\eta + \frac{4}{3}\eta^d)}}$ according to Stroganov (1963)

the following simple analytical expression showing the dependence of the zero isochrone on the viscosity as well as on the thickness of the layer was obtained:

$$\frac{u_0}{q} = 1 - \frac{\cosh z\sqrt{\left(\frac{\gamma_w}{k(\eta + \frac{4}{3}\eta^d)}\right)}}{\cosh h\sqrt{\left(\frac{\gamma_w}{k(\eta + \frac{4}{3}\eta^d)}\right)}} \tag{11.60}$$

For the definition of h and z see Fig. 10.3. By limiting $\eta \to \infty$ the equation reduces to the expression $u_0 = q$ for the zero isochrone in agreement with Terzaghi's theory. When $\eta \neq \infty$ the maximum ordinate of the isochrone can be obtained by inserting $z = 0$:

$$\frac{u_{0m}}{q} = 1 - \frac{1}{\cosh h \sqrt{\left[\frac{\gamma_w}{k(\eta + \frac{4}{3}\eta^d)}\right]}} \tag{11.61}$$

In Fig. 11.20 the term u_{0m}/q is plotted against the dimensionless parameter $\vartheta = h\sqrt{[\gamma_w/k(\eta + 4/3\ \eta^d)]}$. The diagram proves that the consolidation of a thin oedometer sample can start with certain additional effective pressures ($u_{0m}/q < 1$). With increasing layer thickness u_{0m}/q approaches unity and the consolidation curve becomes closer to that resulting from the Terzaghi theory. For the linear Kelvin body similar effects occur in Taylor's (1942) and Florin's (1961) solutions. When interpreting the consolidation process by the isotache method (see section 11.1) these effects are shown to appear in soils of moderate permeability and high viscous structural resistance.

11.4 SOIL AS A KELVIN BODY WITH NON-LINEAR VISCOSITY

Laing Barden (1965b) expressed the axial effective stress–strain relationship of a cylindrical, laterally confined, body of soil by a Kelvin rheological model of non-linear viscosity. In the equation for one-dimensional consolidation

$$\frac{\partial}{\partial z}\left[\frac{k(1+e)}{\gamma_w} \cdot \frac{\partial u}{\partial z}\right] = \frac{\partial e}{\partial t} \tag{11.62}$$

the coefficient of permeability k and the factor $(1 + e)$ are assumed to be constant over a small pressure increment. The compressibility of the skeleton is also assumed to be constant. Consequently, Equation (11.62) reduces to:

$$c\frac{\partial^2 u}{\partial z^2} = \frac{\partial e}{\partial t} \tag{11.63}$$

where c denotes a constant coefficient

$$c = \frac{k(1+e)}{\gamma_w} \tag{11.64}$$

Referring to Fig. 11.21 the relation can be stated as:

$$\frac{e_1 - e_2}{a} = \frac{e_1 - e}{a} + \tau + u$$

Thus:

$$\frac{e - e_2}{a} = \tau + u \tag{11.65}$$

where a is the assumed linear compressibility, τ is the viscous resistance of the skeleton of the clay and u is the excess pore-pressure.

The non-linearity of the viscous resistance has been expressed by the power law of Ostwald in the form:

$$\tau = - b\left(\frac{\partial e}{\partial t}\right)^{\frac{1}{n}} \tag{11.66}$$

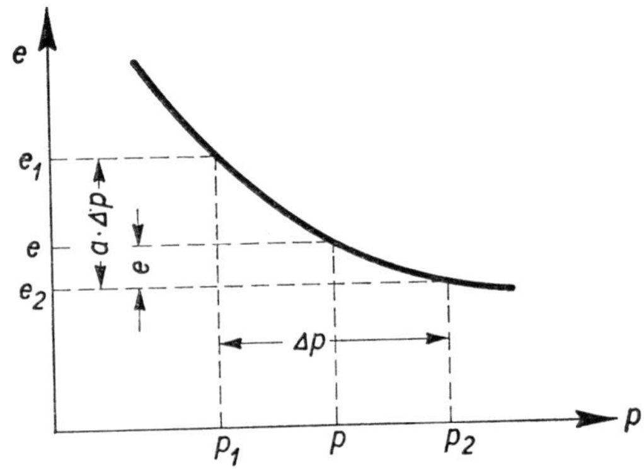

Fig. 11.21. Notation concerning the compression curve in Laing Bardens consolidation theory

where $n > 1$ and b is assumed constant over a pressure increment. Equation (11.65) can therefore be rewritten:

$$e - e_2 = a\left[u - b\left(\frac{\partial e}{\partial t}\right)^{\frac{1}{n}}\right] \tag{11.67}$$

The dimension of the coefficient of viscosity b depends on the value of n. The case $b = 0$ leads to the Terzaghi relation $de = a \,.\, du$.

Equations (11.63) and (11.67) have now to be solved for the usual boundary conditions of one-dimensional consolidation. The solution is obtained in dimensionless form by using the following substitutions:

(*a*) $e' = e - e_2$ and $\dfrac{\partial e'}{\partial t} = \dfrac{\partial e}{\partial t}$;

(*b*) $x = z/H$ where H is the length of the drainage path;

(*c*) $u' = u/\Delta p$ where $\Delta p = p_2 - p_1$ is the pressure increment shown in Fig. 11.18;

(*d*) $(e_1 - e)/(e_1 - e_2) = \mu =$ the degree of consolidation and

$$\lambda = 1 - \mu = \frac{e - e_2}{e_1 - e_2} = \frac{e'}{a\Delta p}$$

(*e*) $T = (c/a) \,.\, (t/H^2) =$ Terzaghi's time factor.

By using these symbols the equation of consolidation (11.63) can be written in the form:

$$\frac{\partial^2 u'}{\partial x^2} = \frac{\partial \lambda}{\partial T} \tag{11.68}$$

and Equation (11.67) as follows:

$$\lambda = u' - \frac{b}{\Delta p}\left(\frac{\partial e'}{\partial t}\right)^{\frac{1}{n}}$$

yielding

$$\frac{\partial \lambda}{\partial t} = -\frac{1}{ab^n}\,\Delta p^{n-1}\,(\lambda - u')^n \tag{11.69}$$

Substituting

$$\frac{cb^n}{H^2(\Delta p)^{n-1}} = R$$

and

$$\frac{c}{a} \cdot \frac{t}{H^2} = T$$

Equation (11.69) becomes:

$$\frac{\partial \lambda}{\partial T} = -\frac{1}{R}(\lambda - u')^n \tag{11.70}$$

Two different methods of numerical integration have been applied in order to solve the simultaneous equations (11.68) and (11.70). In both methods the conventional finite difference approximations have been employed

$$\left(\frac{\partial^2 u'}{\partial x^2}\right)_{i,j} = \frac{u'(i-1,j) + u'(i+1,j) - 2u'(i,j)}{\Delta x^2} \tag{11.71}$$

$$\left(\frac{\partial \lambda}{\partial T}\right)_{i,j} = \frac{\lambda(i,j+1) - \lambda(i,j)}{\Delta T} \tag{11.72}$$

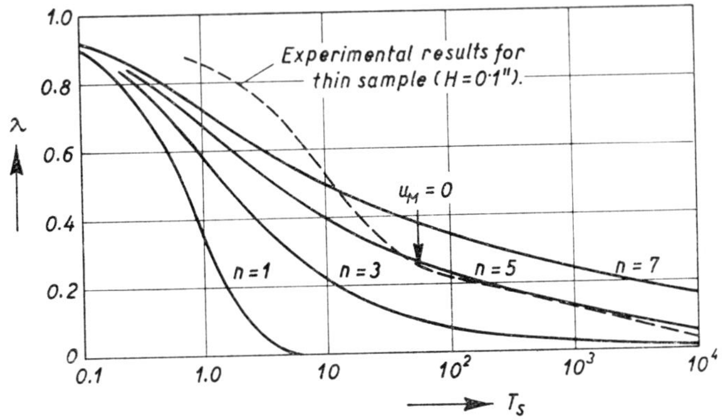

Fig. 11.22. Consolidation curves $\lambda = \lambda(T_s)$ according to Laing Barden for various parameters n; reproduced after Laing Barden (1965b)

The creep behaviour of the soil skeleton when loaded by an increment of effective stress has been obtained from Equation (11.70) by putting $u' = 0$. The equation can then be integrated directly to give:

$$\lambda = \left[\left(\frac{n-1}{R}\right)T + 1\right]^{\frac{1}{1-n}} \tag{11.73}$$

Substituting

$$\frac{T}{R} = \frac{\dfrac{c}{a}\cdot\dfrac{t}{H^2}}{\dfrac{cb^n}{H^2(\Delta p)^{n-1}}} = \frac{t(\Delta p)^{n-1}}{ab^n} = T_s \tag{11.74}$$

the coefficient λ has been plotted against T_s, in Fig. 11.22 (reproduced after Laing Barden), for various values of n.

As an example of the numerical results of integrating Equations (11.68) and (11.70) the isochrones of λ and u' have been plotted in Fig. 11.23 (reproduced after Laing Barden) for the parameters $n = 5$ and $R = 10^{-4}$.

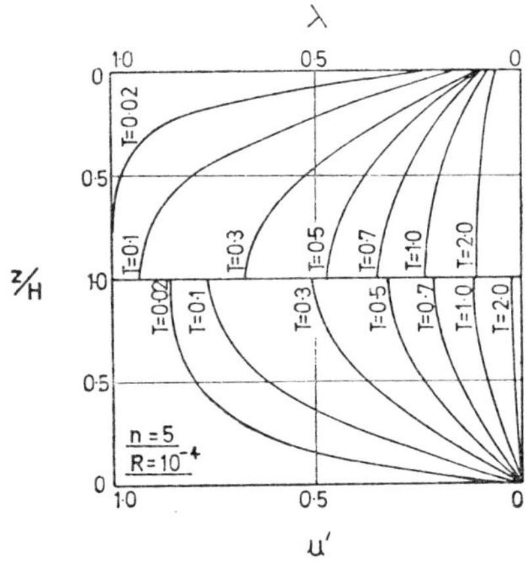

Fig. 11.23. Dissipation (u') and consolidation (λ) ratios plotted versus the depth ratio according to Laing Barden's theory; $n = 5$, $R = 10^{-4}$; reproduced after Laing Barden (1965b)

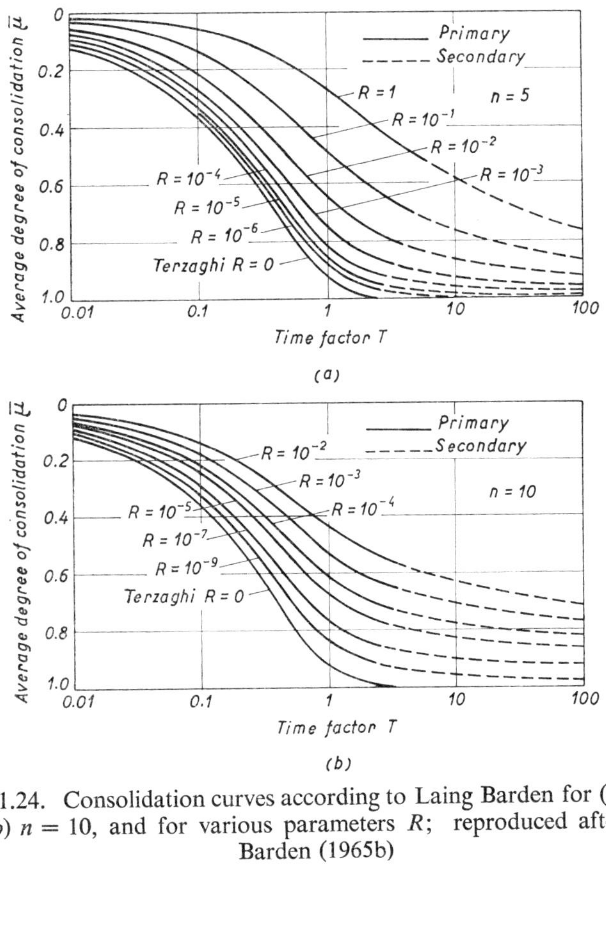

Fig. 11.24. Consolidation curves according to Laing Barden for (*a*) $n = 5$ and (*b*) $n = 10$, and for various parameters R; reproduced after Laing Barden (1965b)

It can be seen that the pore-pressure decays more rapidly than the void ratio, whereas in the Terzaghi solution the isochrones of λ and u' are identical.

'Using the isochrones for λ the mean value $\bar{\lambda}$ can be calculated for any value of T, and hence the average degree of consolidation $\bar{\mu} = 1 - \bar{\lambda}$.

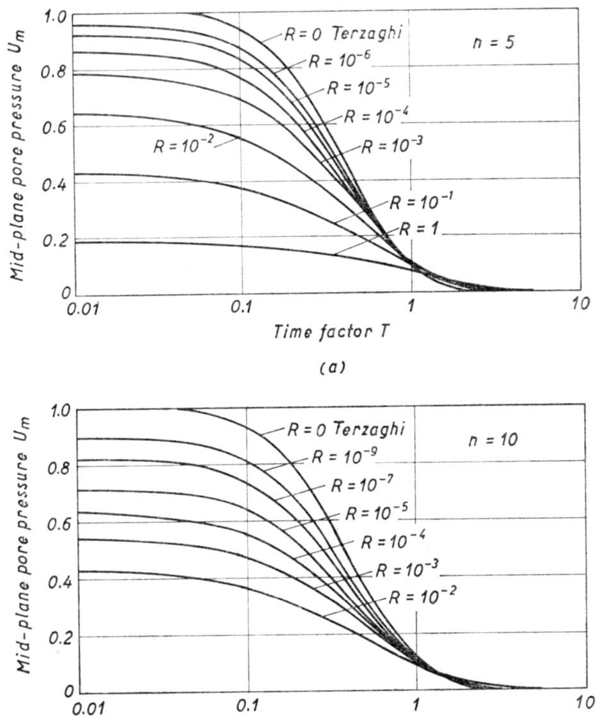

Fig. 11.25. Dissipation curves according to Laing Barden for (*a*) $n = 5$ and (*b*) $n = 10$, and for various parameters R; reproduced after Laing Barden (1965b)

The resulting settlement—log T relations for $n = 5$ and $n = 10$ have been plotted in Fig. 11.24 (*a*) and (*b*) respectively for a range of R values. In these plots the point at which the mid-plane pore-pressure falls below the arbitrary value of $u_m' = 0{\cdot}01$ is taken to be the end of the primary stage and the subsequent secondary or creep deformation is plotted by a broken line. It can be seen that at one limit $R = 0$ gives the Terzaghi solution which is all primary and no secondary. As R increases so does the proportion of secondary.' (Laing Barden, 1965b.)

The rates of dissipation of u_n for $n = 5$ and 10 have been plotted in Figs. 11.25 (*a*) and (*b*) (reproduced after Laing Barden, 1965b).

11.5 GENERAL DISCUSSION OF NON-LINEAR STRESS–STRAIN RELATIONSHIPS IN THE CONSOLIDATION ANALYSIS

The solutions of Stroganov and Laing Barden are two examples of the application of non-linear rheological models in the consolidation analysis. In Stroganov's model the non-linearity concerns the elastic moduli while the coefficients of viscosity are kept constant. On the other hand, Laing Barden's model consists of a linear Hookean spring and a non-linear Newtonian dashpot. Although in both solutions the soil is treated as a laterally confined cylindrical body subjected to one-dimensional drainage along the axis of the cylinder, Stroganov takes the lateral pressures into account while Laing Barden's deformability parameters are related to the axial stresses only.

In Laing Barden's theory the consolidation process is governed by three deformability parameters (a, b, n), by the permeability (k), porosity (e), thickness of the layer (H) and by the load increment (Δp). The theory can explain the main experimental departures from the Terzaghi theory, namely the greater rate of dissipation of pore pressures and the creep effects during and after the dissipation of pore pressures, as well as the dependence of the ratio of primary to secondary compression on the pressure increment ratio and on the thickness of the sample. The theoretical plots presented by Laing Barden (1965b) and their comparison with some experimental data published so far seem, however, to show that in his theory either the viscous effects in the primary consolidation of thick layers are too important or the secondary consolidation is too small and of too short duration.

In Stroganov's solution the consolidation process is governed by eleven independent parameters. Yet, the soil is still considered to be homogeneous and isotropic. Therefore the important effects of structural and deformation anisotropy are not taken into account. Thus, in the author's opinion, the great number of physical parameters to be determined by appropriate tests as well as the mathematical difficulties in solving the stress–strain relationships in non-linear media and in applying the fundamental solutions to technical problems, restrict both the applicability of non-linear models and the possibility of obtaining a general consolidation theory which could be usefully applied in practice.

Brinch Hansen (1961) combined the logarithmic creep function with the Terzaghi consolidation theory. By using reasonable approximations he succeeded in deducing a general expression governing the primary as well as the secondary consolidation. When applying this expression to the consolidation of thicker layers, viscous effects are proved to affect particularly the final value of the primary consolidation. Its development approaches Terzaghi's consolidation curve the final settlement or void ratios of which are close to the extended line of secondary compression observed in oedometer tests. This extension of the observed line of secondary compression becomes the asymptotic line of the primary consolidation curves corresponding to any layer thickness. The applicability of the method, however, is limited to the loading, displacement and drainage conditions of oedometer tests. Furthermore, Brinch Hansen's assumptions do not allow for samples whose consolidation starts with pore pressures less than the total suddenly applied load increment.

The isotache method of the consolidation analysis which was developed by the present author (Šuklje, 1957), allows for any law of development of secondary consolidation provided that the law can be ascertained approximately from extrapolation of the observed consolidation curve of the oedometer sample. The application of this method for the analysis of the pore-pressure dissipation and of the settlement development of thick layers proves that they depend on the amount of the additional pressure in connection with the previous degree of consolidation which is governed by preloading and/or secondary time effects.

In the form presented in Section 11.1 the application of the isotache method is also restricted to uniaxial loading, deformation and seepage conditions. The possibilities of a wider applicability for the method will be dealt with in Sections 12.7, 14.4.2 and 16.3.

12

Linear Compression of Partly Saturated Soils

12.1 PRINCIPLE OF EFFECTIVE STRESS FOR PARTLY SATURATED SOILS

In the preceding chapters the soil was assumed to be fully saturated, i.e. all the voids between the solid particles of the grain skeleton were filled with liquid (the pore water). Most natural deposits lying beneath the ground-water level can commonly be treated as being fully saturated. The layers above the water level, however, are not saturated and all artificially constructed earth embankments such as earth dams supporting storage reservoirs, road embankments or flood dams are built with material whose voids, after compaction, contain both water and air. The ratio between the volume of voids filled with water and the volume of total pore space is called the degree of saturation (S_r). It can be defined by the equation

$$S_r = \frac{e_w}{e} \tag{12.1}$$

e denoting the total void ratio and e_w the void ratio of the voids filled with water.

For saturated soils the effective stress tensor has been defined as that part of the total stress tensor which governs the deformation speed of the soil skeleton (see Section 9.1.1). In homogeneous stress fields the effective stress tensor was proved to be the difference between the total stress tensor and the spherical tensor $u\delta_{rs}$, u denoting the pore pressure and δ_{rs} the delta of Kronecker. In heterogeneous stress fields the effective stress principle was considered to be justified only as an approximation or as a working hypothesis.

For partly saturated soils, even in homogeneous stress fields, the effective stress principle cannot be defined explicitly. The grains can be surrounded partly by water and partly by air bubbles and owing to surface tension the pore water pressure u_w can be less than the pore-air pressure,

u_a. Thus regardless of the discontinuity of the all-round pressure at the contact surfaces between the grains, the boundaries of the single grains are not uniformly loaded by the pore-pressure. Consequently the spherical tensor either $u_w\delta_{rs}$ or $u_a\delta_{rs}$, or any intermediate spherical tensor

$$[u_a - \chi(u_a - u_w)]\delta_{rs},$$

χ being an arbitrary coefficient between 0 and 1, could in principle be subtracted from the total stress tensor to obtain the effective stress tensor. In any case in partly saturated soils with different water and air pore-pressures the effective stresses governing the displacements of the grains are not transferred only by the contact surfaces of the solid grains but also partly by the pore pressures themselves.

In its explicit form the effective stress for partly saturated soils may be expressed as follows:

$$\sigma' = \sigma - [u_a - \chi(u_a - u_w)] \tag{12.2}$$

This equation was first suggested by Bishop (1955). He related the coefficient χ to the effective stress-strain and effective stress-strength relations of saturated soils. In Fig. 12.1 test results are shown from experiments carried out by Donald at Imperial College (reproduced after Bishop, 1960). The values indicated by circles were calculated from two series of triaxial tests on silt, initially placed as a saturated slurry.

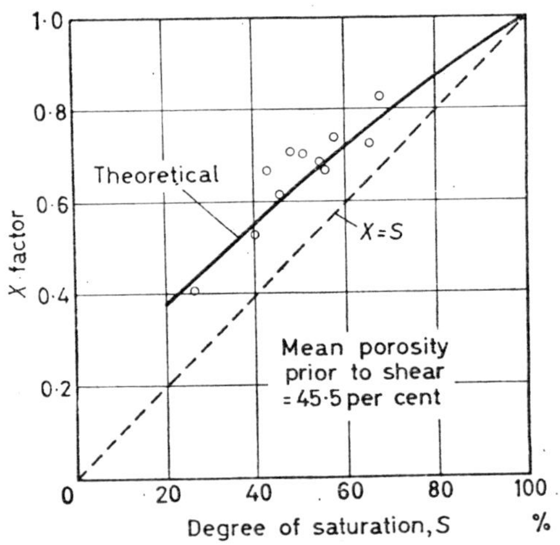

Fig. 12.1. Braehead silt. Values of χ in Equation (12.2) inferred from drained triaxial tests with controlled u_a and u_w (normally consolidated). (After Bishop, 1960)

The first series were run as drained tests without air in the pore space, so that the failure envelope in terms of effective stress could be determined from the usual Terzaghi equation. In the second series drainage was allowed to occur under a controlled negative pore pressure of sufficient magnitude to draw air into the voids. From the observed strength and the values of u_a and u_w the value of the parameter χ could be calculated, using the relationship between effective stress and strength obtained from the tests on the saturated samples.

12.2 DIFFERENTIAL EQUATION OF THE ONE-DIMENSIONAL CONSOLIDATION OF PARTLY SATURATED SOILS

In the section 9.2.2 the differential equation of consolidation of saturated soils was derived from Darcy's law and from the continuity conditions for both the liquid and solid phases of the soil. In order to obtain a general consolidation equation accounting also for partly saturated soils, the continuity condition for the gaseous phase of the soil system must also be considered. This must take into account the compressibility of the air and the solubility of the air in water.

The drainage through the soil may be assumed to occur only in the direction z (Fig. 12.2). The velocity of the air flow may be denoted by w and the density of the air by ρ. If the speed w is assumed to increase with

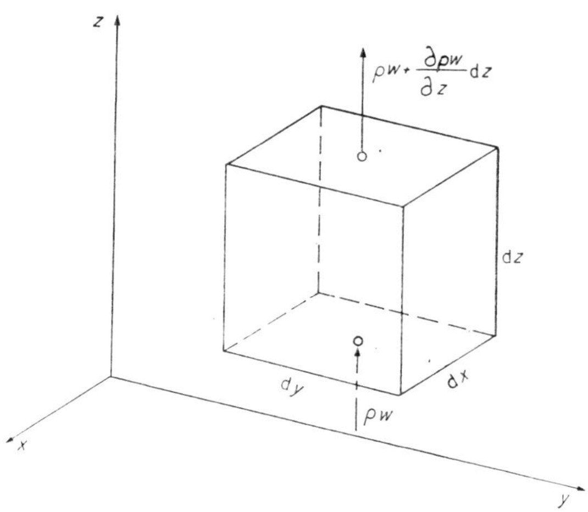

Fig. 12.2. Uniaxial air flow through a soil element

the increasing ordinate z, the mass of air leaving the element $dx\,dy\,dz$ in a time interval dt, is

$$\frac{\partial \rho w}{\partial z}\,dz\,dx\,dy\,dt \tag{12.3}$$

In the same time interval, a further quantity of air becomes dissolved in the water of the element $dx\,dy\,dz$. Denoting the porosity related only to voids filled with water by n_w

$$n_w = S_r n \tag{12.4}$$

and Henry's coefficient of the solubility of air in water by H, the mass of air dissolved in the time dt is:

$$\frac{\partial}{\partial t} H\rho n_w\,dx\,dy\,dz\,dt$$

Assuming an isothermal change of state and a constant coefficient H independent of the change in density, the above expression can be written as follows:

$$H\,dx\,dy\,dz\,dt\,\frac{\partial \rho n_w}{\partial t} \tag{12.5}$$

The sum of the quantities given by expressions (12.3) and (12.5) equals the change of the mass of air in the element $dx\,dy\,dz$ during the time interval dt:

$$\frac{\partial \rho w}{\partial z}\,dx\,dy\,dz\,dt + H\frac{\partial \rho n_w}{\partial t}\,dx\,dy\,dz\,dt = -\frac{\partial n_a \rho}{\partial t}\,dt\,dx\,dy\,dz$$

or:

$$\frac{\partial n_a \rho}{\partial t} + H\frac{\partial \rho n_w}{\partial t} + \frac{\partial \rho w}{\partial z} = 0 \tag{12.6}$$

Thus:

$$\rho\frac{\partial n_a}{\partial t} + n_a\frac{\partial \rho}{\partial t} + H\left(n_w\frac{\partial \rho}{\partial t} + \rho\frac{\partial n_w}{\partial t}\right) + \rho\frac{\partial w}{\partial z} + w\frac{\partial \rho}{\partial z} = 0$$

or

$$\left.\begin{aligned}\frac{\partial n_a}{\partial t} + \frac{n_a}{\rho}\frac{\partial \rho}{\partial t} + H\left(n_w\frac{1}{\rho}\frac{\partial \rho}{\partial t} + \frac{\partial n_w}{\partial t}\right) + \frac{\partial w}{\partial z} + \\ + w\frac{1}{\rho}\frac{\partial \rho}{\partial z} = 0\end{aligned}\right\} \tag{12.7}$$

Neglecting the influence of the change in n_w on the quantity of the dissolved air ($H(\partial n_w/\partial t) \rightarrow 0$), Florin (1961, Equation 3.106) presented the continuity condition of the gaseous phase of the soil system in the form:

$$\frac{\partial n_a}{\partial t} + \frac{n_a}{\rho}\frac{\partial \rho}{\partial t} + Hn_w \frac{1}{\rho}\frac{\partial \rho}{\partial t} + \frac{\partial w}{\partial z} + w \frac{1}{\rho}\frac{\partial \rho}{\partial z} = 0 \tag{12.7-a}$$

The continuity conditions for the liquid and solid phases were deduced, for saturated soils, in Section 9.2.2. For partly saturated soils, the continuity condition Equation (9.23) should be transformed by substituting n_w for n:

$$\frac{\partial v}{\partial z} + \frac{\partial n_w}{\partial t} = 0 \tag{12.8}$$

while the continuity condition for the solid phase remains unchanged

$$\frac{\partial \mathring{v}}{\partial z} + \frac{\partial (1-n)}{\partial t} = 0 \tag{9.25}$$

By adding the continuity conditions for the liquid (Equation 12.8), solid (Equation 9.25) and gaseous (Equation 12.7) phases of the soil system, we obtain the following relation:

$$\frac{\partial}{\partial z}(v + \mathring{v} + w) + (n_a + Hn_w)\frac{1}{\rho}\frac{\partial \rho}{\partial t} + \frac{w}{\rho}\frac{\partial \rho}{\partial t} = 0 \tag{12.9}$$

Now, in the further development of the consolidation equation, Florin's (1961) assumption that the air bubbles cannot move between the grains will be taken and that, as a result, the average values of the real velocities of the grains and of the air bubbles are equal throughout the consolidation process:

$$\mathring{v}_e = w_e \tag{12.10}$$

Thus

$$\frac{\mathring{v}}{1-n} = \frac{w}{n_a} \rightarrow w = \frac{n_a}{1-n}\mathring{v} \tag{12.11}$$

$\mathring{v}$ being the apparent velocity of the solid grains related to the gross cross-sectional area of the soil.

By substituting e_w for e, Equation (9.32) can be applied for partly saturated soils in the form:

$$\frac{\partial v}{\partial z} = e_w \frac{\partial \mathring{v}}{\partial z} + \mathring{v}\frac{\partial e_w}{\partial z} - \frac{\partial}{\partial z}\left(\frac{k}{\gamma_w}\frac{\partial \bar{u}}{\partial z}\right) \tag{12.12}$$

Inserting Equations (12.11) and (12.12) simultaneously into Equation (12.9) we obtain:

$$\frac{1}{1-n}\frac{\partial \mathring{v}}{\partial z} + (n_a + Hn_w)\frac{1}{\rho}\frac{\partial \rho}{\partial t} - \frac{\partial}{\partial z}\left(\frac{k}{\gamma_w}\frac{\partial \bar{u}}{\partial z}\right) + \left[\mathring{v}\frac{\partial e_w}{\partial z} + \frac{w}{\rho}\frac{\partial \rho}{\partial t}\right] = 0$$

Neglecting the members in brackets which are products of two very small factors, the above expression reduces to:

$$\frac{1}{1-n}\frac{\partial \mathring{v}}{\partial z} + (n_a + Hn_w)\frac{1}{\rho}\frac{\partial \rho}{\partial t} = \frac{\partial}{\partial z}\left(\frac{k}{\gamma_w}\frac{\partial \bar{u}}{\partial z}\right) \tag{12.13}$$

Equation (9.25) yields

$$\frac{\partial \mathring{v}}{\partial z} = -\frac{\partial(1-n)}{\partial t} = \frac{\partial n}{\partial t} \tag{12.14}$$

By considering the gas state equation:

$$\rho = \frac{1}{rT}p \tag{12.15}$$

r being the specific gas constant, T the absolute temperature and p the gas pressure, we have in the case of an isothermal change of state:

$$\frac{1}{\rho}\frac{\partial \rho}{\partial t} = \frac{1}{p}\frac{\partial p}{\partial t} \tag{12.16}$$

Denoting the initial air pressure in the voids by u_{a0}, and the pore pressure excess by u_a, the pressure p equals:

$$p = u_a^* = u_{a0} + u_a$$

and

$$\frac{\partial p}{\partial t} = \frac{\partial u_a}{\partial t}$$

Thus

$$\frac{1}{p}\frac{\partial p}{\partial t} = \frac{1}{u_{a0} + u_a}\frac{\partial u_a}{\partial t} \tag{12.17}$$

Inserting expressions (12.17) and (12.14) and considering Equation (9.35), Equation (12.13) yields:

$$\frac{\partial e}{\partial t} + \beta_a\frac{\partial u_a}{\partial t} = \frac{1+e}{\gamma_w}\frac{\partial}{\partial z}\left(k\frac{\partial \bar{u}_w}{\partial z}\right) \tag{12.18}$$

$$\beta_a = \frac{e_a + He_w}{u_{a0} + u_a} \tag{12.19}$$

and

$$\frac{\partial \bar{u}}{\partial z} = \frac{\partial u}{\partial z} - \gamma_w i_0 \tag{12.20}$$

i_0 denoting the limiting hydraulic gradient (see Section 9.2.1).

Now, according to Florin (1961) the pore-air pressure excess, u_a, may be assumed to be equal to the pore-water pressure excess, u_w:

$$u_a = u_w = u \tag{12.21}$$

$$u_{a0} = u_{w0} = u_0 \tag{12.22}$$

In this way Equation (12.18) becomes simplified:

$$\frac{\partial e}{\partial t} + \beta \frac{\partial u}{\partial t} = \frac{1 + e}{\gamma_w} \frac{\partial}{\partial z}\left(k \frac{\partial \bar{u}_w}{\partial z}\right) \tag{12.23}$$

where

$$\beta = \frac{e_a + He_w}{u_0 + u} \tag{12.24}$$

(see Florin, 1961, Equation 3.111).

Substituting Equation (9.37), as in Section 9.2.2 when dealing with the consolidation of saturated soils:

$$\frac{\partial e}{\partial t} = \frac{\partial e}{\partial \sigma'} \cdot \frac{\partial \sigma'}{\partial t} \tag{9.37}$$

and

$$\sigma' = \sigma - u \tag{9.38}$$

σ denoting the total and σ' the effective pressure in the drainage direction, Equation (12.20) can be transformed as follows:

$$\frac{\partial e}{\partial \sigma'}\left(\frac{\partial \sigma}{\partial t} - \frac{\partial u}{\partial t}\right) + \beta \frac{\partial u}{\partial t} = \frac{1 + e}{\gamma_w} \frac{\partial}{\partial z}\left[k\left(\frac{\partial u}{\partial z} - \gamma_w i_0\right)\right]$$

yielding:

$$\frac{\partial u}{\partial t} = \frac{1}{\omega}\left\{\frac{\partial \sigma}{\partial t} - \frac{1 + e}{\gamma_w \dfrac{\partial e}{\partial \sigma'}} \cdot \frac{\partial}{\partial z}\left[k\left(\frac{\partial u}{\partial z} - \gamma_w i_0\right)\right]\right\} \tag{12.25}$$

(compare with Florin, 1961, Equation 3.113), where

$$\omega = 1 - \frac{\beta}{\dfrac{\partial e}{\partial \sigma'}} \tag{12.26}$$

In non-viscous soils $\partial e/\partial\sigma'$ is usually assumed to be constant over the observed stress-interval:

$$\frac{\partial e}{\partial \sigma'} = -a \tag{9.43-a}$$

The void ratio e, the parameter β and, consequently, the coefficient ω change with varying pore pressure u. In solving practical consolidation problems these parameters are usually assumed to be constant (Florin, 1961).

12.3 CONSOLIDATION OF UNIFORMLY LOADED LAYERS WITHOUT PLASTIC RESISTANCE

12.3.1 Idealized Classification of the Consolidation Process for Partly Saturated Soils

Because of the complexity of the continuity condition (12.7), Laing Barden (1965a) sub-divided the consolidation process of compacted and unsaturated clays into a number of less general treatments, each being relevant to a particular range of saturation values.

(1) Extremely dry clays ($S_r < 0{\cdot}5$). The water is firmly attached to the skeleton by capillary forces. The air voids are completely connected and only air will flow from the consolidating clay. Because of the low values of S_r the parameter χ in Equation (12.2) will be small and hence Equation (12.2) can be applied as $\sigma' = \sigma - u_a$.

(2) Clay dry of optimum ($0{\cdot}5 < S_r < 0{\cdot}9$). The water still does not flow to any appreciable degree, possibly because the capillary difference $(u_c - u_w)$ is still large enough to ensure that u_w rarely exceeds zero. The air voids are still continuous and air is again the only fluid to flow from the compressed clay.

(3) A transition stage between the flow of air only in process (2) and the flow of liquid in process (4).

(4) For clay wet of optimum ($S_r > 0{\cdot}9$) the air exists in its occluded state and cannot flow as a separate continuous fluid. Certain smaller air bubbles may be free to flow with the free pore water forming a homogeneous compressible fluid flowing under the gradient $\partial u_w/\partial z$. The majority of the air will be static, trapped by the skeleton. Because of the high value of S_r the parameter χ in Equation (12.2) will approach unity and hence Equation (12.2) can be applied as $\sigma' = \sigma - u_w$.

(5) For very wet clays ($S_r > 0{\cdot}95$) it can be concluded that the small amount of air present will be trapped by the skeleton and the fluid flowing from the clay will be fairly incompressible.

12.3.2 Consolidation of Non-saturated Wet Clays

In order to analyse process (4) as described above, L. Barden, (1965a) assumed that the mobile fluid is homogeneous to the extent that its compressibility characteristics do not vary throughout the mass of the clay and do not change during the consolidation process as a possible consequence of Henry's law.

It is also assumed that since the mass of air is negligible compared with the mass of water in the homogeneous compressible fluid, then the effects of Henry's law can be omitted from the following equation for continuity of the mass of the homogeneous fluid.

$$\frac{\partial}{\partial z}(\gamma_e v_e) = \frac{\partial}{\partial t}(\gamma_e n_e) \tag{12.27}$$

where the suffix 'e' refers to the effective mobile fluid (homogeneous, compressible). Flow is caused by gradients in u_w above

$$v_e = k_e \frac{\partial u_w}{\partial z} \tag{12.28}$$

For k_e, γ_e and n_e the following approximate relations can be assumed:

$$k_e = k_{ef}(1 + bu_w) \tag{12.29}$$

$$\gamma_e = \gamma_0(1 + du_w) \tag{12.30}$$

$$n_e = n_{ef}(1 + au_w) \tag{12.31}$$

(For the justification of these assumptions, see L. Barden, 1965a.) The suffix f in k_{ef} and n_{ef} denotes the final values of k_e and n_e respectively when the excess pore pressure u_w has finally dissipated to zero. γ_0 is the density of the fluid at $u_w = 0$. The parameters b, d, a are assumed to be constant throughout the process analysed.

Substitution of relations (12.28), (12.29), (12.30) and (12.31) into the continuity Equation (12.27) yields:

$$\frac{\partial}{\partial z}\left[\gamma_0(1 + du_w)k_{ef}(1 + bu_w)\frac{\partial u_w}{\partial z}\right] = \frac{\partial}{\partial t}[n_{ef}(1 + au_w)\gamma_0(1 + du_w)] \tag{12.32}$$

To reduce to dimensionless form, put

$$z = \zeta H,\ u_w = uu_0,\ \alpha = au_0,\ \beta = bu_0,\ \delta = du_0 \tag{12.33}$$

where H is the length of the vertical drainage path. u_0 is the initial value of the excess pore-pressure at $t = 0$, and α, β and δ are characteristics which establish the degree of variation of n_e, k_e and γ_e respectively. On expanding and rearranging Equation (12.32) gives

$$(\beta + \delta + 2\delta\beta u)\frac{\partial^2 u}{\partial \zeta^2} + 2(1 - \delta\beta u^2)\frac{\partial^2 u}{\partial \zeta^2} = 2\left(1 + \frac{\delta}{\alpha} + 2\delta u\right)\frac{\partial u}{\partial T} \qquad (12.34)$$

where the time factor

$$T = \frac{k_{ef}u_0 t}{n_{ef}\alpha H^2} \qquad (12.35)$$

Since $(\alpha n_{ef})/u_0$ is equivalent to the conventional coefficient of volume decrease m_v, Equation (12.35) can be written

$$T = \frac{k_{ef}t}{m_v H^2} \qquad (12.35\text{-a})$$

which is identical in form to the Terzaghi expression.

For an incompressible fluid $\delta = 0$ and for a constant value of permeability $\beta = 0$ and so Equation (12.34) reduces to

$$\frac{\partial^2 u}{\partial \zeta^2} = \frac{\partial u}{\partial T} \qquad \text{as required.}$$

As proved by Laing Barden (1965a) appropriate simplifying assumptions lead, in analysing the consolidation process for extremely dry clays ($S_r < 0{\cdot}5$), to an equation identical to Equation (12.34). The only difference is due to the difference in magnitude of the parameters α, β and δ, and so the suffixes *w* and *e* in Equations (12.27) to (12.33) have to be replaced by the suffix *a* (air).

The consolidation process (5) for very wet clays can be treated as process (4) with *d* negligibly small. The main difference between the consolidation of this type of clay and a fully saturated one lies in the disproportionately large effect of the small amount of air on the variation of permeability throughout the process.

12.3.3 The Initial Settlement of Non-viscous Partly Saturated Soils

If the viscous properties are small enough to be neglected, the initial settlement of a laterally confined partly saturated soil sample (layer respectively) is to be attributed to thc instantaneous compression of the voids filled with air. This compression is governed by the law of Boyle-Mariotte, by Henry's law of solubility of gases in liquid and by the effective stress–strain relationship which over a limited load interval may be assumed to be linear. The air may be assumed to exist in its occluded state (process

(4) as described in Section 12.3.1). When we denote the void ratios before and after instantaneous initial settlement by e_0 and e, the corresponding void ratios filled with air by e_{a0} and e_a, and the corresponding air pressures in the pore space including the atmospheric pressure, by u_{a0}^* and u_a^*, the Boyle-Mariotte law yields:

$$e_a u_a^* = (e_{a0} - \Delta e_{a0}) u_{a0}^* \tag{12.36}$$

As a linear stress–strain relationship, governed by the coefficient of compressibility m_v (see Equation 8.24) has been assumed, the void ratio e_a becomes:

$$e_a = e_{a0} - (1 + e_0) m_v \Delta\sigma' \tag{12.37}$$

The instantaneous effective stress difference $\Delta\sigma'$ can be expressed by making use of relation (12.2):

$$\sigma' = \sigma - u_a + \chi(u_a - u_w) \tag{12.2}$$

where u_a and u_w denote the pressures of the air and water respectively in the pore space. The pressures u_a and u_w are expressed relative to the pore pressure u_0 corresponding to the hydrostatic and atmospheric potential at the drainage boundary of the layer:

$$u_a = u_a^* - u_0 \tag{12.38}$$

$$u_w = u_w^* - u_0 \tag{12.39}$$

The initial values of the pressures u_a and u_w are:

$$u_{a0} = u_{a0}^* - u_0 \tag{12.40}$$

$$u_{w0} = u_{w0}^* - u_0 \tag{12.41}$$

Thus the preceding effective stresses are:

$$\sigma_\alpha' = \sigma_\alpha - u_{a0} + \chi(u_{a0} - u_{w0})$$

and the effective stresses corresponding to the instantaneous compression occurring immediately after the load application $\Delta\sigma = \sigma_\beta - \sigma_\alpha$, are

$$\sigma_\beta' = \sigma_\beta - u_a + \chi(u_a - u_w)$$

The instantaneous effective stress increment $\Delta\sigma' = \sigma_\beta' - \sigma_\alpha'$ can be obtained as the difference of the above expressions:

$$\Delta\sigma' = \Delta\sigma - \Delta u_a + \chi(\Delta u_a - \Delta u_w) \tag{12.42}$$

where the following substitutions have been made:

$$u_a - u_{a0} = \Delta u_a \tag{12.43}$$

$$u_w - u_{w0} = \Delta u_w \tag{12.44}$$

Owing to surface tension the initial pore-water pressure u_{w0} is usually less than the initial pore-air pressure, u_{a0}. The pore pressure increments Δu_a and Δu_w are different too. The difference can be expressed by the relation:

$$\Delta u_w = \Delta u_a + \mu_t q \tag{12.45}$$

$$\Delta u_a = u_w - (u_{w0} + \mu_t q) \tag{12.45-a}$$

The second right-hand term of this relation represents that part of the pore-water pressure by which the pressure transferred by the soil skeleton exceeds the simultaneous air pressure increase Δu_a. Inserting Equation (12.45) into Equation (12.42) and substituting q for $\Delta\sigma$ ($q \equiv \Delta\sigma$) we obtain:

$$\Delta\sigma' = q(1 - \chi\mu_t) - \Delta u_a \tag{12.46}$$

With this expression for $\Delta\sigma'$ Equation (12.37) becomes:

$$e_a = e_{a0} - (1 + e_0)m_v[q(1 - \chi\mu_t) - \Delta u_a] \tag{12.47}$$

The void ratio difference Δe_{a0} appearing in Equation (12.36) applies for the air which dissolves in the water due to the change in pore-water pressure from u_{w0} to u_w. Using Henry's law the difference Δe_{a0} may be expressed by the equation:

$$\Delta e_{a0} = e_0 S_{r0}(u_a - u_{a0})H u_{a0}^* \tag{12.48}$$

where S_{r0} denotes the initial value of the degree of saturation S_r defined by the ratio

$$S_r = \frac{\text{volume of water}}{\text{volume of pore space}}$$

and H Henry's coefficient of solubility of air in water. Inserting Equation (12.43), expression (12.48) becomes:

$$\Delta e_{a0} = e_0 S_{r0} H \Delta u_a / u_{a0}^* \tag{12.49}$$

Now, expressing e_a by the relation (12.47), and Δe_{a0} by the relation (12.49) in Equation (12.36), and substituting:

$$u_a^* = u_{a0}^* + \Delta u_a \tag{12.50}$$

a quadratic equation results, the solution of which is:

$$\Delta u_a = \frac{-B \pm \sqrt{(B^2 + 4C)}}{2} \tag{12.51}$$

with the following notation for the constants B and C:

$$B = u_{a0}{}^* - \frac{C}{u_{a0}{}^*} + \frac{e_{a0} + e_0 S_{r0} H}{(1 + e_0) m_v} \tag{12.52}$$

$$C = q(1 - \mu_t \chi) u_{a0}{}^* \tag{12.53}$$

Under the conditions of oedometer tests the initial air pressure $u_{a0}{}^*$ can be taken as equal to 1 atm ($\sim$ 1 kp/cm²) and Henry's coefficient H approximately equals 0·02 cm²/kp. The applied additional applied load q is known and the initial void ratio e_0 as well as the initial degree of saturation S_{r0} can easily be determined. The factors χ and μ_t depend on the actual degree of saturation, soil type and stress history. The available experimental evidence (see Bishop and Henkel, 1962) shows the dependence of χ on S_r varying with soil type and compaction (average values approximately $\chi \simeq S_r$), though a simple common relationship for all soil types is not apparent at this stage. The experimental evidence and our actual knowledge concerning the factor μ_t is not sufficient and requires appropriate investigations. At the present stage a reasonable assumption should be made.

Once the air-pressure increase Δu_a is known Equation (12.46) yields the instantaneous increase $\Delta\sigma'$ of the effective stress and the void ratio e_a can then be found by Equation (12.37).

12.4 ANALYSIS OF CREEP EFFECTS IN PARTLY SATURATED SOILS BY THE ISOTACHE METHOD

12.4.1 Expressions for Consolidation Speeds and Pore Water Pressures of Partly Saturated Soils

By differentiating the equation of the tangent to the consolidation curve $e = e(t)$ with respect to time t the following expression was obtained (see Equations 8.13 and 11.12):

$$\frac{\partial \bar{e}}{\partial t} = -\frac{\bar{\vartheta}_e}{N \,.\, t^m} \tag{12.54}$$

where

(*a*) $N = 2{\cdot}3026$, $\bar{\vartheta}_e = \bar{\alpha}_e$ (see Fig. 8.6), $m = 1$ if the average void ratios e are plotted against the logarithm of time t, and
(*b*) $N = 2$, $\bar{\vartheta}_e = \bar{\beta}_e$ (see Fig. 11.3), $m = \frac{1}{2}$ if they are plotted against the square root of time.

Equation (12.54) holds, naturally, for partly saturated soils too. Assuming a parabolic form for the isochrones with n equalling the degree of the parabola, the following general expressions for the mean excess pore water pressures in saturated soils can be written on the basis of formulae previously derived (see Equations 11.9, 11.13 and 11.22):

(I)

$$\bar{u}_w = \frac{\gamma_w}{N(n+1)} \cdot \frac{\bar{\vartheta}_e'}{(1+\bar{e})} \cdot \frac{h^2}{k \,.\, t^m} \qquad (12.55)$$

when $\bar{u}_w \leqslant [n/(n+1)]q, \quad q \equiv \Delta\sigma$

(II)

$$\bar{u}_w = q - \Delta\bar{\sigma}' = q\left\{1 - \frac{q}{\gamma_w} \cdot \frac{Nn}{n+1} \cdot \frac{1+\bar{e}}{\bar{\vartheta}_e'} \cdot \frac{kt^m}{h^2}\right\} \qquad (12.56)$$

when $\bar{u}_w \geqslant [n/(n+1)]q$

h denoting the thickness of the layer drained from one side only and corresponding to the actual void ratio $\bar{e}$, while n, N, m and $\bar{\vartheta}_e'$ have the meanings explained above.

In partly saturated soils whose air bubbles are occluded (process (4) after Laing Barden), Equations (12.55) and (12.56) can be used by interpreting the gradient $\bar{\vartheta}_e'$ as belonging to an apparent consolidation curve $\bar{e}' = \bar{e}'(t)$ [curve (b) in Fig. 12.3] related to that part of the volume change

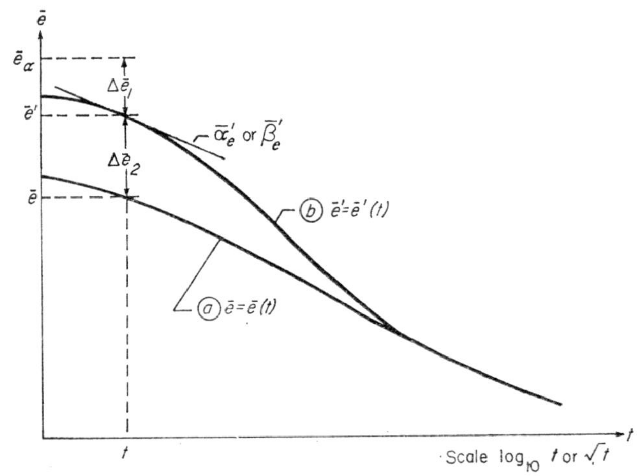

Fig. 12.3. The real (a) and the apparent (b) consolidation curve of a partly saturated soil

corresponding to the escape of water together with some free air bubbles from the voids, the air being assumed incompressible.

In Fig. 12.3 line (*a*) represents the consolidation curve $\bar{e} = \bar{e}(t)$ of an oedometer sample of a partly saturated soil; e_α is the initial void ratio and S_r the initial degree of saturation. Line (*b*) is the apparent consolidation curve $\bar{e}' = \bar{e}'(t)$ corresponding to the assumption of incompressibility of the air in the voids. Consequently the difference $\Delta\bar{e}_2$ between the void ratios $\bar{e}'$ and $\bar{e}$ represents the compression of voids filled with air, which is caused by the increase in pressure of the air in the voids.

The void ratios $\bar{e}$, in Equations (12.55) and (12.56) are related, of course, to the true consolidation curve $\bar{e} = \bar{e}(t)$. The coefficients of permeability that correspond to the entire flow of water and free air bubbles, depend on the void ratios $\bar{e}$ and on the degree of the saturation S_{rt} expressing the volume ratio:

$$S_{rt} = \frac{\bar{e} - (\bar{e}_a)_{tr}}{\bar{e}} \tag{12.57}$$

$(\bar{e}_a)_{tr}$ denoting the void ratio relating, at time t, the volume of occluded air trapped by the soil skeleton to the volume of the solid substance assumed to be compact.

12.4.2 Expression for the Pressure of the Air in the Voids

According to the notation of Fig. 12.3 the volume of the pore-air at the time t can be expressed by the equation:

$$V_a = [e_{a0} - \Delta\bar{e}_1(1 - S_{rf}) - \Delta\bar{e}_2]V_s \tag{12.58}$$

e_{a0} denotes the initial void ratio relating the volume of the air in the voids to the volume V_s of the solid substance assumed to be compact and S_{rf} is a relative coefficient of saturation relating the volume of voids filled with water to the volume of water and that part of the pore-air which, in the form of free bubbles, flows with the free pore-water.

The quantity of air dissolved in the water during the time t under the influence of the excess pore-air pressure $\Delta\bar{u}_a$, may be expressed by the void-ratio decrease corresponding to the initial air pressure $\bar{u}_{a0}{}^*$

$$\Delta\bar{e}_H = H\frac{\Delta\bar{u}_a}{\bar{u}_{a0}{}^*}[\bar{e}' - (\bar{e}_{a0})_{tr}]S_{rf} \tag{12.59}$$

$(\bar{e}_{a0})_{tr}$ denoting the initial value of $(\bar{e}_a)_{tr}$ (at the time $t = 0$), $\Delta\bar{u}_a$ the excess pore-air pressure and H Henry's coefficient of solubility of air in water expressing the ratio between the volume of the dissolved air to the volume of the water solvent at 760 mm Hg air pressure, the volume of the air being reduced to standard conditions (0° C, 760 mm Hg). Thus the volume V_a' of the mass of air at time t related to the initial total air pressure $\bar{u}_{a0}^*$ can be expressed as follows:

$$V_a' = [e_{a0} - \Delta\bar{e}_1(1 - S_{rf}) - \Delta\bar{e}_H]V_s \tag{12.60}$$

The initial volume of air is the sum of the volume of air trapped by the soil skeleton $(e_{a0})_{tr}$ and the volume of free air bubbles flowing with the pore-water. Thus:

$$e_{a0} = (e_{a0})_{tr} + [e_0 - (e_{a0})_{tr}](1 - S_{rf0})$$

or

$$e_{a0} = e_0(1 - S_{rf0}) + (e_{a0})_{tr}S_{rf0} \tag{12.61}$$

According to the definition of the apparent consolidation curve $\bar{e}' = \bar{e}'(t)$ the coefficient S_{rf} in Equations (12.58) and (12.59) equals the initial value S_{rf0}:

$$S_{rf} = S_{rf0}$$

The relation between the volumes V_a (Equation 12.58) and V_a' (Equation 12.60) is governed by Boyle's law of the isothermal change of the gas state. Thus:

$$V_a'\,\bar{u}_{a0}^* = V_a\bar{u}_a^* \tag{12.62}$$

where

$$\bar{u}_a^* = \bar{u}_{a0}^* + \Delta\bar{u}_a \tag{12.63}$$

represents the total pore-air pressure at time t; $\bar{u}_{a0}^*$ is its initial value and $\Delta\bar{u}_a$ the excess pore-air pressure. By inserting Equation (12.60) for V_a' with Equation (12.59) for Δe_H and Equation (12.61) for e_{a0}, Equation (12.58) for V_a and Equation (12.63) for $\bar{u}_a^*$, into Equation (12.62) and rearranging we obtain:

$$\Delta e_2 = \frac{\Delta\bar{u}_a}{\bar{u}_{a0}^* + \Delta\bar{u}_a}\{e_0(1 - S_{rf}) + (e_{a0})_{tr}S_{rf} - \Delta e_1(1 - S_{rf}) + \\ + H[\bar{e}' - (e_{a0})_{tr}]S_{rf}\}$$

yielding:

$$\Delta e_2 = \frac{\Delta\bar{u}_a}{\bar{u}_{a0}^* + \Delta\bar{u}_a}\{\bar{e}' - [\bar{e}' - (e_{a0})_{tr}]S_{rf}(1 - H)\} \tag{12.64}$$

As referred to by Florin (1961) and L. Barden (1965a), the relative degree of saturation S_{rf} is close to unity. The majority of the occluded air bubbles are trapped by the skeleton. With $S_{rf} = 1$ Equation (12.64) reduces to

$$\Delta e_2 = \frac{\Delta\bar{u}_a}{u_{a0}^* + \Delta\bar{u}_a}\{(\bar{e}_{a0})_{tr} + [\bar{e}' - (\bar{e}_{a0})_{tr}]H\} \tag{12.65}$$

or

$$\Delta\bar{u}_a = \frac{\bar{u}_{a0}^*}{\dfrac{(\bar{e}_{a0})_{tr} + [\bar{e}' - (\bar{e}_{a0})_{tr}]H}{\Delta e_2} - 1} \tag{12.66}$$

Drainage of the pore-water begins when the pore-water pressure excess $\bar{u}_w$ reaches a positive value:

$$\bar{u}_w = \bar{u}_w^* - \bar{u}_0 > 0 \rightarrow \bar{u}_w^* > u_0 \quad \text{(see Equation 12.41)}$$

12.4.3 Determination of Pore Pressures

The application of relation (12.45-a) to the average pore-pressure values gives:

$$\Delta\bar{u}_a = \bar{u}_w - (\bar{u}_{w0} + \mu_t q) \tag{12.67}$$

For the initial negative value of the excess pore-water pressure $\bar{u}_{w0}$ and the coefficients μ_t, appropriate assumptions should be made according to experimental evidence. Since experience is lacking, approximate analyses can be based on Florin's assumption (12.6): $u_a = u_w$. Consequently:

$$\bar{u}_{w0} = 0, \mu_t = 0 \qquad \text{and} \qquad \Delta\bar{u}_a = \bar{u}_w.$$

The air and water pressures in the pore space satisfying Equations (12.55) or (12.56) and (12.66) respectively, as well as relation (12.67), can be found by constructing the apparent consolidation curve (*b*) $[\bar{e}' = \bar{e}'(t)]$ in the following way:

At any time t the position ($\bar{e}'$) and the direction ($\bar{\vartheta}_e'$) of the curve (*b*) may be chosen, the corresponding point on the observed consolidation curve (*a*) being ($t,\bar{e}$). The selected values $\bar{e}'$ and $\bar{\vartheta}_e'$ ($\bar{\alpha}_e'$ or $\bar{\beta}_e'$) must satisfy the condition of continuity of line (*b*) with respect to the preceding part of the curve. The values $\bar{e}$ and $\bar{\vartheta}_e'$, as well as the coefficient of permeability k corresponding to the void ratio $\bar{e}$ and to the actual degree of saturation S_{rt}, have to be inserted into Equations (12.55) or (12.56). In this way, the pore-water pressure $\bar{u}_w$ results.

By inserting the same value $\bar{e}'$ and the corresponding difference $\Delta\bar{e}_2$ (see Fig. 12.3) into Equation (12.66) an air pressure increase, $\Delta\bar{u}_a$ is obtained. The calculated pressures $\bar{u}_w$ and $\Delta\bar{u}_a$ must satisfy Equation (12.67). If this relation is not satisfied, new values $\bar{e}'$ and $\bar{\vartheta}_e'$ must be chosen and the whole procedure repeated.

In order to facilitate the procedure two auxiliary graphs may be recommended:

(*a*) The relation (12.55) can be written in the form

$$\frac{\bar{\vartheta}_e'(1+\bar{e})}{k} = \frac{N(n+1)}{\gamma_w h_s^2}\,\bar{u}_w t^m \qquad (12.68)$$

where $h/(1+\bar{e})$ has been replaced by h_s, the latter being the thickness of the solid material assumed to be compact. Now, the plots

$$y = \frac{\bar{\vartheta}_e'(1+\bar{e})}{k} = f(t) \qquad (12.69)$$

can be drawn for various pore pressures $\bar{u}_w$ = const, the values N,n being chosen and the quantities γ_w,h given. If the consolidation curves are plotted against the logarithm of time, it is recommended that the plots (12.69) be drawn to logarithmic scales, for both the time t and the ordinate y, in order to obtain linear graphs.

(*b*) From Equation (12.65) the void ratios $\bar{e}'$ can be plotted against the differences Δe_2 for various constant $\Delta\bar{u}_a$ values and for given values $\bar{u}_{a0}{}^*$, $(\bar{e}_{a0})_t$ and H.

12.4.4 Construction of the Isotache Set

Once the pressures $\bar{u}_w$ and $\bar{u}_a = \bar{u}_{a0} + \Delta\bar{u}_a$ are known, the effective pressures can be determined by Equation (12.2):

$$\bar{\sigma}' = \sigma - \bar{u}_a + \chi(\bar{u}_a - \bar{u}_w) \qquad (12.70)$$

By plotting, on a common diagram, the observed consolidation curve $\bar{e} = \bar{e}(t)$ as well as the corresponding graphs $\partial\bar{e}/\partial t = \omega(t)$ (see Equation 12.54) and $\bar{\sigma}' = \bar{\sigma}'(t)$, the joint values $(\bar{e},\bar{\sigma}')$ can be determined for any consolidation speed $\partial e/\partial t$, the fitting procedure being shown in Fig. 11.10. If various consolidation curves $\bar{e} = \bar{e}(t)$ corresponding to different load increments $\Delta\sigma = \sigma_\beta - \sigma_\alpha$ are available, several values of $(\bar{e},\bar{\sigma}')$ can be obtained for any consolidation speed ω = const and the corresponding isotache can be drawn.

Applying an analogous procedure the isotaches can also be constructed from the consolidation curves corresponding to continuous loading interrupted by some intervals of constant loading. The construction of isotaches becomes simplified considerably if based only on the secondary branches of the consolidation curves. In these regions the pore pressures are small enough to be neglected and the curves (*a*) and (*b*) [see Fig. 12.2] coincide.

Fig. 12.4 represents an isotache set as obtained from the oedometer consolidation curves observed for a lacustrine chalk. The sample was

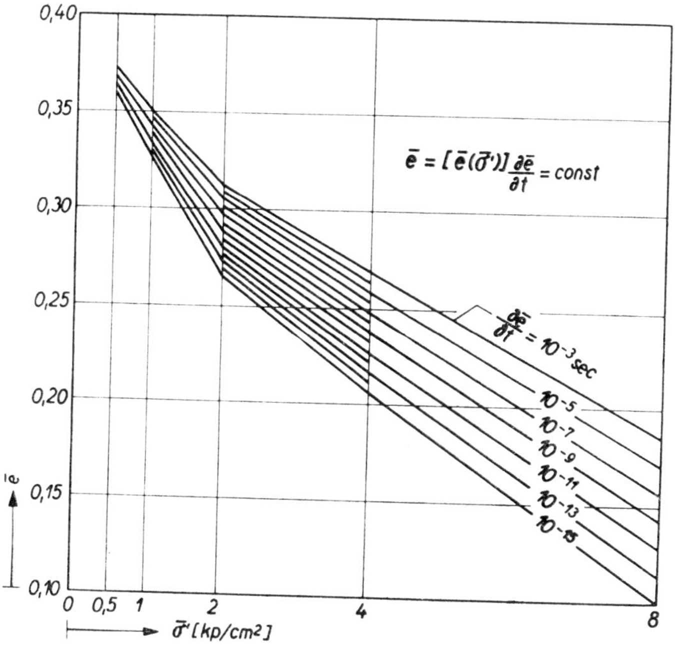

Fig. 12.4. Isotache set obtained from oedometer consolidation curves of a partly saturated precompacted lacustral chalk

precompacted in a Proctor device; the liquid limit of the sample was $w_L = 23{\cdot}6$ per cent, the plasticity index $I_P = 11{\cdot}6$ per cent and the grain size distribution: 17 per cent $< 0{\cdot}002$, 71 per cent 0·002/0·06 and 12 per cent 0.06/2 mm.

12.4.5 Construction of the Consolidation Curve of a Layer of Arbitrary Thickness

Once the isotache set is known, the consolidation curve of a layer of arbitrary thickness corresponding to an arbitrary pressure increment $q = \sigma_\beta - \sigma_\alpha$ can be determined by a procedure analogous to that previously explained for saturated soils (see Section 11.1.6). The need for

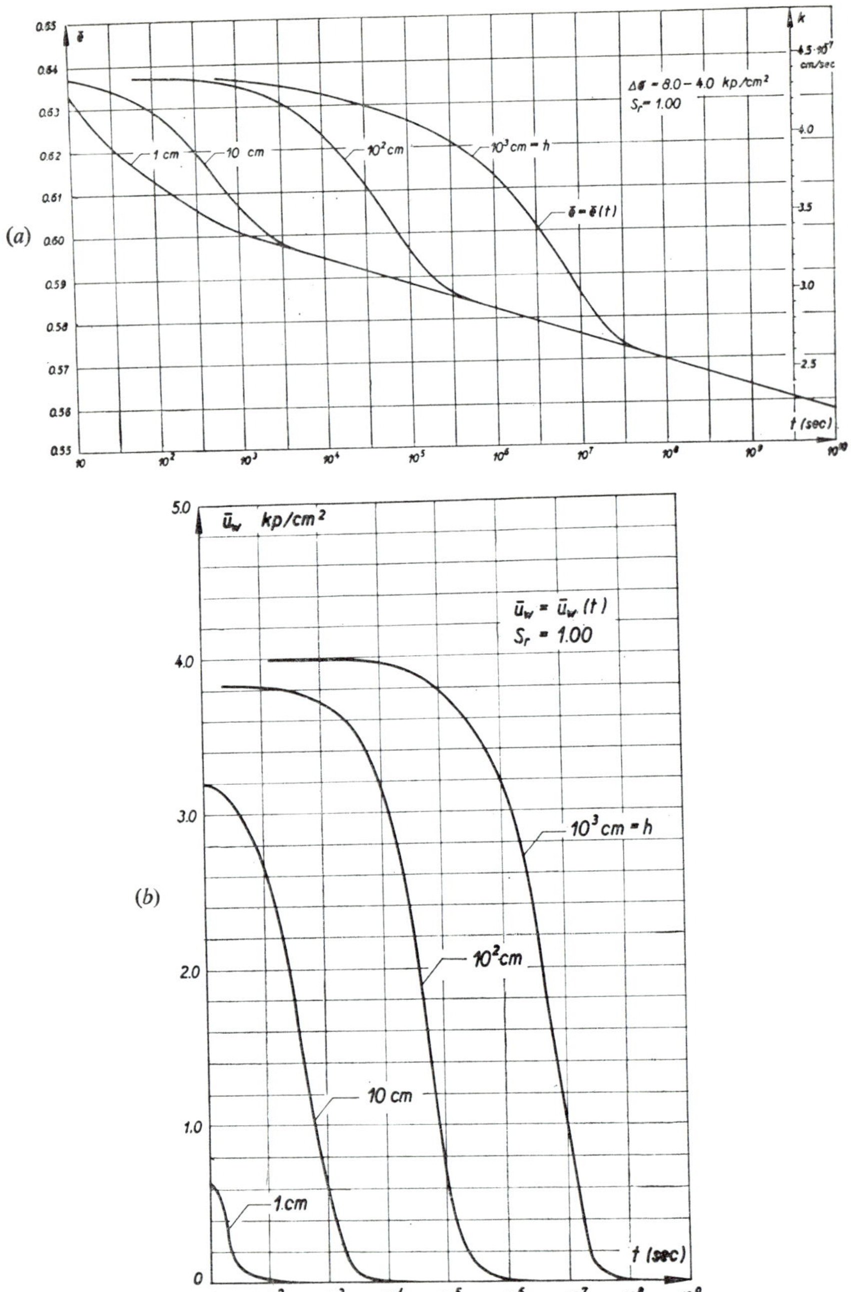

Fig. 12.5. (*a*) Consolidation curves of a saturated, normally consolidated lacustrine chalk, derived from the isotache set presented in Fig. 11.11–*c*, corresponding to seepage paths (layer thicknesses) of 1, 10, 100 and 1000 cms; (*b*) the corresponding pore pressure dissipation

simultaneous control of the air pressure, however, complicates the procedure. The method is the reverse of that explained in the previous section for the construction of isotaches from the consolidation curves.

The apparent (*b*) and the true (*a*) consolidation curves must be constructed simultaneously. At any time t the construction should satisfy the following conditions:

(1) The pore-water pressure $\bar{u}_w$ obtained by Equations (12.55) or (12.56) and the air-pressure increase $\Delta\bar{u}_a$ determined by Equation (12.66) must satisfy the relation (12.67). In these equations the quantities $\bar{\vartheta}_e'$ and $\bar{e}'$ are related to the apparent consolidation curve (*b*).

(2) The consolidation speed $\partial\bar{e}/\partial t$, corresponding in the isotache set to the void ratio $\bar{e}$ of curve (*a*) at time t and to the effective pressure $\bar{\sigma}'$ obtained by Equation (12.70), must equal the value given by Equation (12.54).

(3) The consolidation curves (*a*) and (*b*) must be continuous.

The initial void ratio $\bar{e}_{a0}$ and the initial degrees of saturation S_{r0}, $(S_{rt})_0$ and S_{rf} should be known. The relative degree of saturation S_{rf} is assumed to have the same value at the beginning and at the end of the primary consolidation occurring after the application of a certain load increment $q = \Delta\sigma$. Commonly S_{rf} can be taken equal to unity.

The initial total air pressure $\bar{u}_{a0}^*$ (which includes the atmospheric pressure) and the initial total pore-water pressure $\bar{u}_{w0}^*$ depend on the hydrostatic potentials at the boundaries of the layer and on the preconsolidation or precompaction conditions (stress history). If the hydrostatic pressures at the draining boundaries of the layer are zero and if the soil is not pre-consolidated, the air pressure $\bar{u}_{a0}^*$ is commonly assumed to be $\bar{u}_{a0}^* = 1\ \text{atm} \cong 1\ \text{kp/cm}^2$. In this case the initial excess pore-water pressure is negative or zero.

The auxiliary graphs

$$y = \frac{\bar{\vartheta}_e'(1 + \bar{e})}{k} = [f(t)]_{u_w=\text{const}}$$

(Equation 12.69) and $\bar{e}' = \bar{e}'(\Delta e_2)$ [Equation 12.64 or 12.65] can also be usefully applied during the construction of the consolidation curves corresponding to an arbitrary layer thickness. The graphs $y = [f(t)]_{\bar{u}_w=\text{const}}$ related to a certain thickness h_s can be used also for any other thickness $h_s' = mh_s$. If the graphs are presented on the logarithm of time scale, the values t corresponding to the thickness h_s should be multiplied by the factor m^2 in order to comform to the thickness mh_s.

This method of analysis has been applied to a lacustrine chalk whose isotache set is shown in Fig. 11.11 (*c*) and whose coefficients of

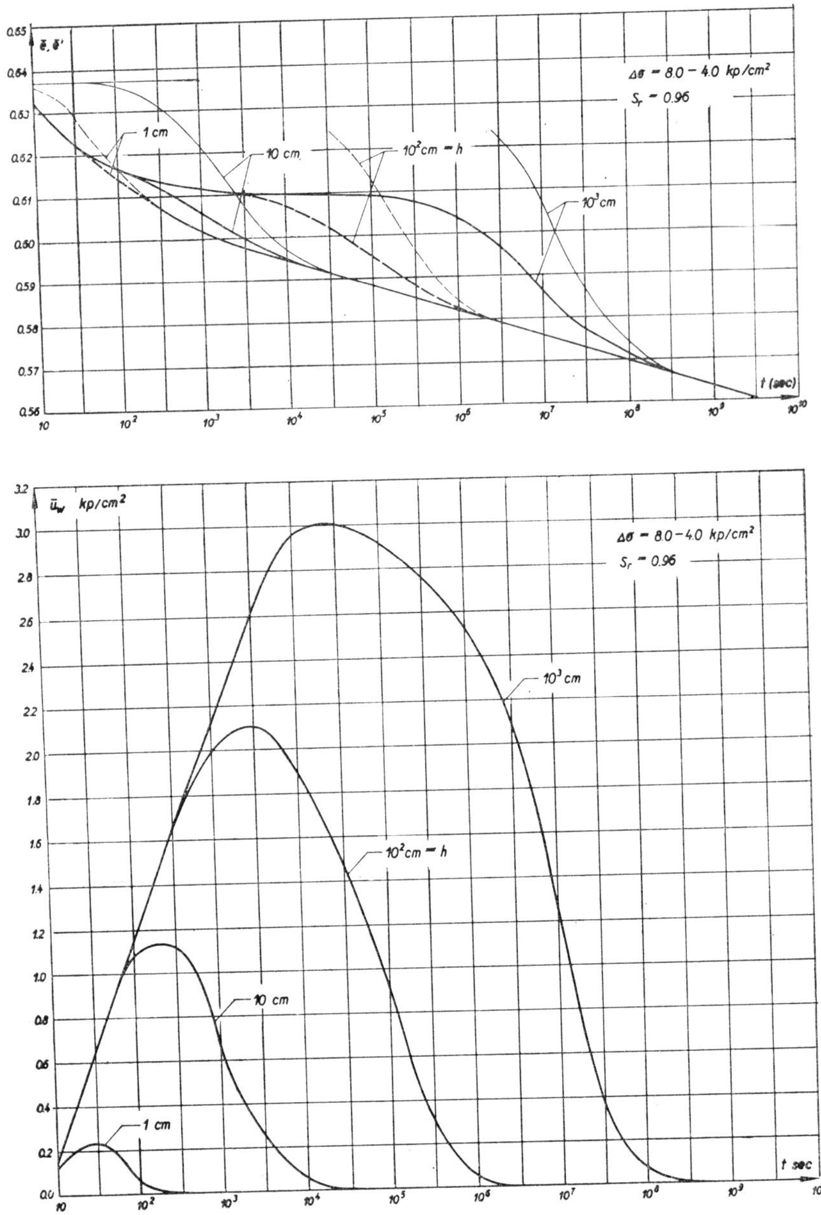

Fig. 12.6. (*a*) Consolidation curves of a partly saturated ($S_r = 0{\cdot}96$), normally consolidated lacustrine chalk, derived from the isotache set presented in Fig. 11.11–*c*, corresponding to seepage paths (layer thicknesses) of 1, 10, 100 and 1000 cm; $\bar{e}_{in} = 0{\cdot}637$, $H = 0{\cdot}02\ \text{kp}^{-1}\,\text{cm}^2$, $\bar{u}_{ao}{}^* = 1\ \text{kp/cm}^2$, $\Delta\sigma = 4 - 2 = 2\ \text{kp/cm}^2$, $\mu_t = 0$; (*b*) the corresponding maximum pore-pressure development

permeability are correlated with the void ratios of saturated soils in Figure 12.5 (*a*). The degrees of saturation $S_r = 1{\cdot}00$, $0{\cdot}96$ and $0{\cdot}92$ have been assumed alternatively and the following initial drainage paths (layer thicknesses) considered: $h = 1$ cm, 10 cm, 1 m, 10 m. The initial void

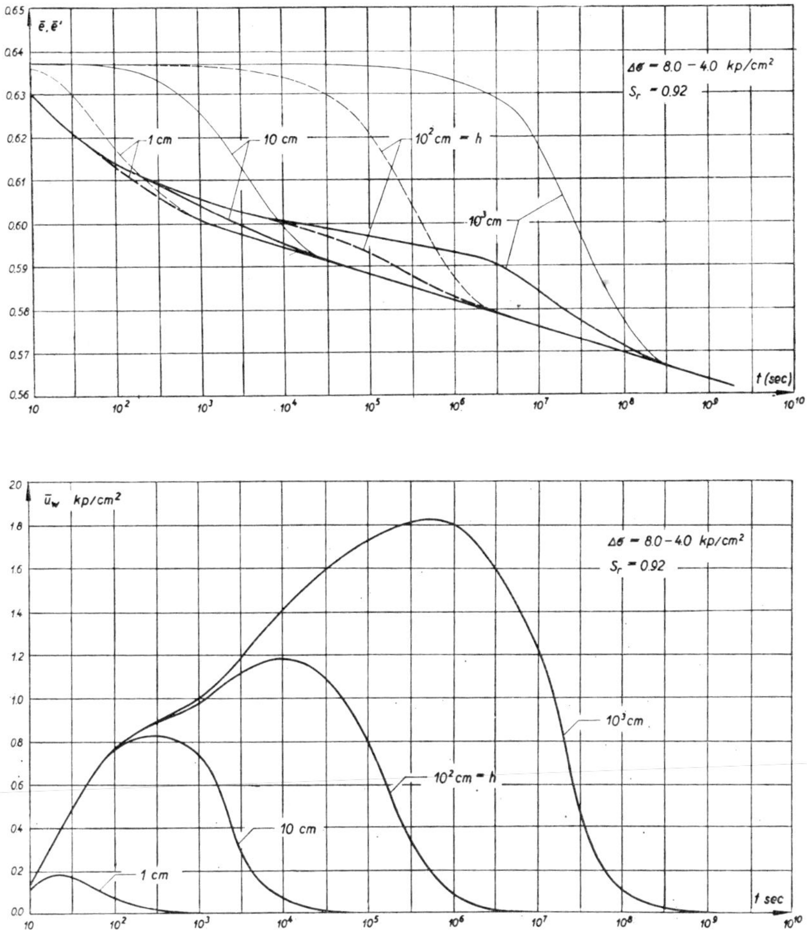

Fig. 12.7. As in Fig. 12.6, for $S_r = 0{\cdot}92$

ratio has been taken as large as $\bar{e} = 0{\cdot}637$ and the initial axial stress $\sigma = 2$ kp/cm². Furthermore the following values have been assumed: $H = 0{\cdot}02$ kp^{-1} cm², $\bar{u}_{w0} = 0$, $\bar{u}_{a0}{}^* = 1$ kp/cm², $\Delta\sigma = 4 - 2 = 2$ kp/cm², $S_{rf} = 1$, $\mu_t = 0$.

The resulting consolidation curves $\bar{e} = \bar{e}(t)$ are presented, in Figs. 12.5, 12.6 and 12.7 parts (*a*), by thick lines; the thin lines represent the corresponding apparent consolidation curves (*b*) facilitating the construction of the true consolidation curves (*a*). The corresponding pore-pressure development has been presented in the same figures in parts (*b*).

The example proves the important influence of the degree of saturation and of the layer thickness on the development of settlements and pore-pressures.

Part C

EXPERIMENTAL AND THEORETICAL FUNDAMENTALS OF THE THREE-DIMENSIONAL CONSOLIDATION ANALYSIS

13

Three-Dimensional Consolidation Analysis Neglecting the Seepage Resistance

13.1 STRESS–STRAIN RELATIONSHIPS IN LIMITED STRESS- AND TIME-INTERVALS

13.1.1 Spherical and Distortional Components of Creep

When in Sections 10.3 and 11.5 the basis and results of the analysis of one-dimensional consolidation were discussed, the possibilities and limitations of their application in computing settlements and their development were established. In most methods of one-dimensional consolidation analysis the strains in the drainage direction are related to stresses acting in the same direction. In some of the methods dealt with, however, the stress–strain relationships were expressed by elastic and viscous parameters governing the displacements in a homogeneous, isotropic soil medium. These general effective stress–strain relationships are the subject of the present and following sections.

The behaviour of soil under an additional loading will be studied in relation to triaxial samples subjected to equal stresses in the radial direction. Nevertheless, the influence of the mean principle stress will also be observed and considered when deciding the general stress–strain relationships.

In order to separate the spherical and distortional creep effects from the influence of the seepage resistance, triaxial tests neglecting the seepage resistance will be treated first. Thus, dry soil samples will be the subject of the present section. Saturated or partly saturated soil samples of which the permeability is high enough or the drainage paths short enough to exclude any noticeable influence of seepage resistance on the consolidation speed will be considered as well. Finally, with some limitations, undrained triaxial samples of saturated soils remaining at constant volume when exposed to additional loading, and drained triaxial samples during the secondary phase of consolidation will be studied.

In Fig. 13.1 the consolidation curves $\Delta\varepsilon_1 = \Delta\varepsilon_1(t)$ and $\Delta\varepsilon_v = \Delta\varepsilon_v(t)$ of a cylindrical sample of the dry powder of a lacustrine chalk are presented. The sample of diameter $d = 10$ cm and of height $h = 20{\cdot}5$ cm was tested in a triaxial cell. After pre-consolidation at an all-round pressure of $\sigma_1 = \sigma_2 = \sigma_3 = 0{\cdot}5$ kp/cm² the sample was exposed to a load increment of $\Delta\sigma_2 = \Delta\sigma_3 = 0$ and $\Delta\sigma_1 = 0{\cdot}5$ kp/cm². The development of the axial linear strain $\Delta\varepsilon_1$ and the volume strain $\Delta\varepsilon_v$ was observed. When determining the volume strain, the influence of temperature was

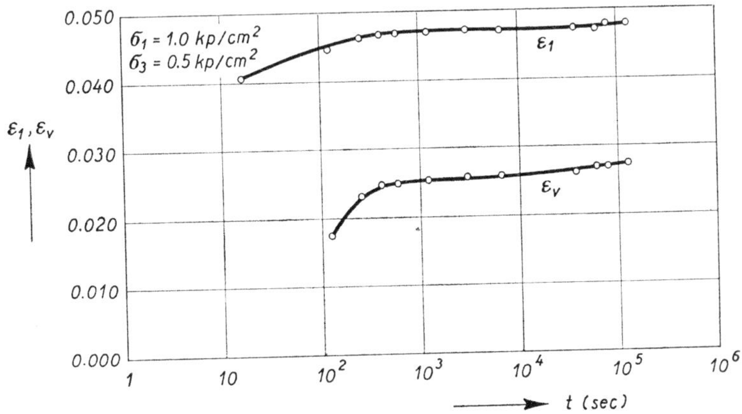

Fig. 13.1. Triaxial test creep curves of the dry powder of a lacustrine chalk; $\Delta\varepsilon_1$ . . . axial strain increase, $\Delta\varepsilon_v$ . . . volume strain increase

taken into account as well as the effects of the deformability of the apparatus and of the measuring device, recording the volume of water in the triaxial cell.

The linear axial strain $\Delta\varepsilon_1$ observed at a given time during consolidation can be separated into its spherical component $\Delta\varepsilon_1{}^0$:

$$\Delta\varepsilon_1{}^0 = \frac{\Delta\varepsilon_v}{3} \tag{13.1}$$

and its distortional component $\Delta\varepsilon_1{}^d$:

$$\Delta\varepsilon_1{}^d = \Delta\varepsilon_1 - \frac{\Delta\varepsilon_v}{3} \tag{13.2}$$

Homogeneity and isotropy of the sample may be assumed together with a linear effective stress–strain relationship in the load interval $\Delta\sigma_1$, $\Delta\sigma_2 = \Delta\sigma_3$. Thus, the deformation constants K and G (see Section 4.1.3) corresponding to the strains at time t are:

(*a*) the modulus of compression:

$$K = \frac{\Delta\sigma_{oct}'}{\Delta\varepsilon_v} = \frac{\frac{1}{3}(\Delta\sigma_1' + 2\Delta\sigma_3')}{\Delta\varepsilon_v} = \frac{\Delta\sigma_1' + 2\Delta\sigma_3'}{3\Delta\varepsilon_v} \tag{13.3}$$

(*b*) the modulus of shear:

$$G = \tfrac{1}{2}\cdot\frac{\Delta\sigma_1'^d}{\Delta\varepsilon_1^d} = \tfrac{1}{2}\cdot\frac{\Delta\sigma_1' - \Delta\sigma_{oct}}{\Delta\varepsilon_1 - \frac{1}{3}\Delta\varepsilon_v} = \frac{\Delta\sigma_1' - \Delta\sigma_3'}{3\Delta\varepsilon_1 - \Delta\varepsilon_v} \tag{13.4}$$

As the seepage resistance is assumed to be small enough to be neglected, the effective pressures can be taken equal to the total pressures; hence:

$$K = \frac{\Delta\sigma_1 + 2\Delta\sigma_3}{3\Delta\varepsilon_v} \tag{13.3-a}$$

Since the difference $\Delta\sigma_1' - \Delta\sigma_3'$ is not affected by pore pressures,

$$\Delta\sigma_1' - \Delta\sigma_3' = \Delta\sigma_1 - \Delta\sigma_3 \tag{13.5}$$

and

$$G = \frac{\Delta\sigma_1 - \Delta\sigma_3}{3\Delta\varepsilon_1 - \Delta\varepsilon_v} \tag{13.4-a}$$

Using the relations (4.14) and (4.17) the modulus of linear deformation E and the Poisson ratio ν can then be found by the equations:

$$E = \frac{9KG}{3K + G} \tag{13.6}$$

$$\nu = \frac{3K - 2G}{2(3K + G)} \tag{13.7}$$

If the creep is observed over a sufficiently long period the consolidation curves can be extrapolated fairly accurately in several cases. Frequently the final observations indicate a logarithmic dependence for the compression with time. When plotted against the logarithm of time the spherical and distortional consolidation curves are straight lines and can easily be extended throughout the life-time (t_e) of the construction under investigation. The restriction of the logarithmic law of secondary compression has already been discussed in Section 8.5.

13.1.2 Influence of Anisotropy on the Spherical and Distortional Stress–Strain Relationships

Suppose the consolidation has been observed long enough to enable the determination of the 'end strains' for the construction in question.

Furthermore, the sample may be assumed to exhibit an orthogonal symmetrical anisotropy the axes of which coincide with the principal stress axes. The deformation modulus in the direction 'i' will be denoted by E_i, and the Poisson ratio governing the influence of the stress σ_k on the strain ε_i, by ν_{ik}.

For the sample defined above the equivalent elastic constants, corresponding to an isotropic sample exhibiting, under the same stress state, equal volume strain and equal distortional strains, may be found. By equating the corresponding strains of the anisotropic and isotropic samples, the following relationships can be obtained (Šuklje, 1963-a):

$$K = \frac{E_1}{3(1 - 2\nu_{31})} \cdot \frac{1 + 2\alpha'}{1 + 2\alpha' \dfrac{(1 - \nu_{23} - \nu_{13})E_1}{(1 - 2\nu_{31})E_3}} \tag{13.8}$$

$$G = \frac{E_1}{2(1 + \nu_{31})} \cdot \frac{1 - \alpha'}{1 - \alpha' \dfrac{(1 - \nu_{23} + 2\nu_{13})E_1}{(1 + \nu_{31})E_3}} \tag{13.9}$$

where

$$\alpha' = \frac{\Delta\sigma_3'}{\Delta\sigma_1'} \tag{13.10}$$

and

$$\nu_{13} = \nu_{31} \frac{E_3}{E_1} \tag{13.11}$$

Equation (13.11) involves the assumption that the work done during deformation does not depend on the sequence applying stresses in the principal directions. The equivalent moduli K and G can of course be found immediately using equations (13.3) and (13.4). If, however, the solutions for the anisotropic half-space are to be applied, equations (13.8) and (13.9) allow the determination of the four unknown independent elastic characteristics E_1, E_3, ν_{31} and ν_{23} in cases of elastic soils with reversible strains. In such cases these constants can be assumed to be independent of the ratio α' and the two pairs of Equations (13.8) and (13.9) corresponding to two values of α' (e.g. $\alpha' = 0$ and $\alpha' \to \infty$) will yield all four elastic constants. Previously the values of K and G had to be calculated from Equations (13.3) and (13.4). In other cases the deformation constants corresponding to increasing principal stress are different from the moduli and Poisson ratios obtained for negative differences $\Delta\sigma_1'$ and $\Delta\sigma_3'$.

13.1.3 Dilatancy

In soil mechanics, the volume strain caused by the distortional components of the stress tensor at constant octahedral normal stress, is called 'dilatancy.' The dilatancy effect is foreign to true Hookean bodies.

Due to the dilatancy effect the volume strain $\Delta\varepsilon_v$, as observed in triaxial tests, is given by Skempton and Bishop (1954) in the form:

$$\Delta\varepsilon_v = C_c\{\Delta\sigma_3' + S_d(\Delta\sigma_1' - \Delta\sigma_3')\} \tag{13.12}$$

where S_d is the structural parameter introduced by Skempton. The parameter S_d is found to have values ranging from slightly more than $+1$ for sensitive clays, to values of less than -1 for dense sands (Skempton and Bishop, 1954). It is evident that $S_d = \frac{1}{3}$ for an isotropic elastic material.

Now, the volume strain of anisotropic soils can be expressed as follows:

$$\Delta\varepsilon_v = \Delta\varepsilon_1 + 2\Delta\varepsilon_3 =$$

$$= \left(\frac{\Delta\sigma_1'}{E_1} - 2\nu_{13}\frac{\Delta\sigma_3'}{E_3}\right) + 2\left\{\frac{\Delta\sigma_3'}{E_3}(1-\nu_{32}) - \frac{\Delta\sigma_1'}{E_1}\nu_{31}\right\} \tag{13.13}$$

$$\Delta\varepsilon_v = \Delta\sigma_1'\frac{1-2\nu_{31}}{E_1} + \Delta\sigma_3'\frac{2(1-\nu_{32}-\nu_{13})}{E_3} \tag{13.14}$$

The comparison of corresponding factors in Equations (13.12) and (13.14) yields:

$$\frac{1}{S_d} = 1 + 2\frac{1-\nu_{32}-\nu_{13}}{1-2\nu_{31}}\cdot\frac{E_1}{E_3} \tag{13.15}$$

$$\frac{1}{C_c} = \frac{E_1}{1-2\nu_{31}}S_d \tag{13.16}$$

Thus, the dilatancy effect can be interpreted also as the effect of the deformation anisotropy.

The dilatancy effect as well as the influence of anisotropy have been proved to depend on the principal stress ratio $\alpha' = \Delta\sigma_3'/\Delta\sigma_1'$. Therefore the equivalent elastic constants K and G have to be determined for a suitable average ratio α' when applying the solutions based on the theory of elasticity. The equivalent elastic constant cannot, however, be used in the case $\alpha' = -\frac{1}{2}$.

13.2 CONSOLIDATION ANALYSIS BASED ON RHEOLOGICAL MODELS

13.2.1 Linear Rheological Model consisting of a Maxwell Body Connected in Parallel with a Hookean Spring in Series with a Saint-Venant Element

(*a*) $p < p_V$.

When dealing with the creep of unsaturated soils Folque (1961) applies the rheological model shown in Fig. 13.2. The model represents the

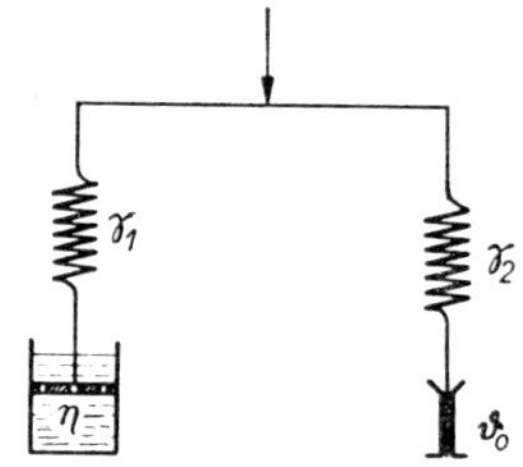

Fig. 13.2. Folque's rheological model

parallel connection of a Maxwell body and a two-element body $\gamma_2 - \vartheta_0$ consisting of a Hookean spring γ_2 and a Saint-Venant element ϑ_0:

$$(\gamma_1 - \eta) \parallel (\gamma_2 - \vartheta_0)$$

The resistance of the Saint-Venant element ϑ_0 may be denoted by p_V and the stress states

$$0 \leqslant p_H < p_V \tag{13.17}$$

may first be investigated. The following subscripts will be used: M for the Maxwell body and H for the Hookean spring. The fundamental stress equation is

$$p = p_{\mathrm{M}} + p_{\mathrm{H}} \tag{13.18}$$

Applying Equation (4.27) in the form

$$\dot{\varepsilon} = \frac{\dot{p}_{\mathrm{M}}}{\gamma_1} + \frac{p_{\mathrm{M}}}{\eta} \tag{13.19}$$

the following expression can be derived:

$$p_{\mathrm{M}} = \eta\left(\dot{\varepsilon} - \frac{\dot{p}_{\mathrm{M}}}{\gamma_1}\right) \tag{13.20}$$

Equation (13.18) can now be written in the form:

$$p = \eta\left(\dot{\varepsilon} - \frac{\dot{p}_M}{\gamma_1}\right) + \gamma_2\varepsilon \tag{13.21}$$

In the case of a constant stress state

$$p = p_M + p_H = \text{const} = p_c \tag{13.22}$$

and the differentiation gives:

$$\dot{p}_M = -\dot{p}_H = -\gamma_2\dot{\varepsilon} \tag{13.23}$$

So Equation (13.21) becomes:

$$p_c = \eta\dot{\varepsilon}\left(1 + \frac{\gamma_2}{\gamma_1}\right) + \gamma_2\varepsilon \tag{13.24}$$

or

$$\frac{d\varepsilon}{p_c - \gamma_2\varepsilon} = A\,dt, \quad A = \frac{\gamma_1}{\eta(\gamma_1 + \gamma_2)} \tag{13.25}$$

Integrating and solving with respect to ε we get:

$$\varepsilon = \frac{p_c}{\gamma_2}\left\{1 - \left(1 - \frac{\varepsilon_0\gamma_2}{p_c}\right)\exp\left(-\gamma_2 A t\right)\right\} \tag{13.26}$$

For $t \to \infty$ this equation reduces to

$$\varepsilon_\infty = \frac{p_c}{\gamma_2} \tag{13.27}$$

as expected (see Fig. 13.2). Due to the viscosity the displacement of the Newton element after the load application initially ($t = 0$) equals zero. Thus, the initial strains are governed only by the deformation moduli of the Hookean springs:

$$\varepsilon_0 = \frac{p_{M_0}}{\gamma_1} = \frac{p_{H_0}}{\gamma_2} \tag{13.28}$$

By inserting p_{M0} and p_{H0}—obtained from the above relation—into Equation (13.18), the following relation is obtained:

$$\varepsilon_0 = \frac{p_c}{\gamma_1 + \gamma_2} \tag{13.29}$$

In Fig. 13.3 the strain ε expressed by Equation (13.26) taking the boundary values ε_0 (Equation 13.29) and ε_∞ (Equation 13.27) is plotted against the time t (Folque, 1961).

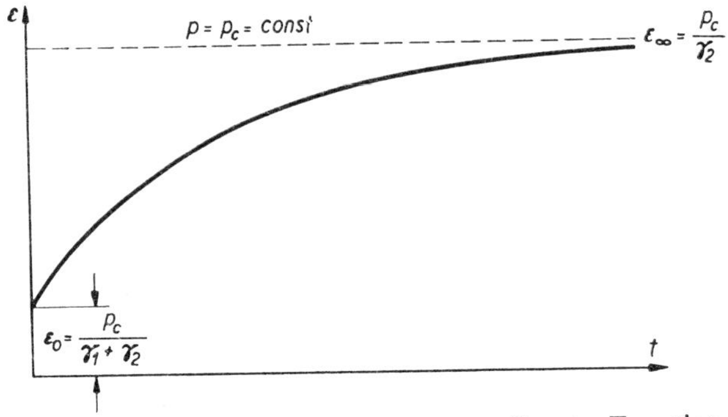

Fig. 13.3. Creep at constant stress state, according to Equation (13.26) (Folque's rheological model at $p < p_v$)

Equation (13.26) holds also for unloading from a load p_a to a constant load $p_c = p_a - \Delta p$. In this case the initial strain ε_0 is given by the expression

$$\varepsilon_0 = \varepsilon_a - \frac{\Delta p}{\gamma_1 + \gamma_2} \tag{13.30}$$

ε_a denoting the preceding strain corresponding to the pressure p_a and the second term denoting the sudden elastic reversible strain due to the pressure decrease $\Delta p = p_a - p$. At complete unloading, $p_c = 0$, $\Delta p = p_a$, Equation (13.26) reduces to

$$\varepsilon = \varepsilon_0 \exp(-\gamma_2 A t) \tag{13.31}$$

The entire strain is recoverable: $\varepsilon_{t\to\infty} = 0$ (see Fig. 13.4).

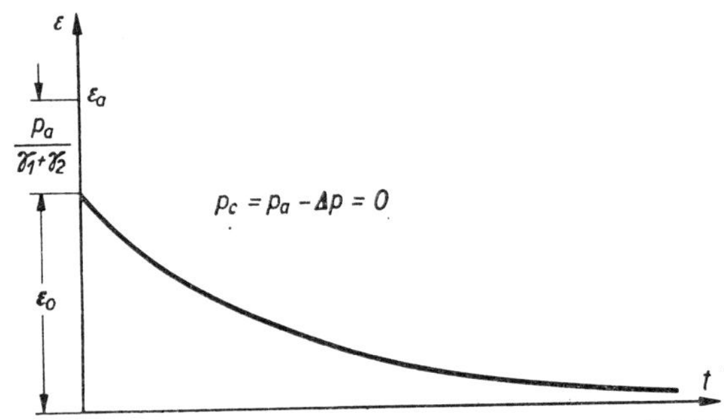

Fig. 13.4. Creep at unloading, according to Equation (13.31) (Folque's rheological model at $p < p_v$)

If the strain is kept constant ($\varepsilon = \varepsilon_c = \text{const}$, $\dot{\varepsilon} = 0$) relaxation of stress appears. Equation (13.21) has the form:

$$p = -\eta \frac{\dot{p}_M}{\gamma_1} + \gamma_2 \varepsilon_c \tag{13.32}$$

As the length of the spring γ_2 does not vary, the stress p_H remains constant and $\dot{p}_H = 0$. Thus, by differentiating Equation (13.18), $\dot{p} = \dot{p}_M$ and Equation (13.32) becomes:

$$\frac{dp}{\gamma_2 \varepsilon_c - p} = \frac{\gamma_1}{\eta} dt \tag{13.33}$$

By integrating from $t = 0$ to t and by rearranging:

$$p = \gamma_2 \varepsilon_c \left\{ 1 + \left(1 - \frac{p_0}{\gamma_2 \varepsilon_c} \right) \exp \left(-\frac{\gamma_1}{\eta} t \right) \right\} \tag{13.34}$$

At the limit $t \to \infty$ this expression reduces to

$$p_\infty = \gamma_2 \varepsilon_c \tag{13.35}$$

In Fig. 13.5 the stress p (Equation 13.34) is plotted versus time t.

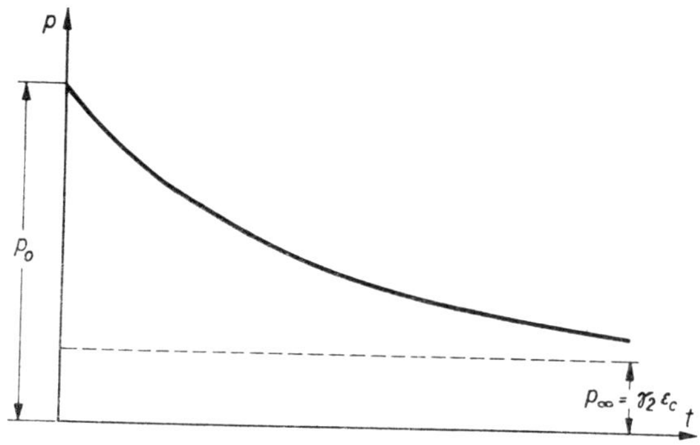

Fig. 13.5. Relaxation of stresses of constant strain, according to Equation (13.34) (Folque's rheological model at $p < p_v$)

In Folque's investigations the soil is treated as a medium having constant elastic and viscous characteristics in any load interval. The rheological model presented in Fig. 13.2 could, however, be applied also to normally consolidated soils the deformations of which are not entirely reversible. In this case different constants γ_1, γ_2 and η should be assumed for different load intervals and each load interval should be analysed separately. Fig. 13.6 illustrates an example when the constants γ_1, γ_2 and η apply for the loading stress interval $\Delta p = p_a$, and the constants $\bar{\gamma}_1$, $\bar{\gamma}_2$ and $\bar{\eta}$ for the unloading interval $\Delta p = -p_a$.

(*b*) $p \geqslant p_V$

After reaching the value of the Saint-Venant resistance p_V, the stress p_H in the Hookean spring no longer increases.

$$p_H = \gamma_2 \varepsilon_H = p_V \tag{13.36}$$

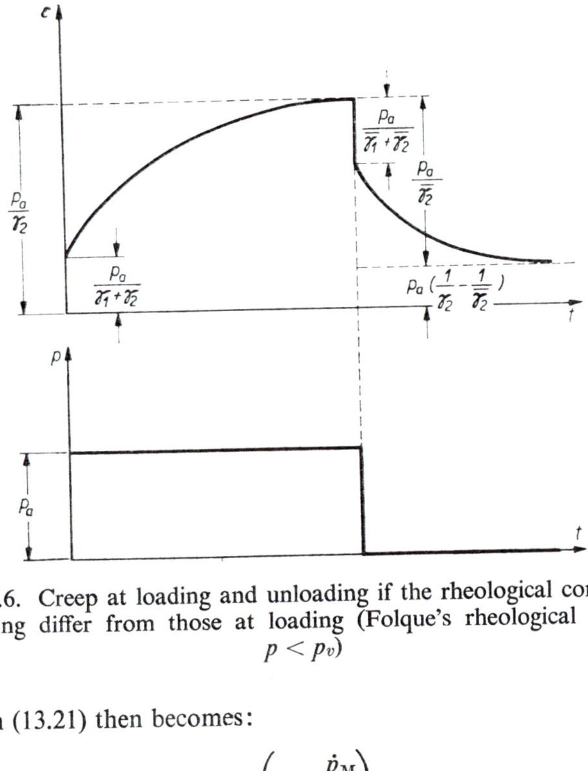

Fig. 13.6. Creep at loading and unloading if the rheological constants at unloading differ from those at loading (Folque's rheological model at $p < p_v$)

Equation (13.21) then becomes:

$$p = \eta\left(\dot{\varepsilon} - \frac{\dot{p}_M}{\gamma_1}\right) + p_V \tag{13.37}$$

If the total stress p is kept constant ($p = p_c = \text{const}$), the relations (13.18) and (13.36) require the stress p_M in the Maxwell body to be constant too. In this case, though, Equation (13.37) reduces to the equation of the Bingham body (Equation 4.34):

$$p_c = \eta\dot{\varepsilon} + p_V \tag{13.38}$$

With the stresses and strains in the Hookean springs γ_1 and γ_2 remaining constant, the development of the stresses and strains is governed only by the Newtonian dashpot η and the Saint-Venant element ϑ_0 (see Fig. 13.2) forming a Bingham body.

After integrating between the limits $t = t_2$ and $t = t_2 + \Delta t$ Equation (13.38) yields:

$$\varepsilon = \varepsilon_2 + \frac{p_c - p_V}{\eta} \Delta t; \tag{13.39}$$

ε_2 denotes the strain at time t_2 when the stress in the Hookean element γ_2 reaches the value p_V. For $t \leqslant t_2$ Equation (13.23) should be applied. The combined graph $\varepsilon = \varepsilon(t)$, $t \leqslant t_3$, is shown in Fig. 13.7.

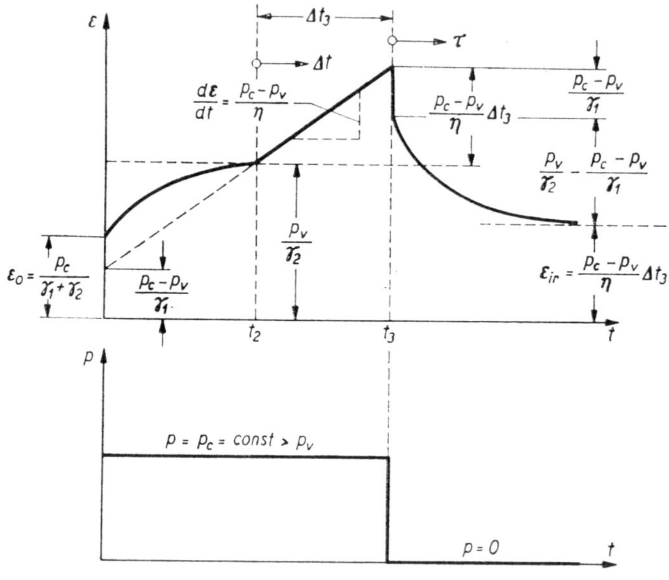

Fig. 13.7. Creep of the soil according to Folque's rheological model: at loading when $p > p_v$ and at subsequent unloading; at the time t_2 the stress in the Hookean element γ_2 has reached the value p_V

The stress p being kept constant ($p = p_c = \text{const}$), the strain ε can be expressed as the strain of the Maxwell body $\gamma_1 - \eta$ as follows:

$$\varepsilon = \frac{p_c - p_V}{\gamma_1} + \frac{p_c - p_V}{\eta} t \tag{13.40}$$

t denotes the time from the beginning of the load application $p_c = \text{const}$. Now, in Equation (13.39) the first term $\varepsilon_2 = p_V/\gamma_2$ represents the total reversible strain, and the second term the irreversible part of the strain. Its magnitude can be obtained by equating Equations (13.39) and (13.40).

$$\varepsilon_{ir} = \frac{p_c - p_V}{\eta} \Delta t = \frac{p_c - p_V}{\eta} t - \frac{p_c - p_V}{\gamma_1} - \frac{p_V}{\gamma_2} \tag{13.41}$$

On unloading, the Hookean spring γ_1 relaxes instantaneously; thus, the part $(p_c - p_V)/\gamma_1$ of the reversible strain occurs immediately. The remaining part of the reversible strain $[(p_V/\gamma_2) - (p_c - p_V)/\gamma_1]$ develops according to Equation (4.38) of the Kelvin body $\eta||\gamma_2$. Using the symbols applied in this section, the strain is given by:

$$\varepsilon = \left(\frac{p_V}{\gamma_2} - \frac{p_c - p_V}{\gamma_1}\right) \exp\left(-\frac{\gamma_2}{\eta}\tau\right) \tag{13.42}$$

where $\tau = t - t_3$ (see Fig. 13.7).

The development of strain on unloading ($p = 0$) is presented in Fig. 13.7, $t > t_3$.

If the strain is kept constant, the stress relaxes. As $\dot{p}_H = 0$ (see Equation 13.36), in Equation (13.37) $\dot{p}$ can be substituted for $\dot{p}_M$ ($\dot{p}_M = \dot{p}$). By integrating and rearranging we get:

$$p = p_V \left\{1 - \left(1 - \frac{p_0}{p_V}\right) \exp\left(-\frac{\gamma_1}{\eta} t\right)\right\} \tag{13.43}$$

p_0 denoting the initial stress at time $t = 0$. At the limit $t \to \infty$ Equation (13.43) yields

$$p_\infty = p_V \tag{13.44}$$

(*c*) $\dot{\varepsilon} = \text{const}$

Finally the case of constant strain velocity $\dot{\varepsilon} = \text{const} = (\dot{\varepsilon})_c$ may be investigated. Under the condition $p < p_V$ and assuming zero initial stress ($p_0 = 0$ at $t_0 = 0$), the stress p_M in Equation (13.18) can be expressed by applying Equation (4.29) of the Maxwell body. Inserting further $p_H = \gamma_2 \varepsilon$ and considering the notation used in this section, Equation (13.18) becomes:

$$p = (\dot{\varepsilon})_c \eta \left\{1 - \exp\left(-\frac{\gamma_1}{\eta} t\right)\right\} + \gamma_2 \varepsilon \tag{13.45}$$

When $p \geqslant p_V$, the initial stress p_0 in Equation (4.29) at time $\Delta t = 0$ (see Fig. 13.7) should be taken as $p_0 = p_c - p_V$ and the second term in Equation (13.18) has the value p_V. Hence:

$$p = [(\dot{\varepsilon})_c \eta + p_V] + [p_c - \{(\dot{\varepsilon})_c \eta + p_V\}] \exp\left(-\frac{\gamma_1}{\eta} t\right) \tag{13.46}$$

(*d*) Determination of the rheological constants

The rheological constants γ_1, γ_2, η and p_V can be determined by applying the appropriate triaxial tests. Folque (1961) proved his rheological scheme by carrying out unconfined triaxial tests (uniaxial loading) where $p_{xx} = p_{yy} = 0$. The axial (ε_{zz}) as well as the lateral strains ($\varepsilon_{xx} = \varepsilon_{yy}$) were measured.

The equations derived under parts (*a*) and (*b*) were applied to express the relationships between the distortional stress and the distortional strain components in the direction of the axis z:

$$p_{zz}^{\ d} = \sigma_1^{\ d} = \tfrac{2}{3}p_{zz} \tag{13.47}$$

$$\varepsilon_{zz}^{\ d} = \varepsilon_1^{\ d} = \tfrac{2}{3}(\varepsilon_{zz} - \varepsilon_{xx}) \tag{13.48}$$

The constant p_V can then be determined by using the relaxation test in the region of plastic deformation ($p > p_V$). The strain should be kept constant ($\dot{\varepsilon}_1^{\ d} = 0$) so long as the plot $\sigma_1^{\ d} = \sigma_1^{\ d}(t)$ approaches the horizontal asymptote, yielding $p_\infty = p_V$ (Equation 13.44).

The coefficient of viscosity η can then be obtained under the following test conditions: $\sigma_1^{\ d} = \text{const}_\sigma > p_V$, $\dot{\varepsilon}_1^{\ d} = \text{const}_\varepsilon$. On the basis of the plot shown in Fig. 13.7, the result can immediately be derived:

$$\eta = \frac{\sigma_1^{\ d} - p_V}{\dot{\varepsilon}_1^{\ d}} \tag{13.49}$$

Once the rheological constants p_V and η are known, the shear modulus $\gamma_1 \equiv G_1$ can be found by carrying out a test at constant strain velocity ($\dot{\varepsilon} = \text{const}$) under the condition $p > p_V$. Inserting the known values η and p_V, and the stress p observed at an arbitrary time t into Equation (13.46), the constant γ_1 is the only unknown quantity.

Finally, by using Equation (13.45) the shear modulus $\gamma_2 \equiv G_2$ can be found from the first phase of the same test, i.e. in the region $\sigma_1^{\ d} < p_V$.

Only the distortional stress–strain relationships can, however, be governed by the constants p_V, η, γ_1 and γ_2 determined in the way described above. The rheological constants governing the spherical strains must be found separately by applying an analogous procedure for the spherical components of the stress and strain tensors. Furthermore, the rheological characteristics vary with the consistency of the soil. For the distortional viscosity coefficient η Folque (1961) assumes the law

$$\eta = \eta_0 \exp{(K\varepsilon_V)} \tag{9.50}$$

If in the domain $p < p_V$ different rheological constants are supposed to express the behaviour of soils during loading and unloading respectively, the number of the parameters determining the consolidation process increases further. Thus, in the author's opinion, the possibility of practical application of the model is rather limited. The anisotropic effects cause further complications.

(*e*) Pre-consolidated soils

In order to adapt the rheological model shown in Fig. 13.2 to the rheological behaviour of pre-consolidated soils, Folque (1961) has added a Hookean spring (γ_3) connected in series with the Saint-Venant element ϑ_0.

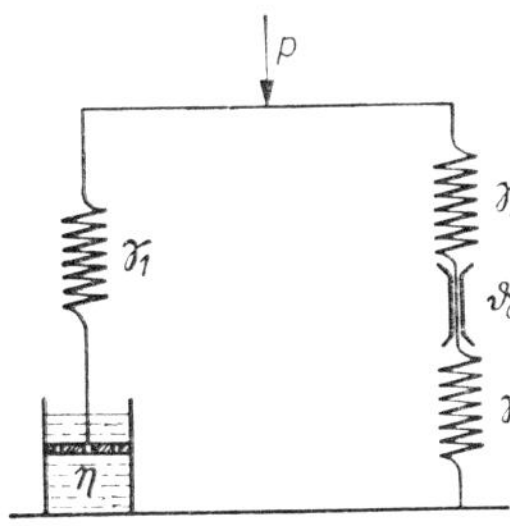

Fig. 13.8. Folque's rheological model for preconsolidated soils

The improved model is presented in Fig. 13.8. Its fundamental rheological equation can be deduced from the strain condition:

$$\varepsilon = \varepsilon_2 + \varepsilon_3 \tag{13.50-a}$$

and the stress condition:

$$p = p_M + p_H \tag{13.50-b}$$

Considering the relations:

$$p_H = \gamma_2 \varepsilon_2, \qquad \gamma_3 \varepsilon_3 = p_H - p_V \tag{13.50-c, d}$$

the strain condition gives:

$$p_H = \frac{\gamma_2 \gamma_3 \varepsilon + \gamma_2 p_V}{\gamma_2 + \gamma_3} \tag{13.50-e}$$

and the stress condition can then be written in the form

$$p = \eta \left(\dot{\varepsilon} - \frac{\dot{p}_M}{\gamma_1} \right) + \frac{\gamma_2 \gamma_3 \varepsilon + \gamma_2 p_V}{\gamma_2 + \gamma_3} \tag{9.50-f}$$

where the expression (13.20) has been taken into account for p_M.

Proceeding from the above equation Folque (1961) has given analytical interpretations of the rheological phenomena of pre-consolidated soils.

13.2.2 Rheological Model Consisting of a Hookean Spring Connected in Series with a modified Voigt element

Murayama and Shibata (1961) investigated the rheological behaviour of soils on the basis of the model shown in Fig. 13.9. The Hookean spring E_1 is connected in series with a modified Voigt element consisting of the Hookean spring E_2, the Saint-Venant resistance σ_0 and the Newtonian

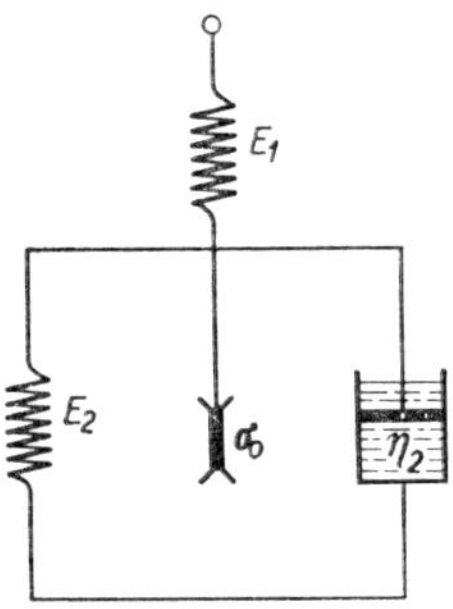

Fig. 13.9. Murayama and Shibata's rheological model

dashpot η_2 all connected in parallel. The coefficient of viscosity η_2 of the latter is considered to vary according to the expression

$$\eta_2 = A_2 \frac{\sigma - \sigma_0}{\sigma_2} \sin\mathrm{h} \frac{B_2 \sigma_2}{\sigma - \sigma_0} \tag{13.51}$$

A_2 and B_2 being constants. This expression which is based on a formula developed by H. Eyring (1941) for the viscosity of certain liquids by the methods of statistical mechanics, will be inserted in the fundamental equation of the Newtonian dashpot:

$$\sigma_2 = \frac{\dot{\varepsilon}_2}{\eta} \tag{13.52}$$

Solving for σ_2 we have:

$$\sigma_2 = \frac{\sigma - \sigma_0}{B_2} \operatorname{arc\,sin\,h} \frac{\dot{\varepsilon}_2}{A_2(\sigma - \sigma_0)} \tag{13.53}$$

This expression may be inserted into the stress condition of the Voigt element:

$$\sigma = \varepsilon_2 E_2 + \sigma_0 + \sigma_2 \tag{13.54}$$

If

$$0 < \varepsilon_2 < \frac{\sigma - \sigma_0}{2B_2E_2}(2B_2 - 1) \tag{13.55}$$

ε is given approximately as

$$\varepsilon = \frac{\sigma}{E_1} + \frac{\sigma - \sigma_0}{E_2}\left(1 + \frac{1}{B_2}\log\frac{A_2}{2}B_2E_2t\right) \tag{13.56}$$

If ε_2 is beyond the upper limit of relation (13.55), at $t \to \infty$

$$\varepsilon_{t\to\infty} = \frac{\sigma}{E_1} + \frac{\sigma - \sigma_0}{E_2} \tag{13.57}$$

as required by the model (Fig. 13.9).

Equations (13.56) and (13.57) show that the creep strain of a clay (ε) is at first proportional to the logarithm of time t, but that it should approach a limit $\varepsilon_{t\to\infty}$ as shown in Fig. 13.10. Murayama and Shibata

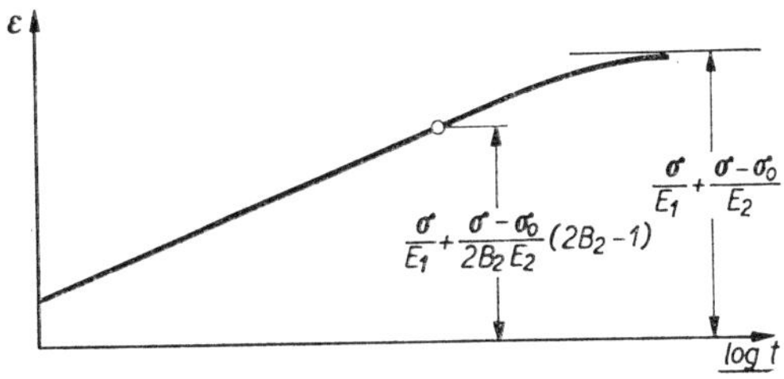

Fig. 13.10. Creep at constant stress according to Murayama and Shibata's (1961) rheological model

performed long term consolidation tests with a compression plastometer consisting of a triaxial testing unit and a recording and control unit. The axial displacement, the pore water pressure at the base of the specimen and the volume of water expelled from it were measured. After the pore pressures had dissipated the secondary consolidation curve had the shape of the logarithmic straight line of Equation (13.56). Similar results have been obtained by the author in most cases of triaxial consolidation tests. The transition towards a horizontal asymptote as shown in Fig. 13.10, however, did not occur during the authors' long term laboratory tests nor is it indicated in the consolidation curves presented by Murayama and Shibata.

When investigating distortional creep the symbol σ in the preceding expressions denotes the distortional stress component. In triaxial tests its magnitude is

$$\sigma = \sigma_1{}^d = \tfrac{2}{3}(\sigma_1 - \sigma_3) \tag{13.58}$$

Differentiating expression (13.56) with respect to the logarithm of time we get:

$$\frac{d\varepsilon}{d \log t} = \frac{\sigma - \sigma_0}{B_2 E_2} \tag{13.59}$$

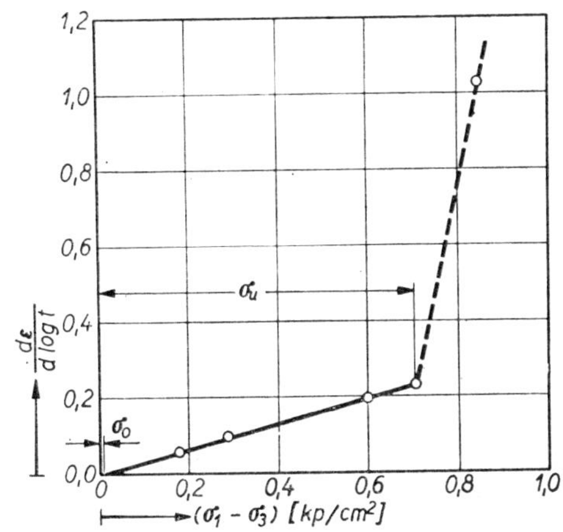

Fig. 13.11. The plot of $\dfrac{d\varepsilon}{d \log t}$ versus $\sigma_1 - \sigma_3 = \frac{3}{2}\sigma$ according to Equation (13.59); reproduced after Murayama and Shibata (1961)

The plot of $d\varepsilon/d \log t$* versus $(\sigma_1 - \sigma_3)$ is a straight line so long as Equation (13.56) is valid. An example of such a plot according to Murayama and Shibata is given in Fig. 13.11. The Saint-Venant resistance σ_0 is represented by the intercept σ_0 on the axis $(\sigma_1 - \sigma_3)$. The upper limit (σ_u) of the validity of the relationship (13.59) determines the distortional stress which if exceeded may lead to rupture after long term creep.

According to the fundamentals of this theory, the coefficient B_2 is inversely proportional to the temperature. Thus, the derivative $d\varepsilon/d \log t$ (see Equation 13.59) must be linearly proportional to the temperature, if the other test conditions are kept constant. Tests made by Murayama and Shibata at eight different temperatures, varying from 5° to 40°, prove

* In Murayama and Shibata's tests the volume strains decreased—in the secondary phase of consolidation—to very small values which were neglected.

the correctness of this theoretical result, a conclusion based on the applied rheological scheme and on relationship (13.51) governing the coefficient of viscosity η_2.

Relaxation

With the strain kept constant ($\varepsilon = \varepsilon_0$), Equations (13.53) and (13.54) lead, according to Murayama and Shibata, to the following approximate solution:

$$\int_{y_0}^{y} \frac{1}{y} \exp(y)\, dy = Ct \tag{13.60}$$

where

$$y = \frac{B_2 E_2}{\sigma - \sigma_0}\left(\varepsilon_0 - \frac{\sigma_0}{E_1}\right) \tag{13.61}$$

$$y_0 = B_2 \frac{E_2}{E_1} \tag{13.62}$$

$$C = \frac{A_2}{2} E_1 \exp\left[B_2\left(1 + \frac{E_2}{E_1}\right)\right] \tag{13.63}$$

For $t \to \infty$ Equation (13.60) yields:

$$\sigma_\infty = \frac{E_1 E_2}{E_1 + E_2}\, \varepsilon_0 \tag{13.64}$$

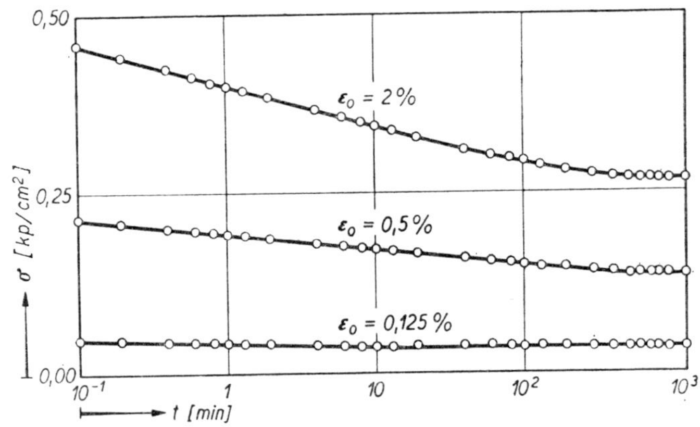

Fig. 13.12. Relaxation of stresses at constant strain, according to Murayama and Shibata (1961)

If, by using approximate test conditions, the rheological constants E_1, E_2 and B_2 are determined, the numerical computation of stresses at various times t and corresponding to various constant strains ε_0 can be carried out using Equation (13.60). Plotting the stress σ against the logarithm of time t, for certain constant strains, the plots $\sigma = [\sigma(t)]_{\varepsilon_0=\text{const}}$ obtained are straight lines over the first part later becoming convex towards the axis t and tending to a horizontal asymptote. Fig. 13.12 shows plots of similar shape based on tests made by Murayama and Shibata with a soil having the following rheological constants: $E_1 = 48{\cdot}0$ kp/cm², $E_2 = 34{\cdot}3$ kp/cm², $B_2 = 5{\cdot}0$.

13.3 THE STATE BOUNDARY SURFACE AND ITS APPLICATION IN STRAIN ANALYSIS

13.3.1 Study of Strains of Normally Consolidated Soils Based on Data from Undrained and 'Anisotropic Consolidation' Tests

13.3.1.1 *The presentation of experimental data*

In Section 13.1 the stress–strain relationships for limited stress- and time-intervals were assumed to be linear for both isotropic and anisotropic deformation conditions. This assumption cannot be sustained for larger stress intervals. When dealing in Part *B* with one-dimensional consolidation, the void ratio was shown to decrease approximately with the logarithm of the axial pressure. Similar relationships can be deduced from triaxial test observations. However, the dilatancy, i.e. the anisotropy effects, requires that the void ratios should be related not only to the spherical stress components, but to the distortional stress components as well.

Since 1958 Roscoe and his colleagues have published several theoretical and experimental studies of the strains in triaxial tests. They presented the stress–strain relationships as observed in triaxial tests by three-dimensional plots relating the void ratio e to the octahedral (spherical) effective stress

$$\sigma_{\text{oct}}' \equiv p \tag{13.65}$$

and to the principal effective stress difference

$$\sigma_1' - \sigma_3' = q \tag{13.66}$$

Although the time factor is not considered in such plots, it can be introduced by preparing the plots at a certain time during the creep process.

For the purpose of this book the paper by Roscoe and Poorooshasb (1963) as well as the study by Roscoe and Schofield (1963) are of special interest. They will be summarized in the present and next sections.

In the (p,q,e) coordinate system, the lines connecting the points representing successive phases of a triaxial test are called state paths. By plotting the complete state paths of conventional undrained ($\Delta\sigma_3 = 0$) and fully drained triaxial compression tests ($\eta = q/p = \text{const}$, $\sigma_1' > \sigma_2' = \sigma_3'$) on normally consolidated samples and analogous paths of tests on over-consolidated samples, the 'yield surface' (Roscoe, Schofield and Wroth, 1958) or 'state boundary surface' (Roscoe and Poorooshasb, 1963) is obtained.

The projections of the state paths on the (p,q) plane are the stress paths. In Fig. 13.13 the line p_0–b–c–d represents the stress path of an undrained test on a non-pre-consolidated sample. The projection of the corresponding state path in the (e,p) plane is the straight line a''–d''. The end point d of the state path represents the yield state when the distortional strains begin to increase without further load.

Further considerations will be related only to normally consolidated soils, whose state paths form the boundary surface between the critical state line and the normal consolidation line.

The stress paths associated with constant volume tests on normally consolidated samples form a series of similar curves which may be represented by the equation

$$\sqrt{(p^2 + q^2)} = p_0 F\left(\frac{q}{p}\right) \tag{13.67}$$

Substituting

$$\frac{q}{p} = \eta \tag{13.68}$$

and

$$\frac{p_0}{p} = \xi \tag{13.69}$$

Equation (13.67) becomes:

$$(1 + \eta^2)^{\frac{1}{2}} - \xi F(\eta) = 0 \tag{13.70}$$

By subjecting non-pre-consolidated samples in drained triaxial tests to all-round pressures $p_0 = \sigma_{\text{oct}} = \sigma_1 = \sigma_2 = \sigma_3$ and by plotting the resulting void ratios e_0 against the pressures p, the 'normal consolidation curve' is obtained (Fig. 13.13). Its empirical equation is [cf. Equation (8.18) for oedometer tests]:

$$e_0 = e_a - \lambda_e \log_e p_0 \tag{13.71}$$

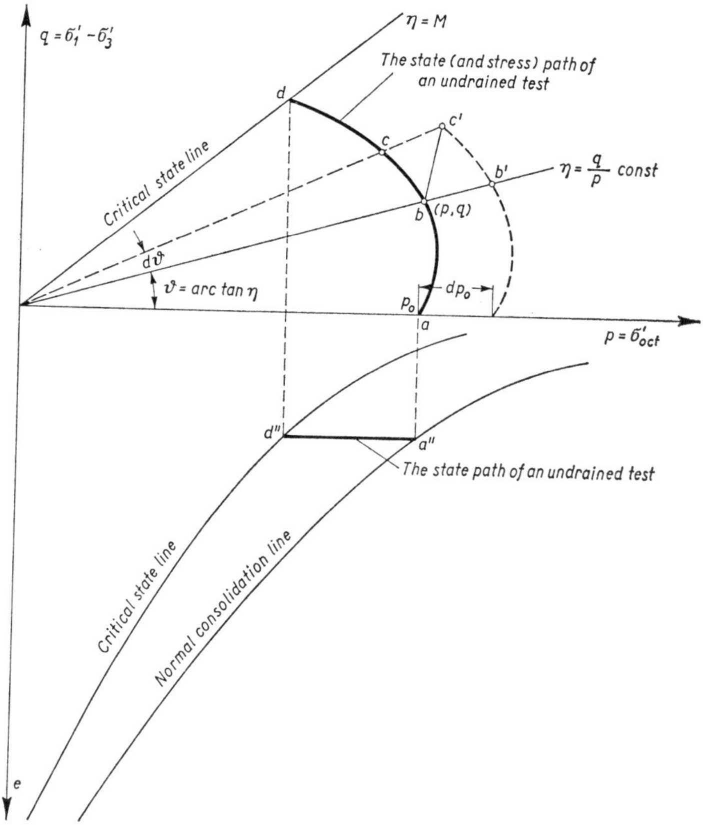

Fig. 13.13. State boundary surface in (q,p) and (e,p) planes. (According to Roscoe and Poorooshasb, 1963)

λ_e is the slope of the consolidation line if the void ratios e_0 are plotted against $\log_e p_0$, and e_a is the void ratio of a normally consolidated saturated sample under unit isotropic stress ($p_0 = 1$). Hence:

$$p_0 = \exp \frac{e_a - e_0}{\lambda_e} \tag{13.72}$$

During an undrained test on a saturated soil the volume does not change.

Thus, when inserting expression (13.72) in to the relation (13.69) for undrained tests, the suffix 0 of the void ratio can be omitted:

$$\xi = \frac{1}{p} \exp \frac{e_a - e}{\lambda_e} \tag{13.73}$$

Equation (13.70) represents the 'normally consolidated' portion of the boundary state surface in the (ξ,η) system of coordinates. The line $\eta = \eta(\xi)$ (Fig. 13.14) can be obtained by plotting the data of the conventional undrained tests ($\Delta\sigma_3 = 0$) and of the normal consolidation test

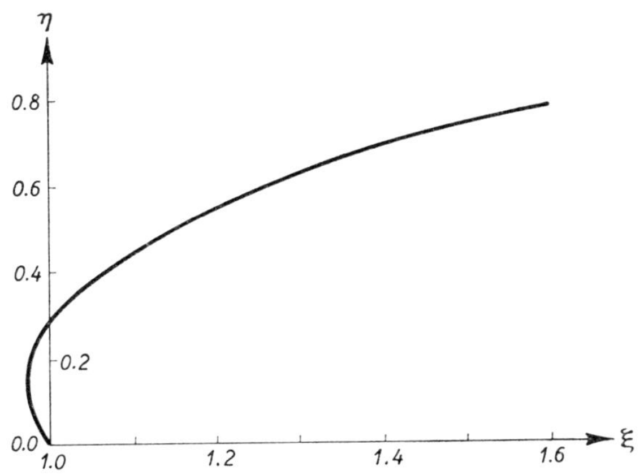

Fig. 13.14. Two-dimensional presentation of state boundary surface for normally consolidated clays according to Equation (13.70) (Roscoe and Poorooshasb, 1963)

($q = 0$). The experimental evidence proves that a continuous line can be drawn through all points (ξ,η) of all such tests, and that the state points (ξ,η) of 'anisotropic consolidation tests' (η = const) of normally consolidated samples lie on the same boundary state surface.

There are two further important results given by the experimental evidence:

(*a*) The contours of equal 'undrained axial strain' $(\varepsilon_1)_{v0}$ (i.e. the total natural axial strain since beginning the conventional undrained process) coincide with the straight lines η = const (Fig. 13.15 (*a*)); consequently:

$$(\varepsilon_1)_{v0} = f(\eta) \tag{13.74}$$

(The suffix 0 indicates that the initial state corresponded to a point on the normal consolidation line, and the suffix v indicates that the volume

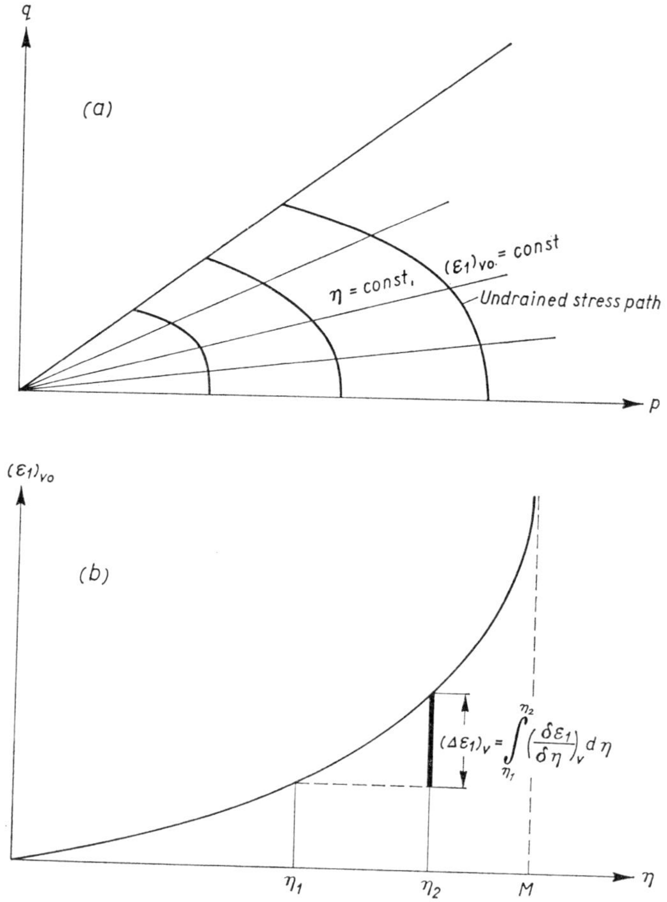

Fig. 13.15. (*a*) Stress paths for undrained tests. (*b*) Variation of axial strain with η. (According to Roscoe and Poorooshasb, 1963)

remains constant throughout the straining process.) This unique relationship between the axial strain in an undrained process and the parameter η is represented by the curve $(\varepsilon_1)_{v0} = f(\eta)$ in Fig. 13.15 (*b*).

(*b*) In 'anisotropic consolidation tests' (drained tests) on normally consolidated samples corresponding to stress paths η = const, the relationship between the total volumetric strain ε_v and the axial strain ε_1 is linear:

$$\left(\frac{d\varepsilon_v}{d\varepsilon_1}\right)_\eta = \text{const} \tag{13.75}$$

(Fig. 13.16 (*a*)). The suffix η indicates the constancy of the parameter η. In the coordinate system $[\eta, (d\varepsilon_v/d\varepsilon_1)_\eta]$ a smooth curve can be drawn through the points *i* relating the gradients $(d\varepsilon_v/d\varepsilon_1)_\eta$ of the straight lines

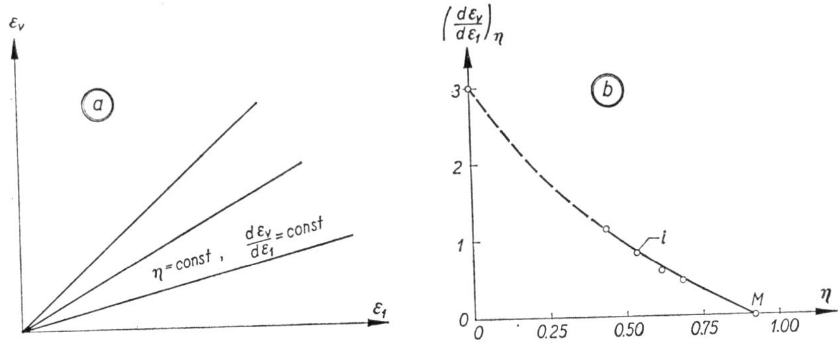

Fig. 13.16. (*a*) Variation of ε_v with ε_1 and (*b*) of $\left(\frac{d\varepsilon_v}{d\varepsilon_1}\right)_\eta$ with η during 'anisotropic consolidation tests'. (According to Roscoe and Poorooshasb, 1963)

$\varepsilon_v = \varepsilon_v[(\varepsilon_1)_\eta]$ and the corresponding quotients η. The extreme points of this curve are:

(I)* $d\varepsilon_v/d\varepsilon_1 = 3$ when $\eta = 0$.

(II) $d\varepsilon_v/d\varepsilon_1 = 0$ when $\eta = \eta_{cr} = M, (d\varepsilon_1/d\varepsilon_v \to \infty)$.

13.3.1.2 *The basic differential equation and its application*

Let an element of soil (clay) traverse a short stress path *bc'* (Fig. 13.13). The incremental strain $d\varepsilon_1$ associated with this change of state is taken to be the sum of two components:

$$d\varepsilon_1 = \left(\frac{\partial\varepsilon_1}{\partial\eta}\right)_v d\eta + \left(\frac{\partial\varepsilon_1}{\partial\varepsilon_v}\right)_\eta d\varepsilon_v \tag{13.76}$$

The first component corresponds to the stress path *b–c* of an undrained test occurring at constant volume (indicated by the suffix *v*); the second one corresponds to the stress path *c–c'* of an 'anisotropic consolidation test' occurring at constant ratio η (indicated by the suffix η).

* In case (*I*) the hydrostatic state of stress is assumed to coincide with the spherical state of strain: $d\varepsilon_1 = d\varepsilon_2 = d\varepsilon_3$. This assumption requires that the deformability of the soil is isotropic and is not compatible, in the present author's opinion, with the dilatancy effects causing the curved form of the undrained stress paths.

If the stress path bc' has finite length, the corresponding strain ε_1 can be obtained by integrating Equation (13.76):

$$\varepsilon_1 = \int_{\eta_1}^{\eta_2} \left(\frac{\partial \varepsilon_1}{\partial \eta}\right)_v \mathrm{d}\eta + \int_{\eta_1}^{\eta_2} \left(\frac{\partial \varepsilon_1}{\partial \varepsilon_v}\right)_\eta \mathrm{d}\varepsilon_v \tag{13.77}$$

The first term of the above expression is given by the difference of the $(\varepsilon_1)_{v0}$ values corresponding, in the plot $(\varepsilon_1)_{v0} = f(\eta)$ (Fig. 13.15 (*b*)), to η_1 and η_2 respectively.

$$\int_{\eta_1}^{\eta_2} \left(\frac{\partial \varepsilon_1}{\partial \eta}\right)_v \mathrm{d}\eta = \int_{\eta_1}^{\eta_2} \mathrm{d}(\varepsilon_1)_{v0} = (\Delta\varepsilon_1)_v \tag{13.78}$$

The second term can be obtained from the plot

$$\left(\frac{\partial \varepsilon_1}{\partial \varepsilon_v}\right)_\eta = \phi(\varepsilon_v)$$

as the area between the curve $\phi(\varepsilon_v)$ and the axis ε_v in the interval from $\varepsilon_v = (\varepsilon_v)_{\eta_1}$ to $\varepsilon_v = (\varepsilon_v)_{\eta_2}$. The curve $\phi(\varepsilon_v)$ can be constructed on the basis of the following considerations:

If the initial void ratio is e, the volumetric strain $\mathrm{d}\varepsilon_v$ corresponding to the void ratio change $\mathrm{d}e$ (compression being positive) can be expressed by the equation

$$\mathrm{d}\varepsilon_v = \frac{-\mathrm{d}e}{1+e} \tag{13.79}$$

The volumetric strain corresponding to the state change from (e_0, p_0) to any arbitrary state (e, p) can then be obtained by integrating the above expression between the limits e_0 and e:

$$(\varepsilon_v)_0 = \Big[-\log_e(1+e)\Big]_{e_0}^{e} \tag{13.80}$$

Inserting, for e_0 and e, the expressions given by Equations (13.72) and (13.73), the above equation yields:

$$(\varepsilon_v)_0 = \log_e \frac{1 + e_a - \lambda_e \log_e(p_0)}{1 + e_a - \lambda_e \log_e(p\xi)} \tag{13.81}$$

where p_0 denotes the all-round pressure corresponding, on the normal consolidation curve, to the initial void ratio e_0, and $p\xi$ (cf. Equation 13.69)

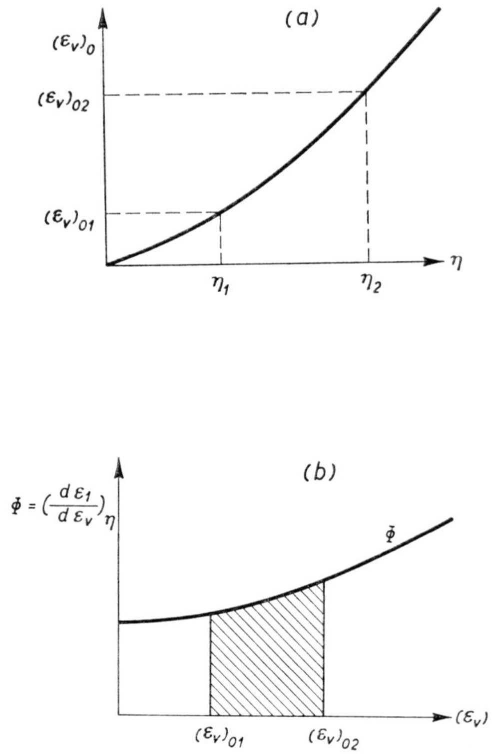

Fig. 13.17. (*a*) $(\varepsilon_v)_0$ versus η plot. (*b*) Determination of $\int_{\eta_1}^{\eta_2} \left(\frac{\partial \varepsilon_1}{\partial \varepsilon_v}\right)_\eta d\varepsilon_v$ (see Equation 13.77). (According to Roscoe and Poorooshasb, 1963)

the all-round pressure p_0' corresponding, on the same curve, to a subsequent void ratio e appearing at the end of the state path.

Now, the graph $\eta = \eta(\xi)$ (Fig. 13.14) representing the boundary state surface enables us to find for any η value the corresponding value of ξ. In this way, Equation (13.81) facilitates the plotting of the volumetric strain $(\varepsilon_v)_0$ against the parameter η (Fig. 13.17 (*a*)). Further, using the plot $(d\varepsilon_v/d\varepsilon_1)_\eta$ against η (Fig. 13.16 (*b*)), the reciprocal differential quotients

$$\left(\frac{d\varepsilon_1}{d\varepsilon_v}\right)_\eta = \frac{1}{\left(\frac{d\varepsilon_v}{d\varepsilon_1}\right)_\eta}$$

can be plotted against the volumetric strains $(\varepsilon_v)_0$ (Fig. 13.17 (*b*)). As stated before, the area between the curve ϕ and the axis ε_v in the interval from $(\varepsilon_v)_{01}$ and $(\varepsilon_v)_{02}$ represents the second term of the expression (13.77).

In view of the lack of reliability of observed strains, and also stresses, in existing test data of soils, the theory developed is assumed, by the authors (Roscoe and Poorooshasb, 1963), for the present to apply only if the following three requirements are simultaneously satisfied:

(*a*) Samples are subjected to triaxial compression tests in which

$$\sigma_1' > \sigma_2' = \sigma_3'.$$

(*b*) The state paths for such tests lie wholly on the boundary surface. (This excludes consideration of tests which entail unloading processes or which have been performed on over-consolidated clay or conducted under conditions implying expansion.)

(*c*) The value of η does not decrease at any stage during the test.

13.3.2 Study of Strains Based on a Deformation Work Hypothesis

13.3.2.1 *Fundamental assumptions*

The side of the boundary state surface which is extended between the normal consolidation line and the critical state line (see Fig. 13.13) has been termed, by Roscoe, Schofield and Wroth (1958), the 'wet' side. Roscoe and Schofield (1963) analysed the mechanical behaviour of idealized 'wet clays' by considering their deformation work. The analysis is based on the following fundamental assumptions:

(1) The distortional strains ε^d are purely plastic:

$$d\varepsilon^d = d\varepsilon_e{}^d + d\varepsilon_p{}^d = d\varepsilon_p{}^d \tag{13.82}$$

$$d\varepsilon_e{}^d = 0 \tag{13.83}$$

(2) The elastic compression increment $d\varepsilon_e{}^v$ satisfies the relationship

$$d\varepsilon_e{}^v = \frac{\kappa}{1+e}\cdot\frac{dp}{p} \tag{13.84}$$

e being the void ratio, p the octahedral normal stress and κ a constant. Considering the well known relationship:

$$d\varepsilon^v = \frac{-de}{1+e} \tag{13.85}$$

Equation (13.84) can be expressed as follows:

$$de_e = -\kappa \frac{dp}{p} \tag{13.86}$$

de_e being the void ratio change corresponding to the elastic compression.

(3) Denoting the deformation work per unit volume corresponding to plastic strains only by W, the differential increment dW is given by the equation:

$$\begin{aligned} dW &= p\, d\varepsilon_p{}^v + [\sigma_1{}^d\, d\varepsilon_1{}^d + \sigma_2{}^d\, d\varepsilon_2{}^d + \sigma_3{}^d\, d\varepsilon_3{}^d] = \\ &= p\, d\varepsilon_p{}^v + [\tfrac{2}{3}(\sigma_1 - \sigma_3)\tfrac{2}{3}(d\varepsilon_1 - d\varepsilon_3) + 2\{-\tfrac{1}{3}(\sigma_1 - \sigma_3)\} \\ &\quad \{-\tfrac{1}{3}(d\varepsilon_1 - d\varepsilon_3)\}] \end{aligned}$$

yielding, taking into account assumption (1):

$$dW = p\, d\varepsilon_p{}^v + q d\varepsilon^d \tag{13.87}$$

At the critical state line

$$q_{cr} = Mp \tag{13.88}$$

(see Fig. 13.13) and the distortional strains increase with the volume remaining constant. In this case the first term on the right hand side of Equation (13.87) vanishes and we get the relation:

$$\frac{dW}{d\varepsilon^d} = Mp \tag{13.89}$$

This relation is assumed to be valid during any slip at the grain contacts.

From Equations (13.83), (13.89) and (13.87) it is possible to derive the relation:

$$\frac{d\varepsilon_p{}^v}{d\varepsilon^d} = M - \frac{q}{p} \tag{13.90}$$

13.3.2.2 *General expression for the yield deviatoric stress*

The projection of the critical state line ($q = Mp$) in the (e,p) plane has the equation

$$e = \Gamma - \lambda_e \log_e p \tag{13.91}$$

(Roscoe, Schofield and Wroth, 1958). The comparison with Equation

(13.71) of the normal consolidation line proves that in the $(e, \log_e p)$ plot both lines are parallel to each other (Fig. 13.18).

Consider a sample of clay having a void ratio e_1 when subjected to a spherical stress p_1 and let the relationship between e_1 and p_1 satisfy Equation (13.91). The state of the soil may then be represented by point A in Fig. 13.18. If this material is subjected to an increasing stress difference q at constant p it is assumed to remain rigid until q reaches the critical

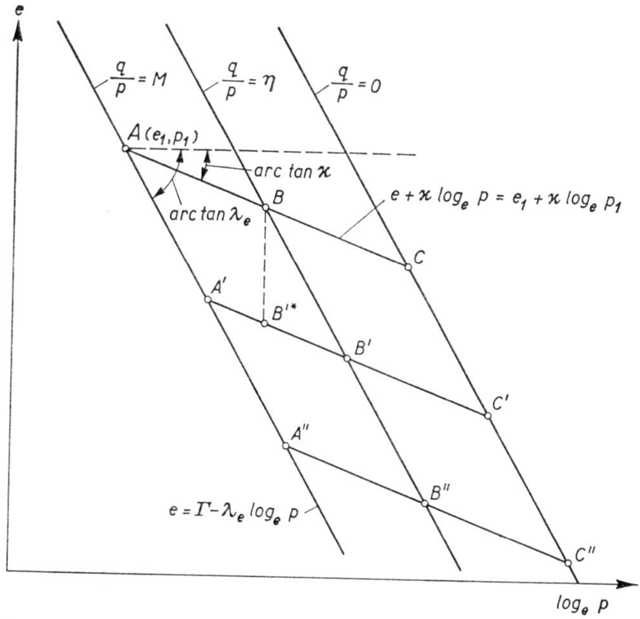

Fig. 13.18. Projection of consolidation lines and critical state line in e-$\log_e p$ plane. (According to Roscoe and Schofield, 1963)

value specified by Equation (13.88) when the material will flow without further change of state.

Alternatively, suppose the same soil is subjected to an additional spherical stress Δp which brings it into a state represented by point B on the elastic compression line (Fig. 13.18). The elastic compression line AB satisfies condition (13.85) and has the equation:

$$e + \kappa \log_e p = e_1 + \kappa \log_e p_1 \tag{13.92}$$

If the material is then subjected to an increasing stress difference q, it remains rigid until q attains a limiting value at which yield occurs. On yielding there is an increment of distortion and a corresponding increment of plastic compression given by Equation (13.90). The grains slip at their contacts and move into a closer packing. The new state of the sample corresponds to a new elastic compression line which lies closer to the origin in Fig. 13.18 (e.g. the line A'–B'–C'). The difference in ordinates between the two elastic compression lines represents the plastic component of the change in void ratio ($B - B'^* = \Delta\varepsilon_p{}^v$ in Fig. 13.18).

The magnitude of the principal stress difference q at which material of any state (p,e) will yield, can be derived from Drucker's principle (1959) that the external agency must never be able to extract work from a stable system of sample and loading apparatus. Thus, for stability,

$$\mathrm{d}p\,\mathrm{d}\varepsilon_p{}^v + \mathrm{d}q\,\mathrm{d}\varepsilon^d \geqslant 0 \tag{13.93}$$

The equality sign in the above equation applies for stress states just before the yield surface is attained. Thus the limiting value of q should satisfy the equation:

$$\frac{\mathrm{d}q}{\mathrm{d}p} = -\frac{\mathrm{d}\varepsilon_p{}^v}{\mathrm{d}\varepsilon^d} \tag{13.94}$$

Combining Equations (13.90) and (13.94) we obtain:

$$\frac{\mathrm{d}q}{\mathrm{d}p} = \frac{q}{p} - M \tag{13.95}$$

or, after integration,

$$\frac{q}{Mp} + \log_e p = \text{const} \tag{13.96}$$

For any given value of this constant, Equation (13.96) refers to a particular state of packing, and hence to one elastic compression line in Fig. 13.18, e.g. the line AB through A. As the point A with coordinates $e_1\,p_1$ lies also on the critical state line (Equation 13.88), the relation (13.96) can be written:

$$\frac{q}{Mp} + \log_e p = \frac{q_1}{Mp_1} + \log_e p_1 = 1 + \log_e p_1$$

yielding:

$$\log_e p_1 = \frac{q}{Mp} + \log_e p - 1 \tag{13.97}$$

Now, in Equation (13.92), e_1 can be expressed by using Equation (13.91), and expression (13.97) can be substituted for $\log_e p_1$. In this way, Equation (13.92) yields:

$$q = \frac{Mp}{\lambda - \kappa}(\Gamma + \lambda - \kappa - e - \lambda \log_e p) \tag{13.98}$$

This is the general equation of the yield surface (state boundary surface).

13.3.2.3 *Anisotropic consolidation*

Anisotropic consolidation may be defined as a process in which the clay yields under principal stresses which increase at a constant ratio:

$$\frac{q}{p} = \eta \tag{13.99}$$

In Fig. 13.19 (*c*) an anisotropic consolidation line is represented, in axonometric view, by the line D–D'–D''. Its projection in the (p,q) plane is the straight line $\eta = q/p = \text{const}$ (Fig. 13.19 (*b*)), and its projection in the (e,p) plane is the logarithmic line B–B'–B'' (see the straight line B–B'–B'' in the semi-logarithmic plot, Fig. 13.18, and the corresponding curved line B–B'–B'' in the linear scale plot, Fig. 13.19 (*a*)).

Inserting the relation (13.99) into the general equation of the boundary state surface (13.98), the equation of the anisotropic consolidation line is obtained:

$$e = \Gamma + (\lambda - \kappa)\left(1 - \frac{\eta}{M}\right) - \lambda \log_e p \tag{13.100}$$

Equations (13.84), (13.85) and (13.100) give the relation:

$$d\varepsilon^v = \frac{\lambda}{1+e}\cdot\frac{dp}{p} = \frac{\lambda}{\kappa}\, d\varepsilon_e{}^v \tag{13.101}$$

The plastic part of the volume strain increment is then:

$$d\varepsilon_p{}^v = d\varepsilon^v - d\varepsilon_e{}^v = \left(1 - \frac{\kappa}{\lambda}\right) d\varepsilon^v \tag{13.102}$$

Inserting this expression for $d\varepsilon_p{}^v$ into Equation (13.90) we obtain:

$$d\varepsilon^v = \frac{M - \eta}{1 - \dfrac{\kappa}{\lambda}}\, d\varepsilon^d \tag{13.103}$$

Thus, in anisotropic consolidation the ratio between the volume and distortional strains remains constant.

For isotropic consolidation ($\eta = 0$) Equation (13.103) gives:

$$d\varepsilon^d = \frac{1 - \dfrac{\kappa}{\lambda}}{M}\, d\varepsilon^v \tag{13.104}$$

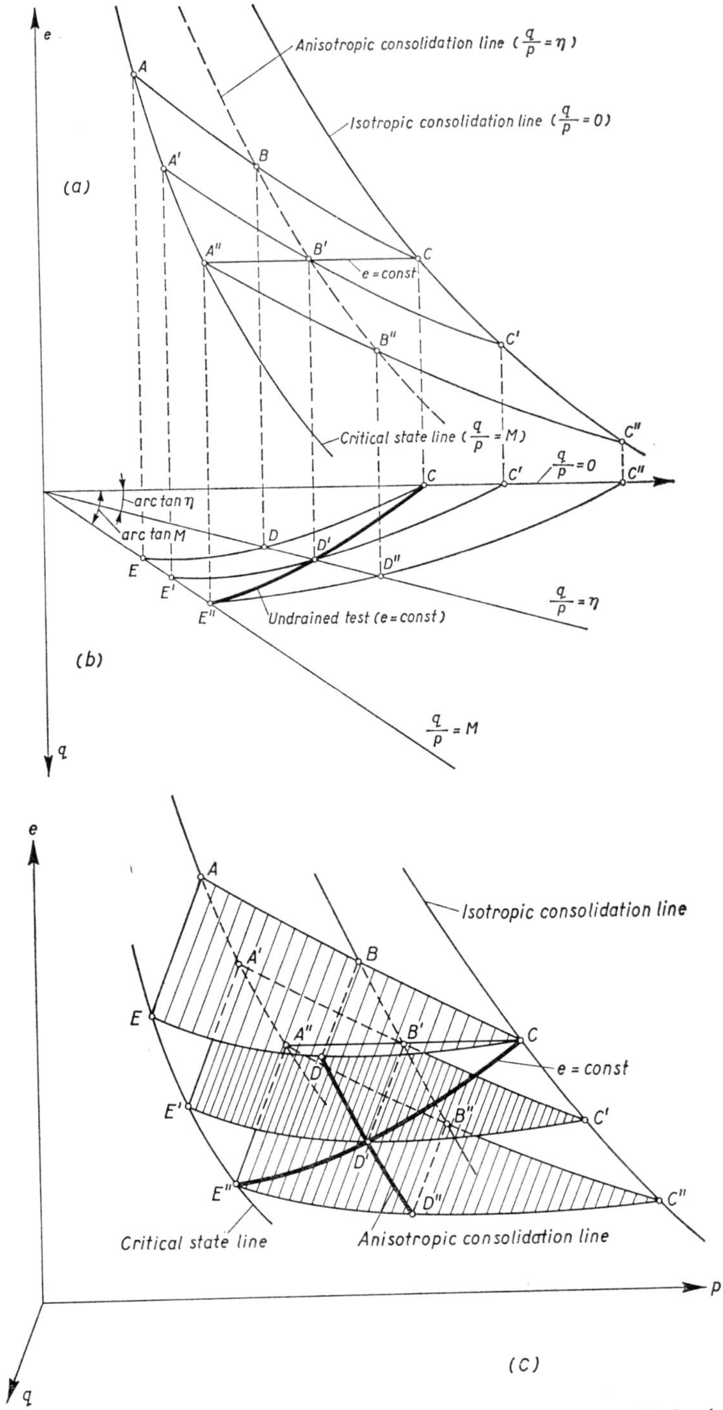

Fig. 13.19. 'Anisotropic' consolidation: (*a*) in (*e*,*p*) plane, (*b*) in (*q*,*p*) plane, (*c*) in axonometric view. (According to Roscoe and Schofield, 1963)

In Roscoe and Schofield's (1963) interpretation the above equation predicts that a distortion of a certain magnitude can take place without specifying the direction because there are no principal axes of stress. The smallest possible deviator stress would suffice to impose definite directions for the principal stresses, but in the absence of any deviator stress it is conceivable that the local distortions in different parts of a sample are such that no overall distortion of the whole sample appears to take place.

13.3.2.4 *Undrained tests*

In undrained tests on saturated soils the void ratio remains constant: i.e. $e =$ const. In Fig. 13.19 (*c*) the line C–D'–E'' represents an example of an undrained state path, A''–B'–C being its projection in the (p,e) plane (see also Fig. 13.19 (*a*)). Its projection in the (p,q) plane is shown in Fig. 13.19 (*b*) by the stress-path C–D'–E''.

In order to obtain the equation of the undrained state path, the fact that the void ratio e remains constant should be considered in the general equation of the boundary state surface (13.98). By taking $q = 0$, Equation (13.98) reduces to the equation of the isotropic consolidation line:

$$e = \Gamma + \lambda - \kappa - \lambda \log_e p_0 \qquad (13.105)$$

p_0 denoting the all-round pressure in isotropic consolidation. In an undrained test whose stress path starts at p_0, the void ratio does not change throughout the test. Thus, expression (13.105) can be substituted for the void ratio e in Equation (13.98). In this way Equation (13.98) yields the equation of the undrained state path:

$$q = \frac{Mp}{1 - \dfrac{\kappa}{\lambda}} \log_e \frac{p_0}{p} \qquad (13.106)$$

In undrained tests the volume strain is zero; thus:

$$d\varepsilon^v = d\varepsilon_e{}^v + d\varepsilon_p{}^v = 0 \qquad (13.107)$$

Considering the assumption (13.84) and Equation (13.90), Equation (13.107) yields the distortional strain increment:

$$d\varepsilon^d = \frac{\kappa}{M(1+e)} \cdot \frac{(-1)}{1 - \dfrac{q}{Mp}} \cdot \frac{dp}{p} \qquad (13.108)$$

Substituting q/Mp from (13.106), Equation (13.108) becomes:

$$\mathrm{d}\varepsilon^d = \frac{\kappa\left(1 - \frac{\kappa}{\lambda}\right)}{M(1 + e)} \cdot \frac{(-1)}{\left(1 - \frac{\kappa}{\lambda}\right) - \log_e \frac{p_0}{p}} \cdot \frac{\mathrm{d}p}{p} \tag{13.109}$$

After integration this becomes:

$$\varepsilon^d = \frac{\kappa\left(1 - \frac{\kappa}{\lambda}\right)}{M(1 + e)} \log_e \frac{1 - \frac{\kappa}{\lambda}}{\left(1 - \frac{\kappa}{\lambda}\right) - \log_e \frac{p_0}{p}} \tag{13.110}$$

13.4 A GENERAL THEORY OF DISTORTIONAL CREEP ACCOUNTING FOR THE INFLUENCE OF THE MEAN PRINCIPAL STRESS

13.4.1 Graphical Presentation of the Stress–Strain Relationships at a given Time and their Analytical Expressions

Goldenblat (1960) suggested that the general stress–strain relationships in a soil medium be expressed by the equations:

$$\varepsilon_{\text{oct}} = f_{\text{oct}}(T, \sigma_{\text{oct}}) \quad \text{or} \quad \sigma_{\text{oct}} = \varphi_{\text{oct}}(\varepsilon_{\text{oct}}, T) \tag{13.111-a,b}$$

$$\Gamma = f(T, \sigma_{\text{oct}}) \quad \text{or} \quad T = \varphi(\Gamma, \sigma_{\text{oct}}) \tag{13.112-a,b}$$

where the following notation has been used:

for stresses:

$$\sigma_{\text{oct}} = \frac{I_1}{3} \tag{2.19}$$

$$T = \sqrt{\left(\frac{3}{2}\right)}\,\tau_{\text{oct}} = \sqrt{\left(\frac{I_1^2}{3} - I_2\right)} \tag{13.113}$$

and for strains:

$$\varepsilon_{\text{oct}} = \frac{J_1}{3} \tag{3.21}$$

$$\Gamma = \sqrt{\left(\frac{3}{2}\right)}\,\gamma_{\text{oct}} = 2\sqrt{\left(\frac{J_1^2}{3} - J_2\right)} \tag{13.114}$$

I_1 and I_2 are the invariants of the stress tensor and have been defined by Equations (2.7) or (2.12) and (2.8) or (2.13) respectively. T is called the intensity of the shear stresses. J_1 and J_2 are the invariants of the strain tensor and have been defined by Equations (3.8) and (3.9) respectively. Γ is called the intensity of the shear strains. (For σ_{oct}, τ_{oct}, ε_{oct}, γ_{oct} see Equations 2.19, 2.21, 3.21 and 3.22.)

Using the general relations (13.112-a,b) as a basis and introducing the time factor in order to account for creep phenomena, Vyalov (1955, 1957, 1959, 1961, 1962, 1963) developed a general theory of distortional creep accounting for the influence of the mean principal stress. This theory was originally related to frozen soils at low temperature in which volumetric strains are not important. Thus in Vyalov's theory the volumetric stress–strain relationships (13.111-a,b) are not considered. As for distortional strains, Vyalov's theory is also of interest for unfrozen clay soils since the creep processes are governed by similar stress–strain relations.

For a given time $t = t_i$ during the creep process, the relations (13.112) can be presented either in a three-dimensional coordinate system (σ_{oct},Γ,T) (Fig. 13.20 (*a*)) or in a plane coordinate system: (Γ,T), (σ_{oct},T), (σ_{oct},Γ) (Fig. 13.20 (*b*)). The various curves in the (T,Γ) system correspond to various constant octahedral pressures σ_{oct}, the various curves in the (σ_{oct},T) plane to various constant intensities of shear strains Γ and the various plots in the (σ_{oct},Γ) system to various constant intensities of shear stresses T.

Considering the mutual relations shown in Fig. 13.20, the general Equation (13.112-b) accounting for the curves in the (σ_{oct},T) plane can be written as follows:

$$T = \varphi_1(\Gamma) + \varphi_2(\Gamma)\Omega(\sigma_{oct}) \tag{13.115}$$

The function $\varphi_1(\Gamma)$ represents the value of T at $\sigma_{oct} = \sigma_{oct}{}^0 = 0$ which depends on the shear strain intensity. The dependence is given by the $\sigma_{oct}{}^0$ curves in the (Γ,T) system. The function $\varphi_2(\Gamma)$ expresses the dependence of further shear stress intensities on the Γ values as given by the $[\sigma_{oct} = \text{const}]$ curves in the (Γ,T) system.

Solving Equation (13.115) for Γ, we have:

$$\Gamma = f_1(T) - f_2(T)\omega(\sigma_{oct}) \tag{13.116}$$

This equation refers to curves in the (σ_{oct},Γ) planes. The second right-hand term expresses the reducing influence of the mean principle stress on the magnitude of the shear strain intensity. If the stress state is purely distortional, the above relationship becomes:

$$\Gamma_0 = f_1(T) \tag{13.117}$$

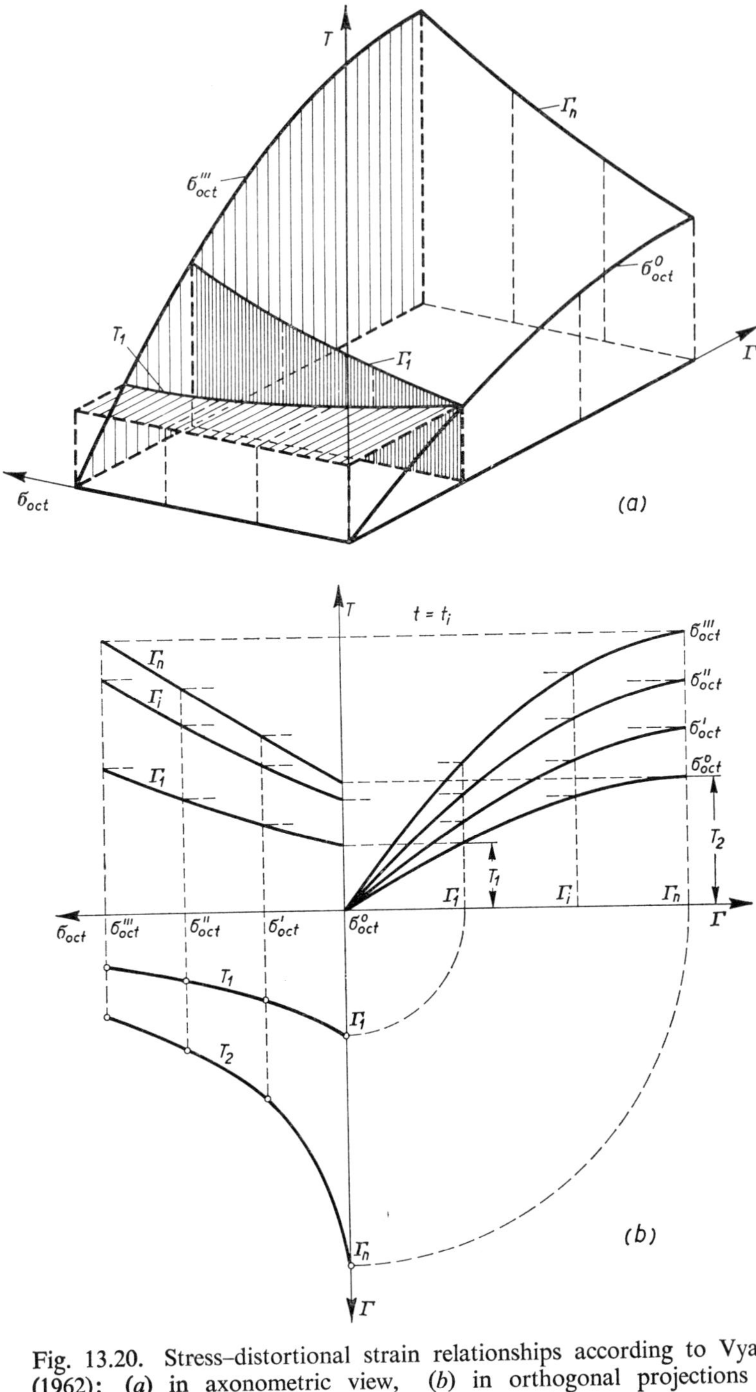

Fig. 13.20. Stress–distortional strain relationships according to Vyalov (1962): (*a*) in axonometric view, (*b*) in orthogonal projections on (σ_{oct}, Γ), (σ_{oct}, T) and (*T*, Γ) planes

The relations (13.115) and (13.116) apply for constant Γ or T values respectively. If both the distortional and the spherical stresses change during loading, the corresponding relations have the form of the integral equations of Volterra-Boltzmann:

$$T = \varphi_1(\Gamma) + \int_0^{\sigma_{oct}} S(\sigma_{oct} - \xi)\varphi_2[\Gamma(\xi)]\,d\xi \tag{13.118}$$

where

$$S(\sigma_{oct}) = \frac{\partial\Omega(\sigma_{oct})}{\partial\sigma_{oct}} = \frac{1}{\varphi_2(\Gamma)}\frac{\partial T}{\partial\sigma_{oct}} \tag{13.119}$$

denotes the change of the stress T with increasing σ_{oct} at unit shear strain intensity ($\Gamma = 1$).

$$\Gamma = f_1(T) - \int_0^{\sigma_{oct}} P(\sigma_{oct} - \xi) f_2[T(\xi)]\,d\xi \tag{13.120}$$

where

$$P(\sigma_{oct}) = \frac{\partial\omega(\sigma_{oct})}{\partial\sigma_{oct}} = -\frac{1}{f_2(T)}\frac{\partial\Gamma}{\partial\sigma_{oct}} \tag{13.121}$$

denotes the change of the shear strain Γ with increasing σ_{oct} at unit shear stress intensity ($T = 1$).

According to experimental evidence the functions $\varphi_1(\Gamma)$ and $\varphi_2(\Gamma)$ can be taken as exponential functions (Vyalov, 1963):

$$\varphi_1(\Gamma) = \bar{A}\Gamma^m \tag{13.122}$$

$$\varphi_2(\Gamma) = \bar{B}\Gamma^n \tag{13.123}$$

Generally the function Ω is non linear. In the first approximation, however, it can be considered to be linear. Thus:

$$\Omega(\sigma_{oct}) = \sigma_{oct} \tag{13.124}$$

and, consequently,

$$S(\sigma_{oct}) = 1 \tag{13.125}$$

Equation (13.118) then reduces to:

$$T = \bar{A}\Gamma^m + B\int_0^{\sigma_{oct}} \Gamma^n(\xi)\,d\xi \tag{13.126}$$

For $\Gamma = \text{const}$ the following simpler form results which can also be derived directly from (13.115):

$$T = A\Gamma^m + B\sigma_{oct}\Gamma^n \tag{13.127}$$

Similar solutions can be written for Γ. A graphical presentation of the family of straight lines the analytical expression of which is Equation (13.127) is shown in Fig. 13.21. Substituting:

$$\varphi_2(\Gamma) = B\Gamma^n = \tan \psi(\Gamma) \tag{13.128}$$

$$\varphi_1(\Gamma) = A\Gamma^m = T_0(\Gamma) = H(\Gamma) \tan \psi(\Gamma) \tag{13.129}$$

$$H(\Gamma) = \frac{T_0(\Gamma)}{\tan \psi(\Gamma)} = \frac{A\Gamma^m}{B\Gamma^n} \tag{13.129'}$$

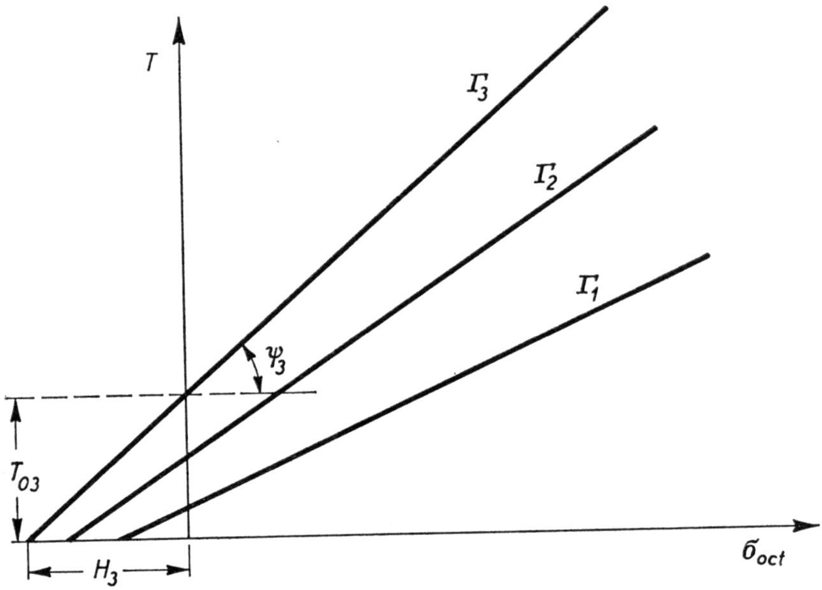

Fig. 13.21. Stress states at various constant values of the shear strain intensity Γ. (According to Vyalov, 1962)

Equation (13.127) can be written in the form:

$$T = T_0(\Gamma) + \sigma_{oct} \tan \psi(\Gamma)$$

or

$$T = [H(\Gamma) + \sigma_{oct}] \tan \psi(\Gamma) \tag{13.130}$$

In the special case $m = n$: $H(\Gamma) = \text{const}$. Equation (13.127) then reduces to:

$$T = (A + B\sigma_{\text{oct}})\Gamma^m = \left(1 + \frac{\sigma_{\text{oct}}}{H}\right) A\Gamma^m \tag{13.131}$$

$$H = \frac{A}{B} = \text{const} \tag{13.131'}$$

13.4.2 General Relations Accounting for Distortional Creep

In the foregoing section all the graphs and equations were related to a given time t during the creep process. General relations accounting for creep effects can be obtained by introducing creep and relaxation functions respectively. In this way, Equations (13.118) and (13.120) may be generalized as follows:

$$\begin{aligned} T = {} & \varphi_1[\Gamma(\sigma_{\text{oct}},t) - \int_0^t R_1(t - \nu)\varphi_1[\Gamma(\nu)]\,\mathrm{d}\nu + \\ & + \int_0^{\sigma_{\text{oct}}(t)} S(\sigma_{\text{oct}} - \xi)\varphi_2[\Gamma(\xi)]\,\mathrm{d}\xi - \\ & - \int_0^t \int_0^{\sigma_{\text{oct}}(t)} R_2(t - \nu)S(\sigma_{\text{oct}} - \xi)\varphi_2[\Gamma(\xi)]\,\mathrm{d}\xi\,\mathrm{d}\nu \end{aligned} \tag{13.132}$$

$$\begin{aligned} \Gamma = {} & f_1[T(\sigma_{\text{oct}},t)] + \int_0^t Q_1(t - \nu)f_1[T(\nu)]\,\mathrm{d}\nu - \\ & - \int_0^{\sigma_{\text{oct}}(t)} P(\sigma_{\text{oct}} - \xi)f_2[T(\xi)]\,\mathrm{d}\xi - \\ & - \int_0^t \int_0^{\sigma_{\text{oct}}(t)} Q_2(t - \nu)P(\sigma_{\text{oct}} - \xi)f_2[T(\xi)]\,\mathrm{d}\xi\,\mathrm{d}\nu \end{aligned} \tag{13.133}$$

The functions φ_1 and φ_2, f_1 and f_2, S and P were defined in the preceding section (Equations 13.122, 13.123, 13.116, 13.119 and 13.121). The creep functions $Q_1(t)$ and $Q_2(t)$, as well as the relaxation functions $R_1(t)$ and $R_2(t)$ are defined by the following equations:

$$Q_1(t) = \frac{1}{f_1(T)}\frac{\partial \Gamma}{\partial t} \tag{13.134}$$

$$Q_2(t) = \frac{1}{f_2(T)}\frac{\partial}{\partial t}(\Gamma_0 - \Gamma) \tag{13.135}$$

$$R_1(t) = -\frac{1}{\varphi_1(\Gamma)}\frac{\partial T}{\partial t} \tag{13.136}$$

$$R_2(t) = -\frac{1}{\varphi_2(\Gamma)}\frac{\partial}{\partial t}(T - T_0) \tag{13.137}$$

The functions $Q_2(t)$ and $R_2(t)$ express the rate of change of the respective supplementary stresses and strains with time by which, under the influence of σ_{oct}, the stresses and strains corresponding to pure shear are exceeded.

The empirical relations (13.122) and (13.123) can be accepted for φ_1 and φ_2 and the approximation (13.125) for S. With these assumptions Equation (13.132), applied for Γ = const, yields:

$$T = \bar{A}\Gamma^m[1 - \int_0 R_1(t)\,\mathrm{d}t] + \bar{B}\Gamma^n\sigma_{oct}[1 - \int_0 R_2(t)\,\mathrm{d}t] \quad (13.138)$$

This equation can be written more simply in the following parameter form:

$$T = A(t)\Gamma^m + B(t)\sigma_{oct}\Gamma^n \quad (13.139)$$

The functions $A(t)$ and $B(t)$ can be determined from experimental observations of the relaxation of the shear stress intensity at constant shear strain intensity Γ. The results frequently approach the following functions:

$$A(t) = \frac{A}{1 + (at)^{\lambda_a}} \quad (13.140)$$

$$B(t) = \frac{B}{1 + (bt)^{\lambda_b}} \quad (13.141)$$

Inserting them into Equation (13.139) gives:

$$T = \frac{A\Gamma^m}{1 + (at)^{\lambda_a}} + \frac{B\sigma_{oct}\Gamma^n}{1 + (bt)^{\lambda_b}} \quad (13.142)$$

When $m = n$ and $a = b$, $\lambda_a = \lambda_b = \lambda$ the above equation becomes simplified:

$$T = \frac{A\Gamma^m}{1 + at^\lambda}\left(1 + \frac{\sigma_{oct}}{H}\right) \quad (13.143)$$

$$H = \frac{A}{B} \quad (13.131')$$

Frequently the second term in Equation (13.138) does not depend on time. In such cases $B(t) = B$ and

$$T = \frac{A\Gamma^m}{1 + at^\lambda} + B\sigma_{oct}\Gamma^n \quad (13.144)$$

The linear strain ε_x in the direction x can be determined on the basis of the following relationship:

$$\frac{\Gamma(t)}{2T(t)} = \frac{\varepsilon_x - \varepsilon_{\text{oct}}}{\sigma_x - \sigma_{\text{oct}}} = \frac{\gamma_{xy}}{2\tau_{xy}} \tag{13.145}$$

Considering zero volumetric strain ($\varepsilon_{\text{oct}} = 0$), Equations (13.139) and (13.145) give:

$$\varepsilon_x = \frac{\Gamma^{1-n}}{2[A(t)\Gamma^{m-n} + B(t)\sigma_{\text{oct}}]}(\sigma_x - \sigma_{\text{oct}}) \tag{13.146}$$

$$\gamma_{xy} = \frac{\Gamma^{1-n}}{A(t)\Gamma^{m-n} + B(t)\sigma_{\text{oct}}} \cdot \tau_{xy} \tag{13.147}$$

In the cases $m = n$ and $H = A(t)/B(t) = \text{const}$, corresponding simplifications to the above equations can be deduced.

13.4.3 Equation of Distortional Creep Related to the Creep Rate

In Equations (13.112-a,b) the intensity of shear strain can be replaced by the intensity of creep rate $\dot{\Gamma}$:

$$\dot{\Gamma} = \frac{\partial \Gamma}{\partial t} \tag{13.148}$$

Then this gives the relations

$$\dot{\Gamma} = F(T,\sigma_{\text{oct}}) \qquad \text{or} \qquad T = \phi(\dot{\Gamma},\sigma_{\text{oct}}) \tag{13.149}$$

The mutual relationships between $\dot{\Gamma}$, T and σ_{oct} can be presented in the coordinate system ($\dot{\Gamma},T,\sigma_{\text{oct}}$), in a manner similar to those for Γ, T and σ_{oct} shown in Fig. 13.20; and analytical expressions can be set out similar to those relating either Γ to T and σ_{oct}, or T to Γ and σ_{oct} (see the previous section). By way of illustration the parameter form of equation (13.112-b), i.e. Equation (13.139), may be replaced by the relation:

$$T = \bar{A}(t)\dot{\Gamma}^m + \bar{B}(t)\sigma_{\text{oct}}\dot{\Gamma}^n \tag{13.150}$$

Taking $m = n$ it reduces to

$$T = \dot{\Gamma}^m[\bar{A}(t) + \bar{B}(t)\sigma_{\text{oct}}] \tag{13.151}$$

$\bar{A}(t)$ determines the change of distortional stress (T_0) which is needed to maintain a given constant intensity of distortional creep rate ($\dot{\Gamma}$). $\bar{B}(t)$ governs the development of that part of the distortional stress ($T - T_0$) which is influenced by the mean principal stress σ_{oct}.

In the special case $\bar{A} = \text{const}_a$, $\bar{B} = \text{const}_b$, $H = \bar{A}/\bar{B} = \text{const}$. We have:

$$T = \bar{A}\dot{\Gamma}^m + \bar{B}\sigma_{\text{oct}}\Gamma^n \tag{13.152}$$

or, for $m = n$:

$$T = \bar{A}\dot{\Gamma}^m\left(1 + \frac{\sigma_{\text{oct}}}{H}\right) \tag{13.153}$$

where $\bar{A}$ denotes the coefficient of distortional viscosity (cf. Equation 4.19). With $m = 1$, Equation (13.153) reduces to Bingham's equation.

13.5 SOME FINAL STRESS–STRAIN RELATIONSHIPS FOR SOILS

By means of a simple assumption concerning the distribution of the deformation states of particle contacts in a potential rupture surface before loading, the following equation was developed by J. Brinch Hansen (1965) for the relationship between shear stresses and strains at first loading:

$$\frac{\tau}{\tau_f} = \left(\frac{\delta}{\delta_f}\right) + \left(\frac{\delta}{\delta_f}\right)^n - \left(\frac{\delta}{\delta_f}\right)^{n+1} \tag{13.154}$$

δ is the sliding or shear deformation corresponding to the shear stress τ, and δ_f, τ_f are the respective deformation and stress values at failure. Empirical relationships can be expressed in the following still simpler form:

$$\frac{\tau}{\tau_f} = \sqrt{[(b + 1)\delta/(b\delta + \delta_f)]} \tag{13.155}$$

(Brinch Hansen, 1965).

According to Brinch Hansen (1965), the results of an isotropic first loading compression can usually be expressed approximately by the equation:

$$\varepsilon_v = \varepsilon_1 + \varepsilon_2 + \varepsilon_3 = 3A\left(\frac{\sigma_m}{M}\right)^m \tag{13.156}$$

where M is a constant 'deformation modulus' for the soil in question, whereas A and m may depend upon the water content w or the void ratio e.

In the conditions of axial symmetry the deviator stress $\sigma_1 - \sigma_3 = \sigma_1 - \sigma_2$ has two effects. First, it produces shear deformations $\varepsilon_1 - \varepsilon_3 = \varepsilon_1 - \varepsilon_2$, and second, it produces a volume dilatation which is usually proportional to the shear deformation. In an oedometer test ($\varepsilon_1 = \varepsilon_3 = 0$) the stress ratio $\sigma_3/\sigma_1 = K_0$ is found to be approximately independent of σ_1 (or σ_m). This means that the parenthesis in (13.156) must also be a factor for the shear deformation and dilatation.

If we define a 'stress angle' v by the equation:

$$\sin v = \frac{\sigma_1 - \sigma_3}{\sigma_1 + \sigma_3 + 2c \cot \varphi} \tag{13.157}$$

we have, according to Coulomb's failure criterion, $v_f = \varphi$.

We can then combine (13.155), (13.156) and (13.157) as follows:

$$\varepsilon_1 - \varepsilon_3 = 3B\left(\frac{\sigma_m}{M}\right)^m \frac{1}{\left(k\dfrac{\sin \varphi}{\sin v}\right)^2 - 1} \tag{13.158}$$

Further we get:

$$\varepsilon_v = \varepsilon_1 + 2\varepsilon_3 = 3A\left(\frac{\sigma_m}{M}\right)^m - \beta(\varepsilon_1 - \varepsilon_3) \tag{13.159}$$

It is then easy to derive:

$$\varepsilon_1 = \left(\frac{\sigma_m}{M}\right)^m \left[A + \frac{B(2-\beta)}{\left(k\dfrac{\sin \varphi}{\sin v}\right)^2 - 1}\right] \tag{13.160}$$

and

$$\varepsilon_2 = \varepsilon_3 = \left(\frac{\sigma_m}{M}\right)^m \left[A - \frac{B(1+\beta)}{\left(k\dfrac{\sin \varphi}{\sin v}\right)^2 - 1}\right] \tag{13.161}$$

The dimensionless quantities m, β, k, A and B, as well as the deformation modulus M must, of course, be determined experimentally. They are, with the exception of M, functions of w or e. (Brinch Hansen, 1965.)

If only the deformations during the triaxial compression test are required the (isotropic) deformations due to the consolidation pressure σ_c must be deduced:

$$\Delta\varepsilon_i = \varepsilon_i - \frac{\varepsilon_v}{3}, \; (i = 1,2,3) \tag{13.162}$$

where the expression (13.160) or (13.161) has to be inserted for ε_i and the expression (13.156) for ε_v.

14

Three-Dimensional Consolidation Analysis of Saturated and Partly Saturated Soils

14.1 GENERAL DIFFERENTIAL EQUATION OF THREE-DIMENSIONAL CONSOLIDATION

14.1.1 Deduction of the Fundamental Equation

When analysing, in the previous chapters, the consolidation process of either saturated or partly saturated soils, the drainage was assumed to occur in one direction only. Based on the respective continuity conditions (12.8), (9.25) and (12.7-a) for the liquid, solid and gaseous phases of partly saturated soils, the general consolidation equation of the following form has been deduced:

$$\frac{\partial e}{\partial t} + \beta \frac{\partial u}{\partial t} = \frac{1 + e}{\gamma_w} \frac{\partial}{\partial z}\left(k \frac{\partial \bar{u}}{\partial z}\right) \tag{12.59}$$

the meaning of the symbols being:

e = void ratio at time t
u = excess pore pressure at time t
z = co-ordinate in the drainage direction
$_w$ = bulk density of the pore water
k = coefficient of permeability at the depth z
$\beta = (e_a + He_w)/(u_0 + u)$ (12.24), H denoting Henry's constant of the solubility of air in water, and e_w, e_a the respective void ratios related to voids filled with water (e_w) and air (e_a)
$\partial\bar{u}/\partial z = (\partial u/\partial z) - \gamma_w i_0$, i_0 being the limiting hydraulic gradient (see Section 9.2.1).

If the drainage is allowed to occur in all directions (see Fig. 9.3), the continuity conditions for the liquid, solid, and gaseous phases respectively, take the following forms:

$$\frac{\partial v_x}{\partial x} + \frac{\partial v_y}{\partial y} + \frac{\partial v_z}{\partial z} + \frac{\partial n_w}{\partial t} = 0 \tag{14.1}$$

$$\frac{\partial \mathring{v}_x}{\partial x} + \frac{\partial \mathring{v}_y}{\partial y} + \frac{\partial \mathring{v}_z}{\partial z} + \frac{\partial(1-n)}{\partial t} = 0 \tag{14.2}$$

$$\frac{\partial w_x}{\partial x} + \frac{\partial w_y}{\partial y} + \frac{\partial w_z}{\partial z} + \frac{\partial n_a}{\partial t} + \beta \frac{\partial u}{\partial t} = 0 \tag{14.3}$$

where the symbol $\boldsymbol{v}$ stands for the apparent (drainage) velocity of the water, $\mathring{v}$ for the apparent velocity of the solid grain movements and w for the apparent velocity of the air, the symbol $\boldsymbol{n}$ for the porosity, the suffixes x, y and z for the directions of the axes of the orthogonal coordinate system (x,y,z) (see Fig. 9.3) and the suffixes w and a for water and air respectively.

In a way analogous to that in which the general equation of one-dimensional consolidation (12.59) was deduced from the continuity conditions (12.8), (9.25) and (12.7-a), the generalized continuity conditions (14.1), (14.2) and (14.3) lead to the following general equation of three-dimensional consolidation:

$$\frac{\partial e}{\partial t} + \beta \frac{\partial u}{\partial t} = \frac{1+e}{\gamma_w}\left[\frac{\partial}{\partial x}\left(k_x \frac{\partial \bar{u}_x}{\partial x}\right) + \frac{\partial}{\partial y}\left(k_y \frac{\partial \bar{u}_y}{\partial y}\right) + \frac{\partial}{\partial z}\left(k_z \frac{\partial \bar{u}_z}{\partial z}\right)\right] \tag{14.4}$$

whereby:

$$\frac{\partial \bar{u}_x}{\partial x} = \frac{\partial u}{\partial x} - \gamma_w i_{0x} \tag{14.5}$$

and similarly for the directions y and z.

In isotropic conditions:

$$k_x = k_y = k_z = \text{const}_k = k, \quad i_{0x} = i_{0y} = i_{0z} = \text{const}_{i_0} = i_0 \tag{14.6-a,b}$$

the Equation (14.4) reduces to:

$$\frac{\partial e}{\partial t} + \beta \frac{\partial u}{\partial t} = \frac{(1+e)k}{\gamma_w} \Delta \bar{u} \tag{14.7}$$

Δ denoting Laplace's operator:

$$\Delta = \frac{\partial^2}{\partial x^2} + \frac{\partial^2}{\partial y^2} + \frac{\partial^2}{\partial z^2} \tag{14.8}$$

The deduction of the Equation (14.7) is to be attributed to Florin (1961, for saturated soils: 1938). Under the assumption that the coefficient β is a constant, and for $i_0 = 0$ it was firstly deduced by Biot (1935 and 1941).

14.1.2 Further Development of the Fundamental Equation

The sum of the excess pore pressure u and the effective octahedral stress σ_{oct}' equals the total octahedral stress σ_{oct}:

$$\sigma_{oct}' + u = \sigma_{oct} \tag{14.9}$$

Substituting:

$$\frac{\partial e}{\partial t} = \frac{\partial e}{\partial \sigma_{oct}'} \cdot \frac{\partial \sigma_{oct}'}{\partial t} \tag{14.10}$$

and considering the relation (14.9), Equation (14.4) can be written in the following form:

$$\frac{\partial u}{\partial t} = \frac{1}{\mathring{\omega}} \left\{ \frac{\partial \sigma_{oct}}{\partial t} - \frac{1 + e}{\gamma_w \dfrac{\partial e}{\partial \sigma_{oct}'}} \Big[\quad \Big] \right\} \tag{14.11}$$

where

$$\mathring{\omega} = 1 - \frac{\beta}{\dfrac{\partial e}{\partial \sigma_{oct}'}} \tag{14.12}$$

and

$$\Big[\quad \Big] = \frac{\partial}{\partial x}\left(k_x \frac{\partial \bar{u}_x}{\partial x}\right) + \frac{\partial}{\partial y}\left(k_y \frac{\partial \bar{u}_y}{\partial y}\right) + \frac{\partial}{\partial z}\left(k_z \frac{\partial \bar{u}_z}{\partial z}\right) \tag{14.13}$$

(cf. Equation 14.4).

In non-viscous soils, a linear stress–strain relationship may be assumed over limited stress intervals. Consequently, the following expression can be substituted for $\partial e/\partial \sigma_{oct}'$:

$$\frac{\partial e}{\partial \sigma_{oct}'} = -\frac{1 + e}{K} \tag{14.14}$$

K denoting the modulus of compressibility. Introducing the expression (14.14) into Equation (14.11) gives:

$$\frac{\partial u}{\partial t} = \frac{1}{\mathring{\omega}} \left\{ \frac{\partial \sigma_{oct}}{\partial t} + \frac{K}{\gamma_w} \Big[\quad \Big] \right\} \tag{14.15}$$

In an oedometer test, the octahedral effective stress σ_{oct}' can be expressed by the equation

$$\sigma_{oct}' = \frac{\sigma_1' + 2\sigma_3'}{3} = \frac{1 + 2\zeta_0}{3}\sigma' \tag{14.16}$$

$\sigma' = \sigma_1'$ being the axial effective pressure in the sample and

$$\zeta_0 = \frac{\sigma_3'}{\sigma_1'} \tag{14.17}$$

the earth pressure coefficient at rest. The zero lateral strain condition yields:

$$\sigma_3' - \nu(\sigma_3' + \sigma_1') = 0 \rightarrow \zeta_0 = \frac{\nu}{1 - \nu} \tag{14.18}$$

ν standing for Poisson's ratio. Now, the coefficient of volume change m_v, defined by Equation (8.24):

$$m_v = -\frac{1}{1 + e} \cdot \frac{de}{d\sigma'} \tag{8.24}$$

can be related to the modulus of compressibility by introducing σ_{oct}' after (4.16) and de from (8.24) into Equation (14.14).

$$K = \frac{1 + 2\zeta_0}{3} \cdot \frac{1}{m_v} \tag{14.19-a}$$

or

$$K = \frac{1 + \nu}{3(1 - \nu)} \cdot \frac{1}{m_v} \tag{14.19-b}$$

Substituting

$$(1 + e)m_v = a \tag{14.20}$$

and inserting, for K, the expression (14.19-a), Equation (14.15) takes the form presented by Florin (1962).

For the conditions of plane strain ($\varepsilon_3 = 0$) $\sigma_3' = \nu(\sigma_1' + \sigma_2')$ and

$$\sigma_{oct}' = \frac{2}{3}(1 + \nu)\frac{\sigma_1' + \sigma_2'}{2} \tag{14.21}$$

Substituting

$$\frac{\sigma_1' + \sigma_2'}{2} = \sigma_{med}' \tag{14.22}$$

and considering the condition

$$\frac{\partial \bar{u}}{\partial y} = 0 \tag{14.23}$$

Equation (14.15) for plane states of strain can be written in the form:

$$\frac{\partial u}{\partial t} = \frac{1}{\mathring{\omega}} \left\{ \frac{\partial \sigma_{\text{med}}'}{\partial t} + \frac{1+e}{2(1-\nu)a} \left[\frac{\partial}{\partial x}\left(k_x \frac{\partial \bar{u}}{\partial x}\right) + \frac{\partial}{\partial z}\left(k_z \frac{\partial \bar{u}}{\partial z}\right) \right] \right\} \tag{14.24}$$

where

$$2(1-\nu) = \frac{2}{1+\zeta_0} \tag{14.25}$$

(cf. Florin, 1962).

Equations (14.4), (14.7), (14.11), (14.15) and (14.24) account for saturated soils with coefficients

$$\beta = 0, \qquad \mathring{\omega} = 1 \tag{14.26-a,b}$$

In partly saturated soils these coefficients are defined by the respective expressions (12.63) and (14.12).

When introducing the expression (14.14) for $\partial e/\partial \sigma_{\text{oct}}'$, the linear relationship between effective spherical stress and volume strain has been assumed without taking into account the influence of seepage forces (see Section 9.1.2). They appear in the fundamental equilibrium conditions of the theory of elasticity in the following form:

$$\left.\begin{aligned} \Delta\xi + \frac{1}{1-2\nu}\frac{\partial \varepsilon_v}{\partial x} - \frac{1}{G}\frac{\partial u}{\partial x} &= 0 \\ \Delta\eta + \frac{1}{1-2\nu}\frac{\partial \varepsilon_v}{\partial y} - \frac{1}{G}\frac{\partial u}{\partial y} &= 0 \\ \Delta\zeta + \frac{1}{1-2\nu}\frac{\partial \varepsilon_v}{\partial z} - \frac{1}{G}\frac{\partial u}{\partial z} &= 0 \end{aligned}\right\} \tag{14.a}$$

where ξ, η, ζ are the displacements in the x, y, z directions respectively of a Cartesian system of coordinates, ε_v the volume strain:

$$\varepsilon_v = \varepsilon_x + \varepsilon_y + \varepsilon_z = \frac{\partial \xi}{\partial x} + \frac{\partial \eta}{\partial y} + \frac{\partial \zeta}{\partial z} \tag{14.b}$$

u the excess pore pressure, G the shear modulus, ν Poisson's ratio and Δ the Laplace operator. Differentiating the first of Equations (14.a) with respect to x, the second will respect to y and the third with respect to z, and adding the equations obtained and reducing the resulting expression we have:

$$\frac{2(1-\nu)}{1-2\nu} G\, \Delta\varepsilon_v = \Delta u \tag{14.c}$$

Expressing ν and G by Lame's coefficients λ and μ (see the relations 4.18-b, 4.18-c and 13.7), Equation (14.c) obtains the form:

$$(\lambda + 2\mu)\Delta\varepsilon_v = \Delta u \tag{14.d}$$

If the soil is saturated ($\beta = 0$) and the limiting hydraulic gradient zero ($\bar{u} = u$), Equation (14.7) reduces to

$$\frac{\partial \varepsilon_v}{\partial t} = \frac{k}{\gamma_w} \Delta u \tag{14.e}$$

where $d\varepsilon_v$ has been substituted for $de/(1 + e)$. Inserting, for Δu, expression (14.d), the above equation can be written:

$$\frac{\partial \varepsilon_v}{\partial t} = \frac{k(\lambda + 2\mu)}{\gamma_w} \Delta\varepsilon_v \tag{14.f}$$

or

$$\frac{\partial \varepsilon_v}{\partial t} = c_v{}' \,\Delta\varepsilon_v \tag{14.f$'$}$$

where the coefficient of consolidation $c_v{}'$ is defined by:

$$c_v{}' = \frac{k(\lambda + 2\mu)}{\gamma_w} \tag{14.g}$$

Equation (14.f′) was firstly deduced by Biot (1935 and 1941).

14.2 THREE-DIMENSIONAL CONSOLIDATION OF SATURATED TRIAXIAL SAMPLES

14.2.1 Three-dimensional Consolidation of Saturated Non-viscous Triaxial Samples

In his thesis Anagnosti (1962) presented the solution of the three dimensional consolidation of isotropic triaxial samples for two different fundamental assumptions. In the first case the sample was assumed to be subjected to a homogeneous constant total stress field with one principal stress axis in the direction of the axis of the cylindrical sample and with two equivalent principal stress axes in arbitrary radial directions. In the second case the sample was assumed to undergo shear stresses in the base of the cylinder. Only the simpler solution for the first assumption will be summarized in this book. Firstly it will be reproduced for non-viscous soils.

The sample is assumed to be saturated ($\mathring{\omega} = 1$) and isotropic ($k_x = k_y = k_z = k$). The linear stress–strain relationship is governed by the constant moduli of compressibility (K) and of shear (G), and the constant state of the total stress field requires $\partial\sigma_{\text{oct}}/\partial t = 0$. Thus the general differential equation of consolidation (14.11) reduces to (see also Equation 14.15):

$$\frac{\partial u}{\partial t} = \frac{kK}{\gamma_w}\Delta u \tag{14.27}$$

Substituting

$$\frac{kK}{\gamma_w} = c_v \tag{14.28}$$

Equation (14.27) can be written in the form:

$$\frac{1}{c_v}\frac{\partial u}{\partial t} = \Delta u \tag{14.29}$$

For cylindrical co-ordinates (r,φ,z) (Fig. 14.1) the Laplace operator Δ is:

$$\Delta = \frac{\partial^2}{\partial r^2} + \frac{1}{r}\frac{\partial}{\partial r} + \frac{\partial^2}{\partial z^2} \tag{14.30}$$

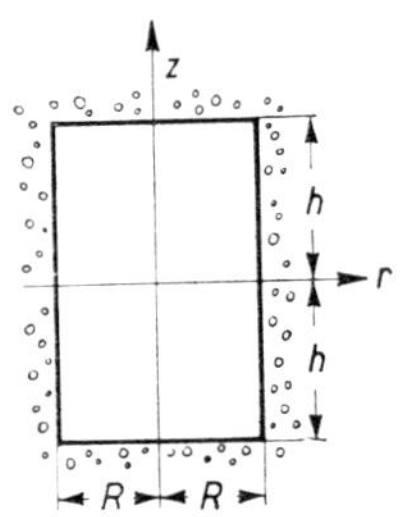

Fig. 14.1. Boundary conditions according to Equations (14.34) and (14.35)

In order to obtain the solution of Equation (14.29), the excess pore pressure may be expressed by

$$u = u_0 U_1(z,t) U_2(r,t) \tag{14.31}$$

u_0 being the excess pore pressure at time $t = 0$, and Equation (14.27) separated into the following two equations:

$$\frac{1}{c_v}\frac{\partial U_1}{\partial t} = \frac{\partial^2 U_1}{\partial z^2} \tag{14.32}$$

$$\frac{1}{c_v}\frac{\partial U_2}{\partial t} = \frac{\partial^2 U_2}{\partial r^2} + \frac{1}{r}\frac{\partial U_2}{\partial r} \tag{14.33}$$

The following boundary conditions have to be respected:

for U_1:

$$\left.\begin{array}{llll} (a) & U_1 = 1 & t = 0 & 0 \leqslant z \leqslant 2h \\ (b) & U_1 = 0 & t \to \infty & 0 \leqslant z \leqslant 2h \\ (c) & U_1 = 0 & 0 \leqslant t \leqslant \infty & z = 0,\ z = 2h \\ (d) & \dfrac{\partial U_1}{\partial z} = 0 & 0 \leqslant t \leqslant \infty & z = h \end{array}\right\} \tag{14.34}$$

for U_2:

$$\left.\begin{array}{llll} (e) & U_2 = 1 & t = 0 & 0 \leqslant r \leqslant R \\ (f) & U_2 = 0 & t \to \infty & 0 \leqslant r \leqslant R \\ (g) & U_2 = 0 & 0 \leqslant t \leqslant \infty & r = R \\ (h) & \dfrac{\partial U_2}{\partial r} = 0 & 0 \leqslant t \leqslant \infty & r = 0 \end{array}\right\} \tag{14.35}$$

Equation (14.32) is identical to Terzaghi's classical equation of consolidation (9.46). For analogous boundary conditions (14.34) its solution has been represented by Equation (9.63). With the notation used in Equations (14.32) and (14.34) it can be rewritten in the following form:

$$U_1 = \frac{4}{\pi}\sum_{m=0}^{\infty}\frac{1}{2m+1}\sin\left\{\frac{\pi}{2}(2m+1)\frac{z}{h}\right\}\exp\left\{-(2m+1)^2\frac{\pi^2}{4}T_v\right\} \tag{9.63-a}$$

$$T_v = \frac{c_v t}{h^2} \tag{9.64}$$

the coefficient of consolidation c_v being defined by Equation (14.28).

The function $U_2(r,t)$ in (14.31) may be expressed in the form:

$$U_2 = \phi(r)F(t) \tag{14.36}$$

The functions $\phi(r)$ and $F(t)$ should satisfy the differential Equation (14.33). Thus:

$$\frac{1}{c_v}\phi\frac{\partial F}{\partial t} = \left(\frac{\partial^2\phi}{\partial r^2} + \frac{1}{r}\frac{\partial\phi}{\partial r}\right)F \tag{14.37}$$

or:

$$\frac{1}{c_v}\frac{1}{F}\frac{\partial F}{\partial t} = \frac{1}{\phi}\left(\frac{\partial^2\phi}{\partial r^2} + \frac{1}{r}\frac{\partial\phi}{\partial r}\right) \tag{14.37'}$$

The solution can be obtained in the customary way:

(1)
$$\frac{1}{c_v}\cdot\frac{1}{F}\frac{\partial F}{\partial t} = -\lambda^2 \tag{14.38}$$

$$\frac{\partial F}{\partial t} + c_v\lambda^2 F = 0 \tag{14.38'}$$

yielding:

$$F = C\exp(-c_v\lambda^2 t) \tag{14.39}$$

where C is an arbitrary constant.

(2)
$$\frac{1}{\phi}\left(\frac{\partial^2\phi}{\partial r^2} + \frac{1}{r}\frac{\partial\phi}{\partial r}\right) = -\lambda^2 \tag{14.39}$$

$$\frac{\partial^2\phi}{\partial r^2} + \frac{1}{r}\frac{\partial\phi}{\partial r} + \lambda^2\phi = 0 \tag{14.39'}$$

The solution of this equation is a Bessel progression of first degree:

$$\phi = J_0(\lambda r) \tag{14.40}$$

Thus the general solution of the differential Equation (14.33) can be written as follows (see Equations 14.36, 14.39 and 14.40):

$$U_2 = C\exp(-c_v\lambda^2 t)J_0(\lambda r) \tag{14.41}$$

The boundary condition (14.35)–(g) requires:

$$J_0(\lambda R) = 0 \tag{14.42}$$

The roots of this equation can be obtained in mathematical tables.

$n =$	1	2	3	4	5
$R\lambda_n =$	2,405	5,520	8,654	11,792	14,931
$n =$	6	7	8	9	>9
$R\lambda_n =$	18,071	21,212	24,353	27,494	$\frac{\pi}{4}(4n-1)+\frac{1}{\pi(4n-1)}$

Consequently, the solution of the differential Equation (14.33) accounting for the boundary condition (14.35)–(g) can be written as follows:

$$U_2 = \sum_1^n C_n \exp(-c_v\lambda_n{}^2 t)J_0(\lambda_n r) \tag{14.43}$$

The coefficients C_n correspond to the arrangement of the function $U_2(r,0)$ in the Bessel progression [see e.g. MacRobert (1954), p. 236]:

$$C_n = \frac{\int_0^R J_0(\lambda_n r)r\,\mathrm{d}r}{\int_0^R J_0{}^2(\lambda_n r)r\,\mathrm{d}r} \tag{14.44}$$

14.2.2 Three-dimensional Consolidation of Saturated Viscous Triaxial Samples

In order to apply the fundamental differential equation of consolidation (14.27) to viscous soil the rheological models of which are linear, the modulus of compressibility should be replaced by the appropriate differential operator. If the transformation of linear viscous-elastic parameters to elastic parameters is related to Lamé's coefficients λ and μ (see Sections 4.1.4 and 4.4.2, Equations 4.18-a and 4.18-b), the relation (4.18-a) should be considered:

$$K = \lambda + \tfrac{2}{3}\mu \tag{14.45}$$

The application of such a procedure may be shown for the linear rheological model which has been proposed and investigated by Anagnosti (1962, 1963). The spherical stress–strain relationship is governed by the equations of Kelvin's body and the distortional strains are related to distortional stresses as in a Maxwell body. The application of this rheological scheme for the analysis of one-dimensional consolidation—after Anagnosti—has been presented in Section 10.2.2. Inserting the differential operators μ and λ as expressed by Equations (10.74) and (10.98), into Equation (14.45) we obtain:

$$K_t = \frac{K + \left(\eta^0 + \frac{K}{G}\eta^d\right)\frac{\partial}{\partial t} + \frac{\eta^0\eta^d}{G}\frac{\partial^2}{\partial t^2}}{1 + \frac{\eta^d}{G}\frac{\partial}{\partial t}} \tag{14.46}$$

By replacing K in Equation (14.27) by the above differential operator we have:

$$\left(1 + \frac{\eta^d}{G}\frac{\partial}{\partial t}\right)\frac{\gamma_w}{k}\frac{\partial u}{\partial t} = \left[K + \left(\eta^0 + \frac{K}{G}\eta^d\right)\frac{\partial}{\partial t} + \frac{\eta^0\eta^d}{G}\frac{\partial^2}{\partial t^2}\right]\Delta u \tag{14.47}$$

The pore pressure u may be expressed again by Equation (14.31). The solution of the above differential equation will be given by assuming:

$$\frac{\partial^2 U_1}{\partial t^2} = 0, \quad \frac{\partial^2 U_2}{\partial t^2} = 0 \tag{14.48-a,b}$$

Considering these assumptions and the expression (14.30) of the Laplace operator Δ, Equation (14.47) can be separated into the following differential equations:

$$\frac{\gamma_w}{k}\frac{\partial U_1}{\partial t} = K\frac{\partial^2 U_1}{\partial z^2} + A\frac{\partial^3 U_1}{\partial t \partial z^2} \tag{14.49}$$

$$\frac{\gamma_w}{k}\frac{\partial U_2}{\partial t} = \left[K + A\frac{\partial}{\partial t}\right]\left(\frac{\partial^2 U_2}{\partial r^2} + \frac{1}{r}\frac{\partial U_2}{\partial r}\right) \tag{14.50}$$

where A is given by:

$$A = \eta^0 + \frac{K}{G}\eta^d \tag{14.51}$$

Anagnosti's (1962) solutions of the above equations may be given in the following (modified) form:

$$U_1 = \frac{4}{\pi}\sum_{m=0}^{\infty}\frac{1}{2_m + 1}\sin\left\{\frac{2}{\pi}(2m+1)\frac{z}{h}\right\}\exp\frac{-(2m+1)^2\frac{\pi^2}{4}T_v}{1 + \frac{A}{Kt}T_v(2m+1)^2\frac{\pi^2}{4}} \tag{14.52}$$

$$T_v = \frac{kKt}{\gamma_w h^2} \tag{14.53}$$

$$U_2 = \sum_1^n C_n \exp\left[\frac{-c_v \lambda_n^2 t}{1 + \dfrac{Ak}{\gamma_w}\lambda_n^2}\right] J_0(\lambda_n r) \tag{14.54}$$

For c_v see Equation (14.28), for λ_n, C_n and $J_0(\lambda_n r)$ see the previous section.

14.3 INITIAL CONSOLIDATION PORE PRESSURES IN SATURATED SOILS

14.3.1 Creep Phenomena Neglected

In an elastic isotropic saturated soil of low permeability the volume strains are linearly proportional to the spherical components of the effective principal stresses. If the compressibility of the solid grains and of the liquid (water) is neglected, the volume of the soil remains unchanged immediately after the load application. Consequently, the spherical component of effective stress does not change either. Thus, the initial pore pressure equals the total increment of the spherical component of total stress. In a triaxial sample $\Delta\sigma_2 = \Delta\sigma_3$ and

$$\Delta u = \Delta\sigma_{\text{oct}} = \frac{\Delta\sigma_1 + 2\Delta\sigma_3}{3} \tag{14.55}$$

The equation can be written also in the form of Skempton's empirical formula:

$$\frac{\Delta u}{\Delta\sigma_3} = 1 + A\left(\frac{\Delta\sigma_1}{\Delta\sigma_3} - 1\right) \tag{14.56}$$

the pore pressure coefficient A having the value:

$$A = \tfrac{1}{3} \tag{14.57}$$

Now, the soil may be assumed to be elastic but anisotropic; the anisotropy is assumed to be orthogonal and one axis of the anisotropy coincides with the specimen axis.* In this case the linear relationship between

* When the deformation parameters in lateral directions (2) and (3) are different from each other, the following considerations apply to rectangulary shaped prismatic samples. The equipment which enables tests with such samples, and the testing procedure, have lately been described in the papers by Ko and Scott (1967), Lomize and Kryzhanovsky (1967).

spherical components of stress and strain tensors no longer holds. The total strain in the direction (1) is given by the expression:

$$\varepsilon_1 = \frac{1}{E_1}(\Delta\sigma_1 - \Delta u) - \frac{\nu_{12}}{E_2}(\Delta\sigma_2 - \Delta u) - \frac{\nu_{13}}{E_3}(\Delta\sigma_3 - \Delta u) \quad (14.58)$$

E_i denoting the deformation modulus in the direction i, and ν_{ik} the Poisson ratio governing the influence of the stress σ_k on the strain ε_i. Analogous expressions will be given for the strains ε_2 and ε_3 and all the expressions inserted into the condition:

$$\varepsilon_v = \varepsilon_1 + \varepsilon_2 + \varepsilon_3 = 0 \quad (14.59)$$

However, the validity of the equation $\varepsilon_v = \varepsilon_1 + \varepsilon_2 + \varepsilon_3$ is limited to cases of very small values of linear strain (see Chapter 3). Rearrangement of the expression obtained produces Skempton's Equation (14.56) but with the following meaning of the pore pressure coefficient A (Šuklje, 1961-c):

$$A = \frac{C_{c1}}{C_{c1} + C_{c2} + C_{c3}} \quad (14.60)$$

where:

$$C_{c1} = \frac{3}{E_1}(1 - \nu_{21} - \nu_{31}) \quad (14.61)$$

and similarly for C_{c3}, C_{c2}. In Fig. 14.2 the pore pressure coefficients A according to Equation (14.60) are plotted against the ratio of anisotropy $\psi = C_{c3}/C_{c1}$.

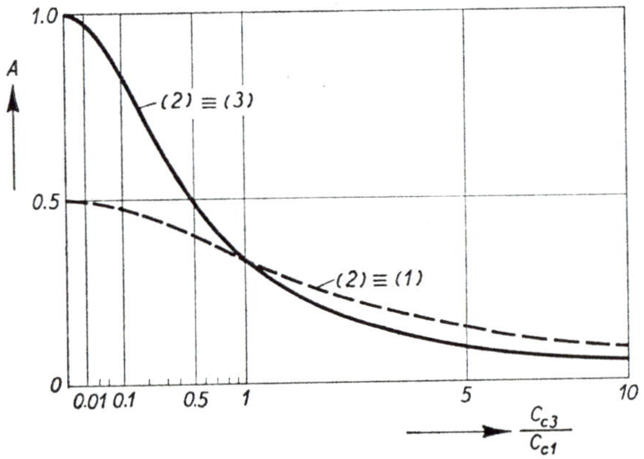

Fig. 14.2. Pore pressure coefficients A of elastic anisotropic soils. (After Šuklje, 1961-c)

In unconfined conditions, assuming $\Delta\sigma_3 < \Delta\sigma_1$, the initial additional effective pressures $\Delta\sigma_3'$ are negative since, according to Equation (14.56), $\Delta u > \Delta\sigma_3$ for any positive value of the pore pressure coefficient A. As the compressibility on unloading is several times smaller than the compressibility at first loading, normally consolidated specimens are to be treated as anisotropic soils with $C_{c3}/C_{c1} > 1$. Consequently, in such soils the coefficient A is expected to be greater than one-third, this being in agreement with experimental evidence (Bishop and Henkel, 1957).

For the conditions of no lateral strain the zero volume strain requires the axial linear strain ε_1 to be zero too. Three equations of the type:

$$\frac{1}{E_1}(\Delta\sigma_1 - \Delta u) - \frac{\nu_{12}}{E_2}(\Delta\sigma_2 - \Delta u) - \frac{\nu_{13}}{E_3}(\Delta\sigma_3 - \Delta u) = 0 \quad (14.62)$$

have non-trivial solutions

$$\Delta u \neq \Delta\sigma_1 = \Delta\sigma_2 = \Delta\sigma_3 \quad (14.63)$$

under the condition that the determinant of the coefficients vanishes. This condition leads to the requirements:

$$2\nu_{13}\nu_{31} = 1 - \nu_{23} \quad \text{in case } (2) \equiv (3) \text{ and}$$
$$2\nu_{13}\nu_{31} = 1 - \nu_{12} \quad \text{in case } (2) \equiv (1)$$

Consequently, one of the Poisson ratios, at least, should be greater than 0·50 if the soil is anisotropic, and $\nu = 0{\cdot}50$ in isotropic soils ($A = \frac{1}{3}$). As even in dense sandy soils the value $\nu_{ik} > 0{\cdot}50$ cannot be expected until the full internal friction—at zero dilatation—has been mobilized (see e.g. Hvorslev, 1960), the theoretical exceptions stated above can be disregarded. Thus, a hydrostatic state of stress appears if the soil is not affected by viscous effects at the very beginning of the consolidation process:

$$\Delta u = \Delta\sigma_1 = \Delta\sigma_2 = \Delta\sigma_3 \quad (10.64)$$

The case of triaxial specimens with zero lateral strain applies also to the initial consolidation state of oedometer tests.

14.3.2 Creep, Relaxation and Thixotropy Effects

In the preceding section the viscous properties of soils were fully neglected. If they were to be taken into account, the speed of spherical creep could be assumed to depend on the porosity and, in cases of deformation isotropy, on the value of the octahedral normal stress (Šuklje, 1963-a).

In the conditions of undrained triaxial tests spherical creep cannot occur. (The deformability of the solid skeleton and of the water may again be neglected.) Thus the pore pressure governed by the coefficient A according to Equation (14.60) is expected to be increased due to relaxation of the intergranular stresses. The greater the creep rate at the preceding state of intergranular stress, the more important the relaxation effect. The effect, can, however, be efficiently retarded by air bubbles. Such relaxation effects account for the observed gradual increase of pore pressures in undrained triaxial samples (Bjerrum, Kringstad and Kummeneje, 1961; Lo, 1961 (a)).

Under drained conditions the relaxation effect does not appear, for the spherical part of creep can develop freely according to the effective pressures which are compatible with the pore pressures governing the pore water drainage. Thereby the initial pore water coefficient A can be less than $\frac{1}{3}$ for isotropic and less than the value given by Equation (14.60) for anisotropic soils. For the conditions of no lateral strain, e.g. in oedometer tests, consolidation can start with pore pressures less than $\Delta\sigma_1$ (Šuklje, 1957).

The increasing pore pressures of saturated undrained samples are accompanied by increasing mobilization of the shearing resistance at a constant total stress state (Bjerrum, 1961). In Fig. 14.3 circle 1 represents

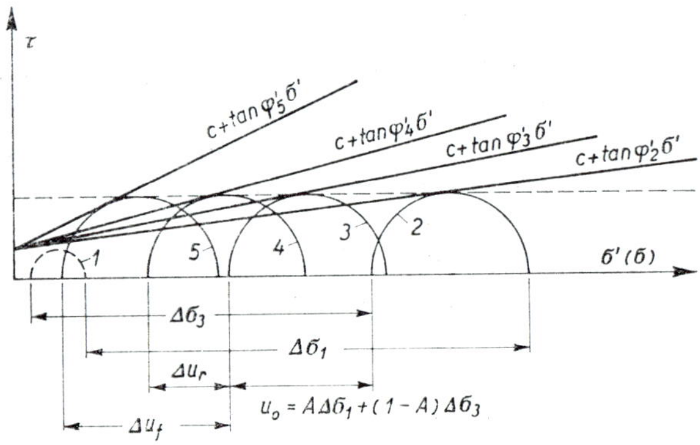

Fig. 14.3. Increase of the strength mobilization in undrained conditions of triaxial tests

the initial effective stress state, circle 2 the new total stress state and circle 3 the corresponding new effective stress state. The distance between circles 2 and 3 is $u_0 = A\Delta\sigma_1 + (1 - A)\Delta\sigma_3$ where A is given by Equation (14.60). The further increase Δu_r of pore pressure due to the relaxation effect mentioned above causes a further increase of the mobilized angle of shearing resistance from φ_3' to φ_4'. If the total stress difference $\Delta\sigma_1 - \Delta\sigma_3$ is large enough, the full effective angle of shear φ_5' (Fig. 14.3) can be mobilized and rupture then occurs. In soils with loose structure rupture can be accelerated by thixotropic phenomena appearing at the partial or complete collapse of the soil skeleton. In these cases very high pore pressures can occur.

If the initial structure of the loose non-cohesive soil is completely destroyed, the intergranular stresses become zero and liquefaction occurs. If, however, the structural destruction is limited, new bonds can be formed which diminish the distortional creep and thus rupture can be avoided.

In natural conditions the influence of the plastic resistance on the initial effective pressure is the smaller the longer the drainage paths. In thick layers the initial pore pressures approach the theoretical values (Equations 14.56 and 14.60) provided that the previous degree of consolidation due to preloading and/or secondary time effects is moderate and the additional pressure big enough.

The relaxation effects do not appear either. Consequently the pore-pressure coefficients governing the initial phase of consolidation of thicker layers are expected to be smaller than observed in undrained triaxial samples and to approach more closely the theoretical values obtained for non-viscous soils. On the other hand, they can exceed the pore pressures as observed at the beginning of the one-dimensional consolidation of drained specimens.

The additional thixotropic effects observed in samples with loose structure and due to the deviatoric stress components, occur in field conditions just as in undrained tests. They can increase considerably the distortional creep rates and influence, though to a lesser degree, the primary consolidation as well. These effects have to be avoided or diminished by introducing sufficiently high safety factors in the bearing capacity analysis and by decreasing the loading rate.

14.4 CONSOLIDATION ANALYSIS BASED ON OBSERVED CONSOLIDATION CURVES

14.4.1 Consolidation Curves of Saturated and Partly Saturated Soils

The oedometer test represents a triaxial test corresponding to the condition of no lateral strain. Consequently the consolidation curves observed in oedometer tests show one of the possible forms of consolidation curves which could be observed in triaxial tests subjected to principal stresses with various relationships between the axial and radial stresses.

Examples of consolidation curves of saturated oedometer samples have been presented in Figs. 9.11 (*a–f*), 11.7 (*A*), 11.11 (*a*), 11.13 and 11.14.

The secondary branch of consolidation curves was shown to be of special interest in the prediction of the consolidation curves of thicker layers, and very frequently the laboratory observations indicated that the logarithmic law of secondary compression held over a long period of time.

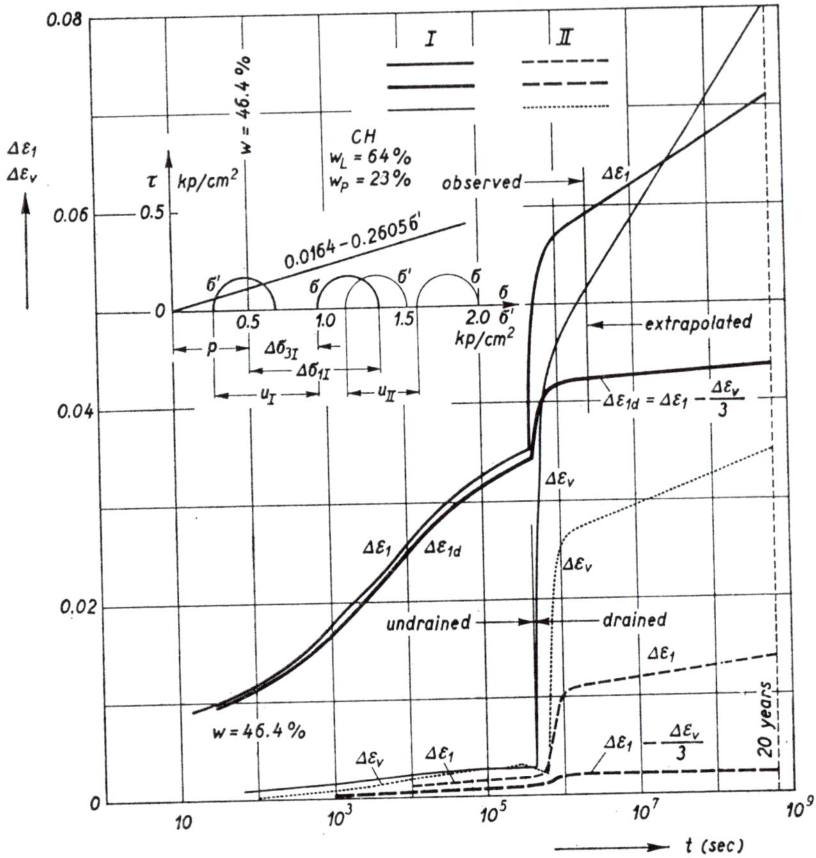

Fig. 14.4. Axial- and volume-strain increase in undrained and in drained conditions during a triaxial test of a normally consolidated clay. (After Šuklje and Sovinc, 1963)

Any secondary consolidation branch of oedometer samples can be split up into its spherical and distortional part. As lateral displacements are not allowed, the total axial strain equals the volume strain ε_v. Thus, according to Equations (3.18) to (3.21), the spherical strain $\varepsilon_1{}^0$ is

$$\varepsilon_1{}^0 = \frac{\varepsilon_v}{3} = \frac{\varepsilon_1}{3} \tag{14.65}$$

the distortional axial strain:

$$\varepsilon_1{}^d = \varepsilon_1 - \varepsilon_1{}^0 = \tfrac{2}{3}\varepsilon_1 \tag{14.66}$$

and the distortional radial strain:

$$\varepsilon_2{}^d = \varepsilon_3{}^d = \varepsilon_3 - \varepsilon_3{}^0 = 0 - \frac{\varepsilon_v}{3} = -\frac{\varepsilon_1}{3} \tag{14.67}$$

If the secondary branch of the total axial strain is a logarithmic straight line, both the spherical and the distortional consolidation components have logarithmic form too.

The consolidation curves ε_1 and ε_v of a dry powder of lacustrine chalk presented in Fig. 13.1 can also approximately be substituted by logarithmic straight lines. For a 'saturated' sample of a normally consolidated clay, Fig. 14.4 shows the observed consolidation curves $\Delta\varepsilon_1$ and $\Delta\varepsilon_v$ in undrained and in drained conditions. That some amount of volumetric strain occurred in undrained condition proves that the saturation was not complete. For this reason the initial pore pressure was less than $\Delta\sigma_{oct}$. Although the relationship between the principal effective stresses was evidently high and the effective stress Mohr circle approached the Coulomb envelope, the distortional strains still did not increase faster than those corresponding to the logarithmic law.

Another example related to a normally consolidated marsh clay of very high compressibility is given in Fig. 14.5.

The distortional consolidation curves of some undrained undisturbed samples of clay presented by Haefeli and Schaerer (1946) follow the logarithmic law of secondary compression as well.

The experiments of Tan Tjong Kie (1954) [Geuze (1953)] made on remoulded samples of a clay of high compressibility ($w_L = 94$ per cent, $w_P = 27$ per cent) subjected in a torsion plastometer to a pure distortional stress tensor prove that in the interval between two limiting shear stresses τ_2 and τ_3 the clay has the properties of a Bingham body. The rate of shear strain $D = \mathrm{d}\gamma/\mathrm{d}t$ is constant at the constant stress state and increases in linear proportion to the increasing shear stress; on unloading only a part of the total deformation is recoverable. Below the lower limit τ_2 of this interval the strains are elastic and in most cases complete recovery might be obtained. Beyond the upper limit τ_3 the flow rate increases with stress to a power more than 1 and also with time indicating that structural disintegration is occurring.

Based on this experimental evidence Geuze (1953) suggested that the admissible shear strength should be determined as a function of the limiting values τ_2 and τ_3 and of the Bingham viscosity governing the creep rate in the interval between τ_2 and τ_3.

The lower limit τ_2 and the coefficient of viscosity during subsequent creep depend on the moisture content of the soil, on the pre-consolidation effects and on the structural resistance associated with the degree of disturbance of the soil. In tests presented by Geuze (1953) and Tan (1954) the lower limit τ_2 is less than one quarter of the total shear strength. It may be expected to be greater in undisturbed samples, in pre-consolidated

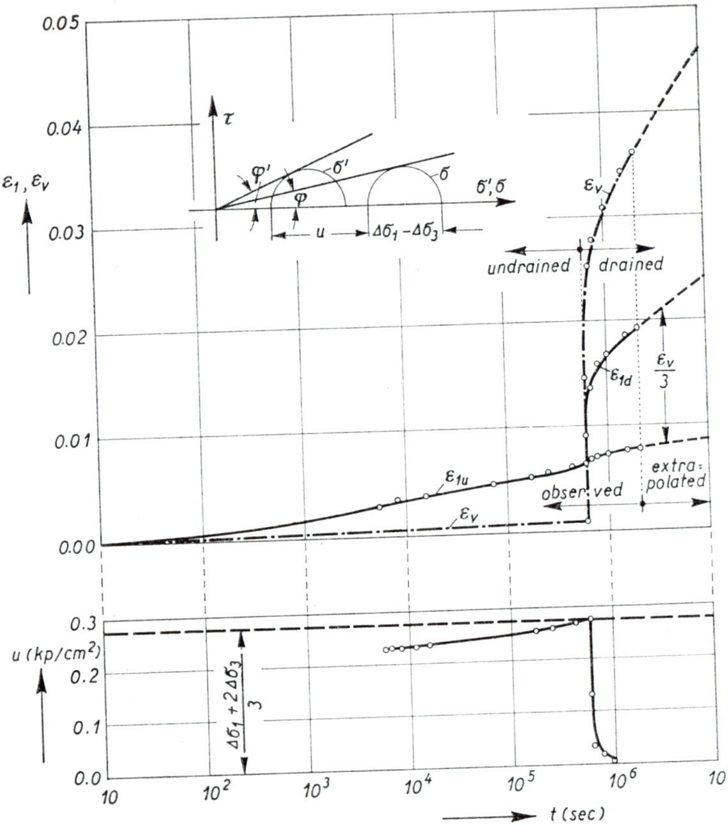

Fig. 14.5. Axial- and volume-strain increase in undrained and in drained conditions during a triaxial test of a normally consolidated lacustrine clay, and the corresponding pore-pressure development. (After Sovinc, 1964)

soils and in soils of smaller compressibility (smaller index of plasticity). Further, the relaxation and thixotropy effects which were discussed in the previous section should be taken into account when applying the test observations to the prediction of the behaviour of foundations and of earth structures.

It is easy to understand that in drained conditions the spherical creep rate decreases as the density increases during the course of secondary consolidation. The increase in the attraction forces of the bonds of adsorbed water between the grains that pass into a denser packing also brings about a decrease in the distortional creep rate. In undrained conditions the hardening effect during incomplete creep recovery could be attributed to the so-called thixotropy. In the present book we are mainly considering the mechanical phenomena and trying to discover the stress–strain relationships in soils without making a more extensive physico-chemical interpretation of them. In the publications of Denissov (1956), Rosenquist (1959, 1961), Geuze (1964), Scott (1963) and others the reader can obtain a better understanding of the observed mechanical behaviour of soils. An unambiguous and complete interpretation of the observed mechanical phenomena can hardly be expected at the present stage of our knowledge.

14.4.2 The Spherical Consolidation Analysis by the Isotache Method

14.4.2.1 *Introductory observations*

In Section 8.3 isotaches were defined as graphs relating the effective pressure σ' to the void ratio e for certain constant consolidation speeds $\partial e/\partial t$. When dealing with one-dimensional consolidation at no lateral strain the isotaches were constructed by plotting the void ratios e against the axial effective stresses σ_1'. In this way the dilatancy effects of distortional stresses were included in the presentation of the void ratio-effective stress relationship. The resulting isotache set can only be applied, however, to stress states corresponding to the zero lateral displacement condition.

When analysing the three-dimensional consolidation the isotaches may represent the plots of the void ratios e against the mean intergranular pressures σ_{oct}' corresponding to certain consolidation speeds $\partial e/\partial t$. In isotropic soils and in isotropic deformability conditions the system of isotaches can be expected to be independent of the distortional stresses. In anisotropic soils or generally in soils exhibiting anisotropic deformability, the distortional stresses do influence the isotache set. The assumption may be made that in anisotropic conditions the same isotache set corresponds to a given principal stress ratio σ_3'/σ_1', (or to the corresponding ratio $\eta = q/p = 3(\sigma_1' - \sigma_3')/(\sigma_1' + 2\sigma_3')$).

The application of the isotache method to arbitrary consolidation problems can be based on the general differential equation of three-dimensional consolidation (14.4) which should be solved by numerical methods. The procedure for this will be dealt with in Section 16.3.

In the present Chapter the application of the isotache method will be given only for isotropic cylindrical samples subject to the conditions of one-dimensional drainage. In terms of linear compressibility the corresponding procedure has been developed for saturated soils in Section 11.1 and for partly saturated soils in Section 12.3. The application of an analogous procedure for triaxial samples subjected to arbitrary controlled principal stress ratios σ_3'/σ_1' is the subject of the following considerations.

14.4.2.2 *Spherical One-dimensional Consolidation of Saturated Soils*

Let us consider a triaxial sample with axial drainage. If the lateral displacements are small enough to allow the change in the drainage cross-section to be neglected, the Terzaghi linear consolidation theory related to octahedral stresses can be applied, provided that the soil is isotropic, the effective stress–strain relationship linear and that secondary time effects do not appear.

The initial isochrone is given by the expression:

$$\Delta u_0 = \Delta\sigma_{oct} = \tfrac{1}{3}(\Delta\sigma_1 + 2\Delta\sigma_3) \tag{14.68}$$

and the coefficient of consolidation is

$$c_v = \frac{k}{\gamma_w m_v} = \frac{k}{\gamma_w} K \tag{14.69}$$

where k = the coefficient of permeability, γ_w = the specific gravity of water and K = the modulus of volume compressibility.

Now, if the spherical creep effects cannot be ignored, one of the basic assumptions of the Terzaghi theory, that is the linear relation between effective spherical stress and spherical strain, no longer holds. Still considering the deformation isotropy of the sample, the hypothesis that the speed of spherical consolidation depends on the mean values of void ratio $\bar{e}$ and total intergranular octahedral stress σ_{oct}' can be assumed.

The spherical consolidation of the sample can be presented by the plot $e = e(t)$ (Fig. 11.10).

If the lateral displacements are small enough, the drainage cross-section can be assumed to remain approximately constant during consolidation. In this case the pore pressures corresponding to the observed consolidation curves can be determined by the same procedure as explained in Section 11.1.2 dealing with the one-dimensional consolidation of confined specimens. The mean values of pore pressures $\bar{u}_w$ are given by the equations (see Equations 11.9, 11.13, 11.22, 12.55, 12.56):

$$\text{(I)} \qquad \bar{u}_w = \frac{\gamma_w}{N(n+1)} \bar{\vartheta}_e (1 + \bar{e}) \frac{h_s^2}{kt^m} \tag{14.70}$$

when $\bar{u}_w \leqslant \dfrac{n}{n+1}\Delta\sigma_{oct}$,

(II) $$\bar{u}_w = \Delta\sigma_{oct}\left(1 - \frac{\Delta\sigma_{oct}}{\gamma_w}\cdot\frac{Nn}{n+1}\cdot\frac{kt^m}{\bar{\vartheta}_e(1+\bar{e})h_s^2}\right) \qquad (14.71)$$

when $\bar{u}_w \geqslant \dfrac{n}{n+1}\Delta\sigma_{oct}$

where the following notation has been used: $h_s = h/(1+e)$ = height of solid in the specimen, assumed to be compact; $\bar{e}$ = mean void ratio and k = the corresponding coefficient of permeability at time t; $\Delta\sigma_{oct} = \frac{1}{3}(\Delta\sigma_1 + 2\Delta\sigma_3)$; γ_w = unit weight of water; n = degree of the parabolic isochrone; further:

(*a*) $N = 1{\cdot}3026$, $\bar{\vartheta}_e = \bar{\alpha}_e$ (see Fig. 8.6), $m = 1$ if the mean void ratios are plotted against the logarithm of time t, and
(*b*) $N = 2$, $\bar{\vartheta}_e = \bar{\beta}_e$ (see Fig. 11.3), $m = \frac{1}{2}$, if they are plotted against the square root of time.

The corresponding mean values of effective octahedral stress can then be found by using the relation:

$$\Delta\bar{\sigma}_{oct}' = \Delta\sigma_{oct} - \bar{u}_w \qquad (14.72)$$

and the simultaneous consolidation speed is given by Equation (12.27′):

$$\frac{\partial\bar{e}}{\partial t} = -\frac{\bar{\vartheta}_e}{Nt^m} \qquad (12.27')$$

Now, by applying an analogous procedure to that shown for the case of no-lateral strain in section 11.1.5, the isotaches $e = e(\sigma_{oct}')$ can be constructed for various consolidation speeds $\partial e/\partial t$ (see Fig. 11.10).

The isotache set can be used to construct the spherical consolidation curve of a thicker layer. The following conditions should be satisfied: (*a*) The sum of the mean effective octahedral stress $\bar{\sigma}_{oct}'$ and the average pore pressure $\bar{u}_w$ must equal the total octahedral stress σ_{oct}; (*b*) the specific consolidation speed $\partial\bar{e}/\partial t$ as well as the mean effective octahedral stress $\bar{\sigma}_{oct}' = \sigma_{oct\alpha} + \Delta\bar{\sigma}_{oct}'$ at a given time t are determined by Equations (12.27′) and (11.8)–(11.10) respectively, by inserting the slope ($\bar{\vartheta}_e$) of the consolidation curve. From these calculated values $\partial\bar{e}/\partial t$ and $\bar{\sigma}_{oct}'$ a certain void ratio e results from the isotache chart $e = e(\sigma_{oct}')$. The void ratio obtained must agree with the value $\bar{e}$ of the consolidation curve $\bar{e} = \bar{e}(t)$ at time t. (c) The consolidation curve must be continuous.

Considering the above conditions the consolidation curve may gradually be constructed.

In isotropic soils the isotache set can be constructed also from consolidation curves obtained at continuous load application between some constant load intervals of long duration. The continuous loading could be recommended especially for normally consolidated soils with loose structure. The observations given in Section 11.1 6 when dealing with the one-dimensional consolidation at no lateral strain also facilitate the interpretation of certain special phenomena when applying the method to the spherical consolidation of triaxial samples.

If the soil exhibits orthogonal anisotropy with respect to its deformability characteristics, the distortional stress components also influence the volume strains. For different principal stress ratios different isotache sets should be constructed and considered when predicting the consolidation of thicker layers.

14.4.2.3 *Spherical One-dimensional Consolidation of Partly Saturated Soils*

All the deductions derived for the one-dimensional consolidation of partly saturated soils (see Sections 12.2 and 12.3) apply also for the spherical consolidation of triaxial samples provided that the drainage is allowed only in the axial direction of the sample and that the radial strains are small enough to allow the change of the drainage cross-section to be neglected. In these equations the following substitutions should be made:

$$\Delta\sigma_{oct} \quad \text{for} \quad q$$

$$\sigma_{oct} \quad \text{for} \quad \sigma$$

$$\sigma_{oct}' \quad \text{for} \quad \sigma'$$

$$\frac{\partial e}{\partial \sigma_{oct}'} = (1+e)\frac{\partial \varepsilon_v}{\partial \sigma_{oct}'} = \frac{1+e}{K} \quad \text{for} \quad \frac{\partial e}{\partial \sigma'} = -a$$

K being the modulus of (volume) compressibility.

14.4.3 Distortional Displacements of Saturated and Partly Saturated Soils

In elastic porous and saturated triaxial samples exhibiting orthogonal symmetrical anisotropy with respect to the sample axis the total axial linear strain ε_1 is given by the equation

$$\varepsilon_1 = \frac{\Delta\sigma_1}{E_1} - 2\nu_{13}\frac{\Delta\sigma_3}{E_3} \tag{14.73}$$

while the immediate distortional strain is:

$$\varepsilon_{1in} = \frac{\Delta\sigma_1 - \Delta u_0}{E_1} - 2\nu_{13}\frac{\Delta\sigma_3 - \Delta u_0}{E_3} \tag{14.74}$$

Expressing the initial pore pressure Δu_0 by Equations (14.56) and (14.60), considering the relation (13.11) and assuming

$$\nu_{23} = \tfrac{1}{2}(\nu_{13} + \nu_{31}) \tag{14.75}$$

the ratio of the above expressions for ε_{1in} and ε_1 yields:

$$\vartheta = \frac{\varepsilon_{1in}}{\varepsilon_1} = 1 - \frac{1 - 2\nu_{31}}{1 - 2\alpha\nu_{31}}\left\{\alpha + (1 - \alpha)\,\frac{1 - 2\nu_{31}}{1 + 2\beta - (5 + \beta)\nu_{31}}\right\} \tag{14.76}$$

where

$$\alpha = \frac{\Delta\sigma_3}{\Delta\sigma_1}, \quad \beta = \frac{E_1}{E_3} \tag{14.76-a,b}$$

For $\alpha = 0$ (simple compression) the ratio between the initial and final axial strain reduces to:

$$\vartheta_{\alpha=0} = \left(\frac{\varepsilon_{1in}}{\varepsilon_1}\right)_{\alpha=0} = 1 - \frac{(1 - 2\nu_{31})^2}{1 + 2\beta - (5 + \beta)\nu_{31}}; \tag{14.77}$$

for $\beta = 1$, $\nu_{31} = \nu_{13} = \nu$, i.e. for isotropic samples, to

$$\vartheta_{\beta=1} = \left(\frac{\varepsilon_{1in}}{\varepsilon_1}\right)_{\beta=1} = \frac{2(1 - \alpha)(1 + \nu)}{3(1 - 2\alpha\nu)}; \tag{14.78}$$

and for $\beta \to \infty$ to

$$\vartheta_{\beta\to\infty} = \left(\frac{\varepsilon_{1in}}{\varepsilon_1}\right)_{\beta\to\infty} = \frac{1 - \alpha}{1 - 2\alpha\nu_{31}} \tag{14.79}$$

In Fig. 14.6 (*a–c*) the coefficient ϑ is plotted versus the principal stress ratio $\alpha = \Delta\sigma_3/\Delta\sigma_1$ for various Poisson's ratios ν and for various anisotropy ratios $\beta = E_1/E_3$. The plots prove the important influence of the coefficients α, β and ν on the distortional strain ratio ϑ. The influence of anisotropy increases with decreasing Poisson's ratio ν, with decreasing principal stress ratio α and with decreasing anisotropy ratio β.

In isotropic non-viscous soils the total strains are governed by two elastic parameters which can be considered constant over a limited load interval. These elastic parameters are either the shear modulus G and the compressibility modulus K or the deformation modulus E and the Poisson ratio ν. As these two combinations of elastic parameters are related to each other by Equations (4.15) and (4.18) or (13.6) and (13.7), the combination of the parameters G and ν can be taken as the basis of the

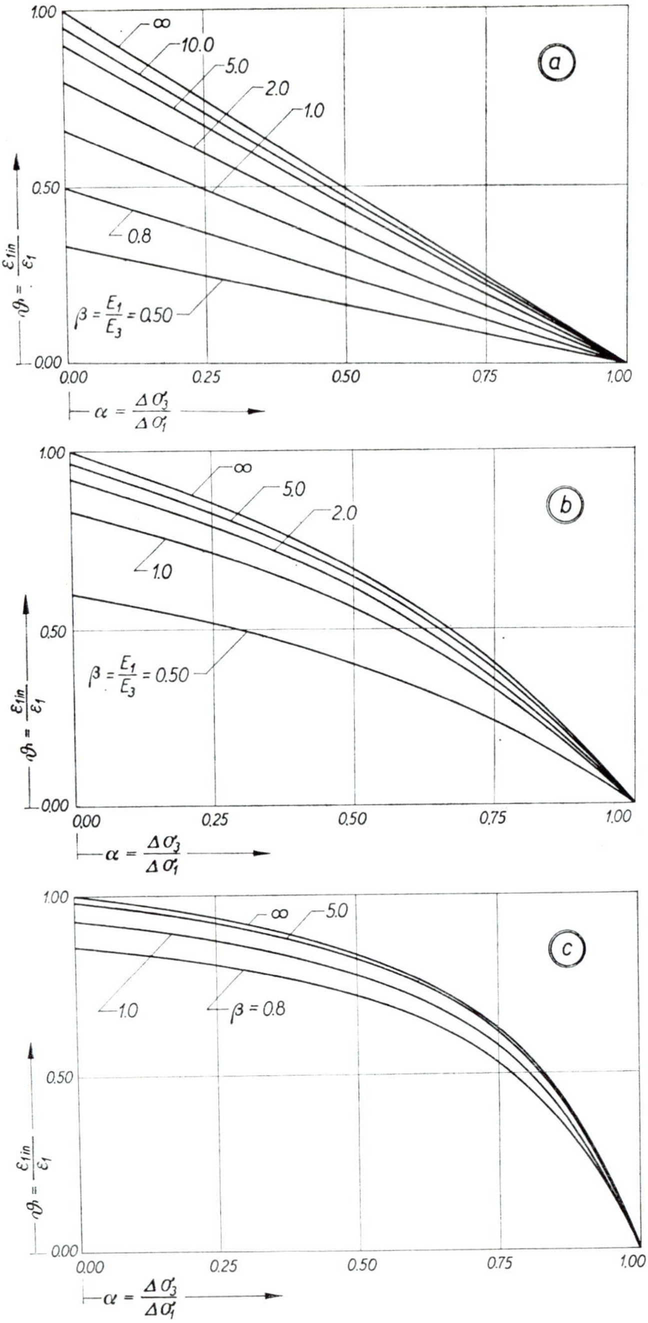

Fig. 14.6. Influence of Poisson's ratio ν, deformation anisotropy (β) and principal stress ratio (α) on the initial strain versus total strain ratio of saturated, non-viscous triaxial samples: (*a*) $\nu_{31} = 0$, (*b*) $\nu_{31} = 0{\cdot}25$, (*c*) $\nu_{31} = 0{\cdot}40$

computations as well. Now, during the primary consolidation of saturated soils the effective compression modulus K_e gradually decreases from infinity to a final value K and the Poisson ratio from $\nu_d = 0{\cdot}50$ to the final value ν given by Equation 13.7. The initial displacements are completely of distortional character and can be computed by considering the given shear modulus G and the Poisson ratio $\nu_d = 0{\cdot}50$.

In anisotropic deformation conditions the above conclusion can only be used for approximate calculations by introducing an equivalent shear modulus G. Using triaxial test data the equivalent modulus can be obtained by Equation (13.4-a). According to Figs. 14.4 and 14.5 the distortional displacements $\Delta\varepsilon_1 - (\Delta\varepsilon_v/3)$ increase with time and consequently the equivalent modulus of shear G decreases. The change due to distortional creep is influenced by the change of principal effective stress ratio α' (Equation 13.10) during primary consolidation.

As shown in Section 13.4.3, the intensity of the distortional creep rate ($\dot{\Gamma}$) depends on the intensity of shear stresses (T) and on the mean principal stress (σ_{oct}). In the coordinate system (T,σ_{oct}) different plots $T = T(\sigma_{oct})$ correspond, in a given time t, to different creep rates $\dot{\Gamma}$. In the state of plane strain ($\varepsilon_3 = 0$) the intensity of the shear stress is (see Equations 13.113, 2.12 and 2.13):

$$T = \frac{\sigma_1 - \sigma_3}{3} \tag{14.80}$$

and the octahedral stress σ_{oct} is

$$\sigma_{oct} = \frac{\sigma_1 + \sigma_3}{2} \tag{14.81}$$

because $\sigma_2 = (\sigma_1 + \sigma_3)/2$ ($\nu = 0{\cdot}50$). The plots $T = T(\sigma_{oct})$ may be assumed to be straight lines, each of them corresponding, at a given time t after load application, to a certain creep rate $\dot{\Gamma}$. The higher the creep rate, the steeper the slope of the straight line $T = T(\sigma_{oct})$ and the greater the degree of strength mobilization.

In constant volume conditions the straight envelopes of the Mohr circles of effective stresses can also be assumed to be isotaches, i.e. the lines corresponding to equal distortional creep rates at a given time t after the load application. In Fig. 14.7 circle 1 is the Mohr circle of the initial consolidated stress state ($\sigma_{1in} = \sigma_{1in}' = p$; $\sigma_{3in} = \sigma_{3in}' = K_0 p$), circle 2 is the Mohr circle of the new total stress state, and circle 3 represents the corresponding effective stress state at the very beginning of the consolidation process. Considering the stress relations shown in the figure and expressing the initial pore pressure $u = \Delta u$ by Skempton's Equation (14.56), the following relation results:

$$\frac{1}{\sin \varphi_m'} = (1 - 2A) + \frac{p_c + (1 + K_0)p}{\Delta\sigma_1 - \Delta\sigma_3} \qquad (14.82)$$

According to this relation the degree of strength mobilization governing the initial creep rate increases with increasing pore pressure coefficient A and with the increasing quotient:

$$\frac{\Delta\sigma_1 - \Delta\sigma_3}{p_c + (1 + K_0)p}$$

The plot of distortional strain versus time presented in Fig. 14.4 shows a step when passing from undrained to drained test conditions. As the

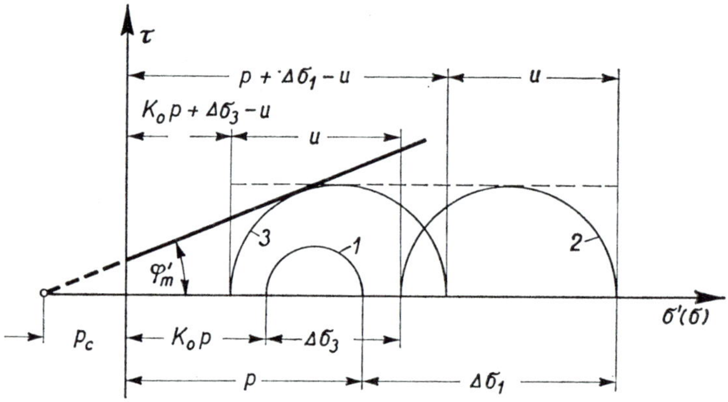

Fig. 14.7. Influence of pore pressures on increasing speed of distortional creep

distortional stresses do not change at this transition, the step can probably be attributed mainly to anisotropic deformability effects. During primary consolidation the pore pressure decreases and the effective stresses increase by the same amount: $\Delta\sigma' = -\Delta u$. Under anisotropic deformation conditions the additional total strains are:

$$\Delta\varepsilon_1 = \Delta\sigma'\left(\frac{1}{E_1} - \frac{2\nu_{13}}{E_3}\right) \qquad (14.83)$$

$$\Delta\varepsilon_3 = \Delta\sigma'\left(\frac{1 - \nu_{32}}{E_3} - \frac{\nu_{31}}{E_1}\right) \qquad (14.84)$$

where a linear stress–strain relationship has been assumed and the creep effect is considered separately. Taking into account the relation (13.11) and the assumption (14.75), the corresponding distortional strain $\Delta\varepsilon_1{}^d$ can be found to be:

$$\Delta\varepsilon_1{}^d = \Delta\varepsilon_1 - \frac{\Delta\varepsilon_v}{3} = \frac{\Delta\sigma'}{E_1}(2 - \nu_{31})\frac{\beta - 1}{3\beta} \tag{14.85}$$

$$\beta = \frac{E_3}{E_1} \tag{14.86}$$

Thus regardless of the creep effect positive additional distortional strains occur during primary consolidation of anisotropic soils with $\beta > 1$. The pre-consolidation effect $E_3 > E_1$ seems to be a consequence of subsequent effective stress states of normally consolidated samples if the load is applied under the following conditions: $\sigma_{3in} \leqslant \sigma_{1in}$, $\Delta\sigma_3 < \Delta\sigma_1$.

The application of the distortional consolidation curves observed in triaxial samples to predict distortional consolidation of thicker layers will be dealt with in Chapter 17.

The possibilities and limitations of determining the independent anisotropic elastic constants E_1, E_3, ν_{13} and ν_{23} have been discussed in section 13.1.2. The same considerations apply to saturated and partly saturated soils in the secondary consolidation stage.

Part D

APPLICATION OF THE CONSOLIDATION ANALYSIS IN THE HALF-SPACE AND IN EARTH-CONSTRUCTIONS

15

Analytical Solutions of the Differential Equation of Consolidation for Non-Viscous Isotropic Soils

15.1 CONSOLIDATION OF THE SATURATED HOMOGENEOUS ISOTROPIC HALF-SPACE SUBJECTED TO A CONCENTRATED SURFACE LOAD

15.1.1 Simplified Solution for the Simple Half-space

The half-space is assumed to be subjected to a suddenly applied concentrated surface load (Fig. 15.1). The soil is assumed to be homogeneous,

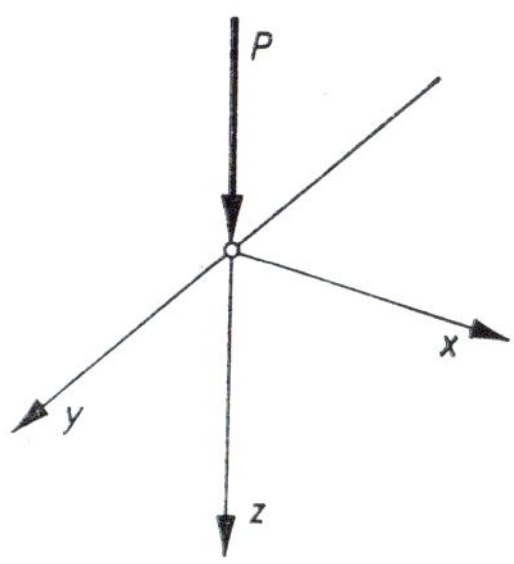

Fig. 15.1. Half-space subjected to a concentrated surface load

isotropic and saturated and the influence of seepage pressures may be neglected. Thus the fundamental equation of consolidation can be applied in its simplified form (14.27):

$$\frac{\partial u}{\partial t} = c_v \Delta u \tag{15.1}$$

u denoting the excess pore pressure and c_v the coefficient of consolidation:

$$c_v = \frac{kK}{\gamma_w} \tag{15.2}$$

where k stands for the coefficient of permeability, γ_w for the bulk density of water and K for the modulus of compressibility.

The solution has been given by Korotkin (1951, 1955). The following boundary conditions were considered:

(*a*) $z = 0, \quad u = 0$

(*b*) $R = \sqrt{(x^2 + y^2 + z^2)} \to \infty, \quad u \to 0$

(*c*) The initial pore pressures ($t = 0$) are equal to the total octahedral pressures σ_{oct} corresponding to the Poisson ratio $\nu = 0{\cdot}50$. Thus, according to Equations (5.19) to (5.21):

$$u_0(x,y,z) = \frac{P}{2\pi} \cdot \frac{z}{(x^2 + y^2 + z^2)^{\frac{3}{2}}} \tag{15.3}$$

The general solution of the differential Equation (15.1) accounting for the boundary conditions (*a*) and (*b*) is given by Florin (1961) in the form:

$$u = \frac{1}{8\sqrt{(\pi c_v t)^3}} \int_0^{\infty} \int_{-\infty}^{+\infty} \int_{-\infty}^{+\infty} u_0(\xi,\eta,\zeta) \left\{ \exp\left[-\frac{(x-\xi)^2 + (y-\eta)^2 + (z-\zeta)^2}{4c_v t} \right] - \right. \\ \left. - \exp\left[-\frac{(x-\xi)^2 + (y-\eta)^2 + (z-\zeta)^2}{4c_v t} \right] \right\} \mathrm{d}\xi \, \mathrm{d}\eta \, \mathrm{d}\zeta \tag{15.4}$$

Inserting Equation (15.3) for u_0 Korotkin obtained, after integration, the following expression:

$$u = \frac{Pz}{\pi} \left[\frac{\exp\left(-\frac{z^2 + R^2}{4c_v t} \right)}{2\sqrt{(\pi c_v t)}(z^2 + R^2)} - \frac{\phi\left(\sqrt{\frac{z^2 + R^2}{2c_v t}} \right)}{(z^2 + R^2)^{\frac{3}{2}}} \right] \tag{15.5}$$

$\phi(x)$ denotes the so-called probability integral:

$$\phi(x) = \frac{1}{2\pi} \int_0^x \exp\left(-\frac{x^2}{2} \right) \mathrm{d}x \tag{15.6}$$

the values of which, for various x, are tabulated in mathematical tables. In Korotkin's paper (1955) and in Florin's book (1961) the reader can find tables and charts giving the values of the quotient $u\pi/P$ for various values of parameters z, $r = \sqrt{(x^2 + y^2)}$ and $c_v t$.

Korotkin (1951, 1955) also presented the analogous solution for the line load q uniformly distributed along the axis y. The resulting expression is

$$u = \frac{q}{\pi} \cdot \frac{z}{r^2}\left[1 - \exp\left(-\frac{r^2}{4c_v t}\right)\right] \tag{15.7}$$

where

$$r^2 = x^2 + z^2$$

The quotients $u\pi/q$ are tabulated and plotted against the parameters x, z and $4c_v t$ (reproduced in Florin, 1961).

In Korotkin's solutions the volume forces due to the heterogeneous hydraulic field are not considered. Furthermore the Poisson's ratio ν related to total stresses changes during consolidation. From being 0·50 at the very beginning of the consolidation process it decreases gradually and at the end of consolidation reaches the value governing the effective stress–strain relationship. The corresponding total stress change during consolidation is not considered in Korotkin's solutions.

Once the pore pressures at a given time t are known, the corresponding settlements of the point $r = r_i = \text{const}$ due to the void ratio change can be computed. The final spherical part of the settlement is given by the integral

$$\rho^0_\infty = \frac{1}{3K}\int_{z=0}^{\infty} u_0(r_i,z)\,\mathrm{d}z \cong \frac{1}{3K}\int_{z=0}^{z_e} u_0(r_i,z)\,\mathrm{d}z \tag{15.8}$$

The area of the plot $u_0 = [u_0(r_i,z)]_{r_i=\text{const}}$ may be determined, in practical computations, to a reasonable depth limit $z = z_e$ and divided by three times the value of the modulus of compressibility, $3K$. The spherical settlement at a time t_k is then given by the equation:

$$\rho^0_t = \rho^0_\infty - \frac{1}{3K}\int_{z=0}^{z_e} u(r_i,z,t_k)\,\mathrm{d}z \tag{15.9}$$

or

$$\rho_t^0 = \frac{1}{3K}\int_{z=0}^{z_e} [u_0(r_i,z) - u(r_i,z,t_k)]\,\mathrm{d}z \tag{15.9'}$$

These settlements should be added to the distortional settlements which can be computed as total settlements corresponding to Poisson's ratio $\nu = 0{\cdot}50$ and to the given shear modulus G.

The shortcoming of the procedure is the incompatibility of the difference between the total and distortional displacements with the assumption of constant Poisson's ratio $\nu = 0{\cdot}50$.

15.1.2 Solution for the Two-layer Half-space

Mandel (1953, 1957, 1961) investigated the consolidation of a half-space of low permeability covered with a very permeable layer of uniform thickness h (Fig. 15.2). The stress–strain relationships in both layers are

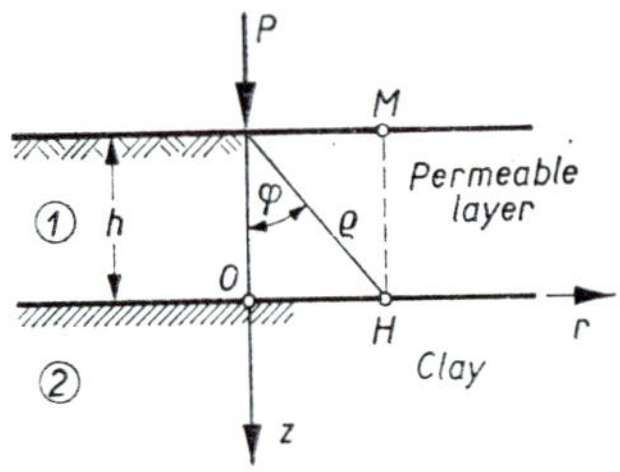

Fig. 15.2. Two-layer half-space subjected to a concentrated surface load

assumed to be linear and non-viscous, ν_1, μ denoting the Poisson's ratio and shear modulus of the upper, permeable layer, and ν_2, μ_2 the corresponding values of the lower layer of low permeability.

If the seepage forces are taken into account the Lamé's equilibrium equations, for the clay layer, are (cf. 14.a):

$$\left.\begin{aligned}(\lambda + 2\mu)\frac{\partial \varepsilon_v}{\partial r} - 2\mu\frac{\partial \gamma_{rz}}{\partial z} &= \frac{\partial u}{\partial r}\\ (\lambda + 2\mu)\frac{\partial \varepsilon_v}{\partial z} + 2\mu\left(\frac{\partial \gamma_{rz}}{\partial r} + \frac{\gamma_{rz}}{r}\right) &= \frac{\partial u}{\partial z}\end{aligned}\right\} \tag{15.10}$$

where:

$$\left.\begin{aligned}\varepsilon_v &= \frac{\partial \xi}{\partial r} + \frac{\xi}{r} + \frac{\partial \zeta}{\partial z}\\ 2\gamma_{rz} &= \frac{\partial \zeta}{\partial r} - \frac{\partial \xi}{\partial z}\end{aligned}\right\} \tag{15.11}$$

denote the volume strain and the double value of the tangential strain respectively; u is the pore-water pressure exceeding the equilibrium pressure; and λ and μ are Lamé's coefficients. Equations (15.10) yield:

$$\Delta u = (\lambda + 2\mu)\Delta\varepsilon_v \qquad (15.12) \equiv (14.d)$$

This expression may now be inserted into the differential equation of the consolidation of saturated isotropic soils (14.7). Taking $\beta = 0$, $\bar{u} = u$ (zero limiting hydraulic gradient) and $de/(1 + e) = d\varepsilon_v$ Equation (14.7) reduces to:

$$\frac{\partial \varepsilon_v}{\partial t} = \frac{k}{\gamma_w} \Delta u \tag{15.13}$$

Substituting Δu from (15.12) we have:

$$\frac{\partial \varepsilon_v}{\partial t} = c_v{}' \Delta \varepsilon_v \qquad (15.14) \equiv (14.f')$$

where

$$c_v{}' = \frac{k(\lambda + 2\mu)}{\gamma_w} \qquad (15.15) \equiv (14.g)$$

Based on Burmister's (1945) solution, Mandel deduced the following expression for the final settlement of an arbitrary surface point M (Fig. 15.2):

$$\zeta = \frac{P}{2\pi} \cdot \frac{1 - \nu_1}{\mu_1} \left[\frac{1}{r} + \frac{F(\varphi)}{\rho} \right] \tag{15.16}$$

where ρ and φ are the polar coordinates of the point H (see Fig. 15.2) and $F(\varphi)$ is the following integral:

$$F(\varphi) = \int_0^\infty \delta(v \cos \varphi) J_0(v \sin \varphi) \, dv \tag{15.17}$$

J_0 is Bessel's function of zero order, $v = m\rho$ and $\delta(x)$ is the function given by the expression:

$$\delta(x) = \frac{N_1 + N_2 + 4N_1(x + x^2) - 2N_1N_2e^{-2x}}{e^{2x} - (N_1 + N_2 + 4N_1x^2) + N_1N_2e^{-2x}} \tag{15.18}$$

where:

$$N_1 = \frac{M - 1}{M + 3 - 4\nu_1}, \quad N_2 = \frac{M(3 - 4\nu_2) - (3 - 4\nu_1)}{M(3 - 4\nu_2) + 1} \qquad (15.19\text{-a,b})$$

$$M = \frac{\mu_1}{\mu_2} \tag{15.20}$$

Mandel (1957) presented charts where the values F are plotted versus φ for various coefficients M between 0 and 40 and for the following two combinations of Poisson's ratios ν_1 and ν_2:

(*a*) $\nu_1 = \nu_2 = 0{\cdot}25$, (*b*) $\nu_1 = 0{\cdot}25$, $\nu_2 = 0{\cdot}50$.

As the distortional creep is neglected, the immediate settlements (ζ_0) can be obtained as the total settlements corresponding to the Poisson's ratio $\nu_2 = 0{\cdot}50$. The interpretation of such a conclusion was given in Section 14.4.3.

The settlement ζ at any time t during consolidation should satisfy the equation of consolidation (15.14), the equilibrium conditions (15.10) and the given boundary conditions. For the corresponding degree of consolidation U at any time t, Mandel (1957, 1961) obtained the following expression:

$$U = \frac{\zeta - \zeta_0}{\zeta_\infty - \zeta_0} = \frac{\int_0^\infty K(mh)[1 - f(mh, m\sqrt{(c_v t)})]J_0(mr)\,dm}{\int_0^\infty K(mh)J_0(mr)\,dm} \tag{15.21}$$

$K(x)$ is the function:

$$K(x) = [\delta(x,\nu_2) - \delta(x,\tfrac{1}{2})]\,\frac{2\mu_2(1 - \nu_1)}{\mu_1(1 - 2\nu_2)} \tag{15.22}$$

where $\delta(x)$ is given by Equation (15.18).

$f(mh, m\sqrt{(c_v t)})$, defined by Equation (18) in Mandel (1957), has a very complicated expression when ν_2 is different from 0·5. For $\nu_2 = 0{\cdot}5$ it has been reduced to the form:

$$f(x) = \frac{2}{\sqrt{\pi}}\left[(1 + 2x^2)\int_0^\infty \exp(-u^2)\,du - x\exp(-x^2)\right] \tag{15.23}$$

where: $x = m\sqrt{(c_v t)}$.

By using this simplified expression the degrees of consolidation were computed for $\nu_2 = 0{\cdot}50$ in spite of the differential settlement for this value of ν_2 being zero. The results obtained can, however, be applied for other values of ν_2 assuming that when $\sqrt{(c_v t)}/(1 - \nu_2)$ remains equal, the degree of consolidation maintains its same value too. The justification of this assumption has been proved by Mandel (1953) for $M = 1$ at U near to 1, and for high values of M. Mandel (1961) presented charts in which the consolidation degrees U were plotted versus φ (see Fig. 15.2) for various coefficients

$$\lambda = \frac{1}{2(1 - \nu_2)}\,\frac{\sqrt{(c_v t)}}{\rho}$$

between 1 and 8, for $\nu_1 = 0{\cdot}25$ and for various values of M between 1 and 20. The values corresponding to $\varphi = \pi/2$ give the consolidation of the simple half-space ($h = 0$).

15.2 CONSOLIDATION OF THE TWO-LAYER SATURATED ISOTROPIC AND HOMOGENEOUS HALF-SPACE SUBJECTED TO UNIFORMLY DISTRIBUTED CIRCULAR LOADS

The settlement of the centre of a uniformly distributed circular load acting on the surface of a two-layer half-space was presented in Section 5.4.1.2 in the form (see Equations 5.45 and 5.46), $G_1 \equiv \mu_1$:

$$w_\infty \equiv \zeta_\infty = qa\frac{1 - \nu_1}{\mu_1}[1 + G(\omega)\sin\omega] \tag{15.24}$$

where $G(\omega)$ is a function of the following parameters: ω, $M = \mu_1/\mu_2$, ν_1 and ν_2 (see Fig. 15.3).

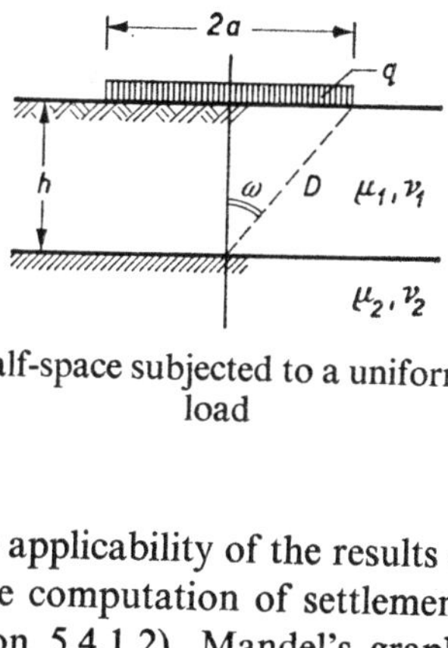

Fig. 15.3. Two-layer half-space subjected to a uniformly distributed circular load

Considering the large applicability of the results obtained for the centre of a circular load to the computation of settlements due to an arbitrary surface load (see Section 5.4.1.2), Mandel's graphs may be reproduced relating $G(\omega)$ to angle ω (see Fig. 15.3) for the values of $M = 0$, 1, 2, 4, 8, 13, 20 and 40, and for two combinations of Poisson's ratios ν_1 and ν_2: (*a*) $\nu_1 = \nu_2 = 0{\cdot}25$, (*b*) $\nu_1 = 0{\cdot}25$, $\nu_2 = 0{\cdot}50$ (Fig. 15.4). The second combination applies to the computation of immediate settlements (ζ_0), and the first to the calculation of final settlements (ζ_∞).

The equilibrium conditions (15.10), the consolidation Equation (15.12) and the boundary conditions corresponding to Fig. 15.3, lead after

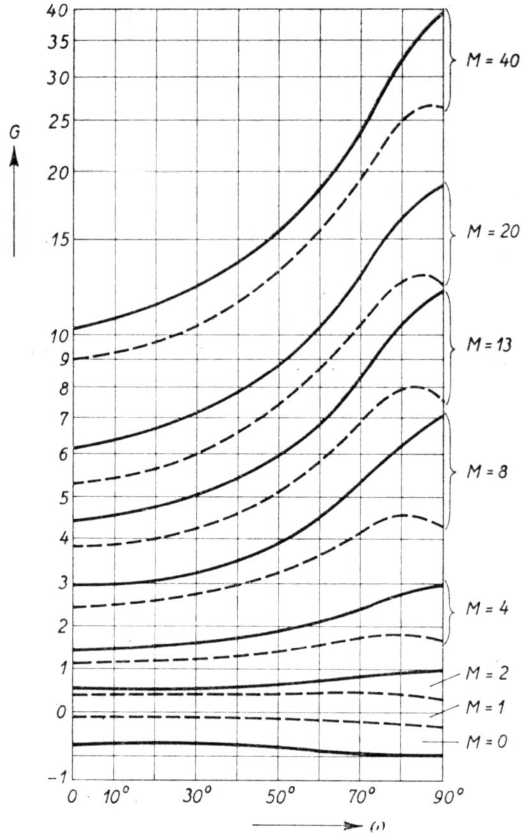

Fig. 15.4. G versus ω plots (see Equation 15.24) for various shear modulus ratios $M = \dfrac{\mu_1}{\mu_2}$ and for the following two combinations of Poisson's ratios: (*a*) full lines: $\nu_1 = \nu_2 = 0{\cdot}25$; (*b*) dashed lines: $\nu_1 = 0{\cdot}25$, $\nu_2 = 0{\cdot}50$. (Reproduced after Mandel, 1957)

cumbersome deductions to the following expression for the degree of consolidation U at a time t during consolidation (Mandel 1957, 1961):

$$U = \frac{\zeta - \zeta_0}{\zeta_\infty - \zeta_0} \tag{15.25}$$

$$\zeta - \zeta_0 = \frac{qa^2(1 - 2\nu_2)}{2\mu_2} \int_0^\infty K(mh) f(mh, m\sqrt{(c_v t)}) \frac{J_1(ma)}{ma} \, dm \tag{15.25'}$$

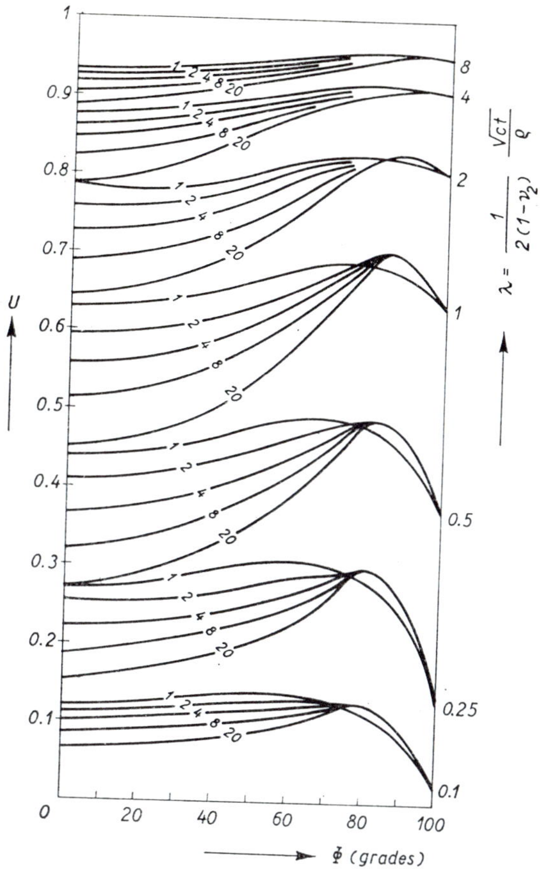

Fig. 15.5. Consolidation ratio U (Equation 15.25) plotted versus angle $\omega \equiv \Phi$ (see Fig. 15.3) for various coefficients λ and for shear modulus ratios $M = 1, 2, 4, 8$ and 20. (Reproduced after Mandel, 1961)

J_1 is Bessel's function of the first order, $K(mh)$ has been defined by Equation (15.22) and $f(x)$, for $\nu_2 = 0{\cdot}50$, by Equation (15.23). The solution obtained for $\nu_2 = 0{\cdot}50$ can be generalized again assuming that the consolidation ratio U obtained for $\nu_2 = 0{\cdot}50$, also suits other Poissons' ratios ν_2 provided that the coefficient

$$\lambda = \frac{1}{2(1 - \nu_2)} \frac{\sqrt{(c_v t)}}{D}$$

(see Fig. 15.3) remains the same. Mandel's plots $U = U(\omega)$ for various coefficients λ and for values of $M = 1, 2, 4, 8$ and 20 are reproduced in Fig. 15.5.

15.3 CONSOLIDATION OF THE SIMPLE SATURATED ISOTROPIC AND HOMOGENEOUS HALF-SPACE SUBJECTED TO A UNIFORMLY DISTRIBUTED LOAD OVER A RECTANGULAR AREA

Based on the differential equation of consolidation (14.f′) and on the equilibrium conditions (14.a) Gibson and McNamee (1957) presented the solution for the immediate and consolidation settlement of the corner of a rectangular uniformly loaded area on the surface of a semi-infinite homogeneous, isotropic, saturated, porous elastic medium. The following boundary conditions were considered:

$$_{xz} = \sigma_{yz} = 0, \qquad u = 0, \qquad \text{on} \qquad z = 0,$$

$$\sigma_{zz} = q = \text{const}$$

over the rectangle $-L < x < L$, $-\lambda L < y < \lambda L$ (see Fig. 15.6) where u stands for the excess pore water pressure and σ_{zz}, σ_{xz}, etc. for the total normal and shear stresses respectively.

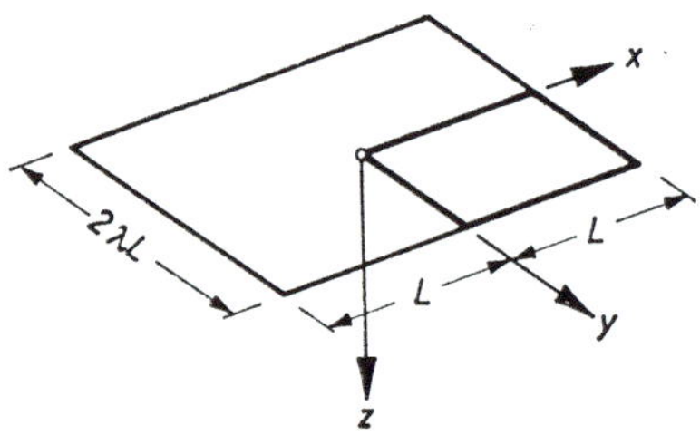

Fig. 15.6. Half-space subjected to a uniformly distributed rectangular load

Taking Poisson's ratio equal to zero ($\nu = 0$) the following results have been obtained:

for the final total settlement $\zeta_c(\infty)$ of the corner of the rectangle ($t \to \infty$):

$$\zeta_c(\infty) = \frac{qL}{2\pi G}[\sinh^{-1}\lambda + \lambda \sinh^{-1}\lambda^{-1}] \tag{15.26}$$

G being the shear modulus;

for the immediate settlement $\zeta_c(0)$ which occurs immediately following the application of the load:

$$\zeta_c(0) = \tfrac{1}{2}\zeta_c(\infty); \tag{15.27}$$

for the degree of consolidation U_c of the corner of the rectangle:

$$U_c = \frac{\displaystyle\int_0^{\frac{ct}{L^2}} \frac{1}{\sqrt{\tau}} \operatorname{erf} \frac{1}{2\sqrt{\tau}} \operatorname{erf} \frac{\lambda}{2\sqrt{\tau}} \, d\tau}{\displaystyle\int_0^{\infty} \frac{1}{\sqrt{\tau}} \operatorname{erf} \frac{1}{2\sqrt{\tau}} \operatorname{erf} \frac{\lambda}{2\sqrt{\tau}} \, d\tau} \tag{15.28}$$

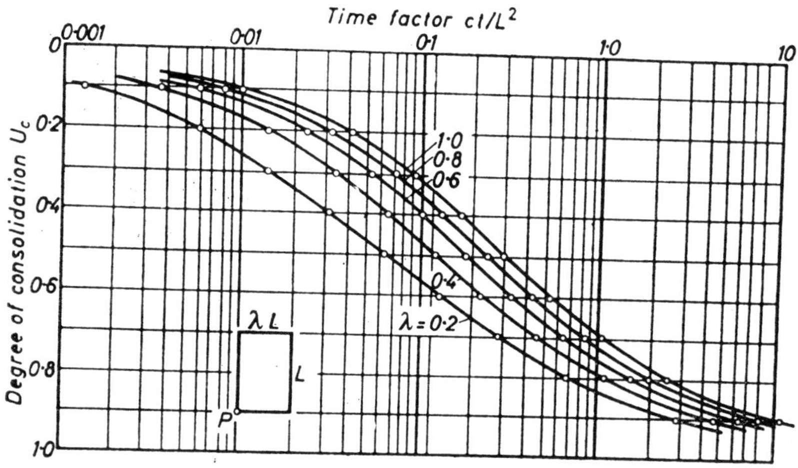

Fig. 15.7. Degree of consolidation U_c versus time factor for corner point P of rectangular uniformly loaded area. (Reproduced after Gibson and McNamee, 1957)

where the error function is defined by

$$\operatorname{erf} y = \frac{2}{\sqrt{\pi}} \int_0^y e^{-u^2} \, du \tag{15.29}$$

and the coefficient of consolidation c by

$$c = \frac{2(1-\nu)}{1-2\nu} G \frac{k}{\gamma_w} = (\lambda + 2\mu) \frac{k}{\gamma_w} \tag{15.30}$$

if λ and μ are Lamé's coefficients (cf. Equation 14.g).

As Poisson's ratio was taken to be zero ($\nu = 0$), the expression for c reduces to:

$$c = 2G \frac{k}{\gamma_w} \tag{15.30'}$$

The numerical results for the degrees of consolidation were presented in the form of graphs which are reproduced in Fig. 15.7. Since the fundamental equations are linear (see Terzaghi, 1943, p. 383), the method of superposition may be applied directly. Thus the settlement of any point of the ground surface due to any number of simultaneously applied rectangular loads may be readily obtained.

16

Numerical Solutions of the Differential Equation of Consolidation

16.1 INTRODUCTORY OBSERVATIONS

The graphs presented in the preceding chapter which facilitate the computation of settlements of the centre of a uniform circular surface load and of the corner of a uniform rectangular surface load, can largely be utilized owing to the possibility of applying the principle of superposition to the solutions obtained. Mandel's charts for the centre of a uniformly distributed circular load allow the computation of the consolidation of an infinite clay layer covered by a permeable layer of finite thickness the surface of which is subjected to a load of any shape and distribution. The procedure to be applied for this purpose is analogous to the procedure explained in Section 5.4.1 dealing with the final settlements of the layered half-space. If the loaded area can be divided into several positive and negative rectangular areas one corner of which coincides with the point whose settlement is to be determined, Gibson and McNamee's solution can be applied in an analogous manner provided that the subsoil can be considered as a homogeneous half-space corresponding to zero Poisson's ratio.

Now, the applicability of the charts mentioned above is restricted by the fundamental assumptions and conditions of their theoretical basis. These are: the homogeneity and isotropy of the medium which concern the deformability as well as the permeability of the soil, the linear effective stress–strain relationship, the infinite depth of the clay layer and the non-viscous soil behaviour. Within certain limits and mutual influences of the preconsolidation and secondary consolidation effects, the magnitude and speed of the overload, and the thickness of the layer, the viscous soil properties affect the primary consolidation mainly by increasing the coefficient of volume compressibility ($m_v = 1/K$). In such cases the solutions obtained for non-viscous soils can be usefully applied for approximate computations of the development of the primary consolidation of soils exhibiting viscous properties. Some instruction on the practical use will be given in Section 17.3.1.

If the thickness of the clay layer is finite and if the soil is anisotropic with respect to permeability, the consolidation can be analysed by applying the differential equation of consolidation in its numerical form. The numerical computation of the consolidation process requires, however, that the total stress increase be known. As shown in Chapter 5 the available solutions facilitating the computation of total stresses are limited even under the condition of linear stress–strain relationship if the stratification, the heterogeneity and the anisotropy of the soil are to be considered. Thus, in most cases, the approximate applications of the solutions valid for the semi-infinite homogenous and isotropic medium are usually applied (see Chapter 5).

If the total stress increase at the beginning and during the consolidation process is assumed to be known, the isotache method permits the approximate prediction of consolidation in non-linear viscous elastic soils. The procedure will be dealt with in Section 16.3.

16.2 NUMERICAL FORM OF THE DIFFERENTIAL EQUATION OF CONSOLIDATION

16.2.1 General Solution for Saturated Soils

For saturated soils ($\beta = 1$) the general equation of consolidation (14.4) reduces to the form:

$$\frac{\partial e}{\partial t} = \frac{1+e}{\gamma_w}\left[\frac{\partial}{\partial x}\left(k_z \frac{\partial \bar{u}_z}{\partial x}\right) + \frac{\partial}{\partial y}\left(k_y \frac{\partial \bar{u}_y}{\partial y}\right) + \frac{\partial}{\partial z}\left(k_x \frac{\partial \bar{u}_z}{\partial z}\right)\right] \tag{16.1}$$

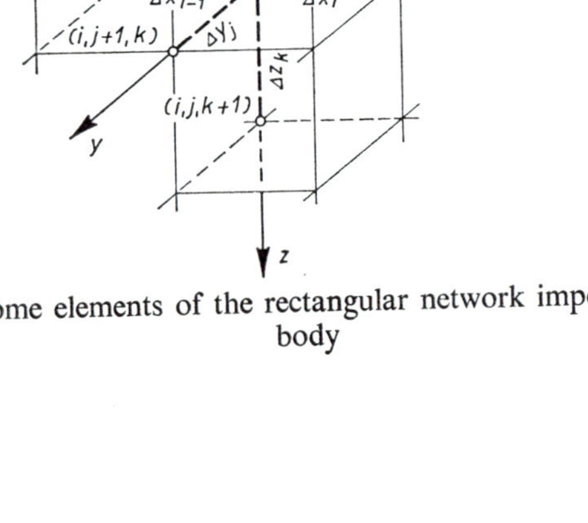

Fig. 16.1. Some elements of the rectangular network imposed to the soil body

where

$$\frac{\partial \bar{u}_x}{\partial x} = \frac{\partial u}{\partial x} - \gamma_w i_{0x}$$

and similarly for directions y and z; u denotes the excess pore pressure, e is the void ratio, k_x, k_y, k_z are the coefficients of permeability and i_{0x}, i_{0y}, i_{0z} the limiting hydraulic gradients in the x, y and z directions, and γ_w is the bulk density of the pore water.

The soil body may have any shape provided that its boundary suits an arbitrary rectangular network the sides of which have the directions of the axes x, y, z of the orthogonal anisotropy of the body with respect to the permeability.

Fig. 16.1 represents some elements of the rectangular network of the soil body. By applying the notation presented in the figure and considering a finite time interval $\Delta t = t' - t$, the derivative on the left side of Equation (16.1) can be substituted, for approximate numerical computation, by the following finite difference quotient:

$$\frac{\partial e}{\partial t} \simeq \frac{\Delta e}{\Delta t} = \frac{e_{t',i,j,k} - e_{t,i,j,k}}{\Delta t} \tag{16.2}$$

where e_t and $e_{t'}$ are the void ratios at the beginning and end of the time interval Δt.

The second order derivatives appearing in brackets on the right-hand side of Eq. (16.1) take the following general numerical form:

$$\begin{aligned}
\frac{\partial}{\partial x}\left[k_x\left(\frac{\partial u}{\partial x} - \gamma_w i_{0x}\right)\right] = \frac{1}{2}\Bigg\{\Bigg[& k_{x,t,i,j,k}\left(\frac{\dfrac{u_{t,i+1,j,k} - u_{t,i,j,k}}{\Delta x_i} - \dfrac{u_{t,i,j,k} - u_{t,i-1,j,k}}{\Delta x_{i-1}}}{\dfrac{\Delta x_i + \Delta x_{i-1}}{2}} - \right. \\
& \left. -\gamma_w\left(\frac{\partial i_{0x}}{\partial x}\right)_{t,i,j,k}\right) + \left(\frac{\partial k_x}{\partial x}\right)_{t,i,j,k}\left(\frac{u_{t,i+1,j,k} - u_{t,i-1,j,k}}{\Delta x_i + \Delta x_{i-1}} - \gamma_w i_{0x,t,i,j,k}\right)\Bigg] + \\
& + \Bigg[k_{x,t',i,j,k}\left(\frac{\dfrac{u_{t',i+1,j,k} - u_{t',i,j,k}}{\Delta x_i} - \dfrac{u_{t',i,j,k} - u_{t',i-1,j,k}}{\Delta x_{i-1}}}{\dfrac{\Delta x_i + \Delta x_{i-1}}{2}} - \gamma_w\left(\frac{\partial i_{0x}}{\partial x}\right)_{t',i,j,k}\right) + \\
& + \left(\frac{\partial k_x}{\partial x}\right)_{t',i,j,k}\left(\frac{u_{t',i+1,j,k} - u_{t',i-1,j,k}}{\Delta x_i + \Delta x_{i-1}} - \gamma_w i_{0x,t',i,j,k}\right)\Bigg]\Bigg\}
\end{aligned} \tag{16.3}$$

where

$$\Delta x_i = x_{i+1} - x_i \tag{16.4}$$

The expressions for the directions y and z are analogous.

The reduction of expression (16.3) will be made assuming that the limiting hydraulic gradient remains constant in each direction of the orthogonal anisotropy:

$$\frac{\partial i_{0x}}{\partial x} = 0 \qquad (x,y,z) \tag{16.5}$$

Furthermore the terms depending on the change of the coefficients of permeability ($\partial k_x/\partial x$) will be neglected although the variation of these coefficients along the axes x, y and z and with time t will be allowed and considered in the remaining terms. In this way expression (16.3) becomes:

$$\frac{\partial}{\partial x}\left(k_x \frac{\partial \bar{u}_x}{\partial x}\right) = \frac{k_{x,t,i,j,k}}{\Delta x_i + \Delta x_{i-1}}\left[\frac{u_{t,i+1,j,k} - u_{t,i,j,k}}{\Delta x_i} - \frac{u_{t,i,j,k} - u_{t,i-1,j,k}}{\Delta x_{i-1}}\right] + \\ + \frac{k_{x,t',i,j,k}}{\Delta x_i + \Delta x_{i-1}}\left[\frac{u_{t',i+1,j,k} - u_{t',i,j,k}}{\Delta x_i} - \frac{u_{t',i,j,k} - u_{t',i-1,j,k}}{\Delta x_{i-1}}\right] \tag{16.6}$$

The above expression can be written more simply in the form

$$\frac{\partial}{\partial x}\left(k_x \frac{\partial \bar{u}_x}{\partial x}\right) = k_{x,t,i,j,k}\left\{\frac{1}{\Delta x_i + \Delta x_{i-1}}\left(\frac{u_{t,i+1,j,k}}{\Delta x_i} + \frac{u_{t,i-1,j,k}}{\Delta x_{i-1}}\right) - \\ - \frac{u_{t,i,j,k}}{\Delta x_i\, \Delta x_{i-1}}\right\} + k_{x,t',i,j,k}\left\{\frac{1}{\Delta x_i + \Delta x_{i-1}}\left(\frac{u_{t',i+1,j,k}}{\Delta x_i} + \frac{u_{t',i-1,j,k}}{\Delta x_{i-1}}\right) - \\ - \frac{u_{t',i,j,k}}{\Delta x_i\, \Delta x_{i-1}}\right\} \tag{16.7}$$

If the distances between the points of the network in the single directions x, y and z are taken equal:

$$\Delta x_0 = \Delta x_1 = \Delta x_2 = \ldots = \Delta x_i = \ldots = \Delta x \tag{16.8}$$

and similarly for Δy_j and Δz_k, the three equations of type (16.7) become simplified:

$$\frac{\partial}{\partial x}\left(k_x \frac{\partial \bar{u}_x}{\partial x}\right) = \frac{k_{x,t,i,j,k}}{2\Delta x^2}\left(u_{t,i+1,j,k} + u_{t,i-1,j,k} - 2u_{t,i,j,k}\right) + \\ + \frac{k_{x,t',i,j,k}}{2\Delta x^2}\left(u_{t',i+1,j,k} + u_{t',i-1,j,k} - 2u_{t',i,j,k}\right) \tag{16.8}$$

and similarly for the directions y and z.

The above numerical expression for the second order derivative

$$\frac{\partial}{\partial x}\left(k_x \frac{\partial \bar{u}_x}{\partial x}\right)$$

is related to the middle of the time interval $\Delta t = t' - t$. By inserting the three expressions of type (16.8), and expression (16.2) into the differential equation (16.1) and taking

$$1 + e = 1 + \frac{e_{t',i,j,k} + e_{t,i,j,k}}{2} \tag{16.9}$$

an equation which can be solved for $e_{t',i,j,k}$ by using relaxation methods is arrived at. An example of such a solution applied to a one-dimensional consolidation problem will be presented in Section 16.2.3.

16.2.2 Simplified Solution for Saturated Soils

In order to simplify the computation the second order derivative

$$\frac{\partial}{\partial x}\left(k_x \frac{\partial \bar{u}_x}{\partial x}\right)$$

(x,y,z) appearing in the equation of consolidation (16.1) will be replaced by numerical values related to the beginning of the chosen time interval Δt (see expression 16.2). The expression (16.8) should then be substituted by the following one:

$$\frac{\partial}{\partial x}\left(k_x \frac{\partial \bar{u}_x}{\partial x}\right) = \frac{k_{x,t,i,j,k}}{(\Delta x)^2}(u_{t,i+1,j,k} + u_{t,i-1,j,k} - 2u_{t,i,j,k}) \tag{16.10}$$

and the equation of consolidation (16.1) becomes:

$$\frac{(\Delta e)_{\Delta t,i,j,k}}{\Delta t} = \frac{1 + e_{t,i,j,k}}{\gamma_w}[\ ] \tag{16.11}$$

$$\begin{aligned}[\ ] = &\frac{k_{x,t,i,j,k}}{(\Delta x)^2}(u_{t,i+1,j,k} + u_{t,i-1,j,k} - 2u_{t,i,j,k}) + \\ &+ \frac{k_{y,t,i,j,k}}{(\Delta y)^2}(u_{t,i,j+1,k} + u_{t,i,j-1,k} - 2u_{t,i,j,k}) + \\ &+ \frac{k_{z,t,i,j,k}}{(\Delta z)^2}(u_{t,i,j,k+1} + u_{t,i,j,k-1} - 2u_{t,i,j,k})\end{aligned} \tag{16.12}$$

In soils of isotropic permeability

$$k_x = k_y = k_z = k \tag{16.13-a}$$

Further, taking equal distances between the points of the network in all directions (x,y,z):

$$\Delta x = \Delta y = \Delta z = \Delta h \tag{16.13-b}$$

the expression (16.12) becomes simplified:

$$[\ ] = \frac{k_{t,i,j,k}}{(\Delta h)^2}(s_{t,i,j,k} - 6u_{t,i,j,k}) \tag{16.14}$$

where

$$s_{t,i,j,k} = u_{t,i+1,j,k} + u_{t,i-1,j,k} + \\ + u_{t,i,j+1,k} + u_{t,i,j-1,k} + u_{t,i,j,k+1} + u_{t,i,j,k-1} \tag{16.15}$$

Thus, $s_{t,i,j,k}$ (Florin, 1961) is the sum of the pore pressures occurring, at time t, at the six points of the network adjacent to the point (i,j,k) for which the void ratio

$$e_{t',i,j,k} = e_{t,i,j,k} + \Delta e \tag{16.16}$$

at the subsequent time $t' = t + \Delta t$ will be determined.

If the relationship between the void ratios e and the effective octahedral stresses σ_{oct}' is known, the effective stresses $(\sigma_{oct}')_{t+\Delta t}$ can be determined corresponding to the void ratios $e_{t+\Delta t}$. The pore pressures $u_{t'}$ to be inserted into Equation (16.11) for the next time interval $\Delta t' = t'' - t'$, can be found as the difference between the total octahedral stresses at time $t' = t + \Delta t$:

$$(\sigma_{oct})_{t'} = (\sigma_{oct})_t + (\Delta\sigma_{oct})_{\Delta t} \tag{16.17}$$

and the effective stresses $(\sigma_{oct}')_{t'}$:

$$u_{t'} = (\sigma_{oct})_{t'} - (\sigma_{oct}')_{t'} \tag{16.18}$$

In Section 16.4 this procedure will be applied to the isotache method of consolidation analysis.

In plane problems (Fig. 16.2) the numerical form of the equation of

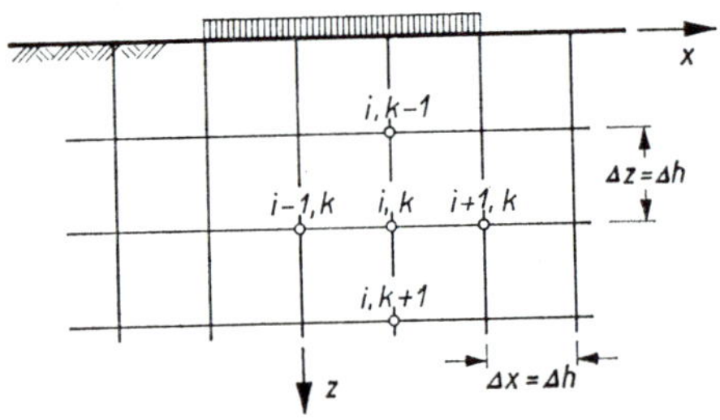

Fig. 16.2. Scheme of the network in plane problems

consolidation (16.11) becomes:

$$\frac{(\Delta e)_{\Delta t,i,k}}{\Delta t} = \frac{1 + e_{t,i,k}}{\gamma_w}\left[\frac{k_{t,i,k}}{(\Delta h)^2}(s_{t,i,k} - 4u_{t,ik})\right] \tag{16.19}$$

The sum $s_{t,i,k}$ contains only four members:

$$s_{t,i,k} = u_{t,i-1,k} + u_{t,i+1,k} + u_{t,i,k-1} + u_{t,i,k+1} \tag{16.20}$$

For the conditions of linear drainage, only two adjacent pore pressures have to be considered in the sum $s_{t,i}$:

$$s_{t,i} = u_{t,i-1} + u_{t,i+1} \tag{16.21}$$

and the factor 4 in (16.19) replaced by the factor 2:

$$[\;] = \frac{k_{t,i}}{(\Delta h)^2}(s_{t,i} - 2u_{t,i}) \tag{16.22}$$

16.2.3 The Application of the Numerical Form of the Equation of Consolidation for Non-viscous Soils with Linear Compressibility

The linear non-viscous stress–strain relationship and full saturation of the soil may be assumed. Thus the differential equation of consolidation can be used in the form (14.15) with $\mathring{\omega} = 1$:

$$\frac{\partial u}{\partial t} = \frac{\partial \sigma^0}{\partial t} + \frac{K}{\gamma_w}\left[\frac{\partial}{\partial x}\left(k_x \frac{\partial \bar{u}_x}{\partial x}\right) + \frac{\partial}{\partial y}\left(k_y \frac{\partial \bar{u}_y}{\partial y}\right) + \frac{\partial}{\partial z}\left(k_z \frac{\partial \bar{u}_z}{\partial z}\right)\right] \tag{16.23}$$

K denotes the modulus of compressibility and the symbol σ^0 has been substituted for σ_{oct}; for explanation of the other symbols see Section 16.2.1.

The numerical form of the expression in brackets [ ] has been developed in the previous two sections. In the case of constant coefficients of permeability k_x, k_y and k_z Equation (16.12) holds. The simpler form (16.14) can be applied if the soil is isotropic with respect to the permeability and if the network is square (Equations 16.13-a,b).

The derivatives $\partial u/\partial t$ and $\partial \sigma_{\text{oct}}/\partial t$ can be substituted by the following difference quotients:

$$\frac{\partial u}{\partial t} \simeq \frac{u_{t',i,j,k} - u_{t,i,j,k}}{\Delta t} \tag{16.24}$$

where Δt is the time interval

$$\Delta t = {}'t - t$$

and u_t, $u_{t'}$ the pore pressures at the beginning and at the end of this interval;

$$\frac{\partial \sigma^0}{\partial t} \cong \frac{\sigma^0{}_{t',i,j,k} - \sigma^0{}_{t,i,j,k}}{\Delta t} \tag{16.25}$$

where $\sigma_t{}^0$, $\sigma_{t'}{}^0$ are the total mean (octahedral) pressures at times t and t' at the limits of the time interval Δt; further suffixes i, j, k indicate the point of the network at which the pressure is considered.

Inserting expressions (16.24) and (16.25) and the expression (16.12) or (16.14) into the equation of consolidation (16.23), produces, after some rearranging, the following form suitable for numerical computation:

$$u_{t',i,j,k} = u_{t,i,j,k} + \Delta\sigma^0{}_{\Delta t,i,j,k} + \frac{\Delta t K}{\gamma_w}[\;] \tag{16.26}$$

where [ ] stands for the expressions (16.12) or (16.14) respectively. With the assumptions (16.13-a,b):

$$k_x = k_y = k_z = k, \qquad \Delta x = \Delta y = \Delta z = \Delta h$$

Equation (16.26) becomes:

$$u_{t',i,j,k} = (1 - 6\alpha)u_{t,i,j,k} + \Delta\sigma^0{}_{\Delta t,i,j,k} + \alpha s_{t,i,j,k} \tag{16.27}$$

where

$$\alpha = \frac{kK}{\gamma_w}\frac{\Delta t}{(\Delta h)^2} \tag{16.28}$$

and $s_{t,i,j,k}$ is the sum of the pore pressures occurring at time t_t at the six points of the network adjacent to the point i, j, k (see Equation 16.15).

In plane problems the sum $s_{t,i,k}$ contains only four terms (see Fig. 16.2) and 4α must be substituted for 6α. For the conditions of linear drainage, only two adjacent pore pressures have to be considered in the sum $s_{t,i}$. When applying Equation (16.27), 6α must be replaced by 2α.

The computation can be further simplified by taking (Florin, 1961):

$\alpha = \frac{1}{6}$ in space problems,

$\alpha = \frac{1}{4}$ in plane problems and

$\alpha = \frac{1}{2}$ in linear problems.

In this case Equation (16.27) becomes:

$$u_{t',i,j,k} = \Delta\sigma^0{}_{\Delta t,i,j,k} + \alpha s_{t,i,j,k} \tag{16.29}$$

Thus the pore pressure at a subsequent time $t' = t + \Delta t$ can be obtained as the sum of the total stress increase in the time interval Δt, and of the sum of the pore pressures at adjacent points of the network at time t at the beginning of the time interval Δt, multiplied by the dimensionless time factor α (see Equation 16.28).

16.2.4 Some Guiding Notes for Practical Computation

In order to facilitate the practical application of the equation deduced in the preceding section, the following instructions may be given (Florin, 1961):

(1) At impervious contacts the condition

$$\frac{\partial u}{\partial n} = 0 \tag{16.30}$$

should be satisfied, u being the excess pore pressure and n the direction of the normal to the contact. Thus the points of the network in the first adjacent row parallel to the contact have to be reflected on the other side of the contact (Fig. 16.3) and at these points the apparent pore pressures should be considered equal to the pore pressures in row (1). In this way condition (16.30) can be fulfilled:

$$\left(\frac{\partial \dfrac{u}{\gamma_w}}{\partial z}\right)_{0,i} = \frac{u_{1,i} - u_{-1,i}}{2\gamma_w \Delta z} = 0$$

$i = 3, 4, 5, 6, \ldots$

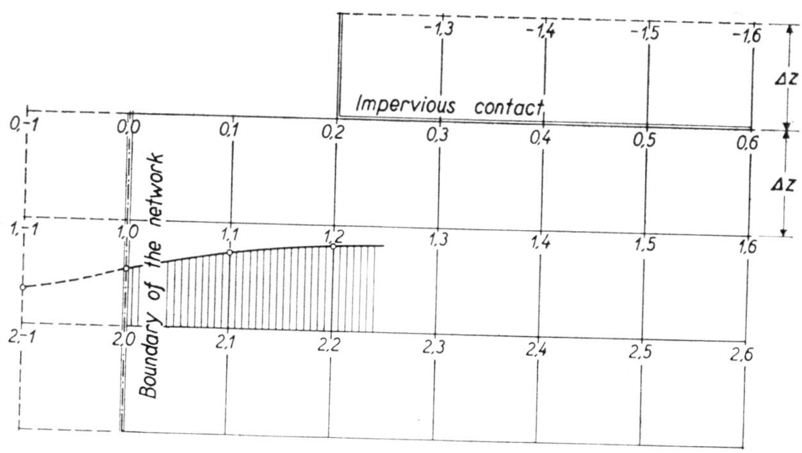

Fig. 16.3. Boundary conditions

(2) The extent of the network has to be limited although the layer considered continues beyond the chosen boundary. When computing the pore pressure development at the boundary points of the network, the preceding pore pressures at the outside points adjacent to the boundary have to be known too. Their values can be estimated by extrapolating the curves through the pore pressure values determined for the points inside the network. In Fig. 16.3 an example is presented.

(3) If the boundary surface of the soil body whose consolidation has to be investigated is oblique or curved, it can be replaced by a broken contour of the rectangular network.

(4) An arbitrary shaped diagram of the total stress increase should be replaced by a steplike diagram (Fig. 16.4).

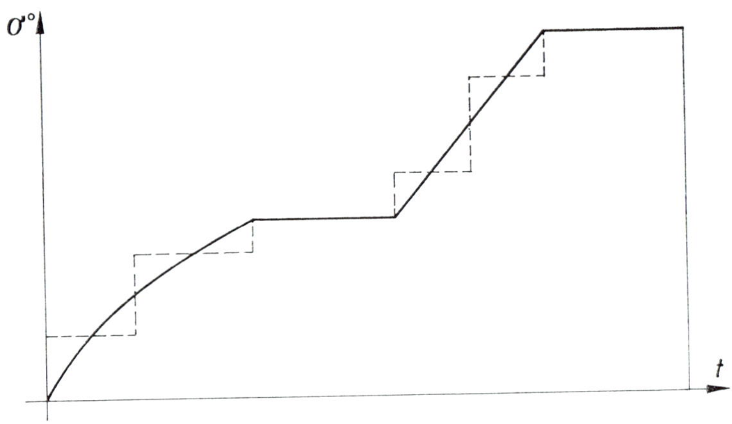

Fig. 16.4. Total stress-time diagram replaced by a steplike diagram

(5) If the soil is composed of several layers with different coefficients of consolidation $c_v = kK/\gamma_w$, each layer can be treated in the same way as in the case of an isolated layer or of the homogeneous half-space. On the contact surface between the layers the excess pore-pressures change approximately according to the relation

$$\frac{k_1(u_{t,i} - u_{t,i-1})}{\Delta x_{i-1}} = \frac{k_2(u_{t,i+1} - u_{t,i})}{\Delta x_i} \qquad (16.31)$$

The points i are assumed to lie in the contact surface between the layers (1) and (2), the points $i + 1$ in the first adjacent row of the upper (2) layer with the coefficient of permeability k_2, and the points $i - 1$ in the first adjacent row of the lower layer (1) with the coefficient of permeability k_1. Denoting

$$\frac{k_2}{k_1} = n \tag{16.32}$$

and taking $\Delta x_{i-1} = \Delta x_i$ Equation (16.31) yields:

$$u_{t,i} = \frac{k_1 u_{t,i-1} + k_2 u_{t,i+1}}{k_1 + k_2} = u_{t,i+1} - \frac{u_{t,i+1} - u_{t,i-1}}{1 + n} \tag{16.33}$$

In order to facilitate the application of Equations (16.27) and (16.29) with the same time factor α, the following condition may be satisfied:

$$c_{v2} \frac{\Delta t_2}{(\Delta h_2)^2} = c_{v1} \frac{\Delta t_1}{(\Delta h_1)^2} \tag{16.34}$$

If $\Delta h_1 = \Delta h_2$, the ratio of the time intervals should be:

$$\frac{\Delta t_1}{\Delta t_2} = \frac{c_{v2}}{c_{v1}} = \vartheta \tag{16.35}$$

This means that ϑ numerical operations made for layer (1), after Equations (16.27) or (16.29), corresponds to one operation made for layer (2).

Alternatively it is possible to take $\Delta t_1 = \Delta t_2$ and the following ratio of the distances between single points in the networks of layers (1) and (2) respectively:

$$\frac{\Delta h_2}{\Delta h_1} = \sqrt{\left(\frac{c_{v2}}{c_{v1}}\right)} = \sqrt{\vartheta} \tag{16.36}$$

Assuming $\vartheta > 1$, the network in layer (1) is denser than in layer (2) the distances between single points being

$$\Delta h_1 = \frac{\Delta h_2}{\sqrt{\vartheta}}.$$

16.3 ONE-DIMENSIONAL CONSOLIDATION OF NON-HOMOGENEOUS CLAY LAYERS

Numerical solutions for the one-dimensional consolidation of a non-homogeneous horizontal clay layer subjected to a suddenly applied uniformly distributed vertical excess pressure $\Delta\sigma$ have been given by Schiffman and Gibson (1964). The coefficient of permeability $k(z)$ and the coefficient of compressibility $m_v(z)$ had been assumed to vary with depth following certain analytical expressions (Fig. 16.5).

Assuming zero total pressure increase ($\partial\sigma/\partial t = 0$) and zero limiting hydraulic gradients ($i_0 = 0$), the general differential equation of one-dimensional consolidation (9.40) reduces to:

$$\frac{\partial u}{\partial t} = -\frac{1+e}{\gamma_w \frac{\partial e}{\partial \sigma'}} \cdot \frac{\partial}{\partial z}\left[k(z)\frac{\partial u}{\partial z}\right] \tag{16.37}$$

This equation can be written in the form:

$$\frac{\partial^2 u}{\partial z^2} + \frac{1}{k(z)}\frac{\mathrm{d}k(z)}{\mathrm{d}z}\frac{\partial u}{\partial z} = \frac{1}{c_v(z)}\frac{\partial u}{\partial t} \tag{16.38}$$

where

$$c_v(z) = -\frac{k(z)\,.\,(1+e)}{\gamma_w \frac{\partial e}{\partial \sigma'}} = \frac{k(z)}{\gamma_w m_v(z)} \tag{16.39}$$

denotes the coefficient of consolidation varying with depth (z).

In order to generalize the solutions, Schiffman and Gibson introduced the following dimensionless parameters:

$$\left.\begin{aligned} \zeta = \frac{z}{h}, \quad v = \frac{u}{\sigma_0} = \frac{u}{\Delta\sigma}, \quad T = \frac{c_0 t}{h^2} \\ k_i = \frac{[k(z)]_i}{k_0}, \quad m_i = \frac{[m_v(z)]_i}{m_0} \end{aligned}\right\} \tag{16.40}$$

u is the excess pore pressure at the depth z and time t, c_0 is the coefficient of consolidation at the upper boundary of the clay layer:

$$c_0 = (c_v)_{z=0} = \frac{k_0}{\gamma_w m_0}$$

For the remaining notation see Fig. 16.5.

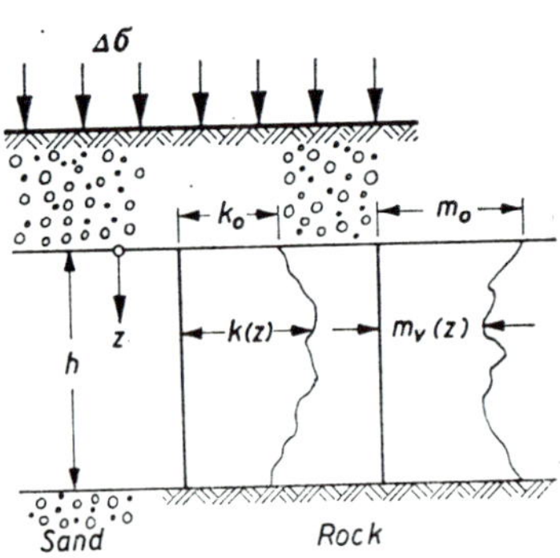

Fig. 16.5. Variation of permeability k and compressibility m_v with depth

The scheme of the numerical computation made by Schiffman and Gibson (1964) is presented in Fig. 16.6. Referring to this scheme the differential equation of consolidation (16.37) can be written in the following form facilitating the numerical computation:

$$\frac{1}{2}\left\{\left[\frac{\dfrac{u_{i+1,j}-u_{i,j}}{\Delta z}-\dfrac{u_{i,j}-u_{i-1,j}}{\Delta z}}{\Delta z}+\frac{1}{k_i}\cdot\frac{\mathrm{d}k_i}{\mathrm{d}z}\cdot\frac{u_{i+1,j}-u_{i-1,j}}{2\Delta z}\right]+\right.$$

$$\left.+\left[\frac{\dfrac{u_{i+1,j+1}-u_{i,j+1}}{\Delta z}-\dfrac{u_{i,j+1}-u_{i-1,j+1}}{\Delta z}}{\Delta z}+\frac{1}{k_i}\cdot\frac{\mathrm{d}k_i}{\mathrm{d}z}\cdot\frac{u_{i+1,j+1}-u_{i-1,j+1}}{2\Delta z}\right]\right\}=$$

$$=\frac{1}{c_v(z)_i}\cdot\frac{u_{i,j+1}-u_{i,j}}{\Delta t} \qquad (16.41)$$

After appropriate rearranging and inserting the parameters (16.40) and the following:

$$\delta=\frac{\Delta z}{h},\quad \tau=\frac{c_0}{h^2}\Delta t \qquad (16.42)$$

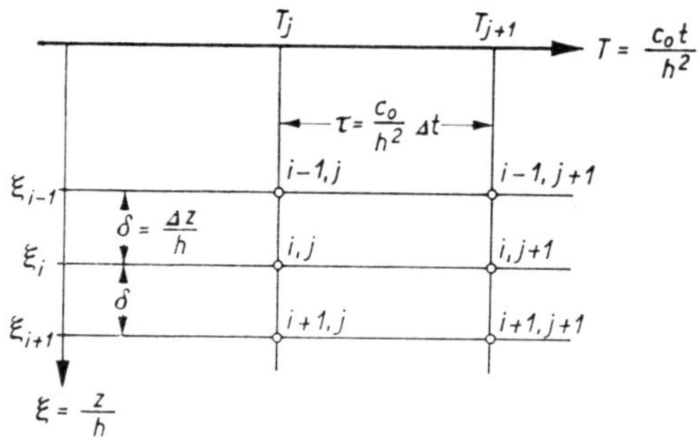

Fig. 16.6. Scheme of the numerical computation. (According to Schiffman and Gibson, 1964)

Schiffman and Gibson obtained Equation (16.41) in dimensionless form:

$$-M_i v_{i-1,j+1}+Q_i v_{i,j+1}-N_i v_{i+1,j+1}=C_{i,j} \qquad (16.43\text{-a})$$

where

$$C_{i,j}=M_i v_{i-1,j}+P_i v_{i,j}+N_i v_{i+1} \qquad (16.43\text{-b}]$$

with the following notation:

$$M_i = \frac{\tau}{\delta m_i}\left(\frac{k_i}{\delta} - \frac{1}{2}\frac{\mathrm{d}k_i}{\mathrm{d}\zeta}\right) \tag{16.43-c}$$

$$N_i = \frac{\tau}{\delta m_i}\left(\frac{k_i}{\delta} + \frac{1}{2}\frac{\mathrm{d}k_i}{\mathrm{d}\zeta}\right) \tag{16.43-d}$$

$$P_i = 2\left(1 - \frac{\tau k_i}{\delta^2 m_i}\right) \tag{16.43-e}$$

$$Q_i = 2\left(1 + \frac{\tau k_i}{\delta^2 m_i}\right) \tag{16.43-f}$$

Schiffman and Gibson (1964) applied Equations (16.43) when investigating the isochrones and time-settlement curves corresponding to the following variations with depth of either the coefficient of permeability (k) or of the coefficient of compressibility (m_v):

(*a*) the polynomial variation of permeability k:

$$k(z) = k_0\left\{1 + (1 + \alpha)^p - \left[1 + \alpha\left(1 - \frac{z}{h}\right)\right]^p\right\}$$

with m_v being held constant;

(*b*) the polynomial variation of compressibility m_v:

$$m_v(z) = m_0\left\{1 + (1 + \beta)^2 - \left[1 + \beta\left(1 - \frac{z}{h}\right)\right]^q\right\}$$

with k being held constant;

(*c*) the hyperbolic decrease of the coefficient of consolidation with depth corresponding (*ca*) to constant permeability k and (*cb*) to constant compressibility m_v;

(*d*) the symmetric sinusoidal distribution of k with constant m_v;

(*e*) the symmetric sinusoidal variation in m_v, the permeability being constant.

In all cases the clay layer is drained from both sides and the initial excess pore pressure (u_0) is constant. A ten-fold permeability or compressibility increase or decrease through the layer has been assumed and isochrones at 50 per cent consolidation compared for either $p = 1$ or $p > 1$ or $p < 1$. The deviations from the conventional theory (constant k and m_v values) have been ascertained and discussed.

16.4 NUMERICAL SOLUTION OF THE DIFFERENTIAL EQUATION OF CONSOLIDATION OF VISCOUS SOILS BY USING THE ISOTACHE METHOD

16.4.1 Spherical Consolidation of Saturated Isotropic Viscous Non-linear Soils

For the conditions of three-dimensional consolidation the isotaches were defined (see Section 14.4.2.1) as the plots $e = e(\sigma_{oct}')$ corresponding to certain consolidation speeds $\partial e/\partial t$. The soil was assumed to be isotropic with respect to deformability. Initially, the following considerations will be based on the same assumption.

Supposing, further, that the soil is fully saturated ($\beta = 0$), the fundamental equation of consolidation can be applied in its numerical form (16.11)–(16.12). For specified conditions the expression in brackets [ ], Equation (16.12), has to be substituted by the respective expressions (16.14), (16.19) or (16.22).

When making use of the isotache method, Equation (16.11) is to be applied in the following way:

The values of the void ratio, coefficients of permeability, pore pressure and total octahedral stress ($\sigma_{oct,t,i,j,k}$) at the beginning of the time interval Δt are assumed to be known throughout the network. Inserting these values into the right-hand side of Equation (16.11), the average consolidation speed

$$v_{e\Delta t,i,j,k} = \frac{\Delta e_{\Delta t,i,j,k}}{\Delta t} \tag{16.44}$$

in the subsequent time inverval Δt at the point (i,j,k) can be obtained. The approximate value of the void ratio at the end of the time interval Δt is then:

$$e_{t+\Delta t,i,j,k} = e_{t,i,j,k} + \Delta e_{\Delta t,i,j,k} \tag{16.45}$$

$$\Delta e_{\Delta t,i,j,k} = v_{e\Delta t,i,j,k} \cdot \Delta t \tag{16.46}$$

Now, in the isotache set $[e = e(\sigma'_{oct})]_{\frac{\partial e}{\partial t} = \text{const}}$ the effective octahedral stress $\sigma_{oct}' = \sigma_{oct}'{}_{(t+\Delta t),i,j,k}$ can be found corresponding to the void ratio $e_{(t+\Delta t),i,j,k}$ on the isotache $\partial e/\partial t = v_{e\Delta t,i,j,k}$. The corresponding pore pressure $u_{t+\Delta t,i,j,k}$ is given by the relation:

$$u_{(t+\Delta t),i,j,k} = \sigma_{oct(t+\Delta t),i,j,k} - \sigma_{oct}'{}_{(t+\Delta t),i,j,k} \tag{16.47}$$

where the total octahedral stress $\sigma_{oct(t+\Delta t),i,j,k}$ at the end of the time

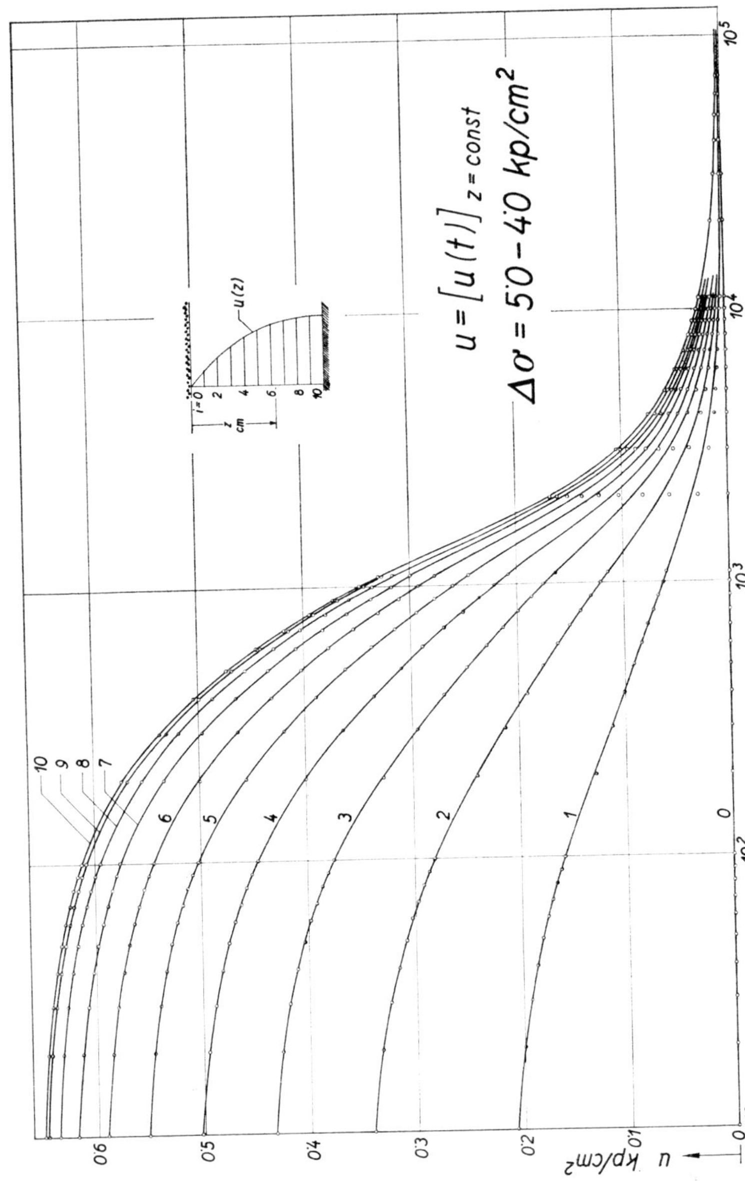

Fig. 16.7. Pore-pressure dissipation plots $u = u(t)_{z=\mathrm{const}}$ corresponding to the is tache set presented in Fig. 11.11–c, $h = 10$ cm

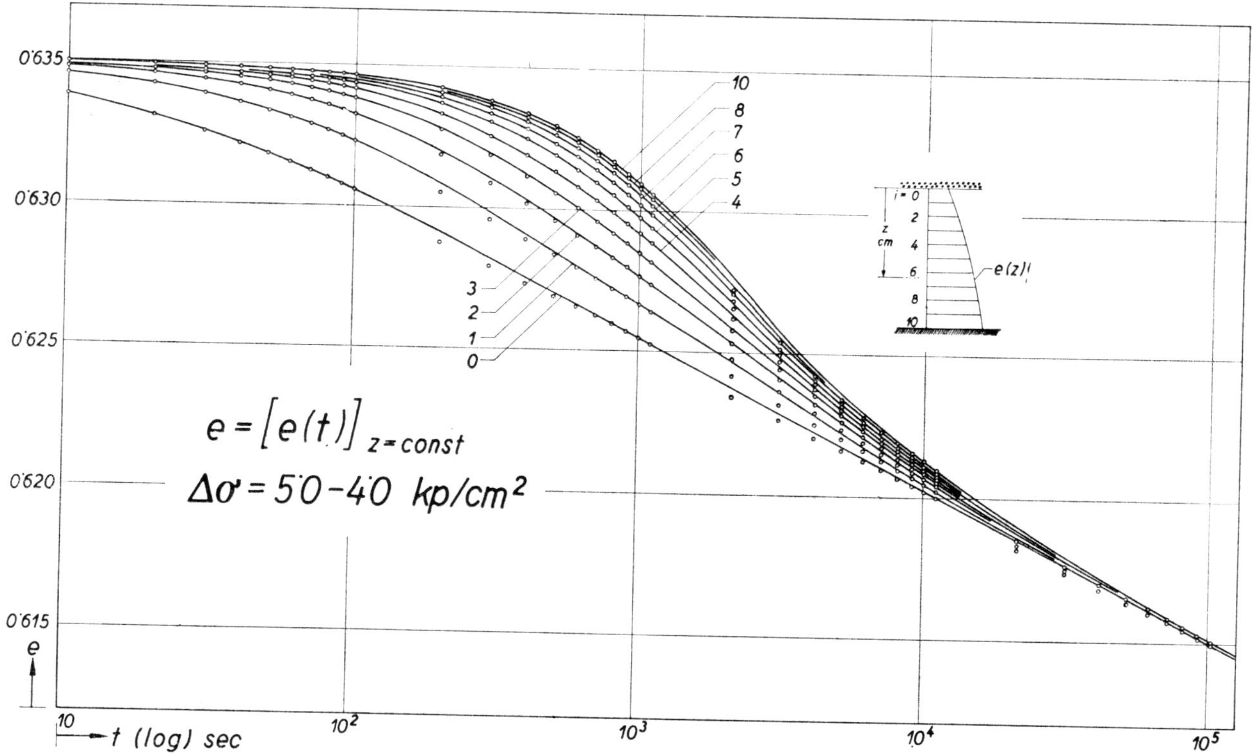

Fig. 16.8. Consolidation curves $e = e(t)_{z=\text{const}}$ corresponding to the isotache set presented in Fig. 11.11–c, $h = 10$ cm

interval Δt is the sum of the preceding total average pressure $\sigma_{\mathrm{oct},t,i,j,k}$ occurring at the beginning of the time interval Δt, and the total pressure increase $\Delta\sigma_{\mathrm{oct},\Delta t,i,j,k}$ during the time interval Δt:

$$\sigma_{\mathrm{oct}(t+\Delta t),i,j,k} = \sigma_{\mathrm{oct},t,i,j,k} + \Delta\sigma_{\mathrm{oct},\Delta t,i,j,k} \tag{16.48}$$

After completing this procedure for all points of the network, the pore pressures obtained can be used in an analogous computation for the next time interval $\Delta t'$ which need not be equal to the preceding one.

The initial values of pore pressures appearing at the beginning of the first time interval Δt_0, i.e. at $t = 0$, should be accorded with the initial total octahedral stress increase $\Delta\sigma_{\mathrm{oct},t=0,i,j,k}$ in such a way that the effective octahedral stresses $\sigma_{\mathrm{oct},t=0,i,j,k}$ and the complementary values of pore-pressures $u_{t=0,i,j,k}$ simultaneously satisfy the isotache set and the equation of consolidation for the initial values of void ratios.

In order to use electronic computers the relations given by the isotache set have to be expressed in analytical form. By way of illustration we can express the isotache set related to axial effective stresses $\sigma_1' \equiv \sigma'$ and presented in Fig. 11.11-c satisfactorily by the equation:

$$\frac{\partial e}{\partial t} = -10^{f(e)+\alpha(e)\frac{1}{\sigma'}} \tag{16.49}$$

where

$$f(e) = a + be = -81{\cdot}11 + 137{\cdot}50e \tag{16.50}$$

$$\alpha_e = (c + de)^{-1} = (0{\cdot}03820 - 0{\cdot}09125e)^{-1} \tag{16.51}$$

and $2 \leqslant \sigma' \leqslant 8\ kp\ \mathrm{cm}^{-2}$.

Using the above relationship the one-dimensional consolidation of a layer of thickness $h = 10$ cm subjected to a suddenly applied overload $\Delta\sigma = (5\text{–}4)kp\ \mathrm{cm}^{-2}$ has been analysed by B. Kogovšek and the author (Šuklje and Kogovšek, 1968). The initial void ratio and coefficient of permeability have been supposed to be constant throughout the layer thickness: $e = 0{\cdot}635$, $k = 3{\cdot}10^{-7}\ \mathrm{cm\ sec^{-1}}$. The lower contact of the layer has been assumed to be impervious and the upper contact pervious. The layer has been divided into $n = 10$ parts and the Equation (16.11) applied in the form:

$$u_{t,i-1} = u_{t,i} + \frac{\gamma_w(\Delta z)^2}{k}\left\{\sum_{j=i}^{n-1}\frac{\left(\frac{\partial e}{\partial t}\right)_{t,j}}{1+e_{t,j}} + \frac{\left(\frac{\partial e}{\partial t}\right)_{t,n}}{2(1+e_{t,n})}\right\} \tag{16.52}$$

where $\Delta z = \dfrac{h}{n}$

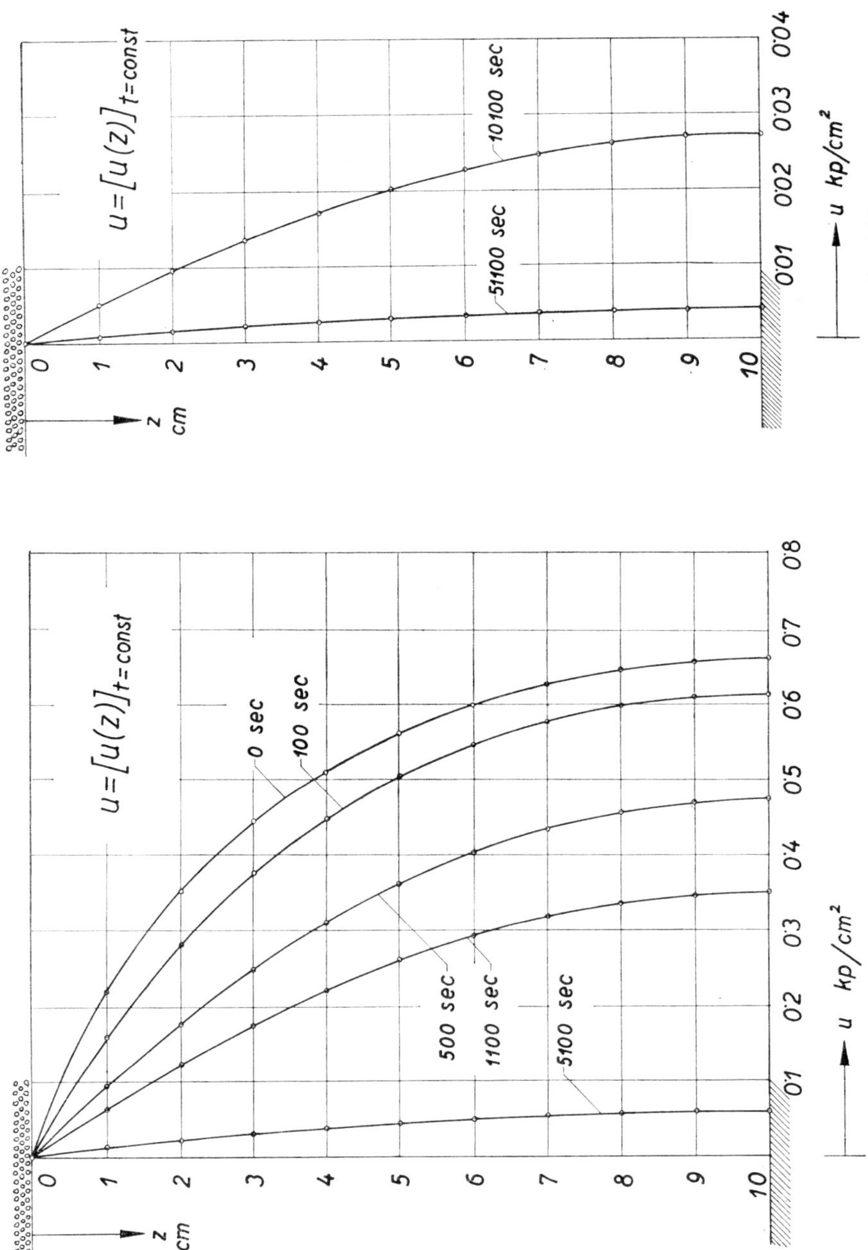

Fig. 16.9. Isochrones corresponding to the plots $u = u(t)_{z=\text{const}}$ presented in Fig. 16.7

The above equation involves the condition $u_{n-1} = u_{n+2}$. The initial and the subsequent isochrones were obtained by trial and error choosing and changing the value of u_n. The right value of u_n is controlled by the condition $u_0 = 0$. The consolidation speeds $\left(\frac{\partial e}{\partial t}\right)_{t,j}$ were determined from the isotache set for the values of void ratios e at the beginning t of the time interval Δt and for the corresponding effective stresses $\sigma' = \sigma - u$. The void ratios at the end of the time interval Δt were then calculated by Equation (16.45).

The pore-pressure dissipation plots $u = u(t)$ and consolidation curves $e = e(t)$ resulting for different depths of the layer are presented in Figs. 16.7 and 16.8. The corresponding isochrones are shown in Fig. 16.9.

16.4.2 Spherical Consolidation of Saturated Viscous Non-linear Soils Exhibiting Orthogonal Anisotropy

In the previous section the soil was assumed to be isotropic with regard to deformability as well as permeability. As proved in Chapter 14, the deformation anisotropy causes dilatancy, i.e. the influence of distortional stresses on spherical strains and spherical stresses on distortional strains.

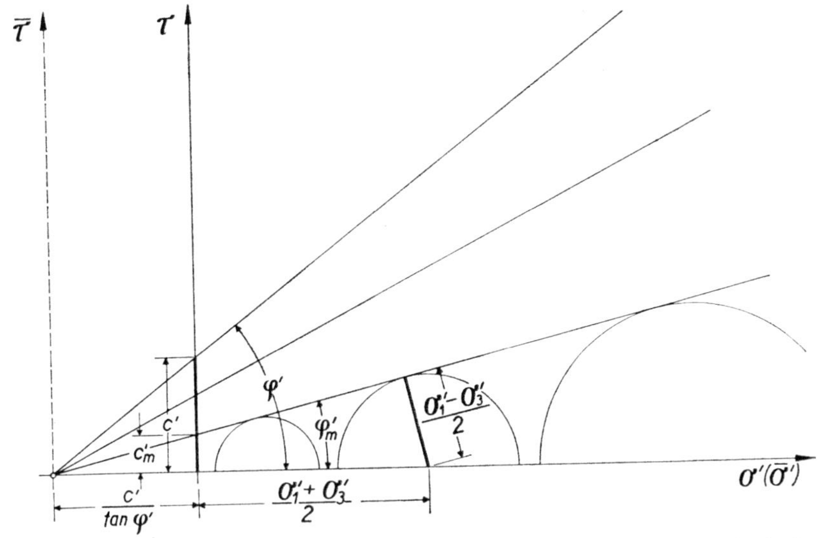

Fig. 16.10. Envelopes of Mohr's circles of stress states whose spherical deformability can be defined by the same isotache set

Thus in anisotropic conditions the isotache set as used in the previous section cannot be considered independent of the distortional stresses. By making several series of triaxial tests different isotache sets can be constructed corresponding to various angles φ_m' appearing in the equation of the straight line envelope to the Mohr effective stress circles:

$$\tau = c_m' + \tan \varphi_m' \,.\, \sigma' \tag{16.52}$$

Referring to Fig. 16.10 the following relation can be determined:

$$\frac{\sigma_1' - \sigma_3'}{2\cot\varphi' + \sigma_1' + \sigma_3'} = \frac{\bar{\sigma}_1' - \bar{\sigma}_3'}{\bar{\sigma}_1' + \bar{\sigma}_3'} = \frac{1 - \dfrac{\bar{\sigma}_3'}{\bar{\sigma}_1'}}{1 + \dfrac{\bar{\sigma}_3'}{\bar{\sigma}_1'}} = \frac{1 - \bar{K}}{1 + \bar{K}} = \sin \varphi_m' \tag{16.53}$$

where

$$\left.\begin{aligned} \bar{\sigma}_1' &= \sigma_1' + c \cot \varphi' \\ \bar{\sigma}_3' &= \sigma_3' + c \cot \varphi' \end{aligned}\right\} \tag{16.54}$$

represent the apparent normal stresses relating to the point $\sigma' = -c \cot \varphi'$ as the new origin of the system of coordinates. Equation (16.53) yields:

$$\bar{K} = \frac{\bar{\sigma}_3'}{\bar{\sigma}_1'} = \frac{1 - \sin \varphi_m'}{1 + \sin \varphi_m'} \tag{16.55}$$

Thus Equation (16.52) can be written:

$$\tau' = \tan \varphi_m' \left(\frac{c}{\tan \varphi'} + \sigma'\right) = \bar{\sigma}' \tan \varphi_m' \tag{16.56}$$

and the straight lines represented by this equation are also lines of equal $\bar{K}$ values. If the soil has zero cohesion, $\bar{\sigma}' = \sigma'$ and $\bar{K} = K = \sigma_3'/\sigma_1'$.

The total stress increase may be assumed to be known. The spherical effective stress development and the corresponding pore pressure dissipation can be found by using an analogous procedure to that explained in the previous section for soils of isotropic deformability. The only difference is that at any time t the effective stress ratio $\bar{K} = \bar{\sigma}_3'/\bar{\sigma}_1'$, has to be considered and that the corresponding isotache set must be applied when determining the effective octahedral stress $\sigma_{oct}' = \sigma_{oct}'{}_{(t+\Delta t),i,j,k}$, corresponding, on the isotache $\partial e/\partial t = v_{e\Delta t,i,j,k}$ (Equation 16.11), to the newly obtained void ratio $e_{(t+\Delta t),i,j,k}$ (Equation 16.45). There the two isotache sets corresponding to the two adjacent $\bar{K}$ values in the interval of which lies the actual $\bar{K}$ value, should in principle be considered and appropriate interpolation made.

17

Development of Distortional and Total Settlements

17.1 THE COMPLEX CONSOLIDATION ANALYSIS BASED ON EXPERIMENTALLY DETERMINED GRAPHS OF THE STRESS–STRAIN RELATIONSHIPS IN SOILS

17.1.1 Strain Characteristics

The computation of the spherical consolidation by using numerical methods of solving the differential equation of consolidation (Chapter 16) was based on the assumption that the total stress distribution was known. With the same assumption the distortional deformations can be computed by applying either Roscoe and colleagues' strain analysis according to the state boundary surface (Section 13.3) or Vyalov's graphical presentation of the distortional stress–strain relationships (Section 13.4). Accounting for creep effects the graphs $\dot{\Gamma} = F(T,\sigma_{oct})$ (Equation 13.149) are of particular interest. Now, according to the considerations presented in Section 13.4.3 the function F is in general time-dependent. The distortional consolidation curves presented in Figs. 14.4 and 14.5 seem to prove this. This time-dependence of the distortional creep, i.e. the initial increase and further decrease of the distortional creep speed, has to be attributed partly to thixotropic effects, partly to the variation in the mean effective stress σ_{oct}' and partly, in drained conditions, to increasing density. The effective stress decrease appearing under undrained conditions has been attributed to relaxation effects (see Section 14.3.2) and the effective stress increase during primary consolidation to the pore pressure dissipation.

For predicting the consolidation of thicker layers the development of both the total and the distortional strains in the secondary consolidation stage of triaxial samples (see Figs. 14.4 and 14.5) is important. In this stage the effective stresses coincide approximately with the total stresses. The isotache sets as well as the sets of graphs $\dot{\Gamma} = F(T,\sigma_{oct}')$ can easily be obtained if a sufficient number of results from different well-chosen triaxial tests is available.

Under the conditions of triaxial tests ($\sigma_2 = \sigma_3$), ($\varepsilon_2 = \varepsilon_3$), the intensity of shear stresses is (see Equations 13.113 and 2.21):

$$T = \frac{\sqrt{3}}{3}(\sigma_1 - \sigma_3) \tag{17.1}$$

and the intensity of shear strains (see Equations 13.114 and 3.22):

$$\Gamma = \frac{2\sqrt{3}}{3}(\varepsilon_1 - \varepsilon_3) \tag{17.2}$$

Using the graphs of additional axial distortional strains $\Delta\varepsilon_1{}^d$ plotted against time t (see Figs. 14.4 and 14.5) the following relation can be taken into account:

$$\Delta\varepsilon_2{}^d = \Delta\varepsilon_3{}^d = -\tfrac{1}{2}\Delta\varepsilon_1{}^d \tag{17.3}$$

Inserting $\Delta\varepsilon_2{}^d$ and $\Delta\varepsilon_3{}^d$ for ε_1 and ε_3 in Equation (17.2), we have:

$$\Gamma = \sqrt{3}\Delta\varepsilon_1{}^d \tag{17.4}$$

and consequently:

$$\dot{\Gamma} = \sqrt{3}\Delta\dot{\varepsilon}_1{}^d \tag{17.5}$$

From the experimental evidence (see, for example, Figs. 14.4, 14.5 and Bishop, 1966, Figs. 34–35), the secondary consolidation of triaxial samples increases for a wide range of stress states, with the logarithm of the time t:

$$\varepsilon_1 = (\varepsilon_1)_1 + \alpha \log_{10} t \tag{17.6}$$

As the spherical strains are governed by the same law:

$$\varepsilon_1{}^0 = (\varepsilon_1{}^0)_1 + \alpha^0 \log_{10} t \tag{17.7}$$

the deviatoric strains also increase with the logarithm of time:

$$\varepsilon_1{}^d = (\varepsilon_1{}^d)_1 + \alpha^d \log_{10} t \tag{17.8}$$

$(\varepsilon_1)_1$, $(\varepsilon_1{}^0)_1$ and $(\varepsilon_1{}^d)_1$, represent the axial strains at time $t = 1$ and α, α^0, α^d the strain increase during a logarithmic time decade increment $\Delta t = 10^{n+1} - 10^n (n = \pm 0,1,2, \ldots)$. Both of these strain characteristics depend on the initial porosity and on the stress and strain history as well as on the actual effective stress state. For triaxial compression tests the latter is fully defined by the octahedral normal stress $\sigma^{0\prime}$ and the slope $\varphi_m{}'$ of the Mohr envelope originating from the point $\sigma' = p_c = -c' \cot \varphi'$ on the axis σ', φ' being the slope of the failure envelope. The slope $\varphi_1{}'$ of the limiting envelope of stress states obeying the logarithmic law of creep is not much less than the failure slope φ' (see Section 19.5.1).

For the domain $\varphi_m{}' \leqslant \varphi_1{}'$, the author suggests that the deformability of soils should be investigated by drained triaxial compression tests carried out at several values of effective spherical stress and at increasing

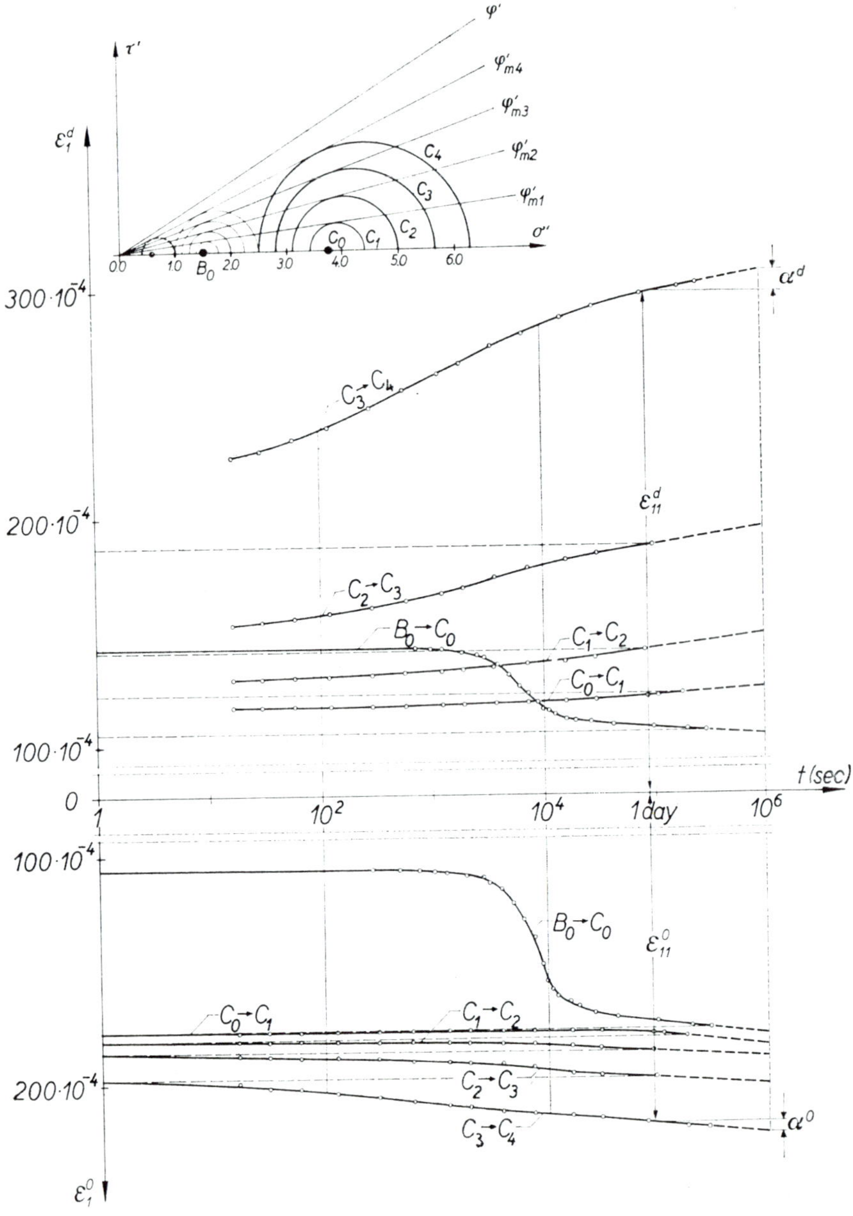

Fig. 17.1. Spheric and deviatoric strain versus time plots derived from triaxial tests of a sandy silt: for the stress states whose Mohr's circles are presented on the top of the figure. Full lines correspond to observed consolidation curves, dashed lines have been extrapolated

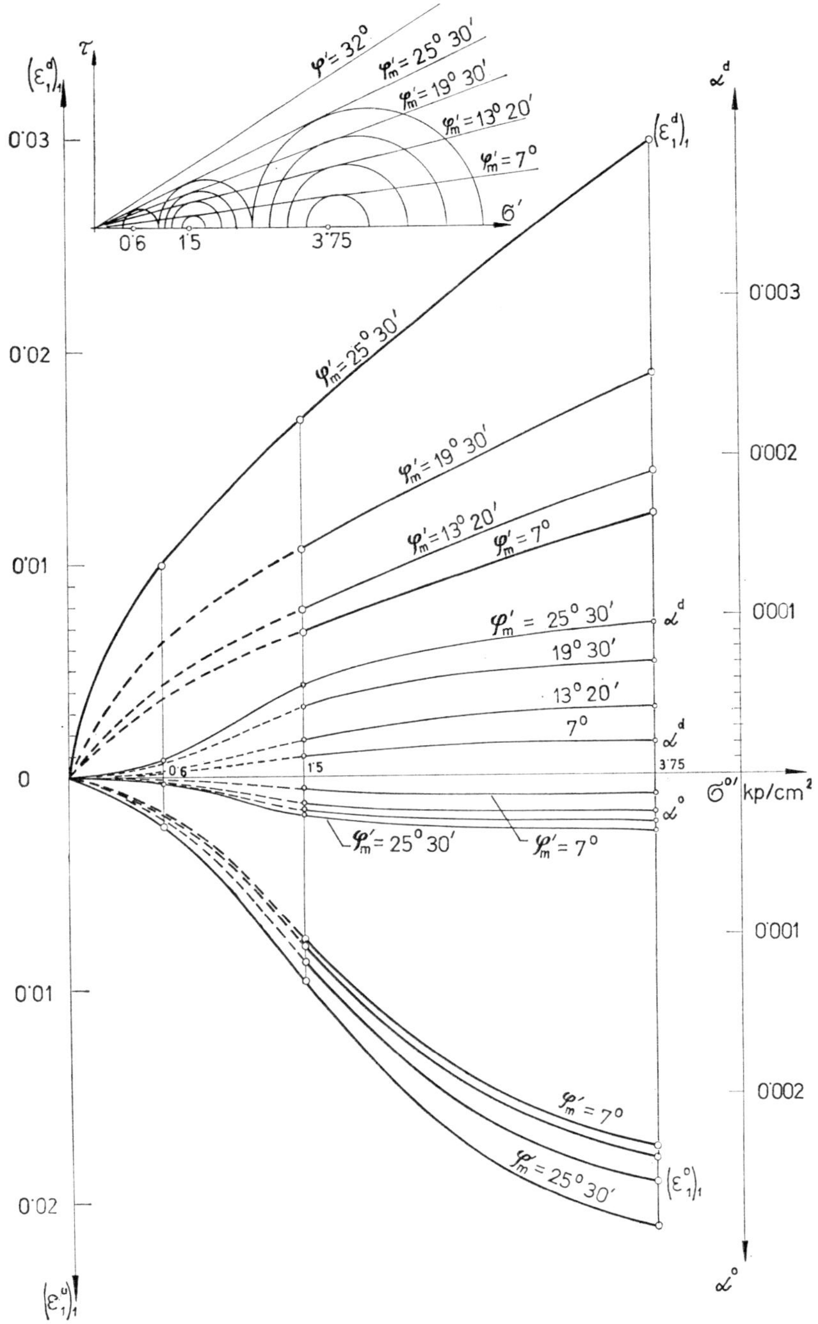

Fig. 17.2. Variation of strain-and strain velocity-parameters with the stress states. The charts are related to the sandy silt whose consolidation lines are presented in Fig. 17.1

deviatoric stresses corresponding to certain stress envelope slopes. In order to avoid a sudden breakdown in grain structure, negative thixotropic effects and temporary over-mobilization of the strength of the soil by pore-pressures, the loading must be increased continuously and slowly from one stress step to the next. Fig. 17.1 shows the method of presenting the results of such a test series by consolidation curves accounting separately for the spherical and the deviatoric strain components and corresponding to successive stress states which are presented in the Mohr-circle coordinate system at the top of the figure.

For saturated soils the time t_s at which the pore pressure dissipate to sufficiently small values (e.g. 0·01 kp/cm^2), can be determined to satisfactory approximation from the average consolidation speed $\dot{\varepsilon}_1{}^0 = \frac{1}{3}\dot{\varepsilon}_v$ by using a simplifying assumption, e.g. by assuming a uniform speed of volume change throughout the sample. The inaccuracy of such an assumption does not affect the order of magnitude of the resulting pore pressures which are small. The coefficient of permeability corresponding to the actual porosity, and the boundary seepage conditions should be known and considered.

The spherical strain characteristics $(\varepsilon_1{}^0)_1$ and α^0 govern the secondary phase of spherical consolidation. The development of spherical consolidation in the primary phase can be based on the isotache sets of volume change as explained in Section 16.3.2.

For the deviatoric strains, the relation (17.8) gives too high values in the early stage of consolidation of the triaxial sample, i.e. at $t < t_s$. During the primary consolidation of thicker soil masses, the deviatoric strains according to Equation (17·8) are somewhat over-estimated if the coefficients α^d increase with increasing spherical effective stresses $\sigma^{0\prime}$, and under-estimated in the opposite instance.

Several series of tests similar to those shown in Fig. 17.1 enable the complete stress–strain relationships of viscous soils to be presented by the strain characteristics $(\varepsilon_1{}^0)_1$, α^0 and $(\varepsilon_1{}^d)_1$, α^d plotted against the spherical effective stresses $\sigma^{0\prime}$ for various values of the parameter $\varphi_m{}'$. The method of presentation of these plots is shown in Fig. 17.2.

For unloading, the families of plots presented in Fig. 17.1 have to be established separately by applying an opposite test procedure. At a constant spherical normal stress $\sigma^{0\prime}$ the deviatoric stress component is continuously decreased but longer pauses are maintained at certain envelopes $\varphi_m{}' =$ const. For over-consolidated soils the unloading plots are expected to be rather close to the loading plots.

The choice of the stress paths and the duration of the preceding load steps certainly influence, to some extent, the values of the proposed strain characteristics. The importance of these effects and the influence of the mean principle stress (compression and extension tests) must be checked by appropriate investigations.

17.1.2 Approximate Strain Fields

If a stress field is known or assumed, the Mohr circle corresponding to the maximum and minimum principal stress can be drawn for any point of the field, and the slope φ_m' of the straight envelope originating from the point $\sigma' = -p_c$ established. The octahedral normal stress $\sigma^{0\prime}$ of the

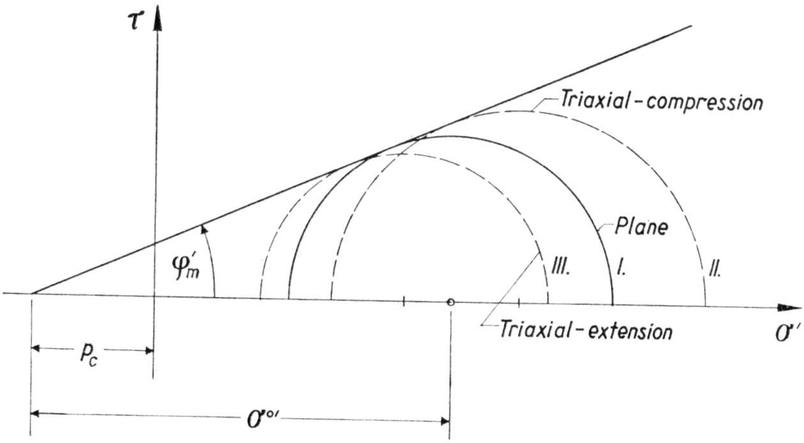

Fig. 17.3. Correlated Mohr's circles of triaxial compression and extension stress states and of the plane state of strain

observed point in the field does not coincide, however, with the value σ^0 of a triaxial compression test corresponding to the same Mohr circle. Thus, an assumption should be made concerning the correlation between the stress state in the soil in question and the stress state in the triaxial compression sample. According to our definition of the radial symmetric stress state in a triaxial compression sample (see the previous section), the slope φ_m' of the Mohr envelope—at a given p_c—and the octahedral effective normal stress $\sigma^{0\prime}$ may be chosen parameters for correlation. Fig. 17.3 represents the correlated Mohr circles: I of a plane stress state with $\sigma_2' = \frac{1}{2}(\sigma_1' + \sigma_3')$, II of a triaxial compression test, and III of a triaxial extension test, corresponding to the assumed principle of correlation.

Further assumptions should be made concerning the correlated strains. For the application of triaxial test results to plane problems, either the main principal strain and the spherical strain, or the intensity of shear strains and the spherical strain can be chosen as correlated quantities. For the application to space problems, further assumption should be made—e.g. the combination of the above assumptions.

By using the plots of Fig. 17.2, the strains, related for a given stress state ($\sigma^{0\prime}$, $\varphi_m{}'$) and time t during secondary consolidation to the zero stress state, can then be computed from Equations (17.7) and (17.8). The strains corresponding to the transition from an initial to final stress state can be obtained as the difference of the strains obtained for both of them. The approximate values of displacements can then be obtained by integrating the diagrams of strain differences in two straight directions normal to each other, and by respecting the boundary conditions.

The development of spherical strains in the primary phase of consolidation can be pursued by the isotache method as explained in Section 16.3. The development of corresponding deviatoric strains can be computed simultaneously. The increase of deviatoric strain $\varepsilon_1{}^d$ in the time interval $\Delta t = t_b - t_a$ may be taken to be equal to

$$\Delta\varepsilon_1{}^d = \{[(\varepsilon_1{}^d)_1]_b + [\alpha^d]_b \log t_b\} - \{[(\varepsilon_1{}^d)_1]_a + [\alpha^d]_a \log_{10} t_a\} \quad (17.9)$$

where the quantities with the suffix a are related to the effective stress state at the beginning of the time interval Δt, and those with the suffix b to the effective stress state at the end of the time interval Δt. If the plots of the strain characteristics were obtained from triaxial tests made on undisturbed samples, the time $t = 1$ day can be taken as the time of commencing the load change.

The computation of the displacements can be started by applying statically correct stress fields, i.e. stress fields which satisfy the equilibrium and the boundary stress conditions. The stress fields obtained using the theory of elasticity or the theory of plasticity can be used in appropriate conditions. (For the application of stress fields corresponding to arbitrarily-chosen "slip surfaces," see Section 19.5.2.)

Usually, the resulting strain fields do not satisfy either the compatibility conditions or the strain boundary conditions and consequently they are not correct. Nevertheless they can give an idea of the displacements to be expected. In principle, iterative procedures could be applied in order to gain agreement between the stress and strain fields satisfying the equilibrium, compatibility and boundary conditions, and obeying the established stress–strain relationships. Yet even at a selected time during secondary consolidation the stress–strain relations can hardly be expressed in an analytical form which is simple enough to be used for the numerical integration of the equilibrium equations.

17.1.3 Simplified Procedures

The determination of complete families of loading and unloading plots of parameters $(\varepsilon_1^0)_1$, α^0 and $(\varepsilon_1^d)_1$, α^d and of corresponding isotache sets is cumbersome and takes considerable time. It requires extensive experimental work and computation. Simplified approximate analyses can be based on only one family of isotaches and on single plots of strain characteristics corresponding to the stress states having a single envelope of Mohr circles $(\varphi_m')_{med}$ = const. The value of $(\varphi_m')_{med}$ that corresponds to the average mobilization of the strength of the soil can be chosen:

$$\tan \varphi_m' = \frac{\tan \varphi'}{F_s}, \quad F_s \text{ being the safety factor.}$$

For non-viscous soils the coefficients α^0 and α^d vanish and all isotaches corresponding to a given value of φ_m' = const condense to a single line $e = [e(\sigma^{0\prime})]_{\varphi_m'=\text{const}}$.

For soils of isotropic deformability a single family of isotaches $e = [e(\sigma^{0\prime})]_{\partial e/\partial t=\text{const}}$ results. The plots $(\varepsilon_1^0)_1 = f_0(\sigma^{0\prime})$, $\alpha^0 = \alpha^0(\sigma^{0\prime})$ condense to single plots applying for all values of the parameter φ_m', and the plots $[(\varepsilon_1^d)_1]_{\varphi_m'=\text{const}} = f_d(\sigma^{0\prime})$, $[\alpha^d]_{\varphi_m'=\text{const}} = \alpha^d(\sigma^{0\prime})$ become parallel to the $\sigma^{0\prime}$ axis.

The isotropic deformability of non-viscous soils can be defined by a single compression curve $e = e(\sigma^{0\prime})$ and one single graph $\varepsilon_1^d = \varepsilon_1^d(\varphi_m')$.

When computing the settlements and their development by the classical Terzaghi theory, the diagram $e = e(\sigma^{0\prime})$ is replaced by the plot of linear compression $e = e(\sigma_z')$. As, in this case, the lateral strains are assumed to be zero, the deviatoric and spherical deformability are simultaneously included in the plot of linear compression because:

$$\varepsilon_x^d = -\varepsilon_x^0 = -\frac{\varepsilon_v}{3}$$

and

$$\varepsilon_z = \varepsilon_z^d + \varepsilon_z^0 = \tfrac{2}{3}\varepsilon_v + \tfrac{1}{3}\varepsilon_v = \varepsilon_v$$

17.2 APPROXIMATE COMPUTATION OF DISTORTIONAL CREEP AND OF TOTAL SETTLEMENT DEVELOPMENT BY USING REPRESENTATIVE VALUES OF THE EQUIVALENT SHEAR AND COMPRESSION MODULI

Owing to non-linear stress–strain relationships, to different values of deformability on loading and unloading, to viscous properties and to

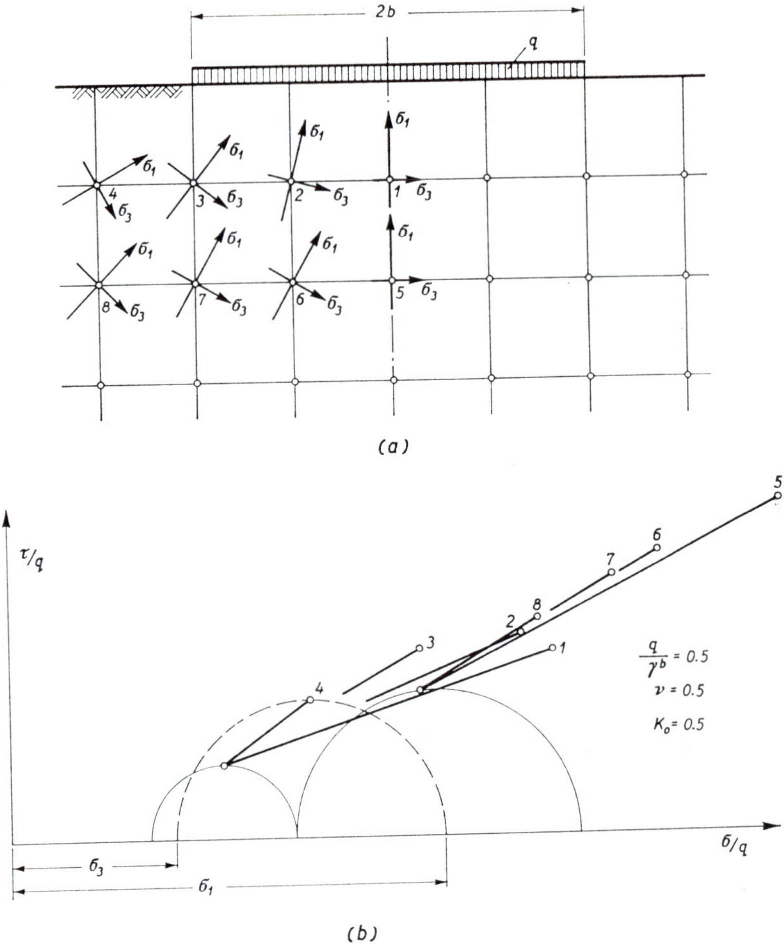

Fig. 17.4. Uniformly distributed strip load: (*a*) directions of the trajectories of principal stresses; (*b*) stress paths for points 1–8

anisotropy, the theory of elasticity can be applied to soil media only for the approximate computation of stresses and displacements. The non-homogeneity and the stratification of soils further complicates the computation.

In order to obtain approximate values of displacements, and the approximate stress distribution by using elastic methods, appropriate equivalent elastic constants can be employed expressing, at the observed time, the average stress–strain relationship in the actual stress field. For this purpose the stresses corresponding to the homogeneous isotropic half-space may first be determined taking a Poisson's ratio of about 0·25. If solutions for the stratified half-space are available in a form facilitating quick computation of the stresses, and if they suit the stratification of the soil in question favourably, appropriate ratios of moduli should be chosen.

In Fig. 17.4 and 17.5 two examples are presented showing the stress path at various points of the half-space subjected respectively to a uniformly distributed load and to a symmetrical triangular strip load. The stress-paths are straight lines connecting the peak points of the Mohr circles corresponding to the stress states before and after the load application.*

The stress paths in Fig. 17.4 correspond to the ratio $q/\gamma b = 0{\cdot}50$, γ being the unit weight of the soil, and those in Fig. 17.5 to the ratio $p/\gamma b = 0{\cdot}50$. In both cases the earth pressure coefficient at rest K_0 has been taken equal to 0·50. The stress paths are denoted by numbers indicating the points of the network below the load-strip (see Figs. 17.4 and 17.5 parts (*a*)).

Now, different consolidation curves result when testing soil samples according to different stress paths. It is to be hoped that sufficient experience, obtained by comparing the results of simplified computations with the results of complete numerical computations accounting for the heterogeneous deformability, as well as with the results of model and field measurements, will, in the future, facilitate the appropriate choice of representative stress paths. In the present state of our knowledge, the paths have to be chosen with some intuition and feeling. Let the stress path 6 be chosen as representative in the case of the uniformly distributed strip load (Fig. 17.4), and the stress path 2 for the triangular strip load (Fig. 17.5).

The consolidation curves $\varepsilon_1 = \varepsilon_1(t)$ and $\varepsilon_1{}^d = \varepsilon_1{}^d(t)$ presented in Fig. 17.6 may correspond to the chosen stress path, ε_1 being the total and $\varepsilon_1{}^d$ the distortional axial strain. From the very beginning the test was carried out as a drained test. In this way the effects of relaxation on the distortional strains (see Section 14.3.2) have been avoided. On the other hand the anisotropy effects appearing at increasing effective stresses (see Section 14.4.3) cannot be separated. Thus the application of the following approximate procedure leads to some overestimation of the immediate distortional strains.

In elastic solutions the instantaneous distortional displacements can be obtained as total displacements corresponding to the apparent Poisson

* This simple presentation of stress state was suggested by Lambe (1964).

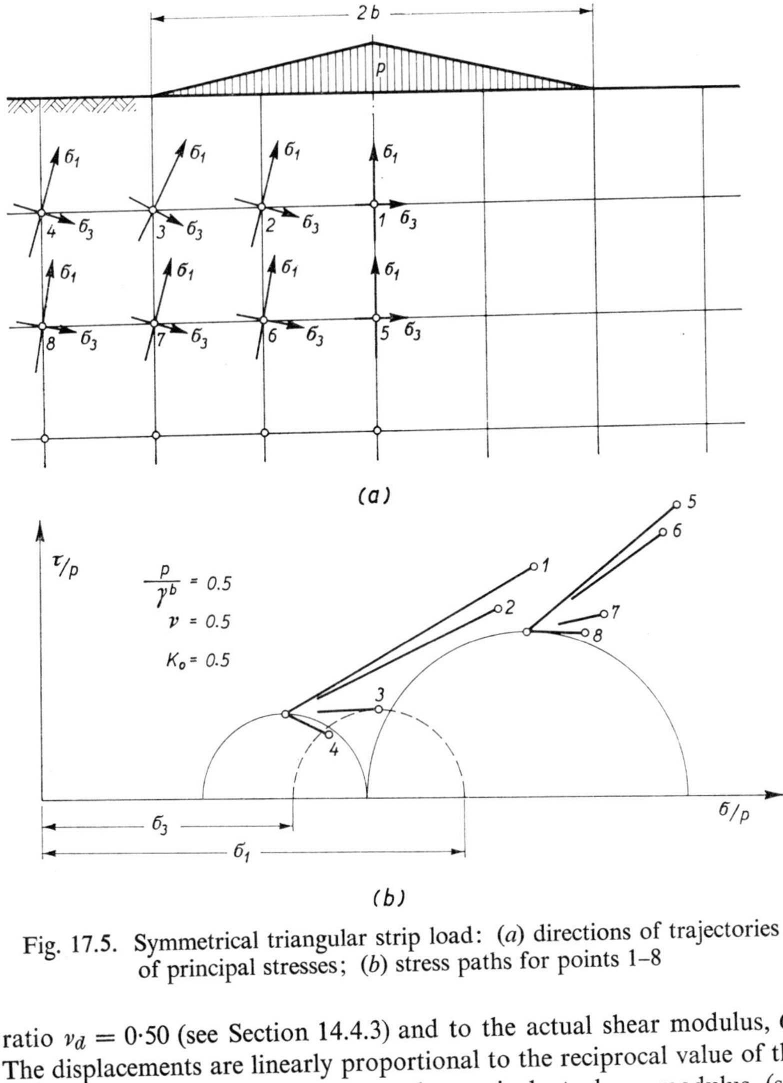

Fig. 17.5. Symmetrical triangular strip load: (*a*) directions of trajectories of principal stresses; (*b*) stress paths for points 1–8

ratio $\nu_d = 0{\cdot}50$ (see Section 14.4.3) and to the actual shear modulus, G. The displacements are linearly proportional to the reciprocal value of the shear modulus. In a triaxial test the equivalent shear modulus (see Section 13.1.2) at any time t is given by the relation

$$2G = \frac{\sigma_1{}^d}{\varepsilon_1{}^d} = \frac{2}{3}\,\frac{\sigma_1 - \sigma_3}{\varepsilon_1{}^d} \tag{17.10}$$

Owing to creep the distortional strains $\varepsilon_1{}^d$ are time-dependent. During the secondary stage of consolidation, i.e. after the pore pressures have dissipated and decreased to very small values, a certain law of distortional strains can be ascertained. Commonly the extrapolation of the observed consolidation gives the logarithmic relationship:

$$\varepsilon_1{}^d = (\varepsilon_1{}^d)_1 + \alpha^d \log_{10} t \tag{17.8}$$

where $(\varepsilon_1{}^d)_1$ denotes the distortional strain at time $t = 1$ and α^d the strain increase over a logarithmic decade increment $\Delta t = 10^n - 10^{n-1}$ (see Section 17.1.1). The reciprocal value of the shear modulus increases obeying the same law:

$$\frac{1}{G} = \frac{3}{\sigma_1 - \sigma_3} [(\varepsilon_1{}^d)_1 + \alpha^d \log_{10} t]$$

$$\frac{1}{G} = \left(\frac{1}{G}\right)_1 + \alpha^G \log_{10} t \tag{17.11}$$

where

$$\left(\frac{1}{G}\right)_1 = \frac{3(\varepsilon_1{}^d)_1}{\sigma_1 - \sigma_3}$$

denotes the distortional deformability at time $t = 1$, and

$$\alpha^G = \frac{3\alpha^d}{\sigma_1 - \sigma_3}$$

the distortional deformability increase over a logarithmic decade increment Δt.

When applying the law of distortional creep, as observed during secondary consolidation, to the primary stage of consolidation of thicker layers, too large distortional displacements result at the very beginning of consolidation and—probably—throughout the primary consolidation. The opposite extreme in evaluating the distortional creep can be obtained by taking the immediate distortional deformability according to the observed immediate distortional strain $(\varepsilon_1{}^d)_1{}^*$ (see Fig. 17.6) and assuming that during primary consolidation the speed of distortional creep is proportional to the speed of the void ratio decrease. The probable curve of distortional creep lies between the extreme curves outlined above.

Similarly the secondary branch of the development of total displacements can be obtained by introducing the same time-dependent shear moduli and the Poisson ratios corresponding to these shear moduli and to the simultaneous compression moduli K, into the elastic solutions. The moduli K are given by the relation

$$3K = \frac{\sigma_{1\,\mathrm{oct}}}{\varepsilon_{1\,\mathrm{oct}}} = \frac{\sigma_1 + 2\sigma_3}{\varepsilon_v} = \frac{\sigma_1 + 2\sigma_3}{3(\varepsilon_1 - \varepsilon_1{}^d)} \tag{17.12}$$

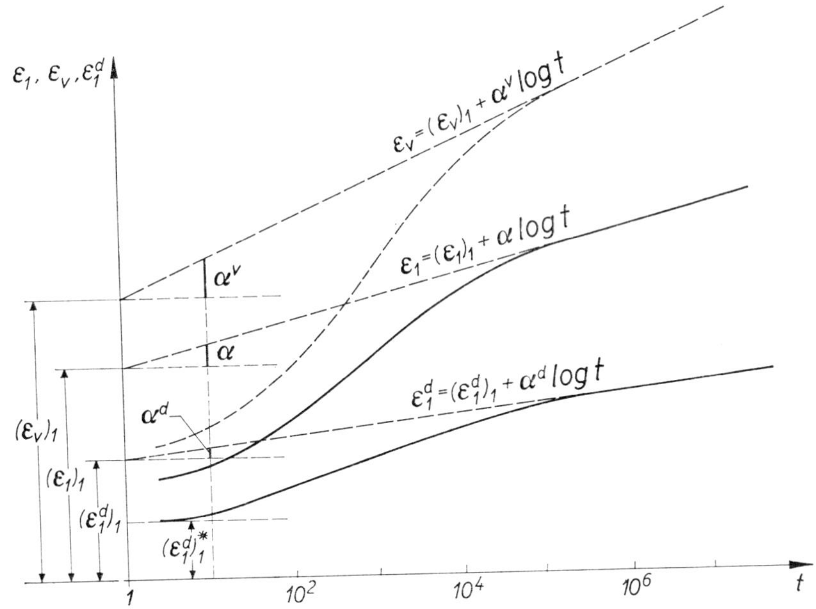

Fig. 17.6. Total (ε_1) and distortional ($\varepsilon_1{}^d$) strain development of a triaxial sample subjected to an average stress path

Commonly the secondary compression can be expressed by the logarithmic law:

$$\varepsilon_v = (\varepsilon_v)_1 + \alpha^v \log_{10} t \tag{17.13}$$

and, consequently, the coefficient of compression m_v by

$$m_v = \frac{1}{K} = \frac{9}{\sigma_1 + 2\sigma_3} [(\varepsilon_v)_1 + \alpha^v \log t]$$

$$m_v = \frac{1}{K} = (m_v)_1 + \alpha^K \log t \tag{17.14}$$

where

$$(m_v)_1 = \frac{9}{\sigma_1 + 2\sigma_3} (\varepsilon_v)_1$$

(for $(\varepsilon_v)_1$ see Fig. 17.6), and

$$\alpha^K = \frac{9\alpha^v}{\sigma_1 + 2\sigma_3}$$

As stated by Rappoport (1948) the solutions obtained for stresses in the homogeneous half-space can be applied with suitable accuracy to the layered half-space if the moduli ratio change remains within moderate limits. Consequently the deformations of a single layer can be obtained by subtracting the displacements of the lower boundary of the layer from the displacements of the upper boundary. Each layer has to be treated separately by applying the solutions for the half-space with corresponding shear and compression moduli. In special conditions, the available solutions accounting for the stratification of the half-space can be applied (see Sections 5.4 and 15.1.2).

The development of the differential settlements can then be predicted approximately by using consolidation ratios corresponding to the time factors according to the known linear (Terzaghi) or three-dimensional solutions (Mandel, Gibson and McNamee). In this the coefficient of compression should be chosen first; it must correspond to the time t_p by which the primary consolidation is expected to be completed. The correct value of the coefficient of compression can easily be found by the iterative procedure; the correct values of the time t_p and of the corresponding coefficient of compression $(m_v)_p$ must give the time factor at which about 99 per cent of the primary consolidation is attained.

17.3 DISCUSSION OF THE SETTLEMENT ANALYSIS BASED ON OEDOMETER TESTS

17.3.1 Settlement Analysis Based Only on Oedometer Tests

The settlement analysis commonly used is based mostly on the classical oedometer test and on Terzaghi's linear consolidation theory. The vertical stress distribution along the vertical line through the point whose settlement is to be computed may first be determined. In Fig. 17.7 line *a* is the diagram of initial vertical pressures due to the self-weight of the soil strata and the water table. Line I represents the influence of temporary unloading ($\Delta\sigma_{\text{I}}$) due to excavation, line II the stresses caused by the load q in the foundation base and line III the stress difference $\Delta\sigma = \Delta\sigma_{\text{II}} - \Delta\sigma_{\text{I}}$ which has to be added to the initial stresses σ_a in order to obtain the final stress diagram c:

$$\sigma_c = \sigma_a + \Delta\sigma$$

Line b determines the transient stresses occurring after the excavation was completed:

$$\sigma_b = \sigma_a - \Delta\sigma_{\text{I}}$$

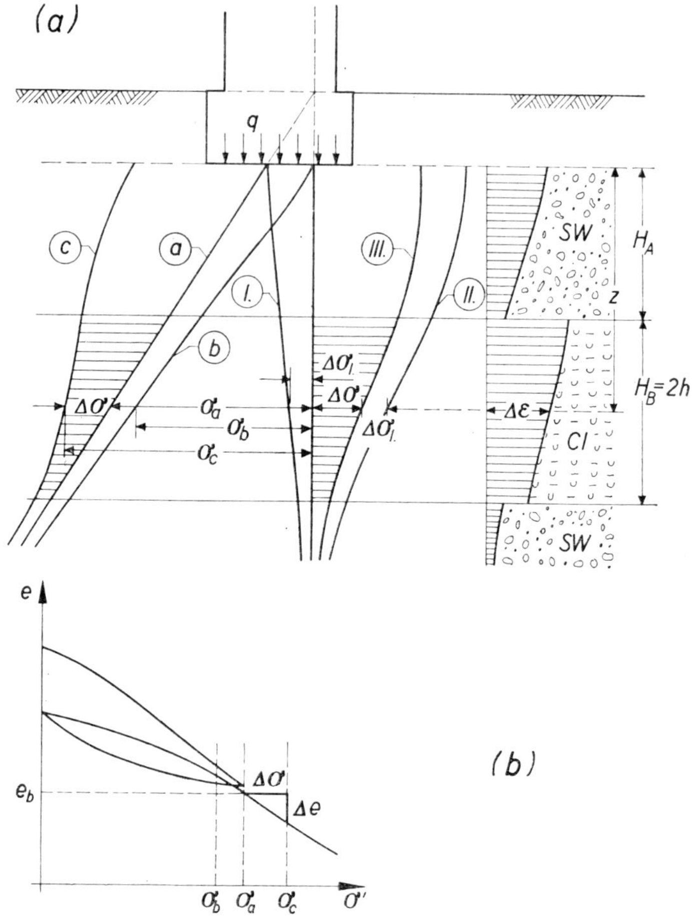

Fig. 17.7. Settlement analysis based on oedometer tests

The vertical stress difference $\Delta\sigma_{I}$ and $\Delta\sigma_{II}$ are usually determined by using integrations of the Boussinesq fundamental equations or Westergaard's charts. For stratified soils the available solutions for multi-layered media can also be applied.

In order to determine the corresponding strain diagrams the $e = e(\sigma')$ plots relating void ratio to effective stress, as obtained by oedometer tests, are to be considered in both cases. In Fig. 17.7 (*b*) the $e = e(\sigma')$ plot represents the results of an oedometer test on an undisturbed sample taken near the upper boundary of the clay layer *Cl*. The strain $\Delta\varepsilon$ set up at the depth z below the foundation base due to the stress increase $\Delta\sigma = \sigma_c - \sigma_a$ is given by the relation:

$$\Delta\varepsilon = \frac{\Delta e}{1 + e_a} \tag{17.15}$$

The plot $\Delta\varepsilon$ versus z is discontinuous at the boundary between layers of different compressibility. The area of the diagram represents the settlement:

$$\rho = \int_{z=0}^{z_e} \Delta\varepsilon \, dz \tag{17.16}$$

The lower limit z_e can be chosen as the depth where the ratio between the additional stress $\Delta\sigma$ and the initial pressure σ_a is sufficiently small (e.g. $\Delta\sigma/\sigma_a = 0{\cdot}05$).

In a similar way the influence of unloading due to excavation can be computed by using the corresponding decompression branches of the oedometer tests. The settlement on reloading in the region of the stress interval $\Delta\sigma_{\mathrm{I}}$ is governed by the recompression branch and is approximately equal to the recovery displacement on unloading. In layers of low permeability only a part of the total swelling displacement can develop as the period of unloading is commonly too short to allow complete swelling. Furthermore, as the deformability of pre-consolidated soils is several times smaller than their deformability at first loading, the influence of unloading and reloading $\Delta\sigma_{\mathrm{I}}$ on the practical settlement computation of normally consolidated soils or pre-consolidated soils of low permeability is often neglected. It cannot be neglected, however, if the stress difference $\Delta\sigma_{\mathrm{I}}$ is very high, if the permeability of the soil is moderate or high, if the duration of excavation is long or if the construction is sensitive to settlement.

Equation (17.16) can also be written in the form:

$$\rho = \int_{z=0}^{z_e} \frac{\Delta\sigma}{\dfrac{\Delta\sigma}{\Delta e}(1 + e_a)} \, dz = \int_{z=0}^{z_e} \frac{\Delta\sigma}{M_{vz}} \, dz \tag{17.17}$$

where M_{vz} is the modulus of compressibility at the depth z:

$$M_{vz} = \frac{\Delta\sigma}{\Delta e}(1 + e_a) \tag{17.18}$$

By introducing a mean value of the modulus M_{vz} for each layer, Equation (17.17) yields:

$$\rho = \sum_{i=1}^{n} \frac{A_i}{M_{vi}} \tag{17.19}$$

where A_i stands for the area of the additional stress diagram along the thickness Δz_i of the layer i:

$$A_i = \int_{z=z_i}^{z_i+\Delta z_i} \Delta\sigma \, dz \tag{17.20}$$

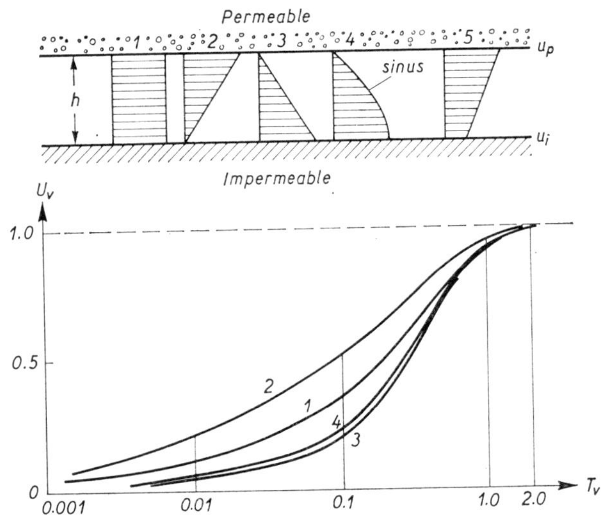

Fig. 17.8. Consolidation ratio versus time factor plots according to Terzaghi's theory, corresponding to different zero isochrones (1–5)

The influence of unloading and reloading can be considered in an analogous way by replacing the compression moduli M_{vi} by corresponding decompression or recompression moduli.

If the foundation contains an isolated saturated clay layer sandwiched between a permeable cover and either a permeable or an impermeable base, the consolidation of this layer can be predicted approximately by applying Terzaghi's one-dimensional consolidation theory to various zero isochrone shapes. The results of some integrations of the differential equation of consolidation (9.76) are presented (after Terzaghi and

Fröhlich, 1936) in Fig. 17.8 where the consolidation ratios $U_v = \rho_t/\rho$ are plotted versus the time factors T_v (see Section 9.3.3):

$$T_v = \frac{c_v t}{h^2} \tag{17.21}$$

In most cases the diagram of additional stress $\Delta\sigma$ occurring across the thickness of an isolated clay layer can be transformed into a trapezium with sufficient accuracy. The consolidation ratios U_{v5} for the trapezoidal zero isochrones can be deduced from the plots U_{v1} and U_{v3} related to rectangular and triangular zero isochrones by using the following relation:

$$U_{v5} = U_{v1} + \frac{u_p - u_i}{u_p + u_i}(U_{v1} - U_{v3}) \tag{17.22}$$

where u_i and u_p are the initial pore pressures at the impermeable and permeable boundaries respectively (see Fig. 17.8).

If the isolated clay layer is drained, at both boundaries, the consolidation ratios U_{v1} related to the rectangular zero isochrone can be used for arbitrarily shaped diagrams of additional stresses. In this case the half-thickness of the layer has to be inserted for h when computing the time factor according to Equation (17.21).

The consolidation of a partly saturated isolated clay layer can be analysed in a similar way by applying the considerations made in Section 12.2.

In the approximate procedure in this section the viscous soil properties have been neglected and the odeometer tests assumed to give final strains. In thick layers of slightly over-consolidated soil exhibiting important secondary effects the above procedure can give useful approximate results if the $e = e(\sigma')$ plots relating to the time t_p at which the end of the primary consolidation of the layers constituting the subsoil occurred, can be provided. This time t_p can be determined with sufficient accuracy by inserting, in the relation

$$t_p = T_v \frac{h^2}{c_v} = T_v \frac{h^2 \gamma_w}{k M_v} \tag{17.23}$$

the value $T_v = 2$. By trial and error the appropriate void ratio e for the extended secondary stages of the oedometer consolidation curves can easily be found. The justification of such a procedure has been given in Section 11.1.6.

The procedure presented makes the assumption that at any point of the subsoil the vertical stresses are principal stresses and that the ratio between the horizontal and vertical principal stresses is the same as in oedometer tests. Furthermore, the horizontal strains are assumed to be zero and the drainage of the pore water is assumed to occur in the vertical direction only. The incompatibility of such assumptions with the actual stress state and with the method applied for computing the vertical stresses is evident. The method is justified by its simplicity and by agreement in several cases of the results obtained with the measured settlements. In the present state of our knowledge this method can be replaced by more precise methods which have been presented and discussed in Chapters 15 and 16 and in the previous sections of the present chapter. The need for applying these more complicated methods and their degree of perfection depend on the character of the soil and the construction. However, no method is perfect and none enables us to get a precise prediction of the settlements and their development.

17.3.2 Settlement Analysis Based on Oedometer Tests and on Simple Unconfined Compression Tests

In the settlement analysis presented in the previous section the initial pore pressures were equal to the full vertical pressure increase $\Delta\sigma_1$. This is the consequence of the assumption of zero lateral strain (see Section 14.3.1) and of neglecting the plastic resistance (see Section 14.3.2). Equally, as in the oedometer test, the vertical displacements are partly of spherical and partly of distortional character. In saturated layers of low permeability, however, the initial distortional displacements are zero because the lateral displacements are assumed to be zero and the vertical displacements cannot occur under a zero initial additional effective pressure in the vertical direction. The initial total additional principal stress ratio

$$\frac{\Delta\sigma_3}{\Delta\sigma_1} = \frac{\Delta\sigma_h}{\Delta\sigma_v} = 1$$

equals unity and gradually decreases to a final value equal to K_0, the earth pressure coefficient at rest. Correspondingly, the distortional displacements increase simultaneously.

This development of the distortional displacements is the logical consequence of the basic assumption of zero lateral strain which is certainly not correct. In order to counteract the consequences of this incorrect assumption the suggestion was made (see e.g. Skempton and Bjerrum, 1957) that the immediate settlements should be determined separately by using formulae resulting from the integration of Boussinesq's expressions for displacements in the half-space subjected to a concentrated surface load (see Chapter 5). In this approach the apparent Poisson's ratio $\nu_d = 0{\cdot}50$ had to be considered while the deformation modulus E or shear modulus G were to be determined by undrained triaxial or by simple unconfined compression tests.

The considerable dependence of undrained triaxial test results on the principal effective stress ratio, on the anisotropy, creep and relaxation effects has been set out and discussed previously (see Sections 14.3.1, 14.3.2). With the exception of the relaxation effect the same influences have to be considered when interpreting analogous results from the

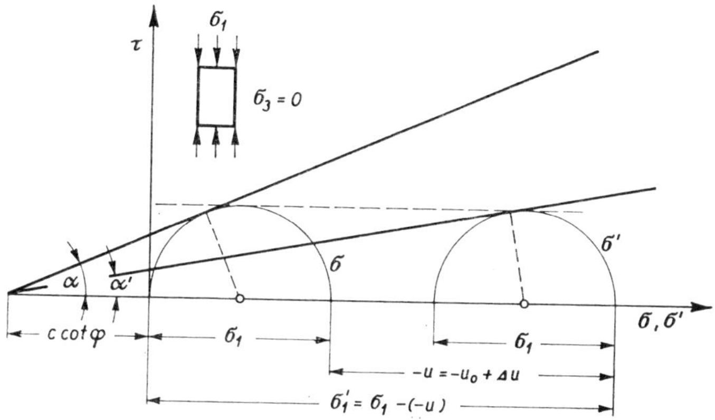

Fig. 17.9. Unconfined compression test. Mohr's circles of total (σ) and effective stresses (σ')

simple unconfined compression test. In this test a cylindrical sample is subjected to a vertical total pressure σ_1, without any lateral loading. Drainage is allowed, but, owing to the short test duration and to the low permeability of the samples, it is assumed that no volume deformation occurs during the test. Owing to the capillary forces the sample is exposed to negative pore pressures $(-u_0)$. After loading these decrease to

$$(-u_0 + \Delta u) \text{ where } \Delta u = A\sigma_1 \tag{17.24}$$

A being the pore pressure coefficient. Now, as stated before (see Sections 13.3, 13.4) the triaxial distortional strains governing the shear modulus G increase with increasing slope α' (see Fig. 17.9). Thus, although the principal distortional stresses might be equivalent to the average stress state in the subsoil, the results would not reflect at all well the role of the mean principal stress, i.e. of the slope α', on the distortional deformations. In most cases the slope α' will be greater than in natural conditions and consequently the resulting distortional deformation overestimated.

The superposition of the immediate settlements obtained by the method explained above and the consolidation settlements according to the assumption of zero lateral strain requires the association of two different assumptions which are not compatible. The resulting end settlements are expected to be overestimated.

As mentioned before, the initial pore pressure equals the total vertical stress increase according to the zero lateral strain assumption. Considering the actual stress state obtained by applying the theory of elasticity the pore pressures are governed by Skempton's equation

$$\Delta u = \Delta\sigma_3 + A(\Delta\sigma_1 - \Delta\sigma_3) \tag{14.56}$$

which can be written in the form

$$\frac{\Delta u}{\Delta\sigma_1} = A + (1 - A)\frac{\Delta\sigma_3}{\Delta\sigma_1}. \tag{17.25}$$

Skempton and Bjerrum (1957) suggested that the pore pressures Δu be subtracted from the additional vertical pressures $\Delta\sigma_1 \equiv \Delta\sigma_v$ when applying oedometer test results to the settlement computation as explained in the previous section. The settlements obtained by the customary procedure (ρ_{oed}) are reduced by the coefficient

$$\mu = \frac{\rho_c}{\rho_{oed}} = A + \alpha(1 - A) \tag{17.26}$$

where

$$\alpha = \frac{\int_0^z \Delta\sigma_3 \, dz}{\int_0^z \Delta\sigma_1 \, dz}; \tag{17.26'}$$

the pore pressure coefficient may be chosen according to the soil character (see Table 17.I), and the mean principal stress ratio α according to the loaded area. For circular and rectangular areas Skempton and Bjerrum have presented the mean values given in Table 17.II.

Table 17.I

Type of clay	*A*
Very sensitive soft clays	>1
Normally consolidated clays	0·50–1
Overconsolidated clays	0·25–0·50
Heavily overconsolidated sandy clays	0–0·25

Skempton and Bjerrum (1957) compared the settlements resulting from the immediate settlements obtained in the manner explained previously added to the reduced settlements computed on the basis of oedometer tests and found satisfactory agreement with the measured values. The

Table 17.II

Values of α in the equation $\mu = A + \alpha(1 - A)$

z/b	*Circular footing*	*Strip footing*
0	1·00	1·00
0·25	0·67	0·74
0·50	0·50	0·53
1	0·38	0·37
2	0·30	0·26
4	0·28	0·20
10	0·26	0·14
∞	0·25	0

incompatibility of the assumptions made in this procedure appears, however, in discontinuities in the void ratios when applying the method for successive load steps (Šuklje, 1963-b). In the two-step loading scheme presented in Fig. 17.10 the total vertical stress increase $\Delta\sigma_{1\text{II}}$ requires, from the oedometer test data, the decrease Δe_{I} of the void ratio and the final ratio $e_{\text{I}} = e_a - \Delta e_{\text{I}}$. The procedure suggested by Skempton and

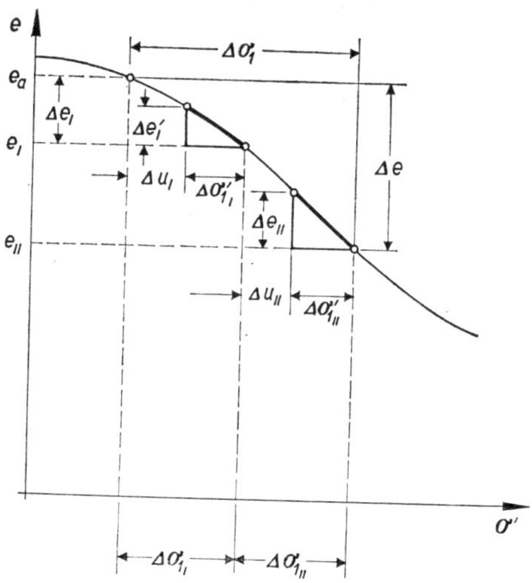

Fig. 17.10. Interpretation of the effect of the reduction of computed settlements by the coefficient $\mu = \frac{\rho_c}{\rho_{\text{oed}}}$ (Equation 17.26)

Bjerrum leads to the final void ratio $e_b' = e_a - \Delta e_{\mathrm{I}}' > e_{\mathrm{I}}$. When computing, following the same procedure, the additional settlement produced by the second load step $\Delta\sigma_{1\,\mathrm{II}}$, it is necessary to start with the void ratio e_{I}. The initial void ratio resulting from the procedure treated above should be e_b'.

17.3.3 Settlement Analysis Based on Oedometer Tests and on Undrained Triaxial Tests

An improved settlement analysis combining oedometer test and undrained triaxial test data was suggested by Lambe (1964). In his analysis the stress paths are presented in the coordinate system (p',q') with the following definition of the coordinates:

$$p' = \frac{\sigma_1' + \sigma_3'}{2} \tag{17.27}$$

$$q' = \frac{\sigma_1' - \sigma_3'}{2} \tag{17.28}$$

Thus the stress paths connect successive positions of the peak points of Mohr circles corresponding to the planes of maximum (σ_1) and minimum (σ_3) principal effective stresses.

Now, in accordance with Roscoe's and Vyalov's investigations (see Sections 13.3 and 13.4) the results of undrained consolidated tests corresponding to several effective stress paths can be expressed, in the system of coordinates p', q', by continuous lines of equal strain ε_1 occurring in the axis (1) of the cylindrical samples. According to experimental evidence these lines are approximately straight lines passing through the origin of the system of co-ordinates p', q'. Thus, they coincide with the straight lines

$$\tan \alpha' = \frac{q'}{p'} = \frac{1 - K}{1 + K} \tag{17.29}$$

connecting the points (p',q') of equal principal stress ratio K:

$$K = \frac{\sigma_3'}{\sigma_1'} \tag{17.30}$$

The straight line

$$\tan \alpha_0 = \frac{1 - K_0}{1 + K_0} \tag{17.31}$$

corresponding to the at-rest earth pressure coefficient K_0 expresses the effective stress state of oedometer tests satisfying the condition of zero lateral strain. The coefficient K_0 can be found either by triaxial tests during which the diameter of the sample is kept constant or in oedometers equipped with peripheral strain-gauges (e.g. Tan's consolidometers—Tan, 1957).

In Fig. 17.11 the straight line K_0 corresponds to the at-rest pressure state, the straight line K_f to the rupture state and the dotted lines to states of equal undrained distortional strain. The initial point A of the

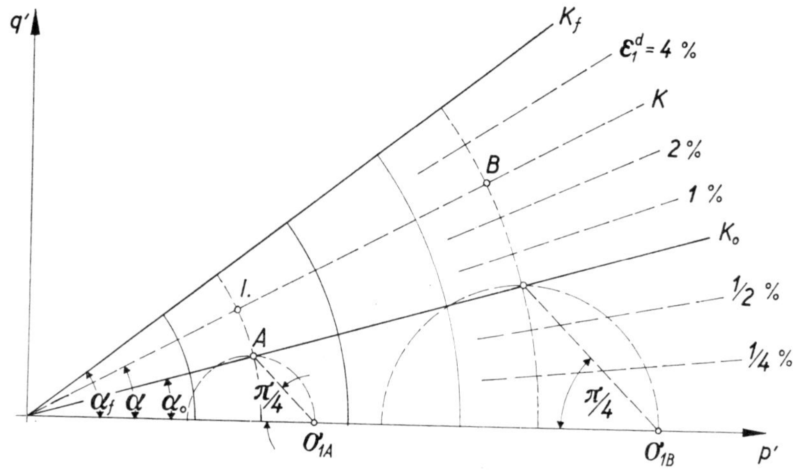

Fig. 17.11. Stress paths according to Lambe (1964)

stress path lies on the K_0 line while line K has been drawn through the final point of the stress path B.

The effective stress path between the initial and end points A and B depends on the speed of the load increase, on the pore pressure coefficient and on the permeability of the soil. It will be assumed that the load application is slow and that the effective stress path does not cross the line K. The path between A and B will follow first the undrained test path A–I and later the K-line between I and B. The distortional settlement corresponding to the component A–I of the path is given by the difference in strain between the lines K_0 and K ($\Delta\varepsilon_1{}^d = 2{\cdot}6 - 0{\cdot}7 = 1{\cdot}9$ per cent in the case presented in Fig. 17.11). The axial strain $\varepsilon_1{}^c$ corresponding to the path component I–B occurs under drained conditions and can be expressed, in triaxial test conditions, by the relation

$$\varepsilon_1{}^c = \frac{\Delta\sigma_1}{E} - 2\nu\,\frac{\Delta\sigma_3}{E} = \frac{\Delta\sigma_1}{E}(1 - 2\nu K) \tag{17.32}$$

while the lateral strain is

$$\varepsilon_3{}^c = \frac{\Delta\sigma_3}{E}(1 - \nu) - \nu\frac{\Delta\sigma_1}{E} = \frac{\Delta\sigma_1}{E}[(1 - \nu)K - \nu] \qquad (17.33)$$

The stress differences $\Delta\sigma_1$ and $\Delta\sigma_3$ are related to the stress path *I–B* (Fig. 17.11). If the linear strains are small, then the volume strain can be considered equal to the sum of the linear strains:

$$\varepsilon_v = \varepsilon_1{}^c + 2\varepsilon_3{}^c = \frac{\Delta\sigma_1}{E}(1 + 2K)(1 - 2\nu) \qquad (17.34)$$

Thus:

$$\frac{\varepsilon_1{}^c}{\varepsilon_v} = \frac{1 - 2\nu K}{(1 - 2\nu)(1 + 2K)} \qquad (17.35)$$

Under zero lateral strain conditions $\varepsilon_1 = \varepsilon_v$, $K = K_0$ and Equation (17.35) yields:

$$\nu = \frac{K_0}{1 + K_0} \qquad (17.36)$$

Inserting this expression for ν into equation (17.35) gives:

$$\frac{\varepsilon_1{}^c}{\varepsilon_v} = \frac{1 + K_0 - 2KK_0}{(1 - K_0)(1 + 2K)} \qquad (17.37)$$

The volume strain ε_v equals the axial strain ε_1 of the oedometer test corresponding to the stress path *A–C* (Fig. 17.11). K_0 can be obtained by the appropriate triaxial test (at $\varepsilon_3 = 0$) and the principal stress ratio K can be obtained from the slope of the line K through the point B:

$$K = \frac{1 - \tan\alpha}{1 + \tan\alpha} \qquad (17.38)$$

In this way the axial strain $\varepsilon_1{}^c$ can be determined from Equation (17.37). The total axial strain ε_1 is the sum of $\varepsilon_1{}^d$ and $\varepsilon_1{}^c$:

$$\varepsilon_1 = \varepsilon_1{}^d + \varepsilon_1{}^c \qquad (17.39)$$

Equation (17.37) has been deduced from the elastic conditions of isotropic soils. Neither the creep nor the anisotropy effects have been considered. As, furthermore, the form of the stress path between the points *A* and *B* was chosen arbitrarily, the procedure suggested by Lambe is only as a rough approach to real strain conditions.

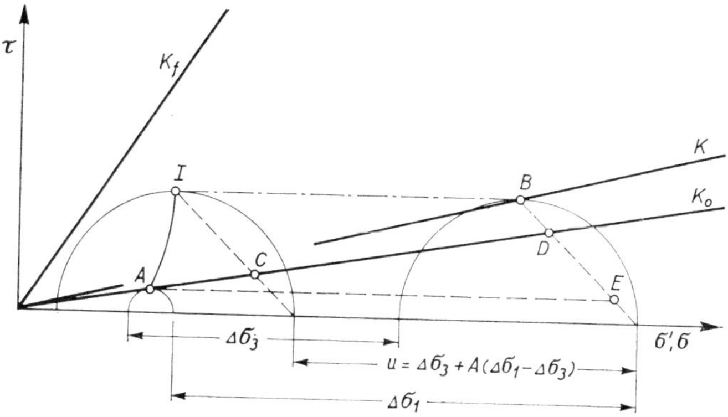

Fig. 17.12. Stress paths corresponding to simplified settlement analysis listed in Table 17.III. (According to Lambe, 1964)

Anyhow the procedure explained above has been deduced only for radial symmetric stress conditions. For plane strain and general stress state conditions the procedure should be correspondingly modified.

A better approach to real strains can be obtained by applying the alternative procedure by Lambe (1964). Following this suggestion the triaxial test will be strictly accorded a given effective stress path. In the settlement computation the resulting total strain ε_1 should be considered. For practical settlement computation this procedure can be recommended in the case of special uniform stress fields. Generally a great number of triaxial tests corresponding to stress states at various points of the subsoil

Table 17.III

	Method	*Stress Path* *Immediate settlements*	*Consolidation settlements*	*Initial pore pressures*
a	Based on oedometer tests only (Section 17.3.1) .	0	A–D	A–E
b	Based on oedometer tests and undrained triaxial tests (Section 17.3.2) .	A–I	A–D	A–E or I–B
c	As under b by using the coefficient of reduction (Skempton and Bjerrum, Section 17.3.2) . .	A–I	C–D	I–B
d	Lambe (1964) (Section 17.3.3) . . .	A–I	I–B	I–B

would be required. Yet, the justification of the chosen stress path remains questionable and the consideration of relaxation, creep and filtration effects open.

Lambe's method of presenting the stress paths allows an interesting comparison with the simplified settlement analyses which were dealt with in Section 17.3. In the case of a sudden load application (*I–B* is then a horizontal line) the comparison is shown, after Lambe (1964), in Fig. 17.12 and Table 17.III.

Part E

LONG-TERM STABILITY OF SLOPES

18

Long-Term Strength of Soils

18.1 GENERAL DISCUSSION OF THE TERMS 'STRENGTH' AND 'STABILITY' OF SOILS

The main problems of soil mechanics are divided into 'settlement' and 'stability' problems. The former are considered to be governed by stress–strain and strain–time relationships* and the latter by the 'shear-strength' of soils.

In the author's opinion there is a close bond between the two groups of problems. The deciding criteria in both groups are the admissible displacement in a given time, the admissible rate of displacement at this time and the corresponding acceleration of the displacement speed. In parts C and D procedures have been presented and discussed expressing general stress–strain and stress–strain–time relationships ranging from the hydrostatic stress-state to the limiting stress-state at which the positive acceleration of the strain speed leads to 'failure' of the soil body.

The only reason for distinguishing between the two groups of problems is our inability to determine the stress states corresponding to non-linear stress–strain relationships and influenced by heterogeneity, various geometrical shapes and various loads of slopes, earth embankments, tunnels and retaining structures. When computing the settlements of a foundation placed on a soil with a more or less horizontal surface, the theory of elasticity is commonly considered to give a satisfactory approximation to the stress state. In such cases the procedure (Fröhlich, 1934), in which the 'stability' criterion is based on the same stress state, seems to be quite justified; yet, the safety factor against failure along continuous rupture surfaces cannot be expressed by means of this procedure.

The available 'elastic' solutions, however, cannot be applied in order to obtain an approximate and satisfactory picture of the stress states in arbitrarily-shaped slopes, dams and the backfills of retaining structures.

* Their experimental and theoretical bases have been dealt with in parts B, C and D of this book.

When computing the safety of such earth masses against failure, the stress states corresponding to the simultaneous mobilization of the same degree of soil strength and satisfying the equilibrium as well as the boundary kinematic conditions are normally considered. The methods available from the theory of plasticity based on Kötter's differential equations of plastic equilibrium have been presented in Chapter 7. Depending on the kinematic conditions, the simultaneous mobilization of the same degree of strength of the soil occurs either in rupture zones or rupture strips or along isolated rupture lines (see Section 7.3). In the 'stability' analyses commonly used the selected rupture lines have either an arbitrary shape or a shape according to the stratigraphy or topography of the slope or dam. The degree of mobilization of soil strength along such potential rupture lines is investigated which satisfies the equilibrium conditions for the mass above the rupture lines. When applying such an approximate procedure the maximum degree of mobilization obtained for the most unfavourable 'rupture line' among all lines investigated, is considered to be the most critical.

Kötter's equations and all the methods of the theory of plasticity based on them which are treated in Chapter 7 involve the Mohr–Coulomb failure criterion which has been dealt with in Chapter 6. This criterion can be applied also with shear parameters c' and φ' reduced by a factor of safety F:

$$c_m' = \frac{c'}{F}, \quad \varphi_m' = \arc\tan\frac{\tan\varphi'}{F} \tag{18.1}$$

The application of the reduced shear parameters c_m' and φ_m' can be interpreted in two different ways—

(*a*) Failure is proved not to be feasible because the envelope of the limiting Mohr stress circles is F-times steeper than the envelope expressing the stress states in accordance with Kötter's equations.

(*b*) The distortional deformations, at the stress states corresponding to the simultaneous mobilization of the same degree of soil strength, are proved not to exceed certain values.

As shown in Sections 13.3.1.1 and 13.4.1, the straight lines issuing from the origin of the coordinate system (σ',τ) or for over-consolidated cohesive soils, from the point

$$\left(-p_0 = -\frac{c'}{\tan\varphi'}, \quad 0\right)$$

on the σ'-axis, coincide approximately with the lines of equal distortional strain occurring in undrained conditions. When the stress states appearing immediately after a sudden loading change are decisive for the degree of the stability of the slope, equal distortional strains correspond to the resulting stress state along the potential failure surface. These 'plastic' stress states certainly do not represent the real stress states. Nevertheless, the distortional strains tend to become uniform and consequently the stress states tend to approach the 'plastic' region the more the greater the degree of strength mobilization.

The Mohr envelope to the critical stress circles can be obtained by carrying out appropriate triaxial tests. Several presentations of triaxial consolidation test results have been given in Sections 13.3, 13.7, 16.3, 17.1 and 17.3.3, the limiting stress states being present in all of them. In the distortional stress-spherical stress planes, the lines of equal undrained distortional strains have commonly been assumed to be straight. Although different systems of presenting the stress–strain relationships have been used, lines of equal ratios $\bar{\sigma}_3'/\bar{\sigma}_1'$ correspond to equal undrained distortional strains in every system; $\bar{\sigma}_3'$ and $\bar{\sigma}_1'$ denote the equivalent stresses:

$$\left.\begin{aligned}\bar{\sigma}_3' &= \sigma_3' + p\\ \bar{\sigma}_1' &= \sigma_1' + p\end{aligned}\right\} \qquad (18.2)$$

(see Fig. 18.1). Thus such straight lines correspond also to the envelopes of the limiting Mohr circles reduced by various 'factors of safety' F, with the limit envelope ($F = 1$) included.

The classical method of 'shear strength' testing is the Krey–Terzaghi 'direct shear test' using square test specimens. Similar 'direct shear' of ring-shaped test specimens can be made for the torsion shear apparatus as constructed by Krey and Hvorslev (1937) and Tiedemann (1937). The stress states in such 'direct shear' apparatuses are not clearly defined

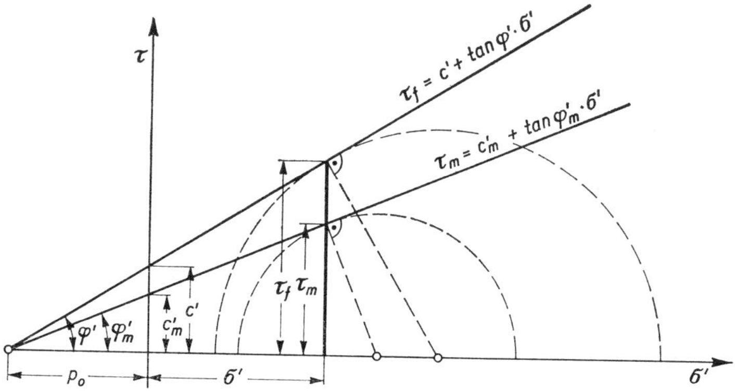

Fig. 18.1. Limit and mobilized strength characteristics of soils

and do not permit the Mohr circles to be drawn with sufficient reliability. The horizontal cross-section separating the lower fixed and the upper movable parts of the apparatus is considered to be the rupture surface and the average normal and shear stresses related to this area the two components of the mobilized strength.

Fig. 18.2 shows the pattern of internal deformation in two direct shear tests carried out and presented by Hvorslev (1960). (The internal deformations in the test specimens were delineated by colouring vertical strips

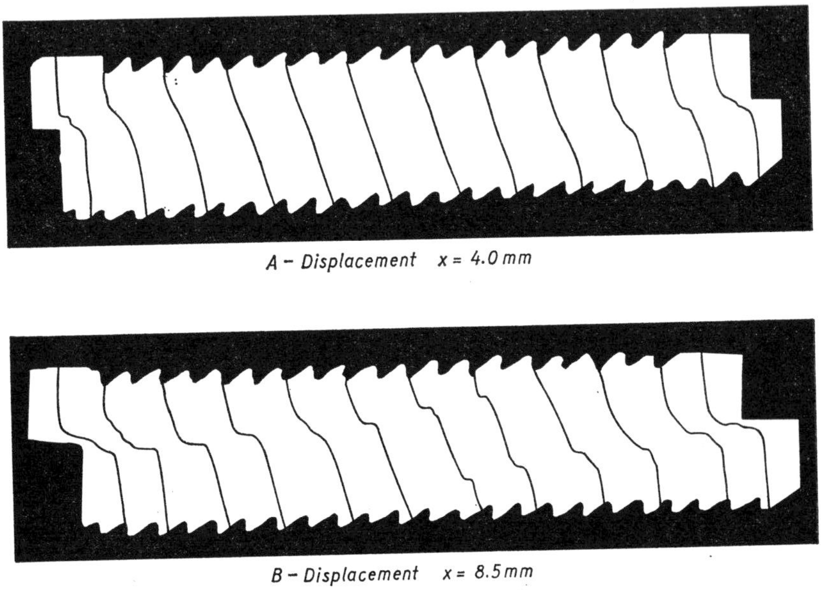

Fig. 18.2. Internal deformations in directs hear tests. (After Hvorslev, 1960)

during consolidation and then slicing and drying the test specimens after completion of the shear test.) The two graphs prove that the initial failure displacement of 3·5 mm did not form a surface of discontinuity along the 'shear surface' between the movable and the fixed parts of the shear box. The shear displacements have the character of continuous angular deformations and at this early stage of failure it is not possible to define the failure surface. (A similar experiment made by the author (Šuklje, 1962) in ring shear apparatuses leads to the same conclusion.) As the complete stress state is not known, the test results do not permit either the envelope of the Mohr critical stress circles or the line connecting the end points of the stress vectors in the surface of failure to be drawn. Nevertheless, in common testing practice the line connecting the end points of the stress vectors related to the horizontal 'shear surface' is considered to be the 'shear strength line' (Fig. 18.3). It is improbable (see Kjellman, 1954) that this line coincides with the envelope of critical Mohr circles. If it is assumed that the normal effective pressure in the

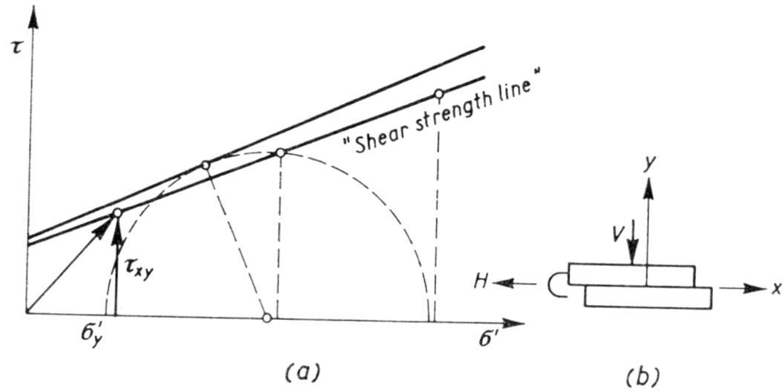

Fig. 18.3. 'Shear strength line' and the limit envelope of Mohr's circles

horizontal surface σ_y' equals the average value of the principal effective stresses σ_1' and σ_3' acting in the vertical *x–y* plane (Fig. 18.4).

$$\sigma_y' = \tfrac{1}{2}(\sigma_1' + \sigma_3')$$

and that the p_y' stress line connecting the end points of the stress vectors related to the horizontal surface with the normal in the *y*-direction

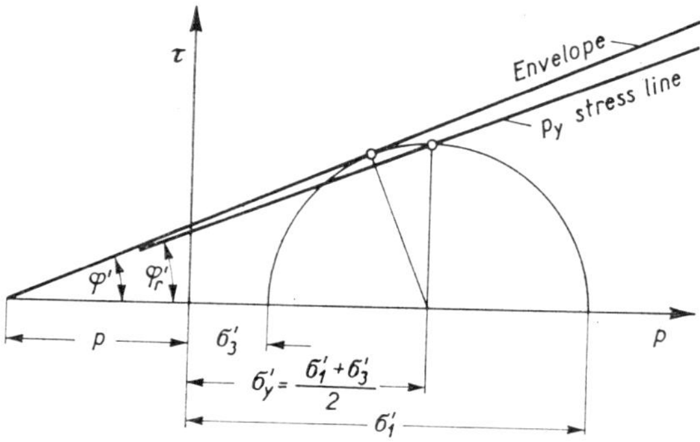

Fig. 18.4. Shear strength line (p_y stress line) and the limit envelope corresponding to the assumption $\sigma_y' = \dfrac{\sigma_1' + \sigma_3'}{2}$

originates from the same point $(-p,0)$ on the σ'-axis as the envelope of limiting Mohr circles, the following relation holds:

$$\sin \varphi' = \tan \varphi_r' \tag{18.3}$$

φ' being the slope of the limiting envelope and φ_r' the slope of the p_y' stress line.

Now triaxial experimental evidence (see Hvorslev, 1960) proves that due to anisotropy effects the failure surface having the slope α_m (Fig. 18.5)

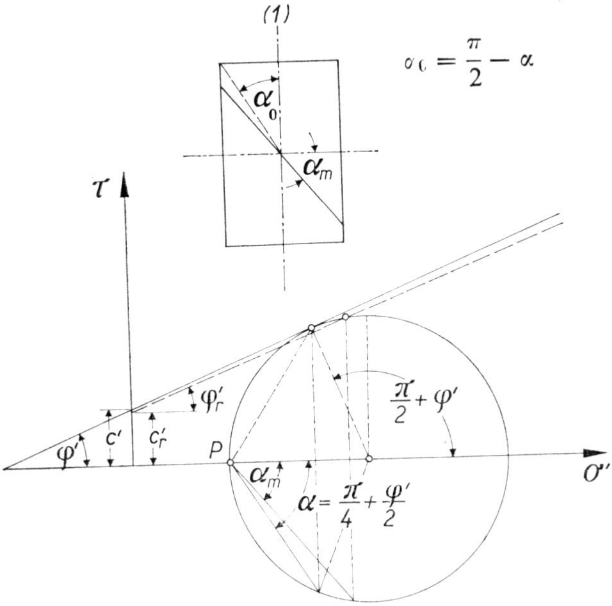

Fig. 18.5. Shear strength line and limit envelope of Mohr's circles of anisotropic triaxial samples. (According to Hvorslev, 1960)

(i.e. the surface of discontinuous displacements, when formed at failure), does not always coincide with the surface corresponding to the point of contact between the envelope and the stress circle of slope α:

$$\alpha = \frac{\pi}{4} + \frac{\varphi'}{2}$$

Thus the 'shear strength line' having slope φ_r' and intercept c_r' on the shear stress axis (Fig. 18.5)

$$\tau_f' = c_r' + \tan \varphi_r' \,.\, \sigma' \tag{18.4}$$

differs from the envelope of the limiting Mohr circles

$$\tau = c' + \tan \varphi' \,.\, \sigma' \tag{18.5}$$

When applying the methods of stability analysis based on Kötter's equations the strength parameters c' and φ' must undoubtedly be considered. On the other hand, it seems that the shear parameters c_r' and φ_r' ought to be introduced into the approximate stability analyses based on arbitrarily chosen potential rupture lines. Now, the strength characteristics c' and φ' as well as the shear characteristics c_r' and φ_r' certainly change somewhat with the orientation of the axis of anisotropy with respect to the principal stress directions and thus they cannot be considered as constant throughout the failure zone or potential rupture surface. If the anisotropy effect is small neither the change of 'strength' and 'shear' characteristics with the orientation of the principal stresses nor the difference between the angles φ' and φ_r' and the cohesion intercepts c' and c_r' is important. In soils with very marked anisotropy, however, the rupture surfaces will probably follow mainly the direction of least resistance. In such cases the shear strength in that direction must be considered in the stability analysis.

In more or less homogeneous and isotropic soils the surface topographic, stratigraphic and kinematic conditions as well as the loading conditions (see Section 20.2) can lead to ruptures along very shallow rupture zones similar to those appearing in a direct shear apparatus. After the angular deformations in these shallow zones have exceeded certain limiting values, a surface of discontinuity appears; further displacements occur mainly only along this 'shear surface.' In pre-consolidated soils, owing to the destruction of the bonds between the particles, to the parallel orientation of the particles and to dilatation effects, the initial strength of the soil is reduced along these surfaces of discontinuity. Further resistance is governed solely by the remaining shear strength along them.

In the author's opinion the use of the term 'shear strength' can be fully justified only in such circumstances. If, however, the failure conditions are governed by the parameters c' and φ' of the critical Mohr–Coulomb envelope, the general term 'strength' of the soil is preferable. The strength may be defined by the stress state at which the displacement speed starts to increase. In common triaxial testing it is often defined by a certain limiting axial strain. In neither case a shear surface of discontinuity needs to appear.

18.2 TIME-DEPENDENT STRENGTH COMPONENTS

The Proceedings of the Research Conference on the Shear Strength of Cohesive Soils sponsored by the Soil Mechanics and Foundations Division, ASCE (issued by the University of Colorado, Boulder, Colorado, June 1960) contain an extremely valuable record of up-to-date research on and knowledge of the strength properties of soils. Excellent reviews of the subject are given in final reports prepared by Casagrande and Wilson (Testing Equipment, Techniques and Errors), by Johnson and Whitman (Shear Strength of Saturated, Remoulded Clays), by Peck and Lowe (Shear Strength of Undisturbed Cohesive Soils), by Seed and Hirschfeld (Shear Strength of Compacted Cohesive Soils) and by Rutledge and Walker (Problems Associated with Practical Applications of Shear Strength Data). This record was largely supplemented by the thorough Conference paper of Hvorslev (Physical Components of the Shear Strength of Saturated Clays) including, in a separate section, investigations of the rheological component of the strength and the rheological properties. The rheological component, c_v, has been defined as the transient part of the effective cohesion component, c_e. The value of c_v converges to zero with increasing time or decreasing rate of deformation, whereas c_e at the same time approaches an ultimate value, c_u, which may be called the 'ultimate cohesion component.' Thus for any given test duration or rate of deformation the following relation exists:

$$c_e = c_u + c_v \tag{18.6}$$

The remaining part of the total effective shear resistance is the effective friction component, τ_φ. It is defined by the shear strength line obtained when varying the effective normal stress with the cohesion components remaining constant; corrections are made for variations of the surface energy component, τ_d. The latter component is due to the volume change during shear and can be found by equating the surface energy to the internal energy. In direct shear tests the resulting relation is

$$\tau_d = \sigma_f' \frac{dy}{dx} \tag{18.7}$$

dy being the change in thickness of a direct shear specimen during a time interval dt at failure and dx the lateral shear displacement during the same time interval (Hvorslev, 1960). For triaxial tests the corresponding relation is (Bishop and Eldin, 1953):

$$\tau_d = \sigma_3' \frac{\dot{\varepsilon}_v}{\dot{\varepsilon}_1} \tag{18.8}$$

where $\dot{\varepsilon}_v$ is the rate of volume strain and $\dot{\varepsilon}_1$ the rate of axial strain at failure.

Thus the total shear strength τ_f obtained in tests and represented, for constant void ratio conditions, by the shear strength line AB in Fig. 18.6, can be considered to be composed of four parts at any time t:

$$\tau_f = \tau_\varphi + \tau_d + c_u + c_v \tag{18.9}$$

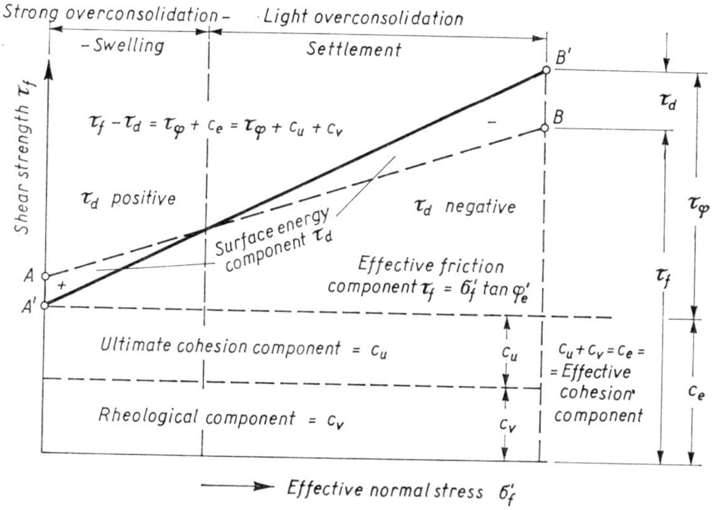

Fig. 18.6. Components of shear strength at constant void ratio. (According to Hvorslev (1960))

By subtracting the surface energy component, τ_d (also called the volume change or the dilatation component) from the observed shear strength τ_f, the so-called effective shear strength $(\tau_f)_e$ is obtained:

$$(\tau_f)_e = \tau_f - \tau_d = \tau_\varphi + c_u + c_v \tag{18.10}$$

expressed, in Fig. 18.6, by the corrected shear line $A'B'$. The general effect of the consideration of the surface energy component is a slight increase in inclination of the shear strength line and a small decrease in its zero intercept. The value of τ_d decreases with increasing test duration and probably reaches zero for tests of very long duration. It is also zero for all undrained or constant volume tests.

The effective friction component τ_φ is usually expressed by the equation

$$\tau_\varphi = \sigma_f' \tan \varphi_e' = (\sigma_f - u) \tan \varphi_e' \tag{18.11}$$

where the pore-water pressure, u, may be a time-dependent variable. According to Hvorslev (1960), the angle φ_e' depends on the composition of the clay and possibly also on the orientation of the flaky mineral particles or zones of stratification with respect to the surface of failure, but it is considered to be independent of the void ratio. Furthermore, according to the same author, it is assumed that φ_e' is independent of time and the rate of deformation. This concept can be a subject of dispute and its full justification requires further investigation.

In the following sections of this chapter the time-dependence of the strength of soil media will be dealt with and several results presented in the previous chapters on consolidation will be referred to.

18.3 CREEP OF SOIL MEDIA PRECEDING FAILURE

18.3.1 Triaxial Tests

When experimental data concerning the creep displacements of unconfined triaxial specimens were presented in Chapters 13, 14 and 17 stress states approaching the failure Mohr envelope were included. The volume deformation was obtained by immediate measurements of the volume change during consolidation and the distortional axial strain was found by subtracting one-third of the volume strain from the observed total axial strain. During the laboratory observations the volume strains usually increased with the logarithm of time. The distortional consolidation curves presented showed a similar logarithmic dependence on time (see Figs. 14.4 and 14.5). (In confined oedometer tests the linear proportionality between spherical and distortional strains is the postulate of the zero lateral strain condition.)

The development of distortional strains discussed in Section 14.4.3 was partly attributed to the decrease in void ratio during secondary consolidation. A similar development of the distortional strain velocity was also observed for undrained triaxial tests at constant volume on saturated soils (Haefeli and Schaerer, 1946); in that case it could be interpreted only by thixotropic hardening effects (see Section 14.4.3). The logarithmic strain–time relationship is included as a special case ($\alpha = 1$) in Goldstein's general equation (Goldstein, 1958) applicable for constant volume unconfined compression tests:

$$\dot{e} = K\sigma^n t^{-\alpha} \tag{18.12}$$

$\dot{e}$ being the axial strain velocity, σ the axial stress, t the time and K, n, α soil parameters (Goldstein, 1958).

Tan's pure torsion tests executed in torsion plastometers (see Section 14.4.1) prove that hardening effects do not appear in every kind of soil. In the clay specimens of high compressibility tested by Tan in torsion plastometers (Tan, 1954) the distortional displacements started developing at constant speed as soon as the limit of recoverable elastic strains had been exceeded; this limit was less than a quarter of the torsion strength. At the upper limit of the Bingham state in which the displacements occur at constant speed, which was quite large for the tested clay, the distortional strains started to increase with positive acceleration and failure occurred.

The large extent of the Bingham state observed in these undrained tests can at least partly be attributed to increasing pore pressures due to the relaxation effects which were treated in Section 14.3.2. For this reason the results of undrained tests must be interpreted with care and the anisotropy effects (see Section 14.3.1) also taken into account.

18.3.2 Direct Shear Tests

At the stress states preceding rupture and in the early stages of rupture drained direct shear tests operate under similar conditions of changing stress and strain as drained triaxial tests. Thus it is not surprising that the observed increase of shear displacements with time obeis analogous laws. The first systematic investigation of the development of these displacements was made by Hvorslev (1937). The test was performed with the torsion shear apparatus and the test specimen was strongly over-consolidated in order to avoid a gradual decrease in void ratio. The total horizontal and vertical displacements for each load increment are shown in Fig. 18.7 (A) (reproduced after Hvorslev, 1960). The relation between the shear strain, γ, and the twist, Θ, of the torsion shear specimen is

$$\gamma = \Theta \frac{2}{3} \frac{R_2}{H_e} \frac{1 - n^3}{1 - n^2} = 4{\cdot}64\Theta \text{ radians}$$

where $n = R_1/R_2$, R_1 and R_2 are the internal and external radii, and H_e is the effective height of the test specimen ($H_e \cong 1$ cm). The angles of twist Θ for each load increment and in the period from 100 to 240 hours after load application are shown in Figs. 18.7 (B and C). The diagram for $\tau_a = 0{\cdot}4$ kp/cm^2 shows the plastic deformation increasing linearly with the logarithm of time. The diagrams for $\tau_a = 0{\cdot}2$ and $0{\cdot}3$ kp/ cm^2 exhibit a slight curvature after a period of 220 hours and the deformation for the stress $\tau_a = 0{\cdot}1$ kp/cm^2 ceased at the same time. The corresponding rates of twist at 120 and 240 hours are presented in Fig. 18.7 (D).

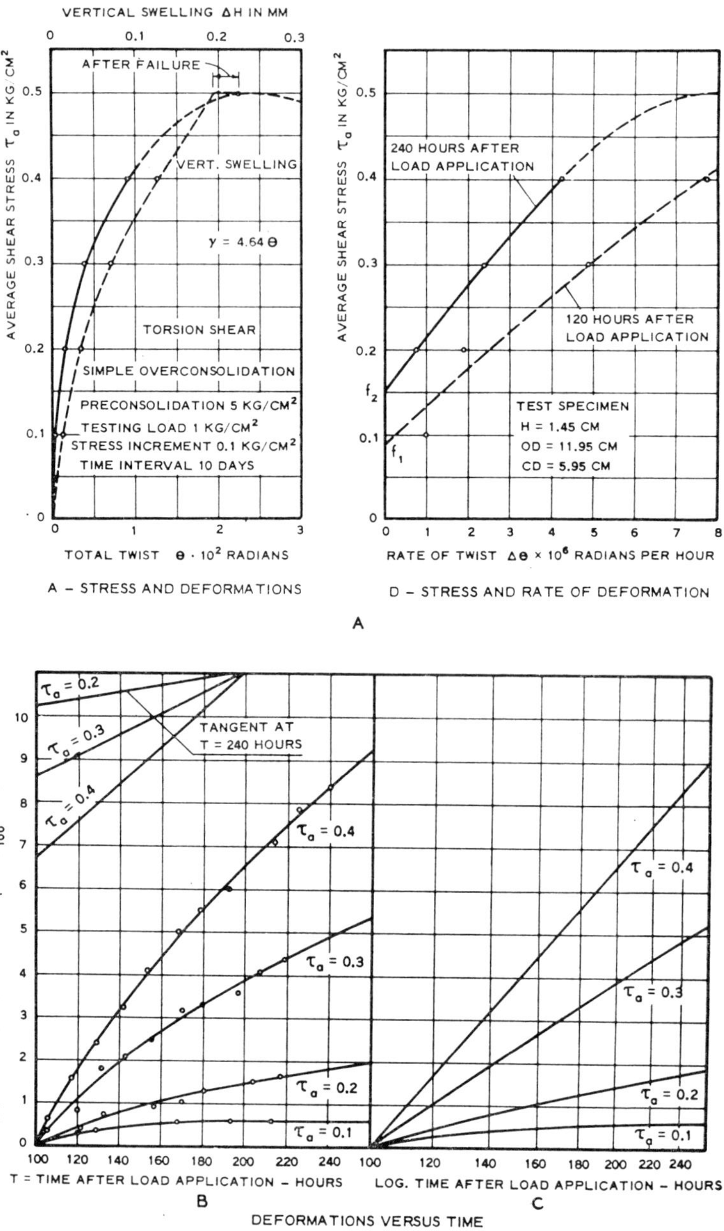

Fig. 18.7. Plastic deformations in torsion shear tests of a strongly overconsolidated clay. (After Hvorslev, 1960)

Similar drained direct shear tests made by Geuze (1948, 1960) and corroborated by field observations also yielded a linear relation between the shear strains and the logarithm of time:

$$\gamma_t = \Delta\tau(\gamma_d + \ddot{\gamma} \log t) \tag{18.13}$$

where γ_t is the total strain at time t, $\gamma_d\Delta\tau$ is the strain one day after placing the stress increment $\Delta\tau$, t is the time in days after placing the stress increment, and $\ddot{\gamma}\Delta\tau$ is the increase in strain for a tenfold increase of t. The rate of strain is obtained by differentiation of Equation (18.13):

$$\frac{d\gamma_t}{dt} = \Delta\tau\ \ddot{\gamma}\ \frac{2{,}3}{t} \tag{18.14}$$

Several long term creep tests of the same type were performed by S. Vidmar and the author (1961-b) with normally consolidated and pre-consolidated silty and clay specimens using torsion shear apparatus with 80/140 mm inside/outside diameters respectively. In Fig. 18.8 an example of the development of the average horizontal displacement of the movable ring is presented by plotting the displacement speed versus time. Because the effective height H_e of the specimen was about 1·5 cm, the corresponding

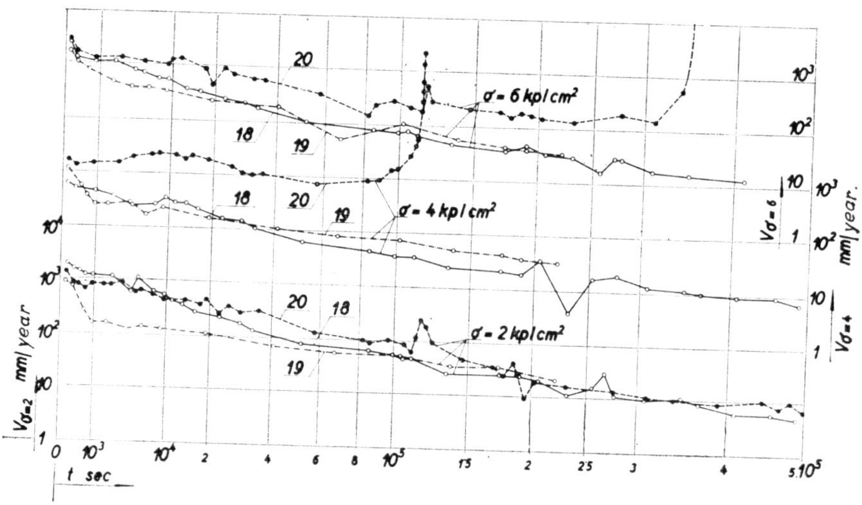

Fig. 18.8. Torsion shear tests of a remolded normally consolidated silt after Šuklje and Vidmar (1961-b): Shear displacement velocities v versus time after load application. The corresponding shear strain velocities:

$$\gamma = \frac{v}{1{\cdot}5 \text{ cm}}$$

shear strain velocities can be deduced by dividing the horizontal displacement velocities by $H_e = 1{\cdot}5$ cm. The example is for a remoulded normally consolidated silt of intermediate compressibility: $w_L = 45$ per cent, and $w_P = 11$ per cent; the axial pressures during shear were $\sigma_1 = 2, 4, 6$ kp/cm^2. The results of this and similar tests prove that in drained conditions the strain velocities of soils similar to those tested decrease even at shear stresses approaching less than 5 per cent of the failure values. The constant velocity of creep is a warning of approaching failure and of an unstable state. Any additional static or dynamic effect can be the cause of an acceleration in creep speed leading to failure. In the early stages of consolidation the renewed decrease in creep speed observed in some experiments after a period of constant or even of increased speed is due to pore-water pressure dissipation. If it appears after the pore pressures have decreased to very small residual values, it is to be attributed to thixotropic strengthening effects.

In slopes, the increase in density due to secondary consolidation depends on the degree of consolidation of the natural deposits and on their preconsolidation. The thixotropic strengthening effects can be reduced by the variation of the stress state caused by erosion, ground-water level oscillation, earthquake shocks and various constructional works. Thus, if the experimental results are to be applied immediately, the test conditions have to be carefully tailored to the natural conditions. The various experimental data and field observations must be considered and interpreted by taking account of all the possible testing errors as well as the different natural stress and deformation conditions.

18.4 LONG-TERM STRENGTH OF COHESIVE SOILS

18.4.1 Triaxial Tests

The strength of the soil can be defined either by a limiting distortional strain or by the speed at which the deformation starts to accelerate or by the appearance of fissures or surfaces of discontinuity. When applying one of the first two definitions the dependence of the long-term strength of the soil on the distortional creep development cannot be discussed. The amount of published investigations of the time-dependence of the strength of cohesive soils is, however, very limited.

Casagrande and Wilson (1951) made the first extensive investigations of the long-term unconfined compressive strength of clays and clay-shales at constant water content and under sustained loads. As the pore pressures were not measured, the effect of the test duration is related to total stresses. Two types of tests were used: 'creep-strength' tests in which loads were built up quickly and maintained constant until the specimen failed; and 'long-time' compression tests in which the specimens were subjected to incremental axial loading, the elapsed time between increments of load varying for different tests. The effect of long-time compression and of sustained loading on the compressive strengths of the nine soils tested is summarized in Figs. 18.9 and 18.10 (reproduced after Casagrande and Wilson, 1951). Fig. 18.9 is a plot of the strength ratio against the duration of sustained load for six undisturbed saturated soils in creep-strength tests, and Fig. 18.10 is a plot of the strength ratio against the time of loading for long-time unconfined compression tests on four soils. Strength ratio is defined as the ratio of the compressive strength obtained in an actual test to the estimated compressive strength corresponding to a normal rate of loading. In creep-strength tests a 'normal' time to failure of 1 minute was used, and for long-time tests a 'normal' time of loading of 10 minutes.

In creep-strength tests at constant water content the strength was reduced to values of between 80 and 40 per cent of its normal value in about 30 days. For long-time unconfined tests, one undisturbed sandy clay and two compacted soils showed a substantial increase of strength with times of loading greater than about 1 day. These three soils were not fully saturated and it is probable that the increase in strength was accompanied by a decrease in void ratio at constant water content.

In Casagrande and Wilson's tests the pore pressures were not measured and it is very probable that, due to the relaxation effects, they increased with time. Thus the effective stresses decreased and the corresponding Mohr stress circles approached the failure envelope faster than they would have done under natural drained conditions.

According to the plots in Fig. 18.9 the strength of saturated soils decreases with the logarithm of time and can be expressed analytically by

$$q_t = q_a - \ddot{q} \log \frac{t}{t_a} = q_a \left(1 - \rho_a \log \frac{t}{t_a}\right) \tag{18.15}$$

or as suggested by Goldstein (1957)

$$q_t = \ddot{q} \log \frac{t_0}{t} \tag{18.16}$$

where t_0 is the time corresponding to zero strength, and $\ddot{q}$ is the decrease

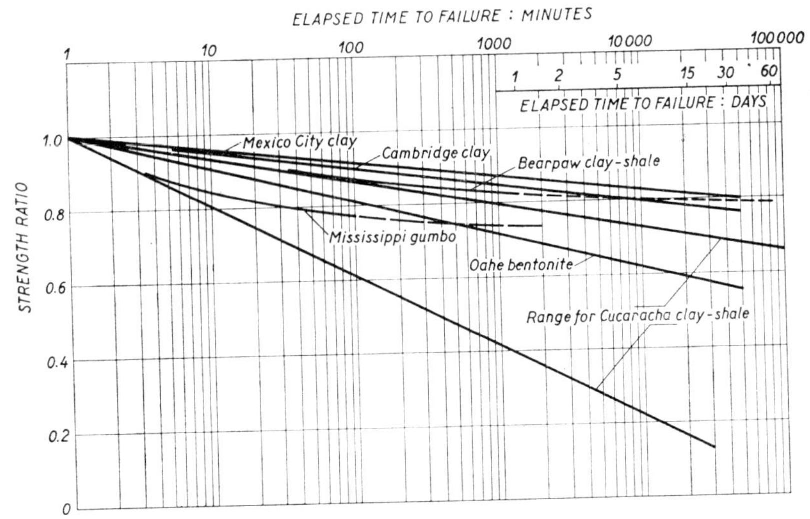

Fig. 18.9. Strength ratio versus elapsed time to failure for six undisturbed fully saturated soils in creep-strength tests. (After Casagrande and Wilson, 1951)

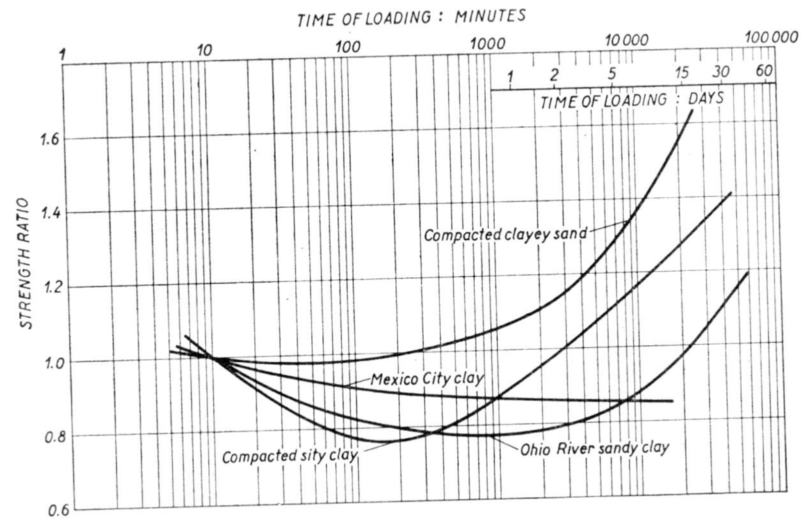

Fig. 18.10. Strength ratio versus time of loading in long-time unconfined compression tests on four soils. (After Casagrande and Wilson, 1951)

in strength for a logarithmic cycle of time. The 'coefficient of rheological decrease in strength' ρ_a (Hvorslev, 1960)

$$\rho_a = \frac{\ddot{q}}{q_a} \tag{18.17}$$

varies with the reference strength q_a and the time t_a. Casagrande and Wilson (1949) use the one-minute strength as a reference value:

$$\rho_1 = \frac{\ddot{q}}{q_1} \tag{18.18}$$

while Hvorslev (1960) suggested that the 1000-minute strength be used as the reference strength, and that the corresponding value of the coefficient of decrease in strength be denoted by ρ_m. According to Equations (18.15) and (18.18) the relation between ρ_1 and ρ_m is

$$\rho_m = \rho_1/(1 - 3\rho_1) \tag{18.19}$$

Except for the Cucaracha clay-shale the plots in Fig. 18.9 yield values of the coefficient ρ_m from 0·05 to 0·13.

The advantage of similar long-term triaxial tests made by Bjerrum, Simons and Torblaa (1958) was in testing at various effective principal stresses and in measuring the pore pressures during the tests. In this way the influence of the strain velocity could be expressed for the whole failure envelope in terms of effective stresses. The tests were made with undisturbed specimens of a normally consolidated marine clay (Fornebu, Oslo). In the first group containing 12 series of consolidated undrained triaxial tests the axial strain rate was varied from 100 per cent to 0·0036 per cent per hour. The ratio between the strengths resulting from long-term tests and from ordinary tests of short duration was about 0·825 ($\rho_m = 0{\cdot}06$ to 0·07). However, in the group of 7 series of drained tests the strength was found to be independent of the test duration when the latter was greater than one day. The authors suggest that the decrease of the rheological component in this case was compensated for by an increase of the cohesion component caused by secondary consolidation and thus a decrease in void ratio.

Goldstein and Ter-Stepanian (1957) presented unconfined compression strength-time data for an unspecified clay which gave values of ρ_m from 0·09 to 0·13. Goldstein found that failure occurred at a constant strain and that this strain was independent of the test duration for a given soil, state of consolidation and testing procedure. Therefore, the time of failure for a given soil can be determined by extrapolating the strain-time curve for this stress to the failure strain. By interpreting the relation

(18.12) he deduced the following expression for the unconfined compression strength:

$$\sigma \equiv q_t = \sqrt[n]{\left[\frac{(e_0 - e_1)(1 - \alpha)}{K(t^{1-\alpha} - 1)}\right]}, \; \alpha \neq 1 \tag{18.20}$$

where e_0 is the axial deformation at failure and e_1 the deformation after 1 day; K, n and α are soil parameters. For $\alpha = 1$ the integration yields:

$$\sigma_t \equiv q_t = \sqrt[n]{\left(\frac{e_0 - e_1}{K \log_e t}\right)}, \quad \alpha = 1 \tag{18.21}$$

(Goldstein, 1958).

Some of the data presented by Casagrande and Wilson (1949, 1951) shows very little influence of the test duration on the failure strain, whereas some of the remainder indicates an increase in failure strain with increasing test duration. Bjerrum, Simons and Torblaa (1958) found a definite decrease in failure strain with increasing test duration. According to Goldstein (1958) this difference of results may be due to the effect of partial restoration of the intergranular bonds during the slow deformation; the intensity of this process must depend upon the physico-chemical nature of the bonds. Whitman (1960) found that the failure strain of plastic soils is nearly independent of the test duration, whereas that of few soils decreases with increasing test duration.

18.4.2 Direct Shear Tests

Hvorslev (1937, 1960) investigated the influence of the time to failure on the shear strength of clays by comparing the cohesion component of the shearing resistance from quick and slow tests performed in torsion direct shear apparatuses. The comparison has been based on Hvorslev's form of the shear law

$$\tau_f = \tan \varphi_e \, . \, \sigma_f' + \kappa \sigma_e' \tag{18.22}$$

which was presented in Chapter 6. σ_f' is the effective normal stress at failure and σ_e' the 'equivalent consolidation pressure' which is defined as the pressure at a point on the virgin branch of the consolidation diagram with ordinate e corresponding to the void ratio e occurring in the shear plane after failure; φ_e and κ are constant shear characteristics of the soil. It was assumed that the shear parameter $\tan \varphi_e$ is not affected by the test duration and that the effective pressure remains constant throughout shearing. Thus the comparison for normally consolidated specimens has been related to the cohesion parameter κ. The ratio of κ-values for

Wiener Tegel obtained in times of 600 minutes and 3 minutes to failure was found to be about 0·73. Approximately the same value was found for the ratio between the effective cohesion components $c_e' = \kappa\sigma_e'$ obtained in slow (600 min) and quick tests (3 min) on over-consolidated specimens of the same clay for effective normal stresses σ_f' at which either the surface movement of the specimens was zero or the equivalent effective pressure σ_e' had the same value in both tests. The extrapolation of the results obtained was based on the assumption that the strength decreases linearly with the logarithm of time. Consequently it should be expected that the rheological component c_v would be dissipated in a period of 10 years and the remaining 'ultimate' cohesion component c_u reduced to 32 per cent of the total cohesion c_e obtained in quick tests (3 min).

Hvorslev (1960) himself declared that the above comparison and extrapolation was speculative. They were presented in order to demonstrate the influence of the rheological component together with a method of computing it when adequate data is available. Unfortunately, this data is not extensive.

S. Vidmar and the author Šuklje, (1962) compared the results of some drained tests on remoulded normally consolidated and on remoulded slightly over-consolidated specimens of Tertiary clays and silts and did not find any noticeable influence of the test duration.

The possible influence of the fissures of stiff clays on the long-term strength will be discussed in the next section.

Generally similar influences and divergencies in the test results as those observed in drained triaxial tests are to be expected (see the previous section).

When concluding his review of the investigations available up to 1960 Hvorslev suggested that the decrease in strength with time of natural clay deposits might have a value between those obtained in undrained and drained tests, because a slight migration of pore water might cause a considerable change of the pore-water pressure in the failure zone. A migration of pore water takes place in undrained compression tests but it may not be of the same magnitude as that which occurs in natural deposits over long periods of time.

Laboratory tests in which the duration does not exceed a few days or weeks or maybe months can hardly be considered as real 'long-term' tests. Extrapolation of the results obtained from tests of shorter duration is unreliable if not corroborated with field observations and thorough analyses of the creep and failure of natural slopes. The strength parameters resulting from analyses of creep or failure conditions should be compared with the parameters obtained by laboratory tests of short and long duration. In these, various influential factors have to be considered. Some of them have already been discussed:

(*a*) The relaxation phenomena appearing in undrained tests of saturated soils under conditions which usually do not correspond to natural conditions.

(*b*) The thixotropy effects either connected with these relaxation effects or appearing independently with different intensity according to the rate of the pore-pressure increase and decrease.

(*c*) The structural disturbance or the increase of strength which depend on the displacements due to the change and oscillations in effective stresses and on the physico-chemical processes along the contact surfaces of solid particles and in the surrounding water-bonds.

The effect of displacement on the strength of the soil after failure will be the subject of the next section.

18.5 CHANGES IN SHEAR STRENGTH AFTER FAILURE

When investigating the strength of non-cohesive soils by triaxial tests and plotting the observed deviator stresses or corresponding angles of shear and the observed volume strains against the axial strain, plots similar to those shown in Fig. 18.11 will be obtained. The slope φ_m of the tangent to the Mohr circle corresponding to any axial strain ε_1 is assumed to be composed of the internal friction angle φ_f and of the angle of surface energy (dilatancy angle) φ (Bishop, 1950; Hvorslev, 1960).

$$\varphi_m = \varphi_f + \vartheta \tag{18.23}$$

According to Equation (18.8) the tangent of the dilatancy angle ϑ is

$$\tan\vartheta = \frac{\dot{\varepsilon}_v}{\dot{\varepsilon}_1} = \frac{d\varepsilon_v}{d\varepsilon_1} \tag{18.24}$$

i.e. the slope $\tan\vartheta$ can be obtained as the tangent to the plot $\varepsilon_v = \varepsilon_v(\varepsilon_1)$. Throughout the test on a loose sand the void ratio decreases and the dilatancy angle ϑ is negative (Fig. 18.11 (*a*)). When testing dense sands, the dilatancy is initially negative. The dilatancy angle ϑ becomes zero when the shear angle equals the internal friction angle φ_f (Fig. 18.11 (*b*)) and then the angle ϑ begins to increase being positive throughout the subsequent increase of axial strain. The maximum shear angle $\varphi_{\max}$ corresponds to the point of flexure (maximum slope) of the dilatancy curve $\varepsilon_v = \varepsilon_v(\varepsilon_1)$. Afterwards the mobilized shear angle φ_m decreases and becomes equal

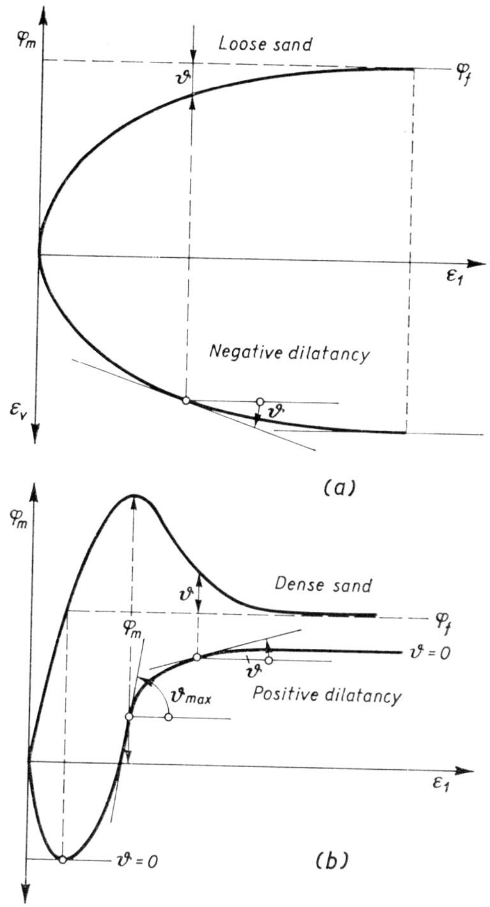

Fig. 18.11. Dilatancy effects in triaxial testing: (*a*) loose sand, (*b*) dense sand

to the pure friction angle φ_f 'after failure' when the dilatancy angle becomes zero.

The plots represented in Fig. 18.11 apply also to direct shear tests if the axial strain is substituted by the horizontal displacement (x) and the volume strain by the vertical displacement (y) (settlement or swelling of the specimen, Rowe, 1958). The slope (tangent) of the dilatancy angle can be deduced from Equation (18.7):

$$\tan \vartheta = \frac{\tau_d}{\sigma_f'} = \frac{dy}{dx} \tag{18.25}$$

When testing normally consolidated or over-consolidated cohesive soils, the effective cohesion component (see Equations 18.6 and 18.9) is a further component of the total shear strength and has to be considered when interpreting the change in strength with increasing deformation. For triaxial tests on normally consolidated clays the total shear strength, τ_f, plotted against axial strain follows curves similar to the curve $\varphi_m = \varphi_m(\varepsilon_1)$ in Fig. 18.11 (*a*); for over-consolidated soils the curves are similar to

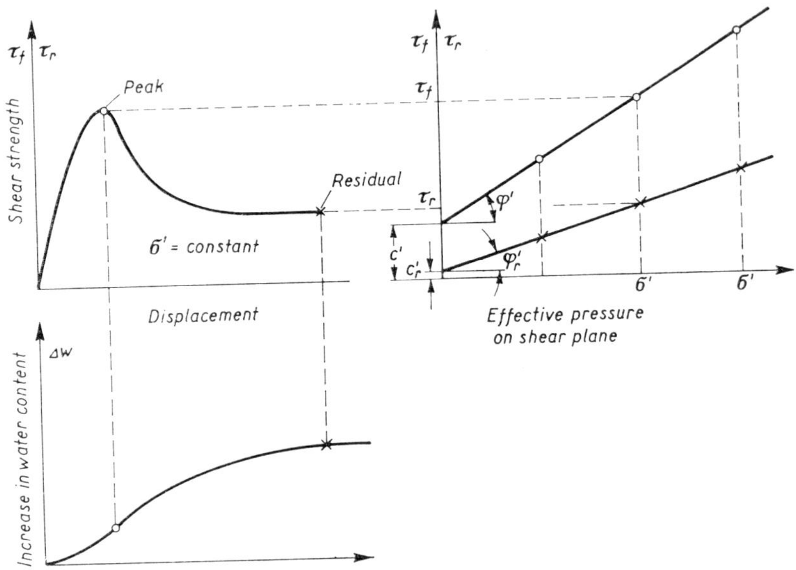

Fig. 18.12. Shear characteristics of over-consolidated clay. (After Skempton, 1964)

those presented in Fig. 18.11 (*b*). The dilatancy effect develops in a similar way in both cases. To the author's knowledge these curves have not yet been subjected to a systematic investigation to determine the character, causes and consequence of the decrease of strength with increasing axial strain. Skempton (1964), however, has made a thorough investigation of the analogous strength-displacement plots obtained in direct shear tests.

A typical plot of the shear strength versus horizontal displacement—as observed in a direct shear test on over-consolidated clays—is presented, after Skempton (1964), in Fig. 18.12. (The shear strength is defined as the total horizontal shear force divided by the area of the square-shaped specimen in the contact plane between the movable and fixed parts of the shear box.) As the clay is strained, so it builds up an increasing resistance. But, under a given effective pressure, there is a definite limit to the resistance the clay can offer, and this is the 'peak strength' τ_f. If, however, the test is continued, then it is found that as the displacement increases so the resistance or strength of the clay decreases. Ultimately a certain 'residual strength' is reached which the clay maintains even when subjected to large displacements.* From the available experimental evidence, the strength falls to the residual value after displacements beyond the peak of several centimetres. Yet, as reported by Skempton, there is field evidence to show that the strength is not less than this laboratory value (or only slightly less) even at displacements of many decimetres.

When carrying out several similar tests under different effective pressures, the peak and residual strength can be plotted against the effective pressure (see Fig. 18.12). Usually both relationships are approximately in accordance with the Coulomb law. The peak strengths can therefore be expressed by the equation:

$$\tau_f = c' + \sigma' \tan \varphi' \tag{18.26}$$

and the residual strengths by the equation

$$\tau_r = c_r' + \sigma' \tan \varphi_r' \tag{18.27}$$

The test results at present available show almost invariably that c_r' is not significantly different from zero. Thus, in moving from the peak to the residual, the cohesion intercept c' disappears. During the same process the angle of shearing resistance also decreases; in some clays by only 1° or 2°, but in others by as much as 10° (Skempton, 1964).

During the shearing process over-consolidated clays tend to expand, especially after passing the peak. According to Denissov (1965) the drop in strength from the peak value is to be attributed to increasing water content and the accompanying destruction of bonds between the solid particles. Skempton (1964) suggested that the development of thin bands or domains in which the flaky clay particles become orientated in the direction of shear is also of comparable importance.

Comparison of the strengths after failure of undisturbed or preconsolidated specimens with those of normally consolidated specimens prepared from slurry of the same clay suggests that the residual strength of a clay, under any given effective pressure, is the same whether the clay has been normally or over-consolidated (Skempton, 1964). Moreover,

* The first determinations of the residual strength were published by Tiedemann (1937) and Hvorslev (1937); the tests were made in ring shear apparatuses. The term 'residual' was introduced by Haefeli (1938). (See Skempton, 1964.)

domains of strongly orientated particles have also been observed when quite soft remoulded clay has been subjected to large strains (Astbury, 1960, cited after Skempton, 1964).

In Fig. 18.13, reproduced after Skempton (1964), the residual angles of shearing resistance of a number of normally and over-consolidated clays are plotted against the clay fraction (percentage of particles by weight, smaller than 2 μ). The angle φ_r' tends to decrease with increasing content

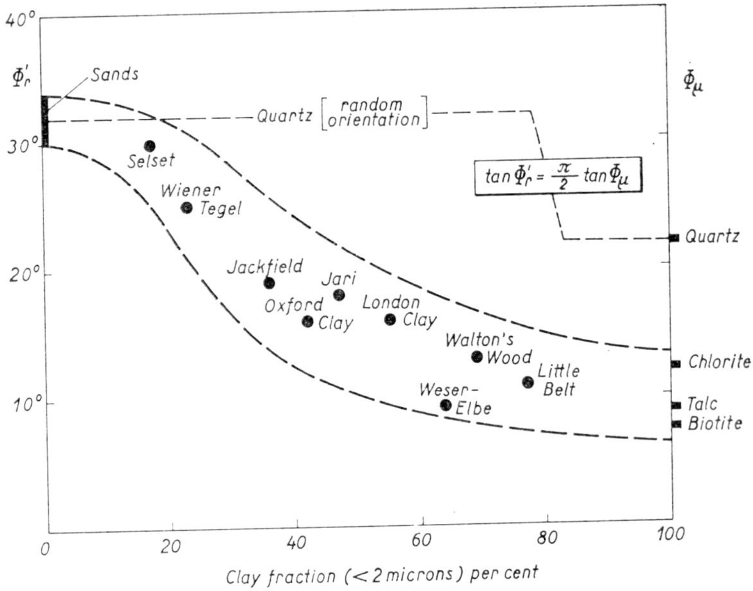

Fig. 18.13. Decrease in $\varphi_p{}^1$ with increasing clay fraction. (After Skempton, 1964)

of clay particles. The residual strength for sands and silts (without a clay fraction) equals the shear angle of spherically shaped particles in random orientation (see Caquot, 1934) and for soil consisting entirely of clay particles it is equal to the angle of friction of layer-latticed minerals.

Skempton (1964) has compared the average shear stress (τ) occurring at failure of several natural slopes and cuttings to the peak (τ_f) and residual (τ_r) shear strengths of specimens from the failure zone (surface) corresponding to the average effective normal stress. Defining the residual factor R by the equation

$$R = \frac{\tau_f - \bar{\tau}}{\tau_f - \tau_r} \qquad (18.28)$$

Skempton found that $R \simeq 0$ for a landslide of a natural slope consisting of non-fissured and unweathered boulder clay; $R \simeq 1$ for two landslides of natural slopes consisting of fissured, jointed and weathered clay (London clay and Coalport Beds respectively); and obtained residual factors ranging from 0·56 to 0·80 for three cuttings in the fissured, jointed and weathered London clay. As a result, Skempton suggests that the presence of fissures and joints can lead to progressive failure in a clay slope and, at its limit, this process can continue until the residual strength is reached. But in clays which are not fissured or jointed the decrease in strength from the peak value is small, or even negligible. Compacted clay fills, as used in embankments and earth dams, may well fall into this latter category. Finally, if a failure has already occurred, any subsequent movements of the existing slip surface will be controlled by the residual strength, no matter what type of clay is involved.

As Skempton (1964) observed: 'Fissures and other physical discontinuities may not be the only explanation of landslips in clays taking place at strengths well below the conventional peak value. The peak strength as measured in laboratory tests lasting, at the most, a few weeks might be appreciably greater than the strength which could be developed in some clays when the stresses are applied over periods of years, decades and centuries. In other words the effects of shear creep have to be considered; though at present very little quantitative information is available to assess these effects so far as the peak strength is concerned.'

Another important factor to be considered when applying laboratory test data to the interpretation of landslides is the character of the displacements of the test specimens. The displacement at which the peak value of the strength is mobilized, when tested in direct shear apparatus certainly has the character of continuous angular deformation (see Fig. 18.3). After passing the peak the angular strain increases gradually and a surface of discontinuity starts to appear. At large displacements, at which the strength falls to the residual value, well-defined slickensides develop.

In natural conditions the absolute amount of displacement required to mobilize the peak strength depends on the thickness of the zone in which the stress–strain relationship approaches the critical state. The thickness of this zone depends on the initial stress distribution as well as on the contacts between strata which influence the pre-stress effects, on tectonic effects, on the consequences of previous slip movements and on the various non-uniformities in the composition of the slope and in the distribution of forces acting on its surface and within it. In connection with these natural conditions Skempton's valuable suggestions will be discussed further in Chapter 20.

19

Common Methods of Stability Analysis

19.1 CREEP EFFECTS IN THE STABILITY ANALYSIS

Conventional methods of stability analysis are based either on various applications of Kötter's differential equations which were dealt with in Chapter 7, or on the concept of potential slip surfaces the general discussion of which was presented in Section 18.1. In both of these two groups of methods static equilibrium conditions and the Mohr–Coulomb condition of limiting plastic equilibrium are applied. The stress–strain relationships are not considered. Nevertheless, approximate values of displacements can be obtained according to the transition from the previous stress state (if known) to the new stress state in the zones or surfaces in which a certain degree of uniform mobilization of strength is assumed to have occurred. The creep speed corresponding to this stress state can also be estimated and the further development of displacements approximately predicted if adequate test data are available.

Besides, the creep effects not contained in the stability analysis itself but considered separately, can be taken into account by basing the stability analysis on the long-term strength of the soil or by reducing the failure envelope $\tau_f = \tau_f(\sigma')$ by a safety factor, F, to $\tau_{fm} = \tau_f(\sigma')/F$ corresponding to the admissible distortional displacement and/or displacement speed (see Section 18.1).

In solving practical problems of the stability and displacements of natural slopes, cuttings and earth dams the potential failure surface method still provides a simple and satisfactory control of the safety and the movements, provided that the method and the strength parameters used suit the conditions of the case being investigated. Although the limited possibilities of considering creep phenomena do not affect the elements of the method a short review of the procedures will be given in the next sections in order to facilitate the discussion of creep effects.

19.2 GENERAL CONCEPT OF THE LIMIT ANALYSIS USING POTENTIAL FAILURE SURFACES

The method of stability analysis using potential failure surfaces was introduced by Peterson (1916) and Hultin (1916) and supplemented by Fellenius (1927). Since the early stages of the application of this method two different procedures have been developed. The first procedure investigates the equilibrium of the earth mass lying between the surface of the slope and the chosen failure surface as a whole. It is assumed that the same degree of strength is mobilized along the potential failure surface simultaneously. The second procedure considers the potential sliding body to consist of several slices with vertical interfaces. The equilibrium of each slice is investigated separately; the internal forces at the interfaces of adjacent slices have to be taken into consideration and the boundary conditions have to be respected.

In the classical, graphical or analytical procedures, circular or straight failure surfaces were used as well as surfaces in the form of logarithmic spirals. Recently procedures for the systematic analytical analysis of arbitrarily shaped failure surfaces have been given (Janbu, 1957; Nonveiller, 1965; Morgenstern and Price, 1965). In the following paragraphs the fundamentals of the graphical and analytical procedures will be given for arbitrarily shaped potential failure surfaces.

Let it be assumed that the gravitational and hydraulic forces acting in the slope presented in Fig. 19.1 are well defined. The vertical cross-section of the slope is assumed to be constant along its length (i.e. along the line normal to the cross-section presented). Both the fields of the gravitational and hydraulic forces are assumed to be constant in any cross-section (i.e. plane problem). When dividing the potential sliding body between the surface of the slope and the potential failure surface *A-B* (Fig. 19.1) into several slices with vertical interfaces, the following forces and their positions are known:

the weight of the slice ΔW,

the resultant water pressures acting on the sides of the slice:

$$P_w \text{ and } P_w + \Delta P_w,$$

the water pressure on the base of the slice: ΔP_b.

The following forces and their positions are not known:

the effective normal pressure $\Delta N'$ and its position—if the law of distribution of normal pressures $\Delta N'$ is assumed, the position of the force $\Delta N'$ is well defined;

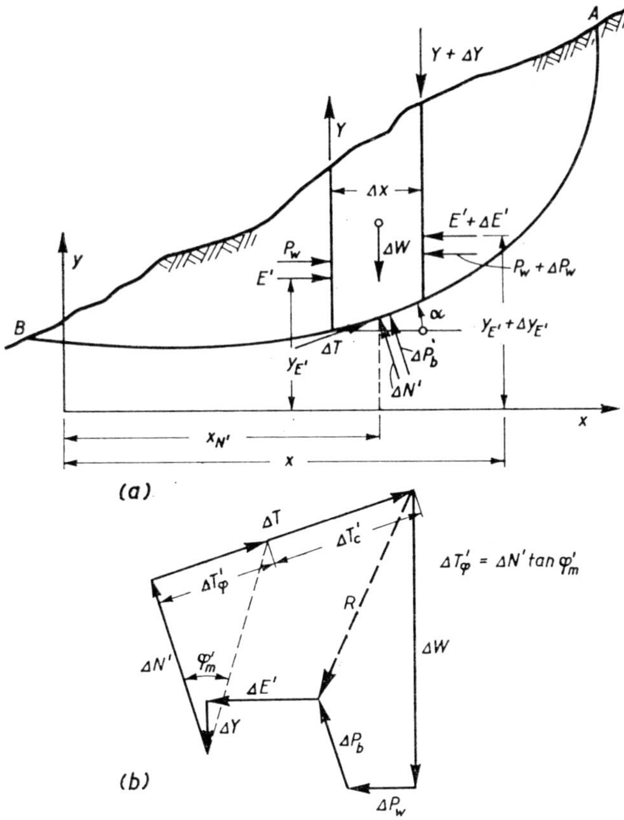

Fig. 19.1. Forces affecting slices of an arbitrary potential sliding body

the shear force acting along the base of the slice ΔT—this force is not an independent unknown quantity for it can be expressed by the Mohr–Coulomb failure criterion in terms of effective stresses as:

$$\Delta T = \frac{1}{F}\left(c' \frac{\Delta x}{\cos \alpha} + \Delta N' \tan \varphi'\right) \tag{19.1}$$

where c' is the cohesion intercept, φ' is the angle of shearing resistance (c' and φ' in terms of effective stresses) and F denotes the factor of safety;

the lateral thrust on the side of the slice in terms of effective stress E' and its position $y_{E'}$; and

the vertical shear force on the side of the slice Y. The corresponding values $E' + \Delta E'$, $y_{E'} + \Delta y_{E'}$ and $Y + \Delta Y$ are unknown quantities of the adjacent slice. In a successive graphical procedure, they are known quantities for the treated slice.

Consequently, in the force polygon in Fig. 19.1 (*b*) the forces $\Delta E'$, ΔY and $\Delta N'$ are unknown. The construction of the polygon involves two equilibrium conditions. The third equilibrium condition expressing the zero summation of the moments about an arbitrary point is not sufficient to determine any one of the three unknown quantities because it contains two further unknown quantities, the ordinate $y_{E'}$ of the position of the thrust E' and the position $x_{N'}$, of the reaction pressure $\Delta N'$. Thus the problem is statically indeterminate. As the stress–strain relationships are complicated and the kinematic conditions difficult to define, it is necessary to invoke two assumptions in order to render the problem statically determinate.

Among the various possible combinations of assumptions the following are used in known graphical or analytical procedures:

(1) The distribution of normal pressures along the potential failure surface can be assumed.

(2) Assumptions may be made regarding the position of the line of thrust and the position of the reaction force $\Delta N'$ (usually at the centre of the base).

(3) Assumptions may be made regarding the position of the line of thrust and the relation between the E' and Y forces, i.e. regarding the inclination of their resultant.

(4) Assumptions may be made regarding the relation between the E' and Y forces, and regarding the position of the reaction forces $\Delta N'$.

(5) The position of the reaction forces $\Delta N'$ may be assumed and the vertical component Y of the internal forces at the interface between the slices may be chosen so as to satisfy the equilibrium conditions.

If the soil mass between the slope surface and the potential failure surface is divided into n slices, then the first of the above group of assumptions involves n boundary conditions concerning the positions of the reaction forces $\Delta N'$ in the bases of the slices, and $n - 1$ boundary conditions concerning the ratios of their magnitudes. Assumptions (2), (4) and (5) yield also n conditions concerning the positions of the internal forces in the bases of the slices and $(n - 1)$ conditions concerning either the positions, directions or magnitudes of the internal forces at the interfaces between the slices. Moreover there are, for all slices of total number n, $3n$ equilibrium conditions and n failure conditions (i.e. n conditions governing the ratios between the shear and normal components of the reaction (internal) forces at failure. Thus, altogether there are

$$[n + (n - 1)] + 3n + n = (6n - 1) \text{ conditions}$$

On the other hand it is necessary to determine $3n$ parameters of the internal forces in the bases of the slices and $3(n - 1)$ parameters determining the internal forces at the interfaces between the slices, as well as the safety factor F_s. So these are $3n + 3(n - 1) + 1 = (6n - 2)$ unknown quantities. Thus the number of conditions exceeds the number of unknown quantities by 1. The problem is therefore statically redundant. The redundant condition restricts the choice of assumptions. The following two requirements should be considered:

(*a*) Attention should be paid to the appearance of tension stresses in the slip-surface and within the sliding body. Generally tension stresses have to be avoided. In order to ensure that this is so, it is recommended that the position of the reaction forces $\Delta N'$ and E' be kept within the middle third of the base and of each interface of the slice respectively.

(*b*) Along the interfaces of the slices the strength of the soil must not be exceeded. As the entire stress state is not investigated, the requirement

$$\frac{Y}{E' + c' \Delta y \, . \, 1} < \tan \varphi'$$

is to be interpreted as a necessary but incomplete condition. Respecting the above condition the slope of the internal forces $(\vec{Y} + \vec{E}')$ is to be estimated intuitively such that the kinematic compatibility of the forces in adjacent interfaces is taken into account.

19.3 RESULTANT METHOD OF STABILITY ANALYSIS

19.3.1 Arbitrarily-shaped Failure Surfaces

When taking the first set of assumptions listed in the previous section, all the surface and volume external forces acting on the sliding body can be summarized vectorially. The resultant active force must be in equilibrium with the resultant of the reaction forces appearing along the potential failure surface. Equilibrium can be obtained if the shear forces along the potential failure surface governed by the strength law and the distribution of both the normal and the shear forces are appropriately chosen.

The method can be applied either in analytical or in graphical form. If the failure surface crosses soils with different strength parameters, different values of the mobilized cohesion c_m' and of the mobilized shear angle φ_m' can be considered in different elements i of the potential failure surface.

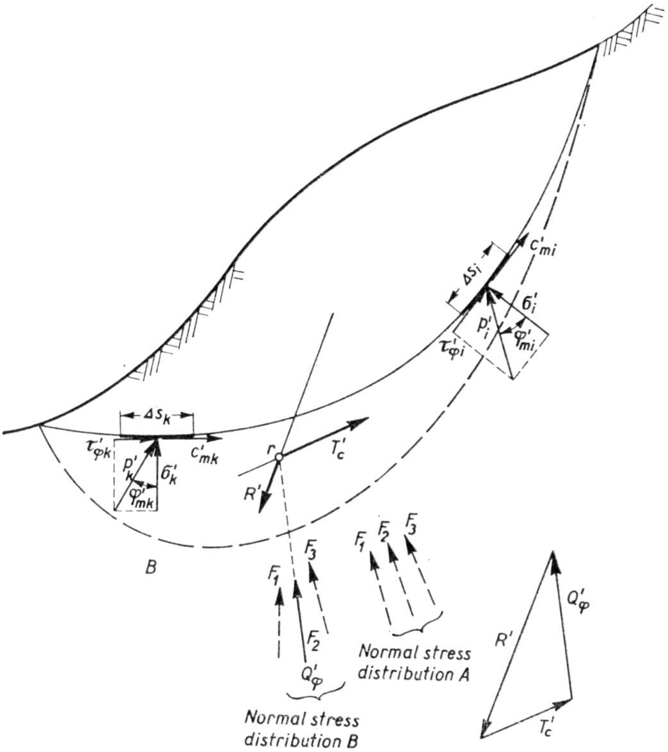

Fig. 19.2. Resultant method of the stability analysis applied to an arbitrarily shaped slip surface

Fig. 19.2 represents the scheme for the corresponding graphical solution. The position and the direction of the resultant cohesion force can be determined as the position and direction of the resultant of the dimensionless vectors $\vec{a}$

$$|\vec{a}| = \frac{c_i'}{c_0'} \cdot \frac{\Delta s_i}{\Delta s_0} \tag{19.2}$$

c_0' and Δs_0 being the arbitrary values of the comparative stress and length. The magnitude of the mobilized cohesion resultant $\vec{T}_c$ is then given by the product

$$|\vec{T}_c'| = T_c' = |\vec{A}| c_0' \Delta s_0 \frac{1}{F_c} \tag{19.3}$$

$\vec{A}$ being the resulting vector of the components $\vec{a}$ and F_c the safety factor.

The resultant $\vec{Q}_\varphi$ of the remaining reactive forces $\Delta\vec{Q}_{\varphi i}$ appearing in the elements Δs_i must be in equilibrium with the force $\vec{T}_c'$ and the resultant $\vec{R}'$ of the active effective forces affecting the sliding body above the potential failure surface. The resultant $\vec{R}'$ involves the surface loads, the gravitational forces $\Delta\vec{W}_i$, and the uplift forces $\Delta\vec{P}_{bi}$ acting in the elements Δs_i (see Fig. 19.1). The force $\vec{Q}_\varphi'$, being in equilibrium with the forces $\vec{R}'$ and $\vec{T}_c'$, must pass through the intersection point r of the forces $\vec{R}'$ and $\vec{T}_c'$. If the slope is stable, this condition can be fulfilled by choosing an appropriate distribution of the normal stresses σ_i' acting on the failure surface, and an appropriate value of the safety factor F. The solution can then be obtained by trial and error.

First a distribution of the normal stresses σ_i' acting in the potential failure surface will be assumed. In Fig. 19.2 the dotted line B represents one of the possible stress distributions. The effective shear angle in the element Δs_i may be denoted by φ_i' and the mobilized shear angle corresponding to a safety factor F by φ_{mi}':

$$\varphi_{mi}' = \text{arc tan} \frac{\tan \varphi_i'}{F} \tag{19.4}$$

Stress vectors $\vec{p}_i' = \vec{\sigma}_i' + \vec{\tau}_{\varphi i}$ may be drawn corresponding to a chosen factor of safety F. The resultants $\vec{B}$ of the vectors $\vec{b}$ having positions and directions of vectors $\vec{p}_i'$ and reduced dimensionless magnitudes

$$|\vec{b}_i| = b_i = \frac{p_i'}{\sigma_0'} \cdot \frac{\Delta s_i}{\Delta s_0} = \frac{\sigma_i'}{\cos \varphi_{mi}' \sigma_0'} \cdot \frac{\Delta s_i}{\Delta s_0} \tag{19.5}$$

may be constructed corresponding to various safety factors $F_\varphi = F_{\varphi 1}$, $F_{\varphi 2}$, $F_{\varphi 3}$ . . .

Several assumptions about the normal stress distribution will be made by using appropriate interpolation of the vectors $\vec{B}$ corresponding to intermediate values of the safety factors F_φ. Some of the vectors $\vec{B}$ from different groups can be found to satisfy the condition that they should pass through the point r. The equilibrium triangle of forces R', T_c' and Q_φ', the last force acting in the direction of the vector $\vec{B}$, yields the magnitude of the forces T_c' and Q_φ'. Among various solutions satisfying the equilibrium conditions the one for which the safety factor F_φ equals the safety factor F_c should be chosen. Furthermore the interface internal forces, which can be obtained by successive construction of the force polygons from slice to slice, should not be inclined by more than φ_i' to the horizontal.

19.3.2 Circular Failure Surfaces

The method presented in the foregoing section becomes simplified if the potential failure surface is circular and if the shear parameters c' and φ' are assumed to be constant throughout the failure surface.

In Fig. 19.3 the vector $\vec{R}$ represents the resultant of the following forces: the weight $\vec{W}$ of the sliding body ($ADBA$); the external loads $\Sigma\vec{P}$

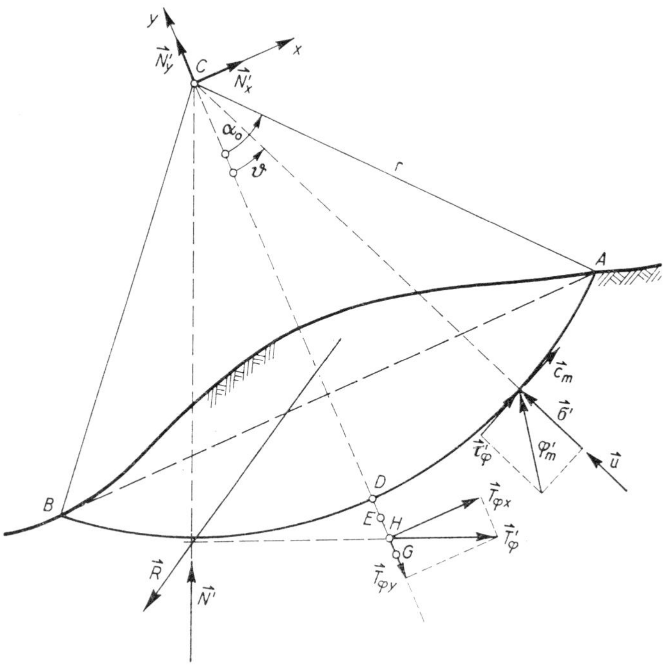

Fig. 19.3. Elementary and resultant reaction forces in a circular slip surface

on the slope surface (AB); the uplift pressures $\vec{u}$ acting along the potential failure surface $\widehat{ADB}$ according to the hydraulic potential distribution; and the cohesion parts c_m' ds of the tangential components of the reaction forces distributed along the potential failure surface:

$$\vec{R} = \vec{W} + \Sigma\vec{P} + \int_A^B \vec{u}\,\mathrm{d}s + \int_A^B \vec{c}_m'\mathrm{d}s \qquad (91.6)$$

Assuming simultaneous mobilization of the same amount of cohesion

$$c_m' = \frac{c'}{F} \tag{19.7}$$

throughout the failure surface, the resultant cohesion force $\vec{T}_c'$ has the magnitude:

$$T_c' = \int_A^B c_m' \cos\vartheta \, ds = 2c_m' r \sin\alpha_0 = c_m' l,$$

$$l = \overline{AB} \tag{19.8}$$

$\vec{T}_c'$ is parallel to the chord AB and intersects the line of symmetry $\overleftrightarrow{CD}$ at the point G defined by the equation:

$$\overline{CG} = a = \frac{1}{T_c} \int_A^B c_m' r \, ds = r \frac{\alpha_0}{\sin\alpha_0} \tag{19.9}$$

The resultant $\vec{R}$ must be in equilibrium with the forces $\vec{N}'$ and $\vec{T}_\varphi'$, $\vec{N}'$ being the resultant of the normal components $\vec{\sigma}' ds$ of the reaction forces along the arc $\widehat{AB}$, and $\vec{T}_\varphi'$ the resultant of the friction parts of the tangential reaction forces:

$$\vec{N}' = \int_A^B \vec{\sigma}' ds, \quad \vec{T}_\varphi' = \int_A^B \vec{\tau}_\varphi' \, ds \tag{19.10}$$

where

$$\tau_\varphi' = \sigma' \tan\varphi_m'; \tag{19.11}$$

φ_m' is the mobilized shear angle:

$$\tan\varphi_m' = \frac{\tan\varphi'}{F} \tag{19.12}$$

if F denotes the factor of safety. As the same degree of mobilization of the shear angle φ' is assumed, the resultant forces $\vec{T}_\varphi'$ and $\vec{N}'$ intersect each other at right angles. At the points of their intersection with the line of symmetry, C and H, respectively, the forces $\vec{N}'$ and $\vec{T}_\varphi'$ may be divided in the directions of the axis of symmetry y and of the axis x perpendicular to the axis y.

$$\left.\begin{aligned} \vec{N}' &= N_x' + \vec{N}_y' \\ \vec{T}_\varphi' &= \vec{T}_{\varphi x}' + \vec{T}_{\varphi y}' \end{aligned}\right\} \tag{19.13}$$

The relations

$$\left.\begin{aligned} T_{\varphi x}' &= N_y' \tan \varphi_m' \\ T_{\varphi y}' &= N_x' \tan \varphi_m' \end{aligned}\right\} \tag{19.14}$$

hold.

As suggested by Caquot (1954), the normal reaction components $\vec{\sigma}' \mathrm{d}s$ in any element $\mathrm{d}s$ may be assumed to be separated into a paired and into an unpaired component in such a way that the resultant force of all the paired components is the force $\vec{N}_y'$ and that of all the unpaired components the force $\vec{N}_x'$. The force $\vec{T}_{\varphi x}'$ is then the resultant of all the elementary forces $(\vec{\tau}_\varphi')_p\, \mathrm{d}s$ corresponding to the paired components $(\vec{\sigma}')_p\, \mathrm{d}s$ of the normal reaction forces. Consequently, the position of the force $\vec{T}_{\varphi x}'$ is completely defined by the way in which the normal stresses $(\vec{\sigma}')_p$ are distributed along the arc $\widehat{ADB}$. If the pressures $(\vec{\sigma}')_p$ are assumed to be uniformly distributed along the arc, the force $\vec{T}_\varphi'$ intersects the axis x at the point G defined by Equation (19.8). If the diagram of the components $(\vec{\sigma}')_p$ is assumed to have the form of a circular segment (ADB), the point of intersection E can be shown to be given by the equation (Caquot, 1954):

$$\overline{CE} = b = 2r \frac{\sin \alpha_0 - \alpha_0 \cos \alpha_0}{\alpha_0 - \sin \alpha_0 \cos \alpha_0} \tag{19.15}$$

The relation

$$\overline{DE} = b - r = \frac{6}{10}\, \overline{DG} = 0.6(a - r) \tag{19.16}$$

holds approximately.

The factor of safety of a slope can be found by trial and error using the following graphical procedure (Fig. 19.4): First the resultant force $\vec{R}'$ (cf. Equation 19.6):

$$\vec{R}' = \vec{W} + \Sigma \vec{P} + \int_A^B \vec{u}\, \mathrm{d}s \tag{19.17}$$

must be constructed and the direction $\vec{T}_c'$ drawn parallel to the chord A-B and passing through the point G. Then a factor of safety F_c may be assumed defining the mobilized cohesion $c_m' = c'/F_c$ (see Equation 19.7). The corresponding cohesion resultant $\vec{T}_c'$ having the magnitude $T_c' = c_m' l$ (Equation 19.8) may now be combined with the force $\vec{R}'$ to give the resultant $\vec{R}$:

$$\vec{R} = \vec{R}' + \vec{T}_c' \tag{19.18}$$

The intersections of the direction of the force $\vec{R}$ with the semi-circles drawn on the diameters $CE = b$ and $CG = a$ (Equation 19.15 and 19.9) determine the angles φ_b' and φ_a' (see Fig. 19.4). The corresponding safety factors are

$$F_{\varphi a} = \frac{\tan \varphi'}{\tan \varphi_a'}, \quad F_{\varphi b} = \frac{\tan \varphi'}{\tan \varphi_b'} \tag{19.19}$$

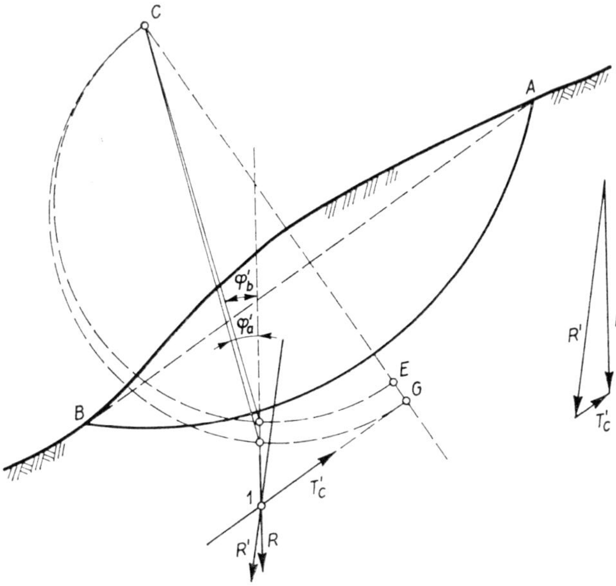

Fig. 19.4. Resultant method of the stability analysis applied to a circular slip surface

If the value F_c chosen initially does not lie in the range between $F_{\varphi a}$ and $F_{\varphi b}$ another value of F_c should be chosen and the construction repeated until this condition is satisfied:

$$F_{\varphi a} \geqslant F_c \geqslant F_{\varphi b} \tag{19.20}$$

When $\varphi' = 0$, the resultant $\vec{R}$ must have the direction $1 - C$ defining the value of T_c' (see the triangle of forces in Fig. 19.4) and the factor of safety F:

$$F = F_c = \frac{c'}{c_m'} = \frac{c'l}{T_c'} \tag{19.21}$$

When $c' = 0$, the intersection points of the force $\vec{R}'$ with the semi-circles define the angles φ_a' and φ_b' and the corresponding range of safety factors.

If the slip wedge (ABC) (Fig. 19.4 (*a*)) is laterally supported by a retaining wall, the resultant of the active and cohesion reaction forces must be in equilibrium with the reaction E of the retaining wall and the resultant Q of the normal and friction reactions along the potential rupture line. Let the point chosen on the axis of symmetry in the interval E-G (Fig. 19.4) be denoted by D and the mobilized shear angle in the backfill of the retaining wall by φ_m'. The reaction Q passes through the point 1 of the intersection of the resultant force R and the earth pressure E, whose

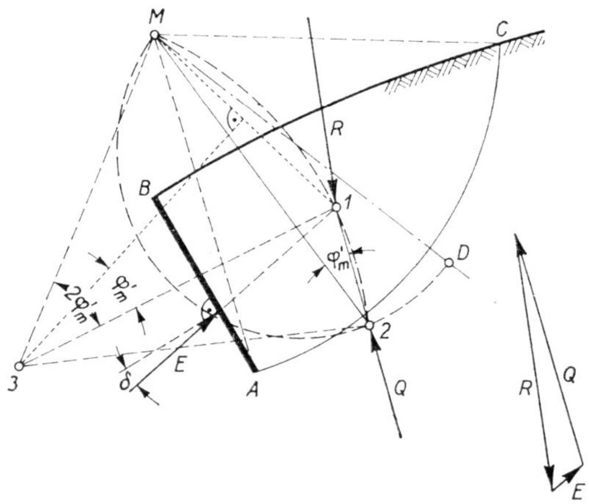

Fig. 19.4*a*. Application of the resultant method of the stability analysis to a circular slip wedge which is laterally supported by a retaining wall

direction and position are assumed to be known, and crosses the semi-circle with diameter MD in the point 2 on condition that the peripheral angle (M − 2 − 1) equals φ_m'. Thus, the point 2 can be found by intersecting the semi-circle with the circular arc whose centre (3) lies on the axis of symmetry of the straight line $M - 1$ and is defined by the central angle ($M - 3 - 1$) equal to $2\varphi_m'$. The magnitude of the earth pressure E corresponding to the chosen potential rupture line $\widehat{AC}$ can then be determined by means of the triangle of equilibrium $\vec{R} + \vec{Q} + \vec{E} = 0$.

19.3.3 Failure Surfaces in the Form of Logarithmic Spirals

If the soil has uniform shear parameters c' and φ', the stability of a slope can simply be analysed by assuming potential failure surfaces having the form of logarithmic spirals with the equation

$$r = r_0 \exp (\alpha_l \mu) \tag{19.22}$$

where $\mu = \tan \varphi_m$, φ_m being the mobilized angle of shear; for the remaining notation see Fig. 19.5. The reaction forces in each element ds can be split up into the cohesion component $d\vec{T}_c{}' = c_m{}'\,\mathrm{d}s$ and the resultant of the normal and friction components $\mathrm{d}\vec{Q}_\varphi$. As all forces $\mathrm{d}\vec{Q}_\varphi$ pass through

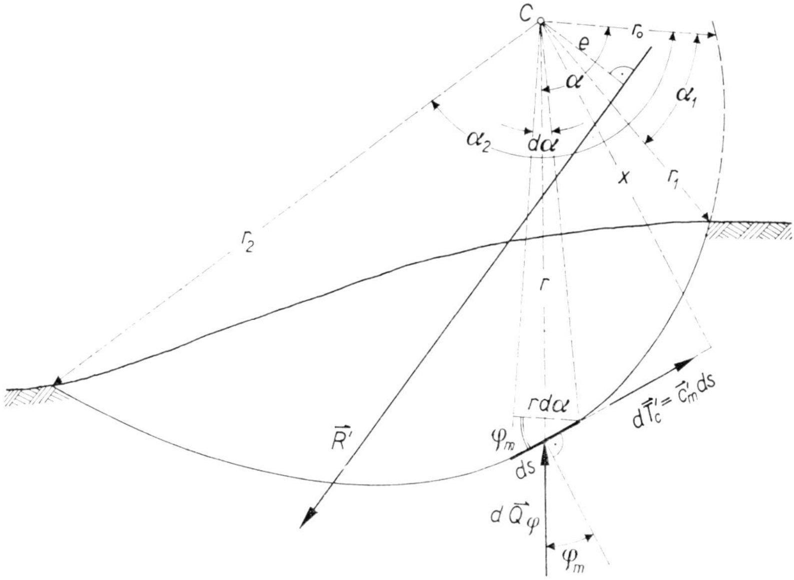

Fig. 19.5. Resultant method of the stability analysis applied to a logarithmic spiral slip surface

the centre C of the spiral, the resultant moment of external forces $M = R'e$ equals the moment of the cohesion forces $\mathrm{d}\vec{T}_c{}'$. This condition yields

$$c_m{}' = R'e\,\frac{2\tan\varphi_m{}'}{r_2{}^2 - r_1{}^2} \tag{19.23}$$

which can easily be proved by simple computation based on Fig. 19.5. The corresponding safety factor

$$F_c = \frac{c'}{c_m{}'}$$

may in principle equal the chosen value F_φ.

19.4 THE METHOD OF SLICES

19.4.1 Graphical Solution

When applying the method of slices in graphical form, either the second or the third group of assumptions mentioned at the end of Section 19.2 can be taken. In Fig. 19.6 the procedure is shown for an arbitrary slice k. The vector $\vec{E}_j^* = \vec{E}_j' + \vec{Y}_j$ (see notation in Fig. 19.1) of the effective reaction force at the interface with the preceding slice j has previously been determined by the successive application of the procedure starting

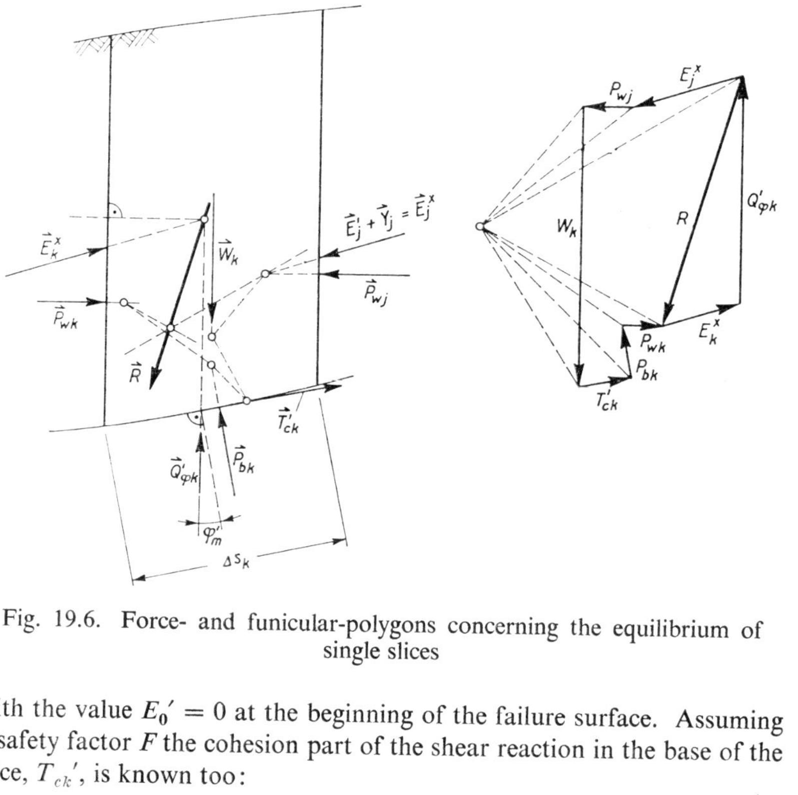

Fig. 19.6. Force- and funicular-polygons concerning the equilibrium of single slices

with the value $E_0' = 0$ at the beginning of the failure surface. Assuming a safety factor F the cohesion part of the shear reaction in the base of the slice, T_{ck}', is known too:

$$T_{ck}' = \frac{c'}{F} \Delta s_k$$

The resultant $\vec{R}$ of all the known forces

$$\vec{R} = \vec{E}_j^* + \vec{P}_{wj} + \vec{W}_k + \vec{T}_{ck}' + \vec{P}_{bk} + \vec{P}_{wk}$$

can be obtained by applying the well known method of the string polygon. (For notation see Figs. 19.6 and 19.1.) The resultant force $\vec{R}$ must be in equilibrium with the remaining part of the reactive effective force, $\vec{Q}_{\varphi k}'$, in the base of the slice, and with the effective reaction at the interface with the next element: $\vec{E}_k^* = \vec{E}_k' + \vec{Y}_k$. The inclination of the force $\vec{Q}_{\varphi k}'$ with the normal to the base is defined by the mobilized shear angle, $\varphi_m' = \arctan(\tan\varphi'/F)$, and its point of application can be assumed to be in the centre or near the centre of the base. As to the thrust $\vec{E}_k^*$, either its position, y_E' (assumption 1), or its inclination (assumption 2) should be assumed. The magnitude of the force $\vec{E}_k^*$ can be determined from the triangle of forces $\vec{R} + \vec{E}_k^* + \vec{Q}_{\varphi k}' = 0$ (see Fig. 19.6). This gives the initial value of $\vec{E}^*$ for the next slice unless the end of the slip surface has been reached. The boundary condition to be satisfied at the end of the failure surface is

$$\vec{E}^* = \vec{E}_n^*$$

where $\vec{E}_n^*$ is usually zero. If the above condition is not satisfied, a new safety factor F has to be chosen and the procedure repeated until the condition is satisfied.

The procedure has to be applied to several potential failure surfaces that are kinematically possible. The smallest safety factor obtained determines the safety of the slope.

The graphical procedure described is, in principle, identical to the procedure that was presented by Fellenius in 1927 for circular failure surfaces.

19.4.2 Analytical Solutions

The analytical solution corresponding to the graphical procedure as explained in the previous section can be obtained by establishing the three equilibrium conditions for the system of forces shown in Fig. 19.1. Furthermore the Mohr–Coulomb failure criterion relating the components $\Delta T'$ and $\Delta N'$ of the reaction effective force in the base of the slice has to be taken into account and one of the groups of assumptions which were listed in Section 19.2 used.

The first analytical solution applying to arbitrarily shaped slip surfaces was elaborated by Janbu (1957). In order to render the problem statically determinate he assumed that the reaction forces dN (Fig. 19.7) acted at

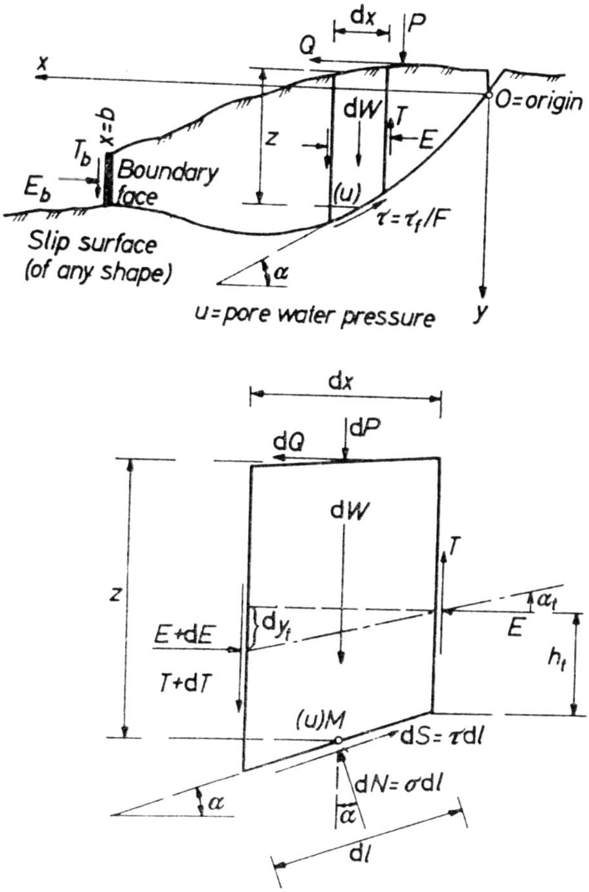

Fig. 19.7. Notation to Janbu's analytical procedure

the centre of the base dl of the slice,* and that the position of the resulting earth pressures E and $E + \mathrm{d}E$ at the interfaces between adjacent slices was known as well. By using the notation of Fig. 19.7 the equations of equilibrium for each slice are:

Vertical: $$\mathrm{d}W + \mathrm{d}P + \mathrm{d}T = \mathrm{d}S \sin \alpha + \mathrm{d}N \cos \alpha \tag{19.24}$$

Horizontal: $$\mathrm{d}E - \mathrm{d}Q = -\mathrm{d}S \cos \alpha + \mathrm{d}N \sin \alpha \tag{19.25}$$

Moment about M: $$T\,\mathrm{d}x + E\,\mathrm{d}y_t - \mathrm{d}Eh_t + \mathrm{d}Qz = 0 \tag{19.26}$$

* If the point of application of the reaction force in the base of the last slice is specified too, the number of conditions exceeds the number of unknown quantities by 1 and the problem is to be interpreted as statically redundant. The redundant condition restricts the choice of assumptions defining the line of thrust.

The effective normal stress $\sigma' = \mathrm{d}N/\mathrm{d}l - u$ can be expressed by the relation

$$\sigma' = (p + t - u) - \tau \tan \alpha \tag{19.27}$$

where

$$p = \frac{\mathrm{d}W}{\mathrm{d}x} + \frac{\mathrm{d}P}{\mathrm{d}x} = \gamma z + q, \quad t = \frac{\mathrm{d}T}{\mathrm{d}x} \tag{19.28-a,b}$$

Inserting Equation (19.27) the relation

$$\tau = \frac{\tau_f}{F} = c_m' + \sigma' \tan \varphi_m' \tag{19.29}$$

yields

$$\tau = \frac{c_m' + (p + t - u) \tan \varphi_m'}{1 + \tan \alpha \tan \varphi_m'} \tag{19.30}$$

When eliminating $\mathrm{d}N$ from conditions (19.24) and (19.25) it is found that:

$$\mathrm{d}E = \mathrm{d}Q + (p + t) \tan \alpha \, \mathrm{d}x - \tau \cos^{-2} \alpha \, \mathrm{d}x \tag{19.31}$$

The horizontal force E at any vertical section is obtained by integrating Equation (19.31) from 0 to x.

The corresponding shear force T is expressed in terms of E and Q on dividing condition (19.26) by $\mathrm{d}x$:

$$T = -E \tan \alpha_t + h_t \frac{\mathrm{d}E}{\mathrm{d}x} - z \frac{\mathrm{d}Q}{\mathrm{d}x} \tag{19.32}$$

If a resultant force at a boundary face, $x = b$, is represented by the components E_b and T_b, the overall directional equilibrium of the free body requires

$$\int_0^b \mathrm{d}T = T_b, \quad \int_0^b \mathrm{d}E = E_b \tag{19.33-a,b}$$

In common cases of slope analysis $T_b = E_b = 0$ has to be considered as the boundary condition. Inserting $\mathrm{d}E$ from Equation (19.31) with the expression (19.30) for τ, into Equation (19.33 (*b*)) we obtain:

$$E_b - \int_0^b u \, \mathrm{d}y = Q + \int_0^b [(p + t - u) + c_m' \cot \varphi_m'] \tan (\alpha - \varphi_m') \, \mathrm{d}x - \int_0^b c_m' \cot \varphi_m' \, \mathrm{d}y \tag{19.34}$$

For the purpose of determining F a working formula is obtained by introducing $\tau_f = F\tau$. For finite differences:

$$F = \frac{\Sigma\tau_f \cos^{-2}\alpha\,\Delta x}{Q - E_b + \Sigma(p + t)\tan\alpha\,\Delta x} \tag{19.35}$$

where

$$\tau_f = \frac{c' + (p + t - u)\tan\varphi'}{1 + \tan\alpha\,\dfrac{\tan\varphi'}{F}} \tag{19.36}$$

By assuming a reasonable position for the line of thrust, accurate values of the corresponding internal forces, E and T, are obtained from Equations (19.31) and (19.32) by means of successive approximation procedures. Initial values of E and T can be calculated for the condition $t = 0$.

Morgenstern and Price (1965) presented a statically determinate* solution by assuming the position of the forces $\Delta N'$ to be known and the slope of the internal forces in vertical sections to be governed by the relation:

$$X = \lambda f(x)E' \tag{19.37}$$

$f(x)$ must be specified and λ, as well as the safety factor F appearing in the failure criterion, must satisfy the appropriate boundary conditions. Eliminating the two unknown quantities ($\Delta N'$ and $\Delta T'$) two governing differential equations remain. They have been put into a form suitable to be programmed for an electronic computer.

An alternative analytical solution has been reached independently by Nonveiller (1965). Whilst, in the solution of Morgenstern and Price, the equilibrium conditions are successively applied from slice to slice and the moment condition related to the centre of the base of each slice, Nonveiller summarized the moment conditions of all slices related to a single pole. He assumed that the reaction force $\Delta N'$ acts at the centre of the slice. The tangential components Y of the effective internal forces are chosen as the remaining unknown quantities. They have to be chosen in such a way that their differences ΔY satisfy the equilibrium conditions. The resulting equations presented by Nonveiller are suitable for computations by digital computer.

An analogous solution restricted to the use of circular failure surfaces has previously been presented by Bishop (1954). Bishop found that a satisfactory approximation could be obtained by disregarding, within

* See the footnote (*) on page 425.

each slice, the differences between the tangential components of the reactive effective forces at the vertical interfaces between the slices:

$$\Delta Y = Y_k - Y_j = 0 \tag{19.38}$$

With this assumption the resulting equation reads:

$$F = \frac{1}{\Sigma \Delta W \sin \alpha} \cdot \Sigma \left\{ [c'b + \tan \varphi' (\Delta W - ub)] \frac{1}{\cos \alpha + \dfrac{\tan \varphi' \sin \alpha}{F}} \right\} \tag{19.39}$$

It has been deduced from the equilibrium equations in the vertical direction, and for the moments, with regard to the centre of the circular failure surface, set up for each slice of finite width b, and summarized for all slices. (For the notation see Fig. 19.1; u = the pore pressure in the base of the slice, $b \equiv \Delta x$). The safety factor F can be found by trial and error. The compatibility of the reaction forces at the interfaces requires that the condition (19.38) should be identical to the condition:

$$Y_k = Y_j = 0 \tag{19.40}$$

When basing the analytical procedure on either group of assumptions listed under (1) to (5) in Section 19.2, the limitations in the choice of assumptions mentioned under (*a*) to (*c*) at the end of the same section should be considered. Thus the resulting positions or inclinations of the internal forces should be checked with respect to their compatibility with the strength and deformation properties of the soils.

Appropriate combinations of the resultant and slices methods described in the present and the foregoing sections (19.4 and 19.3) are suitable especially in cases of heterogeneous profiles of earth bodies with oblique boundaries between regions of different geotechnical properties (e.g. earth dams with cores, backfills of retaining structures).

19.5 ESTIMATION OF DISPLACEMENTS

19.5.1 Characteristic Domains of Creep Velocity

In Section 17.1 a method expressing the anisotropic and viscous deformability of soils by four strain characteristics was presented. The method was based on the assumption that the secondary consolidation is proportional to the logarithm of time. It was stated that the limiting envelope of Mohr circles corresponding to stress states obeying the logarithmic creep law was not far from the failure envelope. Figs. 14.4, 14.5 and 20.1, 20.2 are examples of experimental evidence proving the above relationship. The limiting envelope was assumed to be a straight line originating from the point $\sigma' = p_c = -c' \cot \varphi'$ on the axis σ', φ' being the slope of the failure envelope, and making an angle φ_1' with the axis σ'.

In the domain of the envelopes $\tau = \tan \varphi_m'(\sigma' - p_c)$, $\varphi_m' \leqslant \varphi_1'$, the creep increase with the logarithm of time has been observed, in some laboratory tests, for many months or even years. Comparison of the

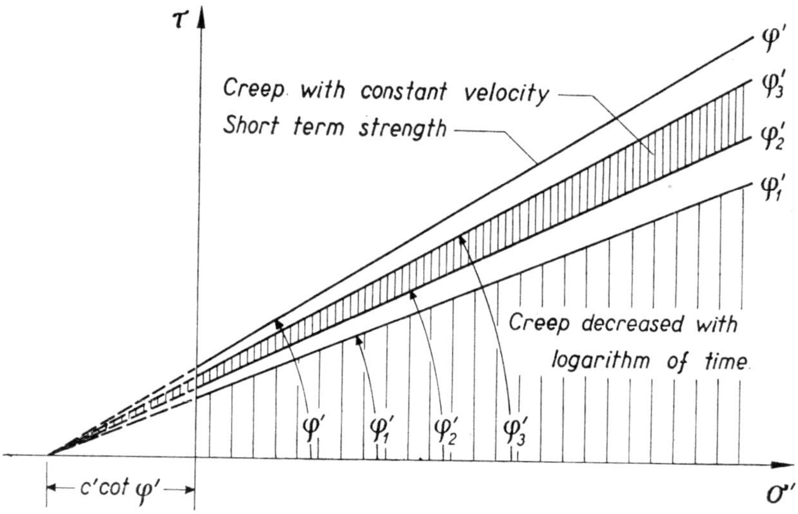

Fig. 19.8. Scheme of characteristic creep ranges as observed in triaxial tests

natural density of some normally consolidated or pre-consolidated soils, whose amount and time of preloading are known, with the density of laboratory specimens (prepared at the liquid limit) under sustained loads corresponding to the natural stress history, proves that creep can increase with the logarithm of time or at a even faster rate for many millenia. Nevertheless it is quite clear that the validity of Buisman's law must be limited and that after a certain time the creep speed tends to zero.

At the limiting slope φ_1' of the Mohr envelope the creep starts to increase at a faster rate than that corresponding to the logarithmic law, yet the acceleration remains negative. At a slope φ_2' which seems to be close to the slope φ_1' (Fig. 19.8), the deviatoric creep starts to develop at constant velocity. The volume creep either stops increasing or occurs at negative speed (dilatation).

At further increase of the slope of the stress envelope the creep speed becomes greater until, at a limiting slope, φ_3', the speed starts to accelerate. At a slope φ' of the envelope, failure occurs. According to the author's experience, the interval $\varphi_3' - \varphi_2'$ is, under drained conditions, rather small and the value φ_3' coincides with the 'short-term strength' value φ'. Tan's experiments (Tan, 1954) prove, however, that the Bingham state of creep can be very extensive for clays of high plasticity under undrained conditions.

According to Maslov (1958) and Vyalov (1961), long-term creep leads to failure at any stress state whose Mohr circle exceeds the envelope φ_2'. Part of the short-term cohesion intercept c' disappears during long-term creep and thus the long-term strength envelope lies below the envelope $\tau = (c' \cot \varphi' + \sigma') \tan \varphi_2'$ (Fig. 19.8).

Only in the domain of damped creep ($\varphi_m' \leqslant \varphi_1'$) can the displacements be approximately predicted. Any creep at constant or increasing velocity should be considered as a process leading to failure. In order to be able to establish the degree of danger, to predict the time to failure and to investigate the effects of remedial measures, field measurements have to be made and interpreted for all creeping slopes, dams and tips which menace human lives and property.

19.5.2 Approximate Strain Fields

In the stability analysis of slopes by the slip surface method the stress states along a potential slip surface are assumed to correspond to a straight

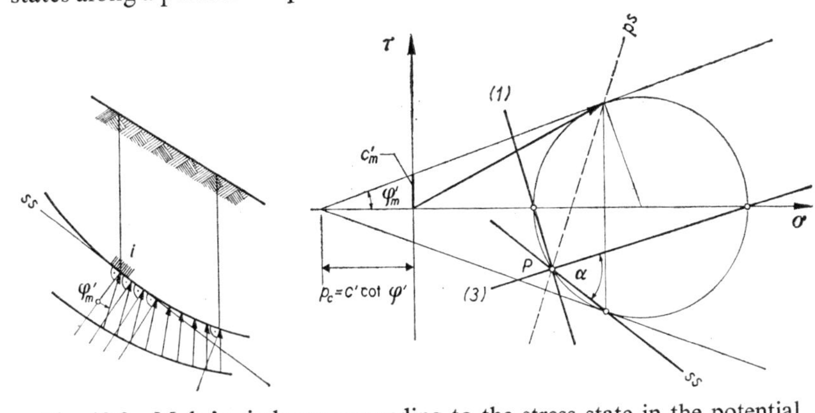

Fig. 19.9. Mohr's circle corresponding to the stress state in the potential slip surface

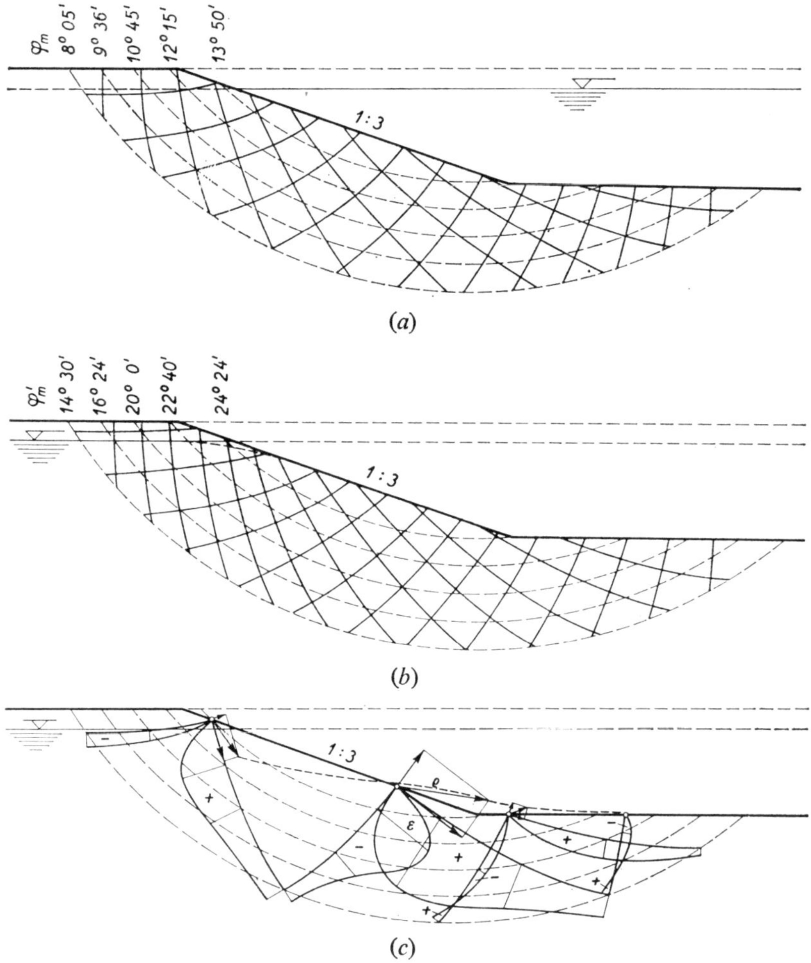

Fig. 19.10. Approximate stress fields (*a*) before and (*b*) after a rapid drawdown of the water in an excavation; (*c*) the corresponding strain diagram (first approximation) along some trajectories of principal stresses according to stress–strain relationships similar to those presented in Fig. 17.2 ($t = 50$ years), and vectors of displacements of some points on the slope surface related to the lowest potential slip surface

Mohr–Coulomb envelope of intercept $p_c \tan \varphi_m'$ on the axis τ and of slope φ_m' with the axis σ'. By using the slices method the full diagram of stresses along the potential slip surface and the Mohr circle for any point on the slip surface can be drawn. In Fig. 19.9 the Mohr circle corresponding to the point i of the slip surface is presented, and the construction of the direction of the potential pseudo-slip surface (*ps*) as well as of the principal stresses (1), (3) is shown. The angle α between the direction of the slip surface and the principal stress direction (3) remains constant along the entire slip surface.

The method of slices can be applied to a set of slip surfaces in a way that the internal forces appearing along any potential slip surface are in equilibrium with the internal forces which were previously obtained for the upper surface, and body forces to which the earth strip between the subsequent slip surfaces is subjected. Consequently, the resulting stress fields are approximately statically correct and can be used in the estimation of displacements by the procedure explained in Section 17.1.2. Fig. 19.10 shows the strain diagrams obtained for the change of stresses corresponding to the drawdown of the water inside an excavation. The strains are plotted along the trajectories of principal stresses of the new stress state and correspond to the stress–strain relationship similar to that presented in Fig. 17.2. The displacement vectors, related to the strain change between the surface and the deepest potential failure surface are presented for three surface points. As the trajectories of principal stresses do not appreciably deviate from the tangents in these surface points, the components of the displacement vectors in the directions of the principal stresses have been obtained by integrating the corresponding strain diagrams. This procedure was discussed in Section 17.1.2.

20

Critical Phases of the Creep of Slopes

20.1 CONDITIONS AFFECTING THE CREEP AND LONG-TERM STABILITY OF SLOPES

The data presented in the previous chapters concerning the creep speed may be summarized in the following statements:

If the distortional creep occurs simultaneously with the spherical (volume) creep, its velocity decreases approximately in inverse proportion to time until the stress state approaches the failure envelope. The lower limit of the Bingham state of distortional creep at constant velocity seems to correspond to about 95 per cent mobilization of the peak strength or even greater so far as the available experimental evidence shows.

On the other hand, pure torsion tests performed on undrained saturated clay specimens of high compressibility (Tan, 1954) showed that the lower limit of the Bingham state of distortional creep can be less than a quarter of the strength if the creep occurs at constant volume and if the mineralogical and physical conditions are unfavourable to allowing strengthening effects of molecular forces between the particles to occur.

When applying the experience obtained from either drained or undrained laboratory tests to the conditions of natural slopes, attention should be paid to different consolidation and relaxation conditions occurring in nature. Under drained conditions the distortional creep of natural slopes may occur at constant volume if the pre-consolidation is high enough. On the other hand, the natural conditions of slow drainage can hardly allow relaxation effects as observed in undrained tests on saturated samples.

There is little or no experimental evidence on sustained load effects of very long duration or, furthermore, on periodic or occasional dynamic effects, or on the effects of a slow, but permanent oscillation of stresses, or on creep effects in contact surfaces between soils of different geotechnical properties. The combination of these effects can lead to more important displacements than expected from extrapolation of the observed deformations of homogeneous laboratory specimens exposed to sustained static loads.

The available experimental evidence indicates that the long-term failure of soils occurs either at the same or at a smaller distortional strain than the short-term failure. However, this conclusion has to be further verified by means of observations of the displacements occurring immediately before slope rupture.

Progressive failure can decisively influence the stability of slopes. Progressive failure is caused either by different resistances of the soil at different parts of the slope or by an unfavourable stress distribution (i.e. by different degrees of strength mobilization at different parts of the soil mass), or by both of these reasons.

The mobilization of a certain degree of strength requires different displacements at different parts of heterogeneous slopes. Thus the more resistant parts are initially stressed to a greater extent than the weaker parts. When the displacements increase sufficiently for the corresponding degree of strength in the less resistant parts to be mobilized, the peak strength of the stronger parts may already have been exceeded and the resistance reduced. During further creep the reduction continues. The successive deterioration of the strength of the more resistant parts can be terminated if, after the reduction in the strength of the stronger parts, the remaining overall resistance is sufficient to withstand the applied stresses. In other cases the reduction continues until angular displacements exceed certain limit values and a slip surface appears across the points of maximum shear strain.

The shear resistance remaining along the slip surface as a continuous surface of the discontinuity of displacements is the residual strength. If the average shear stress along the slip surface is smaller than the residual strength and only temporarily approaches the residual strength, the subsequent creep occurs mainly along the slip surface as a long-term "line creep". If, however, the average shear stress exceeds the residual strength, a sudden rupture occurs after the formation of a continuous slip surface. This can be avoided only by relaxation of distortional stresses which results from remedy measures or changed natural conditions. The thixotropic hardening effect can contribute to the retardation of the creep process.

So far as progressive failure is caused by an unfavourable stress distribution, it also affects normally consolidated sensitive clays and all kinds of over-consolidated soils whose strength decreases after passing a peak value. Bjerrum and Kjaernsli (1957) have presented some interesting case records of landslides in Norwegian quick clays starting with local failure in a part of the slope in which the stress conditions caused a reduction in the strength of the soil to the very small residual value.

In soils with constant peak value the failure develops either by successive mobilization and degradation of the peak value in different parts of the creep zone or as a creep process leading to the decrease of the peak value, or by combination of both effects. In stationary loading conditions the progressive failure caused by successive stress concentrations beyond the peak value is expected to develop with a speed of the order of magnitude close to the rate of the failure process observed on the laboratory specimen. Consequently, the retardation of the failure is to be interpreted either by periodical local stressing of short duration up to the peak strength, with

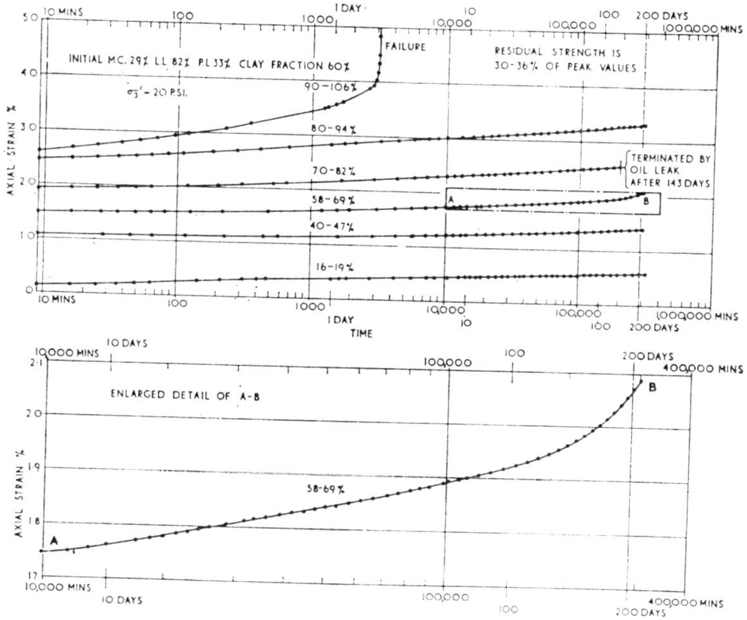

Fig. 20.1. Drained creep tests of undisturbed brown London clay (after Bishop, 1966): Axial strains

long intervals of lower stressing, or by a slow creep leading to the reduction of the peak value with time.

The danger of progressive rupture increases with the degree of heterogeneity of the soil, and the particular disposition of stiff fissured clays to progressive rupture can be attributed to different attractive forces between the bonds surrounding the particles inside and across the fissures and joints. The very slow reduction in the strength of stiff fissured clays (see Fig. 20.8) proves that in such clays progressive failure is usually a long-term process if not accelerated by deteriorated stress states. It appears in the form of a slow creep. The peak value of stiff fissured clays can hardly be expressed by constant strength parameters. In laboratory and field testing different values result; they depend on the testing techniques, on the size of the specimens, on the number and orientation of fissures in the specimen, on the rate of loading, and on the non-uniformity in mineralogical composition (de Beer, 1967; Bishop and Little, 1967; Kenney, 1967). The non-homogeneous resistance of the soil favours the process of progressive failure by stress concentrations; the retardation of

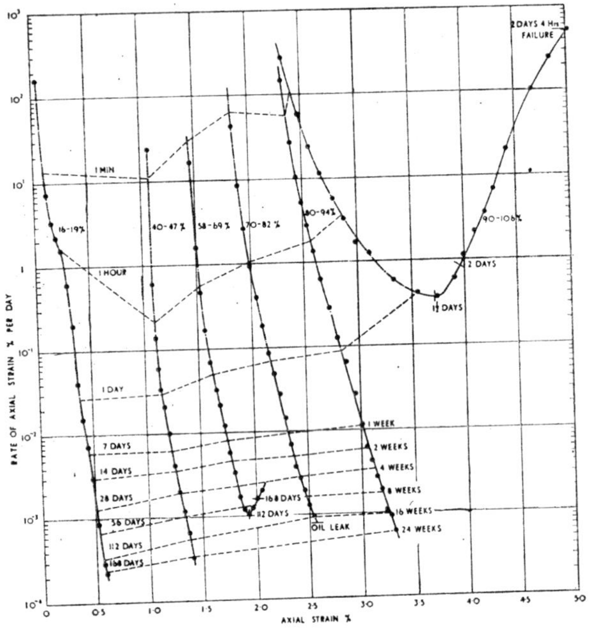

Fig. 20.2. Drained creep tests of undisturbed brown London clay (after Bishop, 1966): Rate of axial strain

the process, however, is to be interpreted similarly as in soils exhibiting constant peak strength (see Šuklje, 1967).

Even if a specimen of a stiff fissured clay is representative enough, the effect of progressive failure due to the heterogeneity of the clay can hardly be expected to reduce appreciably the peak value of the specimen in 'long-term' tests lasting only several months. The drained creep tests on undisturbed brown London clay (from Hendon) presented by Bishop (1966) seems to confirm this statement. The results of these tests made in a special creep cell (detailed description by Bishop, 1966) and running for about seven months are reproduced in Figs. 20.1 and 20.2. The applied principal differences as given in these figures are percentages of the peak values measured in drained triaxial tests of 5 days' duration. The 60 per cent showed a decreasing rate of strain for the first three and a half months (100 days). The rate subsequently increased steadily but at the last reading considered in Figs. 20.1 and 20.2 the axial strain was still only proceeding at two-thousandths of 1 per cent per day and at this rate would have taken another 1200 days to reach the strain at which the 90 per cent sample failed. In the subsequent period of two months the creep rate had not increased and the time to failure still could not be predicted with any certainty.

The stratigraphic contact surfaces of deposits of different geotechnical properties and the lower boundary of the weathered cover of substrata seems to be favourable for the development of failure surfaces of discontinuity. As will be explained in the next section the stress conditions are often favourable to their occurrence.

20.2 DOMAINS OF INTENSIVE CREEP

The extent and the form of the creep domain exhibiting sustained velocity depends on the stress field as well as on the statigraphic conditions and deformation properties of the layers. The role of the stress field may be explained by using the following simplifications of real natural conditions.

Consider a slope of uniform inclination δ (Fig. 20.3) and consisting of homogeneous soil. The ground-water level may be assumed to appear

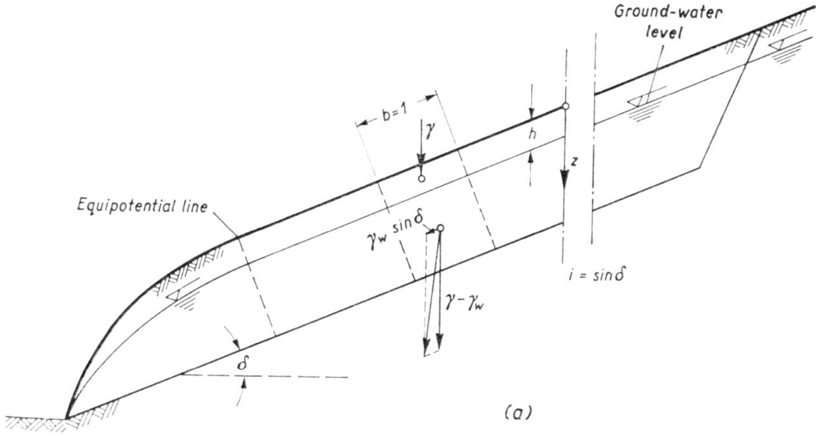

Fig. 20.3. (*a*)

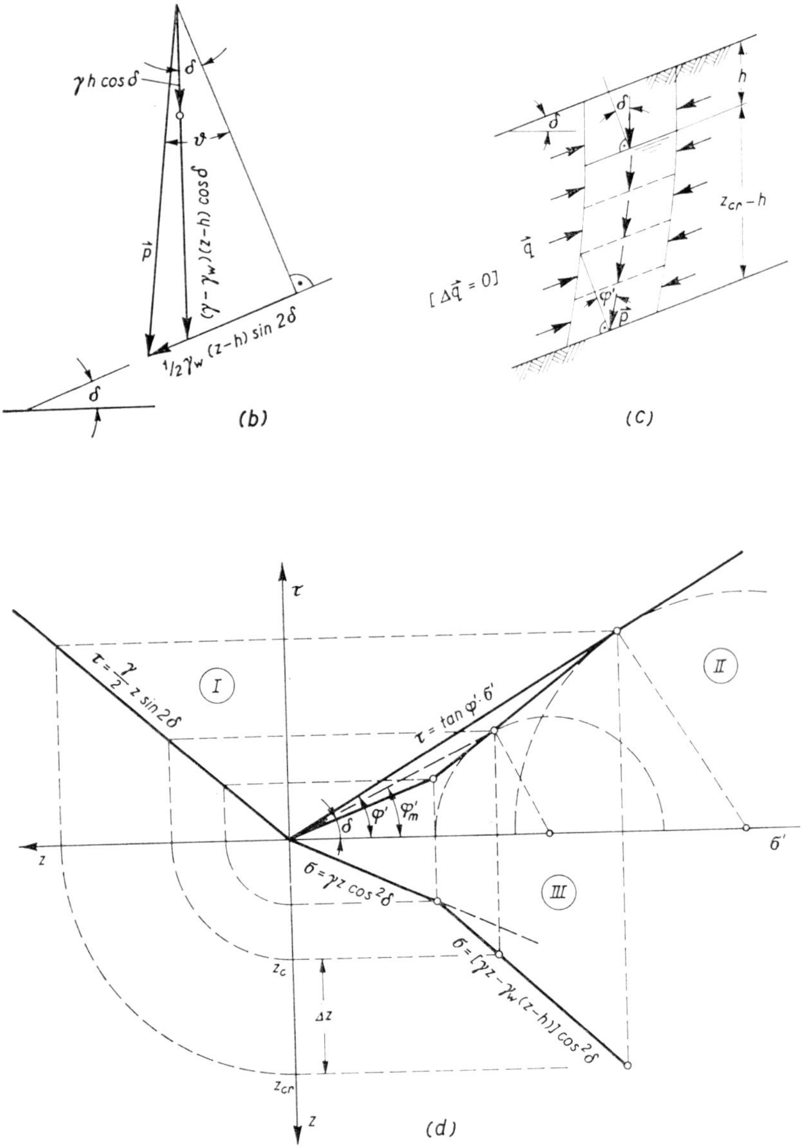

Fig. 20.3. Simplified stress analysis in a long cohesionless slope subjected to gravitational and hydraulic fields: (*a*) hydraulic potential field, (*b*) resultant stress vector *p* acting in the planes with the slope direction, (*c*) conjugate stresses, (*d*) construction of the creep domain

at a depth h below the slope surface parallel to the inclination of the slope. In the main part of the slope the loads due to the gravitational field $\vec{\gamma} z$ above the water level and to the combined gravitational and hydraulic field $[(\vec{\gamma} - \vec{\gamma}_w) + \gamma_w \vec{i}\,](z - h)$ below the water level may be assumed to be transferred entirely to the planes which are parallel to the slope. Thus

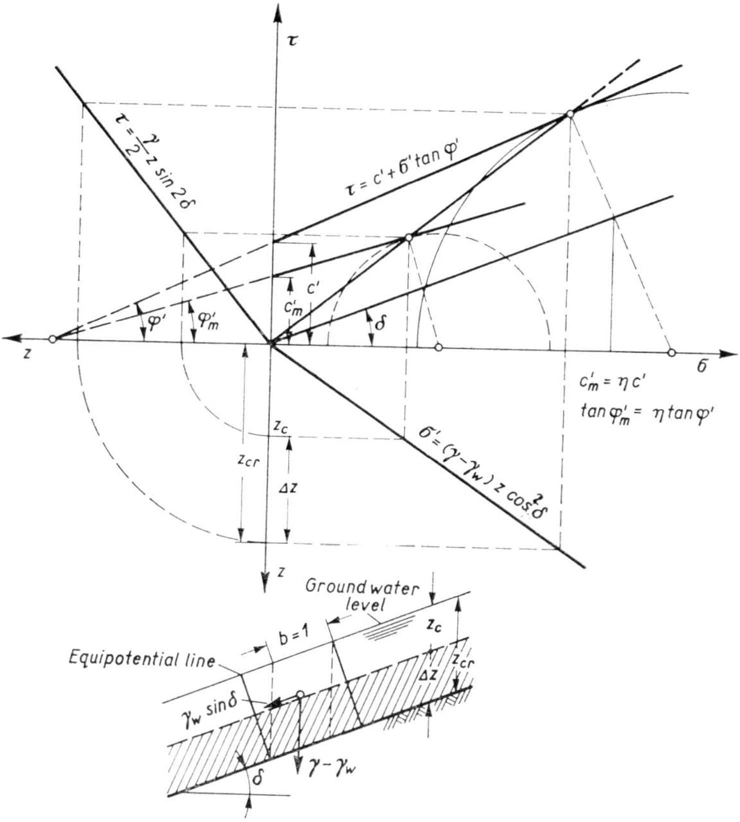

Fig. 20.4. Construction of the creep domain in a cohesive saturated infinite slope

the conjugate stresses at the boundaries of the elements deformed in the direction of the stresses acting on the parallel planes are assumed to originate only from the top of the slide and not to increase in subsequent elements (see Fig. 20.3-*c*).

The resultant stresses $\vec{p}$ acting in the planes with the direction of the slope are shown in Fig. 20.3 (*b*) and the increase of their normal (σ) and tangential (τ) components with depth is presented in the III and I quadrants of Fig. 20.3 (*d*). The corresponding change of the end points of the stress vectors $\vec{p}$ is presented by the broken line in quadrant II of Fig. 20.3 (*d*). The intersection of this broken line with the Mohr–Coulomb envelope determines the depth at which the plane parallel to the slope becomes the failure surface.

If the creep at constant velocity occurs in a domain of stress states defined by the failure envelope $\tau_f = \tau_f(\sigma)$ as the upper limit, and the envelope with ordinates $\tau_\eta = \eta\tau_f$, $\eta < 1$, as the lower limit, then the width of this domain $\Delta z = z_{cr} - z_c$ can be obtained by the graphical construction shown in Fig. 20.3 (*d*). It proves that the smaller the ratio h/z_{cr} for a given value of η, i.e. the nearer the ground-water level to the slope surface, the larger will be the width of the zone with sustained creep speed. If the ground-water level extends to the surface, all planes parallel to the slope are equivalent and the creep process occurs uniformly throughout the slope from the surface down to a firm base or to the parallel plane which intersects the foot of the slope.

The construction of the creep domain shown in Fig. 20.3 (*d*) applies to cohesionless material, but the construction remains the same in principle if the Mohr–Coulomb envelope has a cohesion intercept c' on the τ-axis. In the case of an entirely saturated slope (water-level coinciding with the slope surface) the analogous construction applying to cohesive material is presented in Fig. 20.2. In this case the creep intensity is shown also to increase with the depth z when the water-level reaches the surface of the slope.

Analogous constructions of intensive creep fields can be applied if the slope consists of two or more layers of different strength characteristics. In Fig. 20.5 an example is given for a two-layered slope. The upper non-cohesive layer of thickness H obeys the strength law $\tau_f = \tan\varphi_1' \cdot \sigma'$ while the lower cohesive layer is governed by the failure envelope $\tau_f = c_2' + \tan\varphi_2' \cdot \sigma'$. First the ground-water level is assumed to occur at a depth h below the slope surface. The $\tau = \tau(\sigma')$ solid line connecting the end points of the effective stress vectors acting on planes parallel to the surface of the slope does not intersect the strength diagram $\tau_f = \tan\varphi_1' \cdot \sigma'$ of the upper layer, but does intersect the failure envelope $\tau_f = c_2' + \tan\varphi_2' \cdot \sigma'$ of the lower layer. The point of intersection, A, defines, in the (z,σ') coordinate-system, the critical depth z_{cr}.

If then the ground-water level is assumed to rise up to the depth $z = h'$ below the slope surface, the critical depth in the lower clay layer decreases to z_{cr}°, and simultaneously failure occurs in the upper layer at the depth z_{cr}'. As the strength is exceeded in the entire zones $z_{cr}' \leqslant z \leqslant H$ and $z_{cr}^\circ \leqslant z \leqslant z_{cr}$, a sudden failure is expected to occur.

In the first case (water level at depth h) the development of failure below the depth z_{cr} may be assumed to be prevented by the stratigraphic or kinematic conditions of the slope. The domains of sustained intensive creep can then be ascertained in an analogous way to that demonstrated in the two preceding cases of homogeneous slopes.

The cases explained above prove that, except for entirely saturated non-cohesive slopes, the failure surface tends to be as deep as is allowed

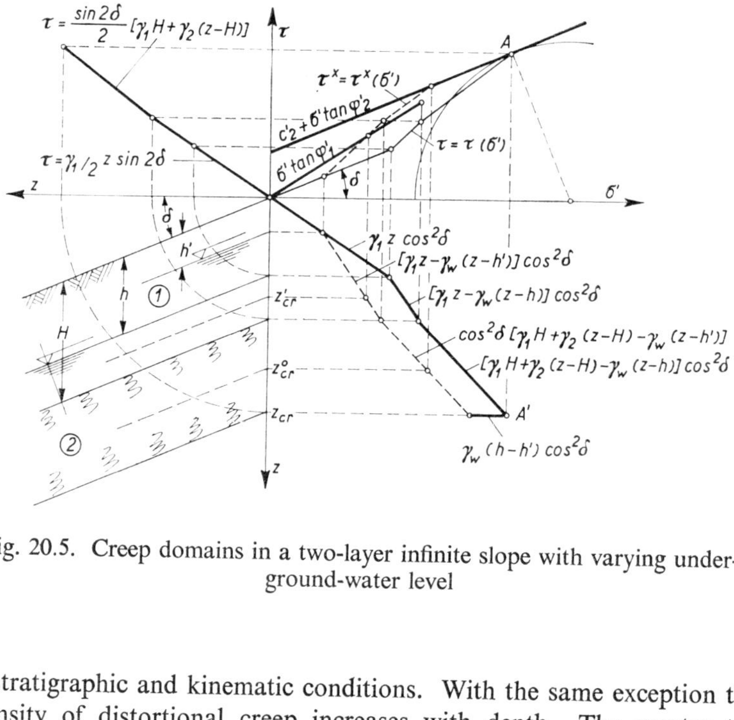

Fig. 20.5. Creep domains in a two-layer infinite slope with varying underground-water level

by stratigraphic and kinematic conditions. With the same exception the intensity of distortional creep increases with depth. The greater the cohesion part of the total shear strength the smaller the zones of intensive creep with sustained velocity. With the rising of the ground-water level the intensity of creep increases. The increase of angular deformations during periodic fluctuations of the ground-water level to the surface can lead to a surface of discontinuity in the corresponding critical depth. After such a surface of discontinuity has been formed, the strength along this surface decreases down to the residual value. Afterwards the creep can also develop with sustained speed during periods of lower water level. During repeated rises of the ground-water level, the creep speed increases and begins to accelerate and if the duration of the high water level is long enough, failure finally occurs.

In several cases investigated by the author (Šuklje, 1953-b; Šuklje and Sovinc, 1957; Šuklje and Vidmar, 1961-b) the surface of discontinuity developed at the contact either of a weaker clay with a firmer substratum or of more resistant and more permeable upper deposits with the lower clay layer. In the first case the appearance of the failure surface at the contact can be attributed to the effects explained above of the cohesion and of the ground-water level. In the second case the development of the failure surface at the upper boundary of the layer was usually caused by the topographic and kinematic conditions, combined with the erosion effects at the foot of the slope. Nevertheless some observations suggest that the pre-consolidation effects at the contact with deposits of different geotechnical properties are smaller than in the rest of the pre-consolidated clay layer. This phenomenon can easily be explained for the contact of a pre-consolidated clay layer with a coarse-grained pervious material. In other cases the above hypothesis needs further investigation.

Various topographic, stratigraphic, tectonic and hydrological conditions influence the development of domains of intensive creep, their size and form, and the nature of the final failure when it occurs. In principle, however, the effects of cohesion and of the ground-water level remain similar to those explained for the simplified conditions which were assumed in the above considerations.

20.3 CASE RECORDS OF THE CRITICAL CREEP OF NATURAL SLOPES

20.3.1 General Observations

By periodic measurements of the displacements of the surfaces of unstable slopes, the directions, speeds and accelerations of the surface creep can be determined. The results facilitate the stability analysis of the character of the creep. This analysis has to be based on the data of geological mapping of the slope, of bore-holes and exploration pits and on data concerning the depths and oscillations of the ground-water levels. More authentic information about the character of the creep can be obtained by supplementary observations of deep measuring points in galleries or boreholes, or by measuring the deformations of the boreholes as a whole. Analogous information can often be obtained by observations of the displacements and damage in tunnels and drainage galleries, as well as of the contour fissures and cracks on the slope and in the structures founded on the slope.

It is not possible to estimate the time to failure by means of the creep speeds of the points on the surface of a slope without knowing the character and the extent of the domain of extensive creep. Based on the considerations of the previous section the following types of extensive creep domains can be distinguished:

(*a*) The plastic zone is extended from the surface of the slope down to a certain depth; (*b*) the plastic domain is large but does not reach the surface of the slope; (*c*) the creep in a shallow plastic zone leads soon to the development of a slip surface of discontinuity. Combinations of these three types can be observed in many cases; e.g. zone creep of the top weathered material combined with deep creep in a limited plastic layer; or simultaneous creep in two creep layers developed at different depths within strata of different strength characteristics; or creep in a large, but limited plastic layer combined with line creep along a failure surface developed at the lower boundary of the layer.

The intensity of zone creep should be expressed by the speed of angular deformation, and that of line creep by the speed of linear displacement along the slip surface. As shown by the considerations of the previous section the creep speed in limited creep zones, which do not reach the surface of the slope, is expected to increase with depth if the values of the strength characteristics do not increase with depth. In such cases the representative value of the creep intensity is obtained by dividing the speed of the displacements at the upper boundary of the plastic zone by the thickness of the plastic zone in the direction of the normal to the boundary. The determination of the increase of the creep speed within the plastic zone requires observation of the displacements of several points within this zone.

At our disposal are numerous thorough descriptions of landslides containing the analyses of the conditions and causes of their occurrence. There is, however, little published information concerning the duration and speed of the creep preceding failure, or concerning the creep speeds of slopes which have been moving for many years without failing. (E.g.: Haefeli, 1953; Fukuoka and Taniguchi, 1961; Moos and Haefeli, 1962. It is to be regretted that the very interesting data published by Moos and Haefeli does not allow judgement to be made of the thickness of the zone of intensive creep.)

Estimates of the danger of creep is in many cases of great economical importance. They represent the bases of decisions concerning the extent of remedial measures or the need for evacuating the threatened slope and building new roads or railways or constructions on safer subsoils. There is an urgent need to collect further data on measurements of displacements of slopes than is available at present. Such data should be accompanied by the geological exploration of the slope, by laboratory testing of the creeping soil and by stability analyses which include the results of all the investigations mentioned above. The following case records do not represent a complete review of the available data. They serve only to demonstrate the possibilities of predicting the critical stage of creep on the basis of field and laboratory investigations.

20.3.2 Zone Creep

20.3.2.1 *Creep of Broken, Partly Disintegrated Rock*

On the basis of unconfined compression and triaxial tests as well as of observations of the displacements of slopes, Saito and Uezawa (1961) suggested that the time to failure t_r could be predicted by the equation:

$$t_r = \frac{0{\cdot}0214}{\dot{\gamma}} \tag{20.1}$$

$\dot{\gamma}$ being the angular strain velocity. The equation agrees with the observations of Goldstein and Ter-Stepanian (1957) that rupture is governed by a critical strain regardless of the duration of the preceding creep corresponding to the applied stress state (see Section 18.4.1).

The above equation will be compared with the observations of E. Nonveiller and the author (1954) made on the big landslide at Zalesina (Yugoslavia). The results of thorough investigations of this landslide were presented to the conference on Landslides in Stockholm (1954) and

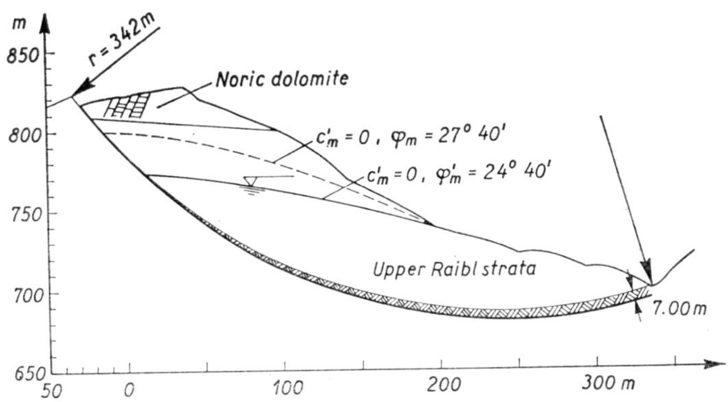

Fig. 20.6. (*a*)

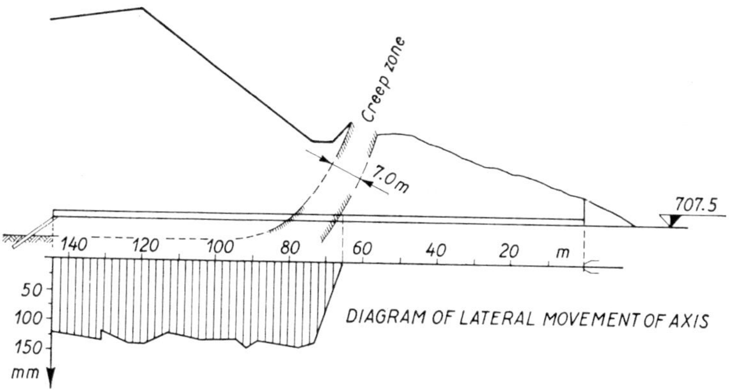

(b)

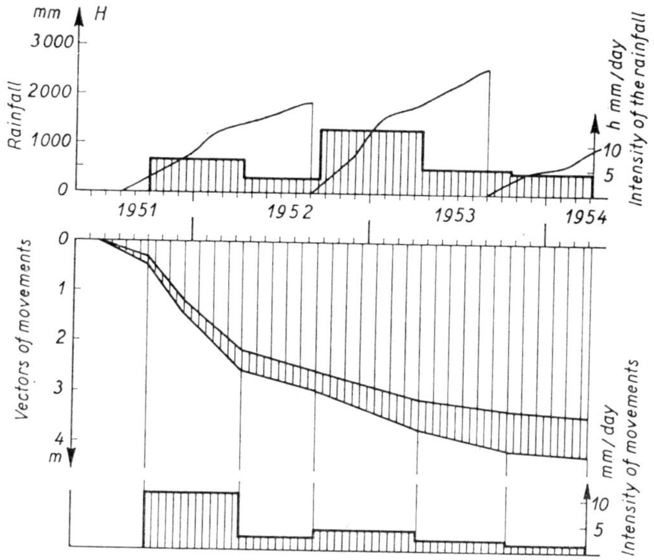

(c)

Fig. 20.6. Zone creep of the Zalesina slope (after Nonveiller and Šuklje, 1954): (*a*) creep zone, (*b*) displacements measured in an exploratory tunnel during the period from November 12, 1953, to March 31, 1954, (*c*) rainfall intensity, displacements and average creep speeds in the period 1951–54

data about further creep development were published in a supplementary report by Nonveiller (1956). Fig. 20.6 reproduced after Nonveiller and Šuklje (1954) shows in (*a*) a typical cross-section of the slope in the direction of the displacements; in (*b*) the displacements measured in an exploratory tunnel which was cut into the creep zone and entered the sliding masses; and in (*c*) the diagram of movements of various points on the surface of the slope for the period from June 1951 to May 1954; the diagrams of the intensity of movements and of the rainfall are also added. Fig. 20.7 shows further development of the creep until the end of 1961.

As proved by the diagram of displacements in the exploratory tunnel and by the boreholes, the creep zone is about 7 m wide and extends within the cross-section in the form of a circular arc down to depths of 80 m below the slope surface. The landslide area is about 500 m wide, and 300 m long in the direction of the movement. The whole creep zone has been

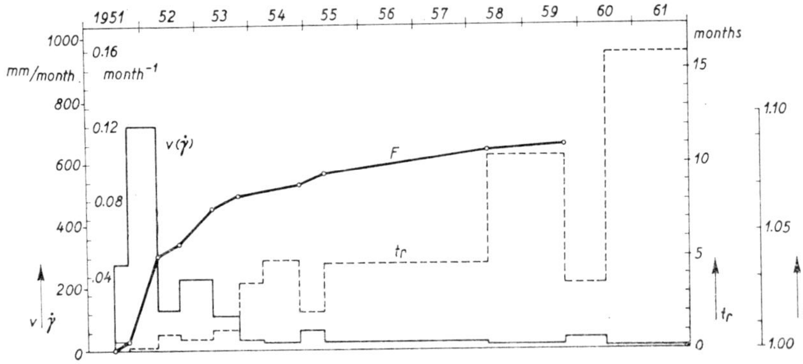

Fig. 20.7. Creep of the Zalesina slope: creep speeds in the period 1951–1961, the corresponding theoretical time to failure (t_r) according to Saito and Uezawa's formula, and the increase of the safety factor with time

developed in upper Raibl strata belonging to the upper Trias and consisting of yellowish siliceous sandstone and yellow-grey dolomitic sandstone, reddish porous slate, sandy schisty shale, and particularly of grey porous dolomite, shale dolomite and dolomitic shale with thin layers of very permeable porous limestone. The strike and dip of the beds changes in the area of the slide and the layers have been badly damaged by tectonic movements. In the creep zone the material was found to be completely broken, partly disintegrated, and fully saturated. By means of a stability analysis the initial mean value of the mobilized shear angle has been found to be about 23° if the observed ground-water level is taken into account, and 26° if the probable maximum level is considered. The intensity of the movements increases and decreases with the intensity of the rainfall and generally decreases when the masses move into a more stable position. This is shown by the relative increase of the factor of safety in Fig. 20.7.

In the same figure the displacement velocities at the upper boundary of the creep zone are plotted versus time. The scale of the angular strain velocities according to the width of the creep zone is also added. Simultaneously the diagram of the corresponding time to failure is shown as obtained by Equation (20.1). It can be seen that in this case the criterion suggested by Saito and Uezawa is on the very safe side.

This conclusion has been corroborated by the application of Equation (20.1) to the displacements observed on slopes consisting of badly damaged and partly disintegrated (due to tectonic movements) schists and sandstones of the lower Trias (Werfen) and creeping towards the Jablanica storage reservoir on the Neretva river (Yugoslavia) (see Šuklje, 1964-b).

20.3.2.2 *Creep of Stiff Fissured Clays*

Typical zone creep with little or no influence of cohesion on the long term stability can be observed in many slopes consisting of younger Tertiary clays (Pliocene, Miocene geological periods). It occurs due to a high ground-water level periodically reaching the surface of the slope. The average inclination ($\tan \delta$) of such slopes corresponds approximately to the formula

$$\tan \delta = \left(1 - \frac{\gamma_w}{\gamma}\right) \tan \varphi' \tag{20.2}$$

which can be deduced from Fig. 20.3 (*b*) by taking $h = 0$, $\vartheta = \varphi'$. The shear angle of specimens of these clays consolidated from slurry at the liquid limit varies from 16° to 23°, the bulk density γ from 1·8 to 2·0 t/m³ and the inclination of natural slopes from 7° to 12°. When excavating exploratory pits and drainage ditches, several slip surfaces parallel to the inclination of the slope can be observed at various depths. In some slopes slickensides appear in the conjugate direction and often secondary slip surfaces or slickensides with various directions can also be found.

Similar zone creep can be observed in the weathered cover of firm layers belonging to various geological periods and consisting of various minerals. Beside secondary slip surfaces appearing throughout the creep zone a very distinct surface of discontinuity can be found at the firm substratum of the weathered material whenever the topographic and kinematic conditions are favourable to its formation at the depth of the contact. The linear creep along such a surface of discontinuity will be considered further in the next section.

The data of stability analyses for slopes in London clay, which is a strongly over-consolidated stiff fissured clay, presented by Skempton (1964), are very instructive. The inclinations of unstable natural slopes in this clay exposed to sustained stresses with certain periodic variations due to ground-water oscillations vary between 10° and 12°. The ground-water level varies between the depths 1·50 and 3·00 m below the slope surface and the contact between the weathered and unweathered clay appears at depths between 7.50 and 13·5 m. The diagram of forces in Fig. 20.1 (*b*) yields the relation:

$$\tan \varphi' = \frac{\tan \delta}{1 - \dfrac{\gamma_w}{\gamma} \cdot \dfrac{z - h}{h}} \tag{20.3}$$

Inserting values of between 10° and 12° for δ, 1·9 t/m³ for γ, and $z = 7{\cdot}50$ to 13·5 *m* and $z - h = 6{\cdot}0$ to 12·0 m ($h = 1{\cdot}50$ m), effective shear angles

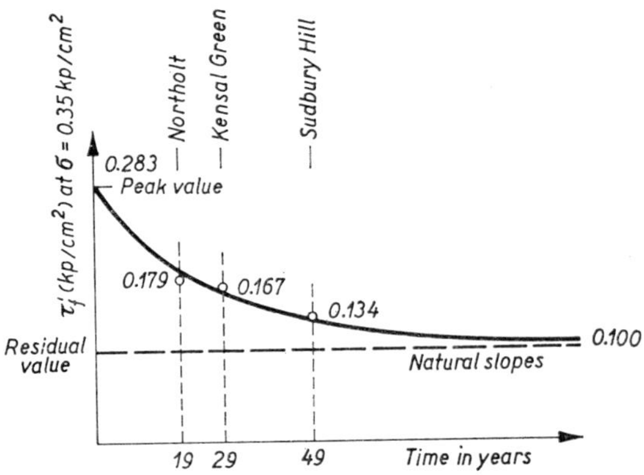

Fig. 20.8. Long-term strength of London clay according to Skempton's analysis (Skempton, 1964)

between 16° 55′ and 21° 50′ are obtained. For the same values of δ, γ, z and for $h = 3{\cdot}00$ m, Equation (20.3) yields $14° 35' \leqslant \varphi' \leqslant 19° 50'$.

When analysing the stability conditions of a limiting natural slope of 10° Skempton (1964) considered alternative values for z and h: 1·7 to 2·5 m for the depth of the slip surface (z) and 0·15 to 0·75 m for the water level (h). With this data Equation (20.3) yields effective shear angles (at zero cohesion) in the range $13° 30' \leqslant \varphi' \leqslant 19° 15'$ (average value: $\varphi' \simeq 16°$).

The 'long-term' strengths of three cuttings have been shown to decrease with increasing time to failure. The stresses at failure of a cutting at Northolt, which occurred 19 years after excavation, correspond to the Mohr–Coulomb envelope with parameters $c' = 0{\cdot}065$ kp/cm² and $\varphi' = 18°$. The rupture conditions of the failure of a retaining wall at Kensal

Green, which occurred 29 years after the construction, were governed by the parameters $c' = 0{\cdot}057$ kp/cm² and $\varphi' = 17° 30'$. Finally, a cutting excavated at Sudbury Hill in 1900 slipped after 49 years of creeping at stresses corresponding to the envelope defined by $c' = 0{\cdot}027$ kp/cm² and $\varphi' = 17°$. Now, in all three cases, the time to failure corresponding to sustained constant load is certainly shorter because in the first years after the excavation the effective stresses increased according to the dissipation of negative pore pressures appearing on the unloading of the slope. But the stability analyses of these three slides were carried out using the effective stress method, and thus the important influence of the creep duration on the long-term strength cannot be doubted.

For an average effective normal stress 0·35 kp/cm² in the potential failure surface of the above cited case records the failure shear stress have been computed and plotted versus time to failure in Fig. 20.8. The plot proves that the degradation of the peak strength of stiff fissured clays is a very slow process. Consequently the laboratory 'long-term' tests lasting, commonly, only a few weeks or months, cannot give but limited information about the time effect on the reduction of the strength of the clay.

20.3.2.3 *Creep of Intact Clays*

The analysis of the enormous Gradot landslide (Macedonia, Yugoslavia, 1956) presented by S. Vidmar and the author (Šuklje and Vidmar, 1961 (*b*)) led to the hypothesis that failure occurred along the slightly inclined contact between the overlying silty deposits and the clayey base. The stratigraphic composition of the slope and the probable form of the slip surface are shown in Fig. 20.9. The basic cause of the landslide has

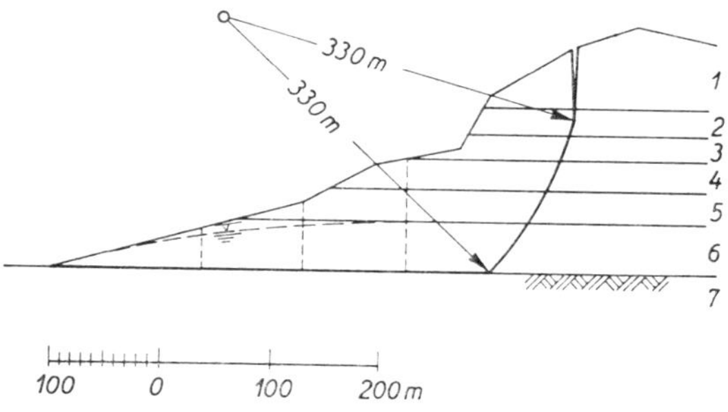

Fig. 20.9. Landslide Gradot, geological cross-section and the slip surface; 1, andesite tuff; 2, sand and gravel, well bound; 3, conglomerate; 4, sand and gravel weakly bound; 5, gravel not strongly bound; 6, silt with silty sand intercalations; 7, clay. (After Šuklje and Vidmar, 1961)

been the erosive action of the Vatasha river deepening its bed through the silty layers into the clay series. Owing to the unequal creep deformation of the underlying cohesive strata, the cohesion of the sandstone, of the conglomerate and of the volcanic overburden, was overcome decades or even centuries before the landslide took place. Later the shear strength of the slightly over-consolidated cohesive lower strata was overcome by creep which occurred over a long period. Dynamic earthquake effects contributed towards cracking of the rigid strata and also helped to sustain the creep in the cohesive base.

Along the upper boundary of the slide a discontinuous series of deep cracks were formed between elevations 664 and 716 over a length of about 270 m. The oldest people had known these since their childhood and thirty years before the failure a man descended one crack by means of a rope to a depth of 30 m but did not reach the bottom. The cracks were separated in the upper part only by loosened and fractured material, but were continuous under the surface. The longest was about 20 m long and 130 cm wide. A year before the failure the lower boundary of the cracks was observed to have subsided 20 cm. Further subsidence appeared some weeks before the failure, and natural bridges, separating single cracks, broke off.

Based on the above data it is possible to estimate that the average rate of creep in the last hundred years must have been smaller than 15 mm per year, but it increased progressively before failure.

Stability analysis made by the authors on four representative cross-sections in the directions of the movements has proved that the creep preceding failure could have occurred at the following combinations of the mobilized shear angles:

Mobilized shear angles φ' in the layer series (see Fig. 20.9):

2–5 sand and gravel	6 silt and silty sand	7 clay	
34° 30′	24° 55′	19° 45′	average values of four cross-sections
38°	28°	17° 30′	
36° 55′	27° 5′	21° 15′	maximum values related to the cross-section f–f (see Šuklje and Vidmar, 1961-b)
38°	28°	20° 30′	

By laboratory tests on remoulded normally consolidated samples the following parameters of failure envelopes have been found:

Layer series	Peak value of the inclination of the tangent to the failure stress circles at lower ($\sigma \leqslant 4$ to $6\ kp/cm^2$) normal stresses	at higher ($\sigma \geqslant 6\ kp/cm^2$) normal stresses	Residual value at lower stresses ($\sigma \leqslant 6\ kp/cm^2$)
6 silt	26° 30′ to 27°	24° 30′	18°
6 silty sand	32°	30° 30′	—
7	20° 30′ to 24° 30′	19°	10° to 15° 30′

The friction angle of pre-consolidated specimens of the clayey layer (*b*) was found to be $\varphi_p' = 17° 30'$ as compared with the shear angle $\varphi' = 20° 30'$ obtained when testing the normally consolidated specimens belonging to the same test series; both values were obtained in the domain of normal stresses $\sigma \leqslant 6\ kp/cm^2$.

The actual normal pressures preceding failure varied, in the clayey part of the slip surface, between zero and 280 t/m^2 while the pre-consolidation pressure is estimated to have been about 270 t/m^2.* Thus if the pre-consolidation is taken into account, the average shear resistance in the clayey part of the slip surface would be about

$$27\ (\tan 20° 30' - \tan 17° 30') + 14\ .\ \tan 17° 30' = 6{\cdot}00\ kp/cm^2.$$

If the over-consolidation is neglected, an average shear strength of

$$14\ .\ \tan 20° 30' = 5{\cdot}23\ kp/cm^2$$

corresponds to the virgin shear angle of 20° 30′. The ratio of the above values $F = 6{\cdot}00/5{\cdot}23 = 1{\cdot}15$ proves that the pre-consolidation effect could not have been important. Nevertheless comparison of the computed shear angles and those obtained by strength tests proves that the long-term creep effects completely eliminated the cohesive part of the shear strength of the clay layers as well as across the silty series, and slightly reduced the friction part corresponding to the actual overburden pressure. The total reduction can roughly be estimated to have been about 20 per cent.

The probable mechanism of the failure corresponds generally to the mechanism as explained in Section 20.1 for clays exhibiting a constant peak value. Periodical overstressing was caused by earthquake shocks. The non-uniform over-consolidation as well as the relative rigidity of the

* In the paper by the author and S. Vidmar the pre-consolidation pressure of 500 t/m^2 has probably been overestimated because it had been computed using the total bulk density of the overburden. Assuming that the water level in the lake deposits did not decrease with increasing erosion, the overburden pressure should be computed using the submerged density yielding the above value of about 270 t/m^2.

overlying, partly cemented layers favoured the progressive failure starting from the inner side of the creep zone. The very low value of the residual strength supports the hypothesis that the long-term creep occurred as zone creep in the form of continuous angular deformations and that the continuous slip surface at the contact of silty and clayey layer probably appeared only as the last stage of the long-term creep followed by a sudden failure.

20.3.3 Line Creep

An instructive example of line creep is the creep of the slopes above the Radovna river (Slovenia) consisting of a stiff fissured Tertiary clay,

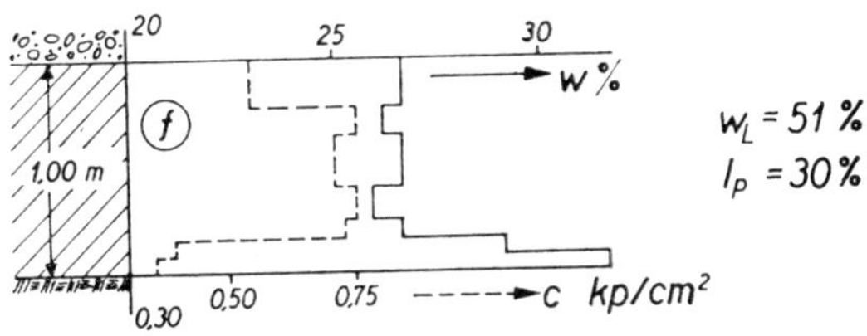

Fig. 20.10. Creep of Radovna-river slopes in a over-consolidated Oligocene clay: water content w and cohesion c varying with depth

belonging to the Oligocene period (Šuklje and Sovinc, 1957). The inclination of the creeping slope is about 14°. The slip surface occurs in the immediate vicinity of the contact between the plastic clay about 1 m thick and its base consisting of an essentially stiffer clay often bound with marl. Over the plastic clay lies a natural capping about 2 to 3 m thick consisting of materials which have previously slid, and deposits of clayey gravel, sand and silt.

When, in 1955, the drainage system in the slope under the railway was reconstructed, the slip surface of discontinuity was clearly visible along the entire main drainage ditch over a length of about 100 m. The shearing strength of remoulded samples was tested in torsion ring shear apparatuses and the results interpreted after Hvorslev (see Section 6.1); in slow tests the shear-strength parameters $\kappa = 0{\cdot}055$, $\varphi_r = 13°\ 55'$ have been obtained. In Fig. 20.10 the moisture content of the clay layer above the slip surface and the cohesion (= half the unconfined compression strength) are plotted versus depth. It can be seen that the moisture content increases considerably in the slip surface near the contact and that the cohesion decreases accordingly.

The effective stress stability analyses has proved that the cohesion for limiting equilibrium equals 0·079 kp/cm², if the angle of shear $\varphi_r = 13° 55'$ (after Hvorslev); it corresponds to an equivalent consolidation pressure of $p_e = 1{\cdot}50$ kp/cm² according to a moisture content of 31·4 per cent. This value agrees with the moisture content of the clay in the slip surface.

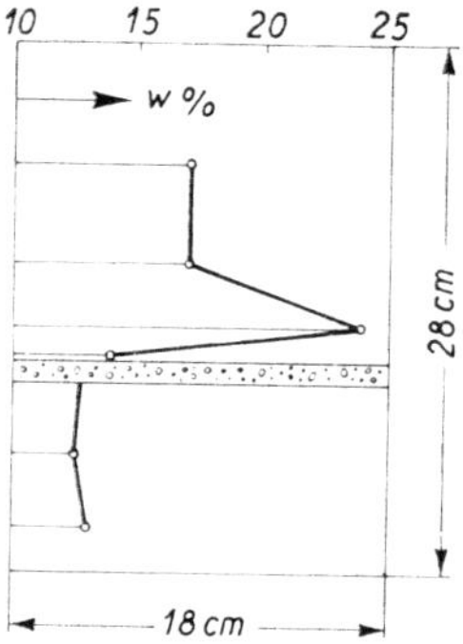

Fig. 20.11. Creep in a Oligocene-clay slope (Moste): variation of the water content in the vicinity of the slip surface

The resulting equivalent consolidation pressure exceeds the actual overburden pressure, however, it is essentially smaller than the equivalent pressure corresponding to the average moisture content of the layer. The $\varphi = 0$ analysis yields $c = 2{\cdot}56$ t/m² which is not far from the value of 3·3 t/m² obtained from the unconfined compression test on the sample from the slip surface. Similar results obtained for samples outside the slip surface are essentially greater. The long-term decrease in the peak value of strength as obtained by unconfined compression tests on small cylindrical samples can be attributed to the combined effects of sustained loading, of progressive failure due to numerous jointing fissures and tectonic slickensides, of free water circulating along the fissures and slickensides, and to the probable diminution of the pre-consolidation effect at the contact with the firmer substratum.

In the same Tertiary clay a clearly developed slip surface running parallel with an inclusion of fine sand some millimeters thick was discovered in 1948 during the excavations of the drainage ditches in the slide at the Moste power station (in the upper Sava Valley, Slovenia). (See Šuklje and

Sovinc, 1957.) The sample containing this inclusion (Fig. 20.11) was sheared directly in the direction of the inclusion. The average moisture content of the clay above the inclusion of about 17 per cent increased to a value of 25 per cent at the slip surface which was developed at a distance of 1–2 cm from the sandy inclusion. The moisture content of the clay below the inclusion was rather smaller—about 13 per cent. The shearing strength obtained of 4·1 t/m^2 agreed satisfactorily with the result of the $\varphi = 0$ stability analysis yielding $c_m = 4{\cdot}57$ t/m^2. The effective stress analysis showed that some pre-consolidation effect has

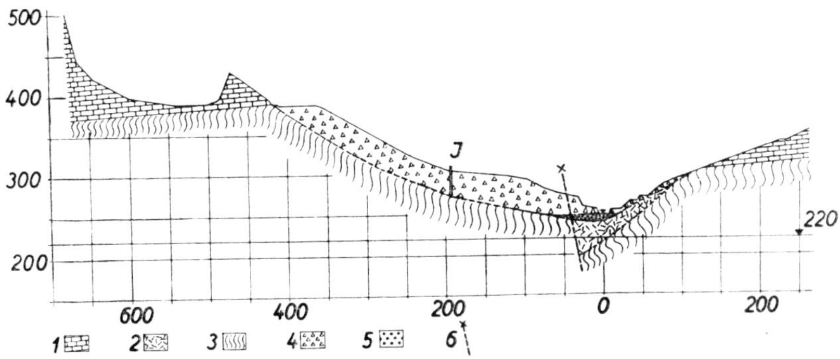

Fig. 20.12. Creep along the clayey contact of detrital limestone and carboniferous shists at Hrastnik (Slovenia): 1, Tertiary (Leitha) limestone; 2, Triassic dolomite; 3, Carboniferous shists; 4, detrital limestone and dump; 5, Alluvion; 6, fault; J, exploratory pit

remained even in long-term stability conditions. It represents, however, only a small part of the full pre-consolidation affecting the density (water content) of the clay.

The inclined contact of a more or less permeable mass with a poorly permeable pre-consolidated clay or clayey substratum is also favourable to the development of a slip surface of discontinuity. Fig. 20.12 presents the cross-section of a slope (Hrastnik, Slovenia) where broken-down blocks of limestone and deposits of a lime-kiln of a total depth up to 40 m are creeping along a thin plastic clayey cover of carboniferous schists. The slip surface was seen in an exploratory pit excavated in the lower part of the moving slope. The creep has been continuing for at least 70 years. The bridge at the toe of the slope was built in 1848 and since 1899 has been reconstructed several times due to the damage caused by the effects of the creep of the slope. After the last repairs in 1958 the displacements have been measured at the bridge and at several points on the slide area. At the toe of the slope the creep speed is about 25 mm per year and it does not increase essentially in the direction of the top of the moving slope in spite of the length of the slip surface being about 450 m. Owing to the additional load of the deposits the speed has probably been increasing since 1948. Before, it was certainly less than 2 cms per year. The mobilized effective shear angle is about 17°.

21

Tensile and Bending Strength of Soils

21.1 TENSILE STRESSES IN SOILS

In the elastic isotropic homogeneous half-space subjected to a concentrated surface load radial tensile stresses (σ_r) appear at certain values of the Poisson ratio ν and direction ϑ (see Equation 5.20 and Fig. 5.2). In the axis of the load ($\vartheta = 0$) the stresses σ_r are positive for any value of ν except $\nu = 0{\cdot}50$.

In a layered half-space with smooth contacts, subjected to circular loads, radial tensile stresses occur at the contacts.

Dangerous tensile fissures can appear in cohesive layers at the base of open excavations subjected to artesian water pressure. Under the influence of bending moments tension occurs on the upper side of the layer in the middle, and on the lower side at the fixed-ends. As shown by the author elsewhere (Šuklje, 1961a, b) the critical hydraulic gradients in the cohesive base depend on the strength (shearing and tension resistance) of the soil, on the dimensions of the excavation and on the surface filter load. The effects of the cohesion decrease with the increase in the ratio between the horizontal dimensions of the excavation and the thickness of the cohesive base, whilst the effect of the seepage pressure decreases with increasing base thickness.

In cohesive slopes which creep, critical stress states with one tensile principal stress appear in the upper parts of the slopes. The corresponding Mohr stress circles approach the failure envelope which may also be extended through the tensile domain of the $\tau = \tau(\sigma)$ diagram.

A thorough analysis of the tensile strain states is needed when constructing clay cores of earth dams. Tensile fissures in such cores can represent a dangerous starting point for the erosive action of seepage water and, therefore, they have to be avoided. They are due either to different settlements and horizontal displacements of the foundation and of the core itself, or to friction forces between the core and the flanks. In the first case vertical fissures parallel to the seepage flow appear at the joints between the core and the valley slopes in the upper part of the core as well as at places of maximum curvature along the contact of the core with the foundation. In the second case horizontal fissures appear as a result of the retention of the settlement of the core.

The influence of creep on the development of tensile strains and on the strength in domains of stress states with tensile stresses is expected to be very important. In the compression stress states the decreasing rate of the spherical component of creep can be interpreted by the increasing density of the soil. In the tensile stress state it can be attributed only to favourable re-arrangement of the clay particles and to increasing molecular forces between the bonds of the bound water surrounding the particles. Thus the intensity of spherical creep decreases with increasing compressive octahedral stress and increases with increasing tensile octahedral stress. The distortional creep is expected to be of similar character as in the domain of compressive normal stresses.

Only pre-consolidated cohesive soils can resist tensile loading. Nevertheless they cannot be treated as media of isotropic deformability. When pre-consolidated soils are subjected to pure tension or to pure compression, different moduli originating from the hysteresis effect are expected to govern their behaviour in the elastic domain. The failure envelope in the domain of tensile normal stresses has to be ascertained by adequate tests.

The uni-axial tensile stress-state is unlikely to appear in natural soils or earth constructions. Nevertheless it is of interest in the study of the tensile deformability and tensile strength of soils because testing of specimens subjected to pure tension enables the problem to be isolated. Now, tests with a pure uni-axial tension stress state cannot easily be effected. For this reason and in order to study the effects of stress states with tension stresses as they really occur, experiments should also be made where cylindrical specimens are subjected to states of plane stress and/or strain with tension and compression stresses, and where beams are subjected to bending moments.

21.2 PURE TENSION TESTS

Pure tension tests have to be carried out in such a way that the distribution of tensile stresses in the test section of the specimen is as uniform as possible. Furthermore the transfer of the tension force from the widened heads to the test section must not cause concentrations of stresses in the transition part of the specimen, and finally, it must also be possible to apply a continuous increase of stress. In order to eliminate the unfavourable influence of the self-weight of the specimen on the initial readings of deformation, the specimens are commonly tested in the horizontal position. The effect of friction should be reduced to a minimum by using adequately smooth supports. In Tschebotarioff and colleagues' (1953) tests (Fig. 21.1) the movable end section of the soil mould was supported on ball bearing rollers and the central section on line supports with ball bearing rollers at their ends. The pull was transferred to the specimen by the compression and shear stresses acting on the lateral flanks of the metal heads. Hasegawa and Ikeuti's (1964) tests were performed on specimens of circular cross-section; the specimens were

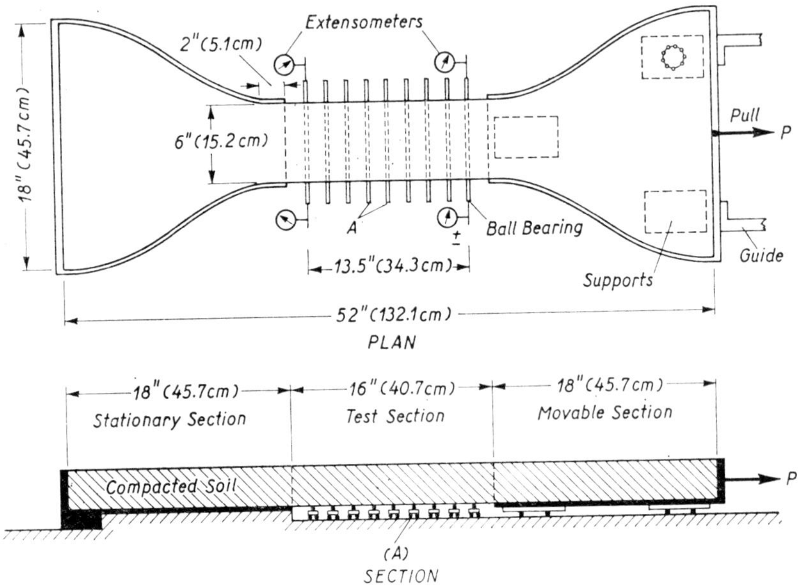

Fig. 21.1. Apparatus for uniaxial tension tests. (After Tchebotarioff, Ward and De Philippe, 1953)

placed on mercury and the pull was transferred by two small anchorage plates built into the egg-shaped heads. Drnovšek and Šubičeva (1966) preferred to use specimens of rectangular cross-section and of the form shown in Fig. 21.2; the pull was transferred to the specimen through metal strips welded to the upper and lower surfaces of the fixed and movable heads, and the central test section as well as the movable end were supported by small cardboard plates lying on smooth line rollers. The strains were computed from the displacements of measuring points arranged on the vertical face of the test section of the specimen in two horizontal and several vertical lines. Photographic recording of the displacements was employed as introduced previously by the author in bending tests (see Section 21.4.1)

Tschebotarioffs and colleagues' (1953) test results show that the tensile strength decreases with the test duration and tends to a final long-term

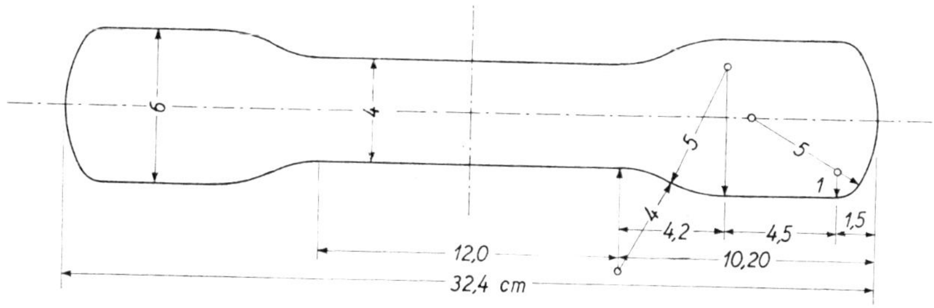

Fig. 21.2. Specimen form for uniaxial tension tests according to Drnovšek and Šubičeva (1966)

value. The decrease of the failure strain with time to failure resulting from the same tests does not agree either with the experiences of Hasegawa and Ikeuti (1964) or with the results of Leonard's and Narain's (1963) bending tests.

When investigating the apparent cohesion of sandy specimens, S. Andrée (1961) eliminated the friction by using equipment the scheme of which is shown in Fig. 21.3. This solution, however, ensures neither a uniform distribution of stresses in the cross-section between both closed

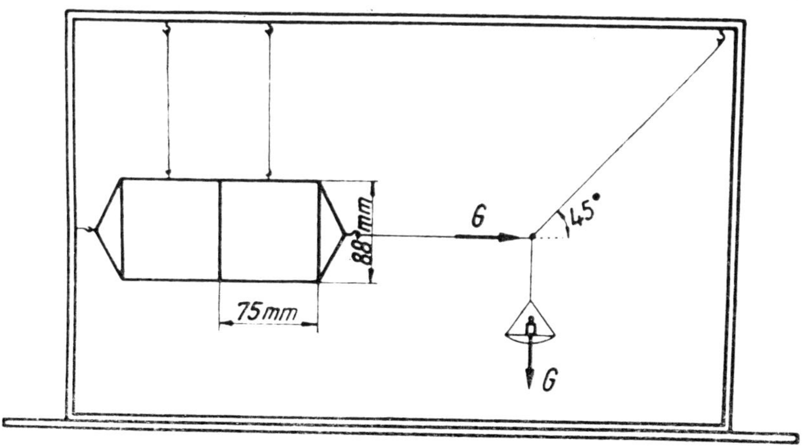

Fig. 21.3. Scheme of apparatus for uniaxial tension tests of sands. (After Andrei (1961))

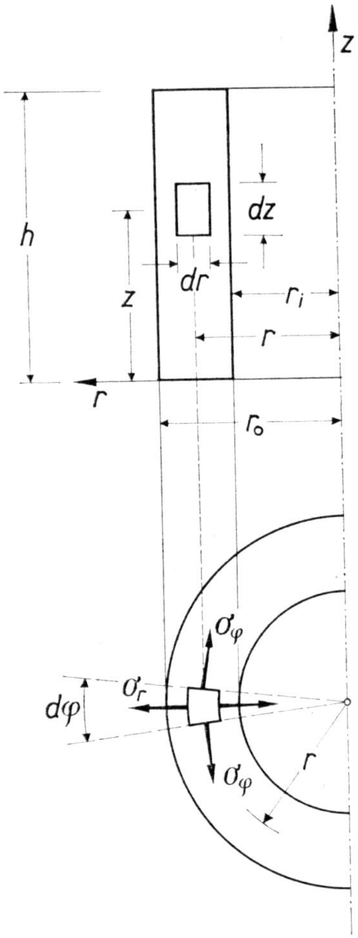

Fig. 21.4. Stresses in a hollow cylinder

parts of the specimen nor does it permit the measurement of the deformation of the specimen.

21.3 TESTS ON HOLLOW CYLINDERS

21.3.1 Elastic Domain

The deformability and the strength of soils in the domain of Mohr stress circles with one positive (tension) principal stress can successfully be investigated on specimens having the form of hollow cylinders and subjected to various internal and external hydrostatic pressures. The advantages of such tests are:

(*a*) Well-defined stress states without uncontrolled stress concentration; (*b*) the possibility of investigating the deformability and strength at various stress states; (*c*) the possibility of taking into account the deformation anisotropy; and (*d*) a more likely possibility of carrying out long-term and drained tests. The analytical expressions enabling the determination of the deformation characteristics in the domain of the linear stress–strain relationship have been derived by Šuklje and Drnovšek (1965); the different deformabilities in the tension and compression directions had been taken into account. Starting from the general equations of the theory of elasticity and considering the boundary conditions, the following solutions have been obtained:

(*a*) For the state of plane strain:

$$\sigma_r = Ar^{x_1} + Br^{x_2} \tag{21.1}$$

$$\sigma_\varphi = A(x_1 + 1)r^{x_1} + B(x_2 + 1)r^{x_2} \tag{21.2}$$

$$\sigma_z = \nu_{rz}\sigma_r + \nu_{\varphi z}\sigma_\varphi \tag{21.3}$$

$$\varepsilon_\varphi = Ar^{x_1}[m(x_1 + 1) - l] + Br^{x_2}[m(x_2 + 1) - l] \tag{21.4}$$

$$\nu_{rz}\left[\frac{A}{x_1 + 2}\,r^{x_1+2} + \frac{B}{x_2 + 2}\,r^{x_2+2}\right]_{r=r_i}^{r_0} + \\ + \nu_{\varphi z}\left[\frac{A(x_1 + 1)}{x_1 + 2}\,r^{x_1+2} + \frac{B(x_2 + 1)}{x_2 + 2}\,r^{x_2+2}\right]_{r=r_i}^{r_0} = \frac{P}{2\pi} \tag{21.5}$$

where the symbols of Fig. 21.4 have been used in addition to the following:

$$A = \frac{p_i r_0^{x_2} - p_0 r_i^{x_2}}{r_i^{x_1} r_0^{x_2} - r_i^{x_2} r_0^{x_1}}, \qquad B = \frac{p_i r_0^{x_1} - p_0 r_i^{x_1}}{r_i^{x_2} r_0^{x_1} - r_i^{x_1} r_0^{x_2}}, \qquad (21.6), (21.7)$$

$$x_{1,2} = -1 \pm \sqrt{\left(\frac{k}{m}\right)} \tag{21.8}$$

$$\left.\begin{aligned} k &= \frac{1}{E_r} - \frac{\nu_{rz}^{\ 2}}{E_z}, \qquad l = \frac{\nu_{\varphi r}}{E_r} + \frac{\nu_{rz}\nu_{\varphi z}}{E_z}, \\ m &= \frac{1}{E_\varphi} - \frac{\nu_{\varphi z}^{\ 2}}{E_z} \end{aligned}\right\} \qquad (21.9\text{–}11)$$

E, E_φ and E_z are linear deformation moduli in the directions r, φ and z; $\nu_{ik}(i,k = r,\varphi,z)$ are the Poisson ratios governing the influence of the stress σ_k on the strain ε_i; p_i is the internal and p_0 the external radial hydrostatic pressure.

(*b*) For the state of plane stress:

$$\varepsilon_\varphi = \frac{1}{rE_\varphi}\{Ar^{\sqrt{(\alpha)}}(\sqrt{(\alpha)} - \nu_{r\varphi}) - Br^{-\sqrt{\alpha}}(\sqrt{(\alpha)} + \nu_{r\varphi})\} \tag{21.12}$$

and Equations (21.1), (21.2), (21.5–7), (21.9–11), all with the following meaning for the parameters x_1, x_2:

$$x_{1,2} = -1 \pm \sqrt{\alpha}, \qquad \alpha = \frac{E_\varphi}{E_r} \tag{21.13}$$

The pressures p_i and p_0 are known. The shear strain ε_φ at $r = r_i$ and $r = r_0$ can be obtained from the volume changes ΔV_i and ΔV_0 of the liquid in the internal and external cells of the equipment (see Šuklje and Drnovšek, 1965) and from the observed linear strain ε_z. Inserting the values obtained for $r = r_i$ and $r = r_0$ in Equation (21.12) two conditions are obtained yielding for various assumed values of the ratio α corresponding combinations of the elastic constants E_φ and $\nu_{\varphi r}$. The correct values can be found either by repeating the procedure for another combination of the pressures p_i' and p_0' within the observed load interval satisfying the condition

$$\frac{p_i'}{p_0'} \neq \frac{p_i}{p_0} \tag{21.14}$$

or by intersecting the obtained plot $E_\varphi = E_\varphi(\alpha)$ with the straight line $E_\varphi = E_r\alpha$. E_r can easily be determined from unconfined compression tests effected in the corresponding interval of the compression stresses.

The tests carried out so far (Šuklje and Drnovšek, 1965) indicate a linear stress–strain relationship for undrained tests up to a yield point at which the stress versus strain plots become curved towards the strain axis. In load steps below the yield point the creep increases with negative acceleration; above the yield point the creep continues at constant rate leading to failure.

If the deformation is assumed to be isotropic and the hollow cylinders are subjected only to internal radial pressure $p_i(p_0 = 0)$, Equations (21.1), (21.3) reduce to:

$$\sigma_r = \frac{p_i}{\rho_i^2 - 1}\,[\rho^2 - 1) \tag{21.15}$$

$$\sigma_\varphi = \frac{p_i}{\rho_i^2 - 1}\,[-\rho^2 - 1] \tag{21.16}$$

$$\sigma_z = -2\nu\frac{p_i}{\rho_i^2 - 1} \tag{21.17}$$

where

$$\rho_i = \frac{r_0}{r_i}, \qquad \rho = \frac{r_0}{r} \tag{21.18–19}$$

Equation (21.17) proves that the stresses σ_z are constant throughout the cross-section of the specimen and along the z-axis. Thus the state of plane strain is also a state of plane stress. As, however, Equation (21.17) requires positive (tension) stresses σ_z for the condition $\varepsilon_z = 0$, the state of plane stress with $\sigma_z = 0$ appears automatically. Equations (21.15) and (21.16) remain unchanged while the expression for the linear strain ε_φ in the tangential direction can be deduced from Equation (21.12):

$$\varepsilon_\varphi = \frac{-p_i}{\rho_i^2 - 1} \cdot \frac{1+\nu}{E} \cdot \left(\frac{1-\nu}{1+\nu} + \rho^2\right) \tag{21.20}$$

(In all the above equations tension stresses and tension strains are considered as positive values.) By inserting the measured strains ε_φ at $r = r_i$ and $r = r_0$, two conditions can be obtained from Equation (21.20) yielding the elastic constants E and ν. The available experimental evidence proves, however, that the deformation anisotropy has to be considered.

21.3.2 Plastic Solutions for Stresses Accounting for the State of Plane Stress ($\sigma_z = 0$)

21.3.2.1 *Solution According to the Mohr–Coulomb envelope of the Failure Stress Circles*

Failure may be assumed to be defined by the Mohr–Coulomb linear

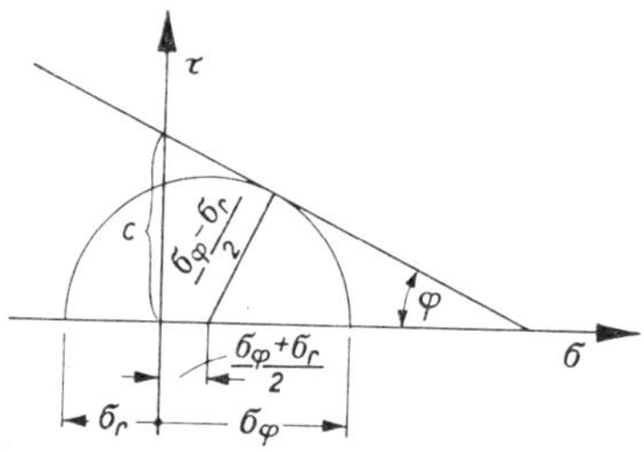

Fig. 21.5. Limit envelope of Mohr's circles in the domain of tensile stresses

envelope to the stress circles (Fig. 21.5). Tensile stresses may again be taken as positive values. After Fig. 21.5 we get the relation:

$$\frac{\dfrac{\sigma_\varphi - \sigma_r}{2}}{c \cot\varphi - \dfrac{\sigma_\varphi + \sigma_r}{2}} = \sin\varphi \tag{21.21}$$

By appropriate rearrangement we obtain:

$$\sigma_\varphi = \sigma_r K_a + 2c\sqrt{(K_a)} \tag{21.22}$$

where

$$K_a = \tan^2\left(\frac{\pi}{4} - \frac{\varphi}{2}\right) \tag{21.23}$$

Expression (21.22) has to be inserted into the equilibrium condition

$$\frac{\mathrm{d}\sigma_r}{\mathrm{d}r} = \frac{\sigma_\varphi - \sigma_r}{r} \tag{21.24}$$

(see Fig. 21.2). The integration yields:

$$\sigma_r = \frac{Dr^{K_a-1} + 2c\sqrt{(K_a)}}{K_a - 1} \tag{21.25}$$

The boundary condition

$$\sigma_{r=r_1} = p_i \tag{21.26}$$

determines the constant D:

$$D = [p_i(K_a - 1) + 2c\sqrt{(K_a)}]r_i^{1-K_a} \tag{21.27}$$

Thus Equations (21.25) and (21.22) yield:

$$\sigma_r = \left(p_i + 2c\frac{\sqrt{(K_a)}}{K_a - 1}\right)\left(\frac{r}{r_i}\right)^{K_a-1} - 2c\frac{\sqrt{(K_a)}}{K_a - 1} \tag{21.28}^*$$

$$\sigma_\varphi = \left(p_i K_a + 2c\frac{\sqrt{(K_a^3)}}{K_a - 1}\right)\left(\frac{r}{r_i}\right)^{K_a-1} - 2c\frac{\sqrt{(K_a)}}{K_a - 1} \tag{21.29}^*$$

* Wu, Loh and Malvern (1963) developed similar expressions for σ_r and σ_φ when $|\sigma_\varphi| > |\sigma_z| > |\sigma_r|$; all normal stresses were taken as compression stresses. Furthermore the same authors deduced equations defining stresses for $|\sigma_z| > |\sigma_\varphi| > |\sigma_r|$. The limiting equilibrium condition was combined with a certain stress–strain relationship. The solutions were given for two different assumptions: (*a*) According to the plastic potential theory which requires the following relationships between the strains in the plastic domain:

$$\dot{\varepsilon}_z{}^p = -\frac{1}{K_p}\dot{\varepsilon}_r{}^p, \qquad \dot{\varepsilon}_\varphi{}^p = 0 \qquad (21.32), (21.33)$$

$$K_p = \tan^2\left(\frac{\pi}{4} + \frac{\varphi}{2}\right) \tag{21.34}$$

(p_i is a compression and consequently has a negative value). By applying Equation (21.28) to $\sigma_{r=r_0} = p_0$ and solving for $2c$ we obtain:

$$2c = \left[p_0\left(\frac{r_0}{r_i}\right)^{1-K_a} - p_i\right] : \left\{\frac{\sqrt{(K_a)}}{1 - K_a}\left[1 - \left(\frac{r_i}{r_0}\right)^{1-K_a}\right]\right\} \tag{21.30}$$

As $K_a \leqslant 1$ and $r_i < r_0$, the denominator is always positive. The condition for $2c > 0$, i.e. for the validity of Equation (4.19), is

$$p_i < p_0\left(\frac{r_0}{r_i}\right)^{1-K} \tag{21.31}$$

or

$$|p_i| > \left|-p_0\left(\frac{r_0}{r_i}\right)^{1-K_a}\right| \tag{21.31'}$$

Failure is defined by the appearance of the first tensile fissure. The corresponding values of p_0 and p_i have to be inserted into Equation (21.30). By plotting $2c$ against φ for two combinations of p_0 and p_i causing failure, the right values of φ and c can be obtained; they are given by the point of intersection of the two plots.

21.3.2.2 *Solution according to the Mises–Hencky failure criterion*

The Mises–Hencky criterion of plastic flow is expressed by the equation:

$$(\sigma_r - \sigma_\varphi)^2 + (\sigma_\varphi - \sigma_z)^2 + (\sigma_z - \sigma_r)^2 = 2\sigma_0^{\,2} \tag{21.37}$$

For the state of plane stress ($\sigma_z = 0$) the above equation reduces to:

$$\sigma_r^{\,2} - \sigma_r\sigma_\varphi + \sigma_\varphi^{\,2} = \sigma_0^{\,2} \tag{21.38}$$

At the beginning of plastic flow, i.e. at the transition from the straight line $\sigma = \sigma(\varepsilon)$ to the curved line (Fig. 21.6) the expressions (21.15) and

---

Consequently the vector of the strain velocity is normal to the failure surface. This condition requires positive dilatation which was not observed in the tests.

(*b*) The volume strains are assumed to be zero. The strains are separated into the elastic and plastic components. Based on general equilibrium conditions and on the Mohr–Coulomb failure hypothesis the following expressions result:

$$\sigma_r = -Ap_i - Bp_0 \tag{21.35}$$

$$\sigma_\varphi = Cp_i - Dp_0 \tag{21.36}$$

where A, B, C and D are parameters depending on the strength characteristics c and φ, on the Poisson ratio ($\nu = 0{\cdot}50$) and on the quotient r/r_i.

(21.16) for σ_r and σ_φ hold. By inserting them into the plasticity condition (21.38) it is possible to obtain

$$\frac{\sigma_0}{p_i} = \frac{\sqrt{\left[3\left(\frac{r_0}{r}\right)^4 + 1\right]}}{\left(\frac{r_0}{r_i}\right)^2 - 1} \tag{21.39}$$

With increasing plastic strain, σ_0 increases. In order to obtain a general solution for stresses, Equation (21.38) may be written in the form

$$\sigma_\varphi = \tfrac{1}{2}\left(\sigma_r + \sqrt{(4\sigma_0^2 - 3\sigma_r^2)}\right) \tag{21.40}$$

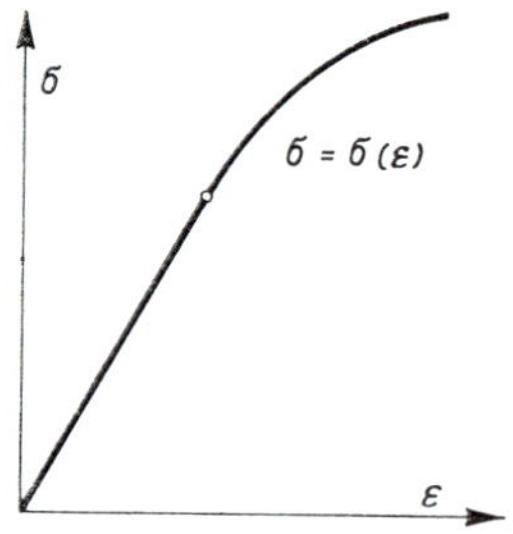

Fig. 21.6. $\sigma = \sigma(\varepsilon)$ diagram

The insertion of the above expression for σ_φ into equilibrium condition (21.24) gives the differential equation:

$$\frac{d\sigma_r}{dr} - \frac{\sqrt{(4\sigma_0^2 - 3\sigma_r^2)} - \sigma_r}{2r} = 0 \tag{21.41}$$

whose solution is (Hoffman and Sachs, 1953, p. 93):

$$\ln r = \frac{\sqrt{3}}{2} \arcsin\left(\frac{\sqrt{3}}{2}\frac{\sigma_r}{\sigma_0}\right) - \frac{1}{2}\ln\left\{\sqrt{\left(1 - \frac{3\sigma_r^2}{4\sigma_0^2}\right)} - \frac{\sigma_r}{2\sigma_0}\right\} + C \tag{21.42}$$

The boundary condition $\sigma_{r=r_0} = 0$ determines the constant C:

$$C = \ln r_0 \tag{21.43}$$

Thus Equation (21.42) can be written in the form:

$$\frac{r}{r_0} = \exp\left[\frac{\sqrt{3}}{2} \arcsin\left(\frac{\sqrt{3}}{2}\frac{\sigma_r}{\sigma_0}\right)\right]\left[\left(1 - \frac{3\sigma_r^2}{4\sigma_0^2}\right)^{\frac{1}{2}} - \frac{\sigma_r}{2\sigma_0}\right]^{-\frac{1}{2}} \tag{21.44}$$

Plate I. (*a*) Initial support of the beam

(*b*) Loading arrangement

[*Facing page* 466]

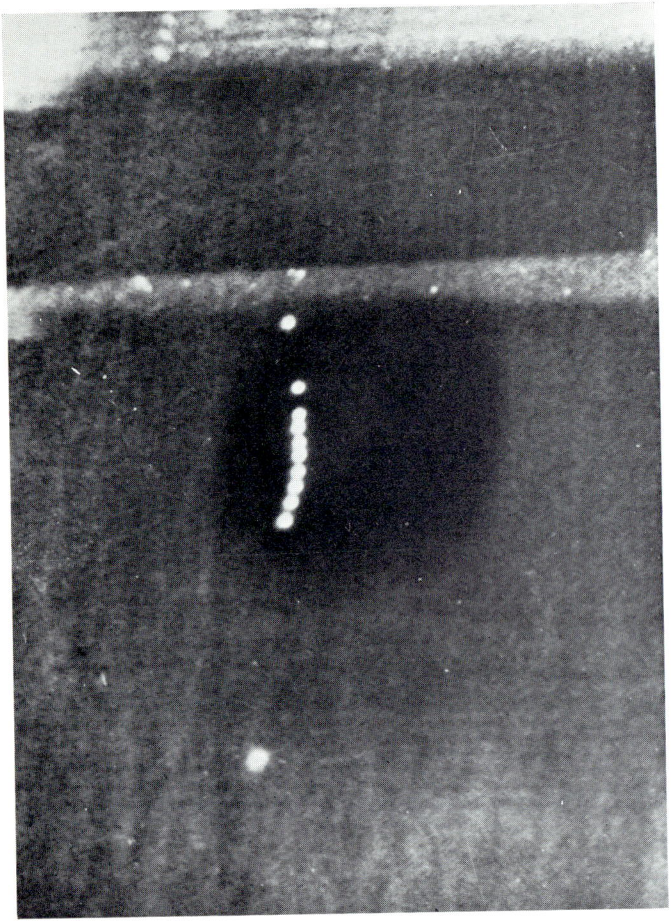

Plate II. (*c*) Enlarged picture of the displacements of a pin

[*Facing page* 467]

For $r = r_i$, $\sigma_{r=r_i} = p_i$, Equation (21.44) becomes:

$$\frac{r_i}{r_0} = \exp\left[\frac{\sqrt{3}}{2} \operatorname{arc\,sin}\left(\frac{\sqrt{3}}{2}\frac{p_i}{\sigma_0}\right)\right]\left[\left(1 - \frac{3p_i^2}{4\sigma_0^2}\right)^{\frac{1}{2}} - \frac{p_i}{2\sigma_0}\right]^{-\frac{1}{2}} \tag{21.45}$$

(p_i is a negative value !). The stress ratio p_i/σ_0 corresponding to a certain ratio r_i/r_0 defines the parameter σ_0 at any stage of plastic deformation, failure included. The stresses σ_φ and σ_r can then be computed from Equations (21.44) and (21.40). The stresses σ_φ and σ_r corresponding to σ_0 at failure determine the failure stress state.

21.4 BENDING TESTS

21.4.1 Linear Stress–Strain Relationship

21.4.1.1 *Critical Strain*

Bending tests are usually carried out on small beams of rectangular cross-section. The strains in bending tests performed on beams of compacted clay can be established from photographic recording of the displacements of measuring points arranged parallel to the axis of the beam.

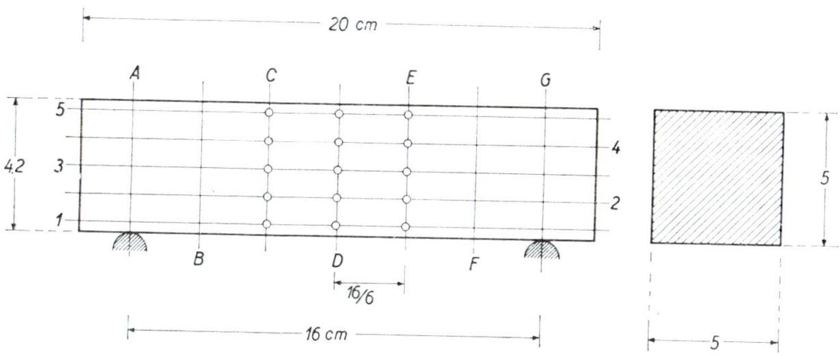

Fig. 21.7. Bending tests: Disposition of measuring pins

By inserting several pairs of measuring pins in adjacent verticals (see Fig. 21.7 and Plates I and II), the diagrams of strain can be derived from the displacements of pins determined from photographs monitored in a photocomparator (Šuklje and all, 1967). By extrapolating the resulting strain diagrams the maximum strains appearing at the edges of the beam can be determined. An example is presented in Fig. 21.8. At the appearance of the tensile crack the boundary tensile strain defines the critical tensile strain corresponding to the linear stress state.

By testing several beams prepared at various moisture contents and with various compaction energies, the resulting critical strains can be plotted against moisture content for different compaction energies. A family of such plots can be used to select a suitable moisture content and density of a clay that has to form the core of a dam.

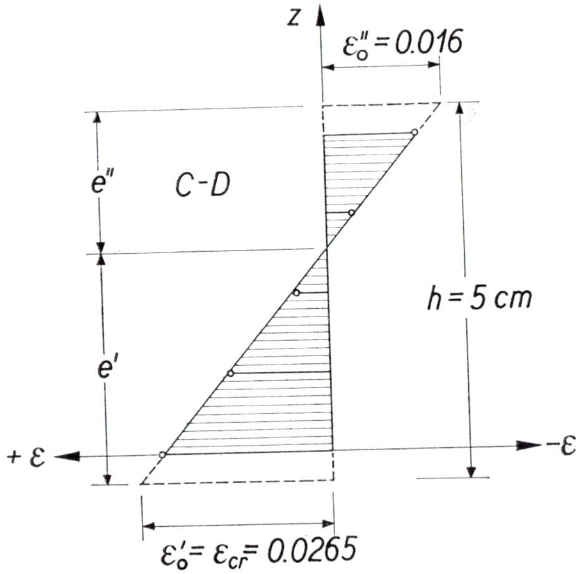

Fig. 21.8. Diagram of bending strains

In tests made so far (Šuklje and all, 1967) the reflections of an electric lamp on the polished round heads of pins provided appropriate measuring points. A uniformly distributed load was applied and the dead weight of the beam was included in the testing load, yet eliminated before commencing with measurement (Plate I (*b*)). In order to reduce the contribution of the dead weight and to obtain a central domain with approximately constant bending moment and negligible shear forces, the support and loading arrangement as shown in Fig. 21.9 can be recommended.

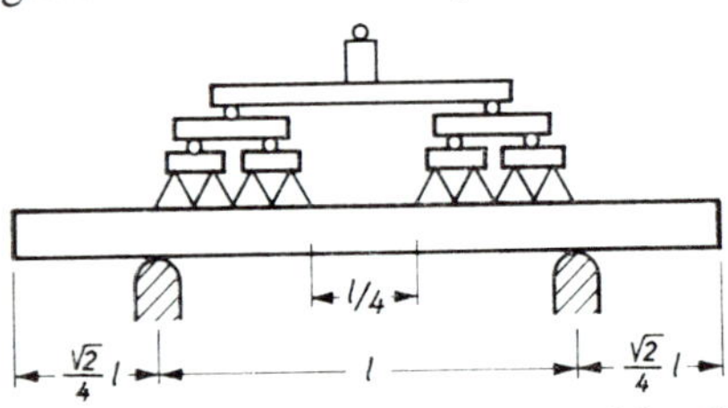

Fig. 21.9. Scheme of recommended support and loading arrangement

21.4.1.2 *Deformation Moduli*

The strain plots $\varepsilon = \varepsilon(z)$ (Fig. 21.8) also render deformation moduli to be determined provided that a certain law of the stress–strain relationship is known or assumed. In the following deductions constant but different values of deformation moduli in the tension (E') and compression (E'') domains of stresses and linear strain plots $\varepsilon = \varepsilon(z)$ (Navier's hypothesis) will be assumed. The results of tests made so far (Šuklje and

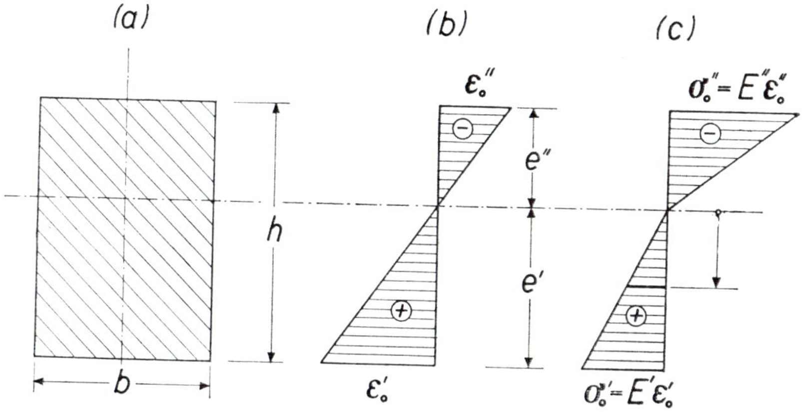

Fig. 21.10. Stress diagram of bending specimens at different but constant values of tension and compression moduli

all, 1967; cf. Fig. 21.8) prove that the strain plots are still approximately linear at the appearance of the first tensile cracks. The neutral axis lies above the centre of gravity of the profile and as the normal forces are zero, the tension and compression parts of the stress plots $\sigma = \sigma(z)$ are equal (Fig. 21.10):

$$b\frac{e'}{2}\sigma_0' = b\frac{e''}{2}\sigma_0'' \qquad (21.46)$$

Inserting the relations

$$\sigma_0' = E'\varepsilon_0', \qquad \sigma_0'' = E''\varepsilon_0'' \qquad (21.47)$$

we have

$$\frac{E'}{E''} = \frac{e''\varepsilon_0''}{e'\varepsilon_0'} = \frac{\varepsilon_0''^2}{\varepsilon_0'^2} = \frac{e''^2}{e'^2} \qquad (21.48)$$

The magnitude of the deformation moduli can then be obtained by considering the moment equilibrium condition

$$b\,\frac{\sigma_0' e'}{2}\cdot\frac{2}{3}h = M \tag{21.49}$$

Thus:

$$\sigma_0' = \frac{3M}{bhe'}, \qquad \sigma_0'' = \frac{3M}{bhe''} \tag{21.50}$$

and, according to the relations (21.47):

$$E' = \frac{3M}{bh^2}\cdot\frac{\varepsilon_0' + \varepsilon_0''}{\varepsilon_0'^2} \tag{21.51}$$

$$E'' = \frac{3M}{bh^2}\cdot\frac{\varepsilon_0' + \varepsilon_0''}{\varepsilon_0''^2} \tag{21.52}$$

An analogous procedure can be applied if the strain diagrams are not linear.

The 'bending modulus' E deduced from the bending line by assuming equal values of tension and compression moduli can differ appreciably from the tension modulus E' deduced from the strain plot. If the strain diagrams $\varepsilon = \varepsilon(z)$ are assumed to be linear, simple deductions (Šuklje and all, 1967) yield the relation:

$$\frac{E'}{E} = \left(\frac{b}{2e'}\right)^2 \tag{21.53}$$

21.4.1.3 *Failure Envelope*

The straight envelope of the limiting stress circles corresponding to the relations (21.50) may be defined by the shear intercept c and the slope φ:

$$\tau = c - \tan\varphi\,.\,\sigma \tag{21.54}$$

(tension stresses have been assumed to be positive and compression stresses negative values). Based on Fig. 21.11 the following relation can be established:

$$\sin\varphi = \frac{\dfrac{\sigma'}{2}}{c\cot\varphi - \dfrac{\sigma'}{2}} = \frac{\dfrac{\sigma''}{2}}{c\cot\varphi + \dfrac{\sigma''}{2}} \tag{21.55}$$

By solving the above relations for σ' and σ'' we obtain the following expression for the tension (σ_0') and compression (σ_0'') stresses at the edges of the profile:

$$\sigma_0' = \frac{2c\cos\varphi}{1+\sin\varphi} = 2c\tan\left(\frac{\pi}{4} - \frac{\varphi}{2}\right) \tag{21.56}$$

$$\sigma_0'' = \frac{2c\cos\varphi}{1-\sin\varphi} = 2c\tan\left(\frac{\pi}{4} + \frac{\varphi}{2}\right) \tag{21.57}$$

By dividing the above expressions and considering the relations (21.50) we obtain:

$$\tan^2\left(\frac{\pi}{4} + \frac{\varphi}{2}\right) = \frac{\sigma_0''}{\sigma_0'} = \frac{e'}{e''} \tag{21.58}$$

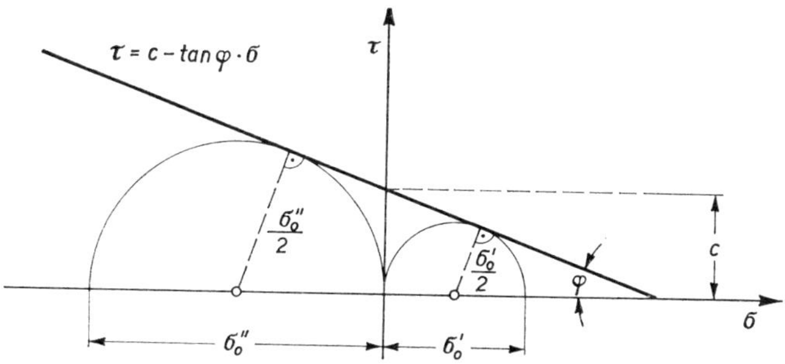

Fig. 21.11. Failure envelope in bending tests

After inserting Equation (21.50) the relation (21.57) yields:

$$c = \frac{3M}{2bhe''}\tan\left(\frac{\pi}{4} - \frac{\varphi}{2}\right) \tag{21.59}$$

As simultaneous rupture at both the tension and compression edges of the profile can hardly be assumed, the resulting parameters c and φ cannot be interpreted as the strength characteristics of the clay. Nevertheless, they govern the 'stability' of the soil body subjected to bending stress states.

21.4.1.4 *Creep Effects*

In bending tests the early creep deformation occurring at a constant stress state increases with the logarithm of time. The creep starts to develop at a faster rate when a certain critical strain or a certain stress yield point is exceeded. In Fig. 21.12 the creep development is presented as observed during a short term bending test performed with a silty clay of low compressibility ($w_L = 25$ per cent, $I_P = 9$ per cent) and with a water content wet of optimum ($w = 13{\cdot}9$ per cent). According to Leonards and Narain (1963) the critical tensile strain as observed in

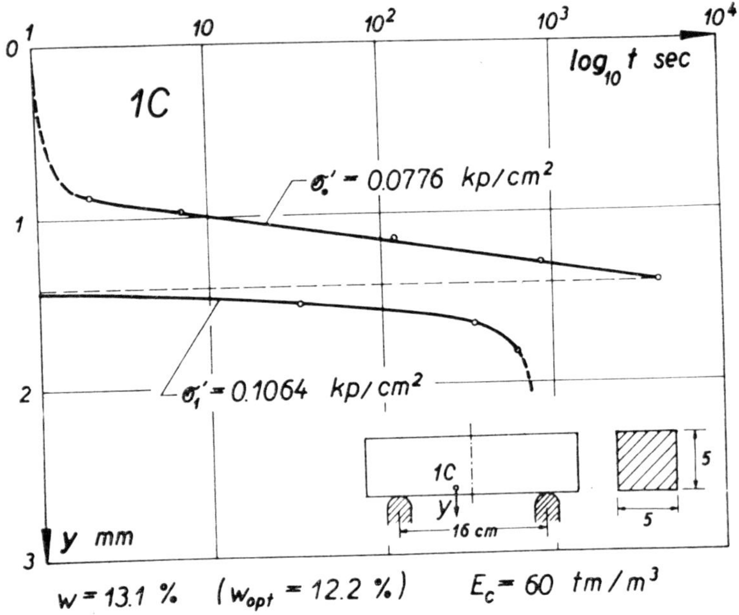

Fig. 21.12. Creep in a short term bending test

bending tests increases with the time to failure. (In long-term tests lasting four weeks the critical tensile strain was about 20 per cent greater than in short-term tests lasting two days.)

21.4.2 Plastic Stress State

In the preceding sections the linear stress–strain relationship was assumed to govern the stress state throughout the loading and deformation process until the tension crack appeared. As the other extreme assumption, the 'plastic' stress–strain relationship according to Fig. 21.13 will be considered.

The equilibrium of forces in the direction of the axis of the beam requires:

$$\frac{\sigma_0'}{\sigma_0''} = \frac{\sigma'}{\sigma''} = \frac{e''}{e'} \tag{21.60}$$

and the moment equilibrium condition:

$$M = \sigma' e' b \frac{h}{2} \tag{21.61}$$

Thus:

$$\sigma' = \frac{2M}{bhe'}, \qquad \sigma'' = \frac{2M}{bhe''} \tag{21.62}$$

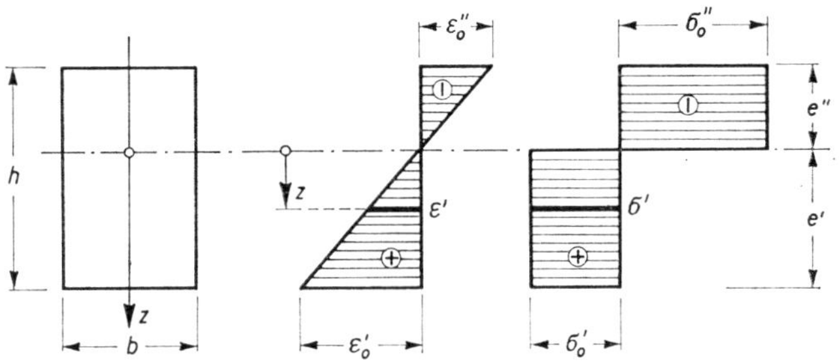

Fig. 21.13. Plastic stress state in bending tests

An analogous computation as applied in Section 21.4.1.3 yields, for the failure envelope parameters c and φ (see Fig. 21.11), the following expressions:

$$\tan^2\left(\frac{\pi}{4} + \frac{\varphi}{2}\right) = \frac{e'}{e''} \tag{21.58'}$$

$$c = \frac{M}{bhe''} \tan\left(\frac{\pi}{4} - \frac{\varphi}{2}\right) \tag{21.59}$$

The interpretation of the parameters c and φ as given in Section 21.4.1.3, holds also in this case.

Part F

CREEP AND RELAXATION EFFECTS ON EARTH PRESSURES

22

Kinematic Conditions of the Magnitude of the Earth Pressure against a Retaining Structure

22.1 THE INFLUENCE OF THE DISPLACEMENT OF THE RETAINING STRUCTURE ON THE MAGNITUDE OF THE EARTH PRESSURE

22.1.1 General Considerations

The theoretical bases for the computation of the earth pressure against retaining structures according to the theory of plasticity have been dealt with in Chapter 7. All the methods presented originate from the Kötter's differential equations, i.e. from the static equilibrium conditions and the plastic limit equilibrium condition according to the Mohr–Coulomb envelope. The stress–strain relationships do not appear in these methods. Only when fixing the rupture figures, according to which Kötter's equations may be applied, are the kinematical conditions considered in a qualitative way.

As suggested in Section 17.1.2, the stress fields obtained by the theory of plasticity can—in appropriate cases—serve as statically correct fields when computing the approximate strain fields. Within a plastic zone the stress state can be considered to have changed from the at-rest state to the active or passive state defined by the envelope that corresponds to the mobilized part of the strength of the soil. Some methods accounting for the spherical and the distortional deformations corresponding to known changes of the stress state were presented in Chapters 13 and 17. The deformations can be computed by integrating the strains along linear paths. If the soil does not exhibit a structural anisotropy, co-axiality of the principal directions of the stresses and strains can be assumed. Thus the displacements perpendicular to the contact surface between the structure and the soil can then easily be found by using—in plane problems—Mohr strain circles.

But even in cases of zone rupture the movement of the wall does not only affect the 'plastic zone', i.e. the zone within which the new stress states reach the envelope corresponding to the mobilized part of the strength. Outside this zone the change of the stress state is not defined by the assumption concerning uniform strength mobilization. A stress change, however, does take place outside the 'plastic zone.' The stresses behind the rupture surface or the envelope of rupture surfaces limiting the plastic domain, and below the surface of the backfill, can be computed approximately by the theory of elasticity. The self-weight of the soil and the pressures along the boundary of the plastic zone should be considered as active loads. There are not, however, solutions available facilitating the computation of such an elastic stress field. Thus, simplifying assumptions must be made in order to obtain a static field in which the equilibrium and the boundary conditions are at least partly satisfied. It seems reasonable to eliminate a domain of the backfill in which the at-rest stress state is assumed to remain unchanged, and to restrict the computation of displacements to the 'plastic field' and to the transient field between the 'plastic' and the 'at-rest' domains.

An example of such an approximate treatment of the problem will be given in the following section.

22.1.2 Some Individual Solutions Applied to Rankine's Stress Condition

22.1.2.1 *The Application of Non-linear Stress–Strain–Time Relationships to Rankine's Case*

In plane state of strain as occurring in Rankine's case (Fig. 22.1) the strains in the mean principal direction are zero. In an elastic medium this condition requires the mean principal effective stress to be:

$$\sigma_2' = \nu(\sigma_1' + \sigma_3'), \tag{22.1}$$

ν being the Poisson ratio. As the rupture is assumed to occur in the plane of principal directions (1), (3), the mean principal stress σ_2' can vary only in the interval

$$\sigma_3' \leqslant \sigma_2' \leqslant \frac{\sigma_1' + \sigma_3'}{2} \tag{22.2}$$

and the octahedral normal stress $\sigma^{o\prime}$ in the interval

$$\frac{\sigma_1' + 2\sigma_3'}{3} \leqslant \sigma^{o\prime} \leqslant \frac{\sigma_1' + \sigma_3'}{2}. \tag{22.3}$$

The assumption shall be made

$$\sigma^{o\prime} = \frac{\sigma_1' + \sigma_3'}{2} \tag{22.4}$$

corresponding to $\nu = 0{\cdot}50$ in the elastic medium. We have found that in the conditions of the following numerical example the choice of the mean principal direction in the interval given by Equation (22.3) does not considerably influence the end-results.

By two conjugate stresses $\vec{p_b}$ and $\vec{p_e}$ (Fig. 22.1) and the assumption (22.4) the stress state is completely defined. The earth pressures $\vec{p_e}$ will now be determined according to the stress–strain–time relationship deduced from triaxial tests and explained in the Section 17.1. The main

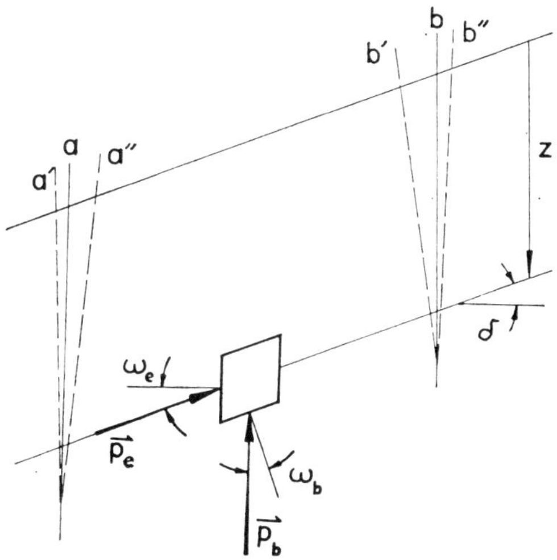

Fig. 22.1. Rankine's case

principal deviatoric strain and the spherical strain will be chosen as correlated strain quantities (see Section 17.1.2). In plane strain conditions ($\varepsilon_2 = 0$) the principal strains ε_1 and ε_3 can be computed according to the relations

$$\varepsilon_1 = \varepsilon_1{}^{\circ} + \varepsilon_1{}^{d} \tag{22.5}$$

$$\varepsilon_3 = 2\varepsilon_1{}^{\circ} - \varepsilon_1{}^{d} \tag{22.6}$$

The spherical strains $\varepsilon_1{}^{\circ}$ and distortional strains in the principal direction (1) $\varepsilon_1{}^{d}$ may be assumed to be governed by Equations (17.7) and (17.8). In the following numerical example the parameters $(\varepsilon_1{}^{\circ})_1$, α° and $(\varepsilon_1{}^{d})$, α^{d} appearing in these equations will be taken from the plots presented in Fig. 17.2*; they are related to a silty soil.

* The charts in Fig. 17.2 suit for the first loading. Thus the displacements necessary to mobilize a certain degree of the active earth pressure will be overestimated.

According to Fig. 22.2 and Equation (22.4) the octahedral strain $\sigma^{o'}$ can be expressed by the equation:

$$\sigma^{o'} = \frac{p_b}{\cos^2 \varphi_m'} \cos \delta \pm \sqrt{(\cos^2 \delta - \cos^2\varphi_m')}] \qquad (22.7)$$

where φ_m' denotes the mobilized shear angle. For various values φ_m' and p_b the corresponding principal strains ε_1 and ε_3 can then be obtained and the strain ε_e in the direction of the slope (δ) determined by using Mohr's circle of strain. For the inclination $\delta = 10°$, time $t = 1$ day and

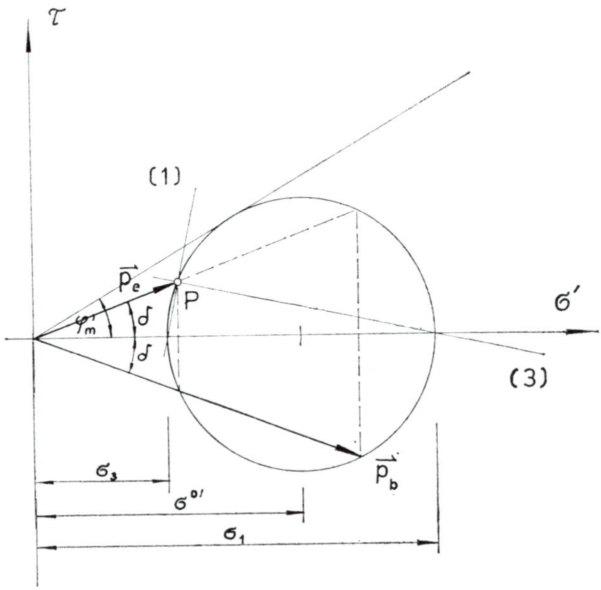

Fig. 22.2. Mohr's stress circle in Rankine's case

various values of γz (γ denoting the bulk density of the soil and z the depth below the slope surface), the resulting plots $[\varepsilon_e = \varepsilon_e(p_e)]_{\gamma z=\text{const}}$ are presented in Fig. 22.3. The asymptotes $p_e = \text{const}$ of these plots are defined by Rankine's active and passive earth pressures.

Similar plots can be obtained for any time t (see Equations 17.7 and 17.8). From various sets of plots

$$[p_e = p_e(\varepsilon_e)]_{\gamma z=\text{const},\ t=\text{const}}$$

the relaxation curves

$$[p_e = p_e(t)]_{\gamma z=\text{const},\ \varepsilon_e=\text{const}}$$

can be deduced. An example is presented in Fig. 22.4.

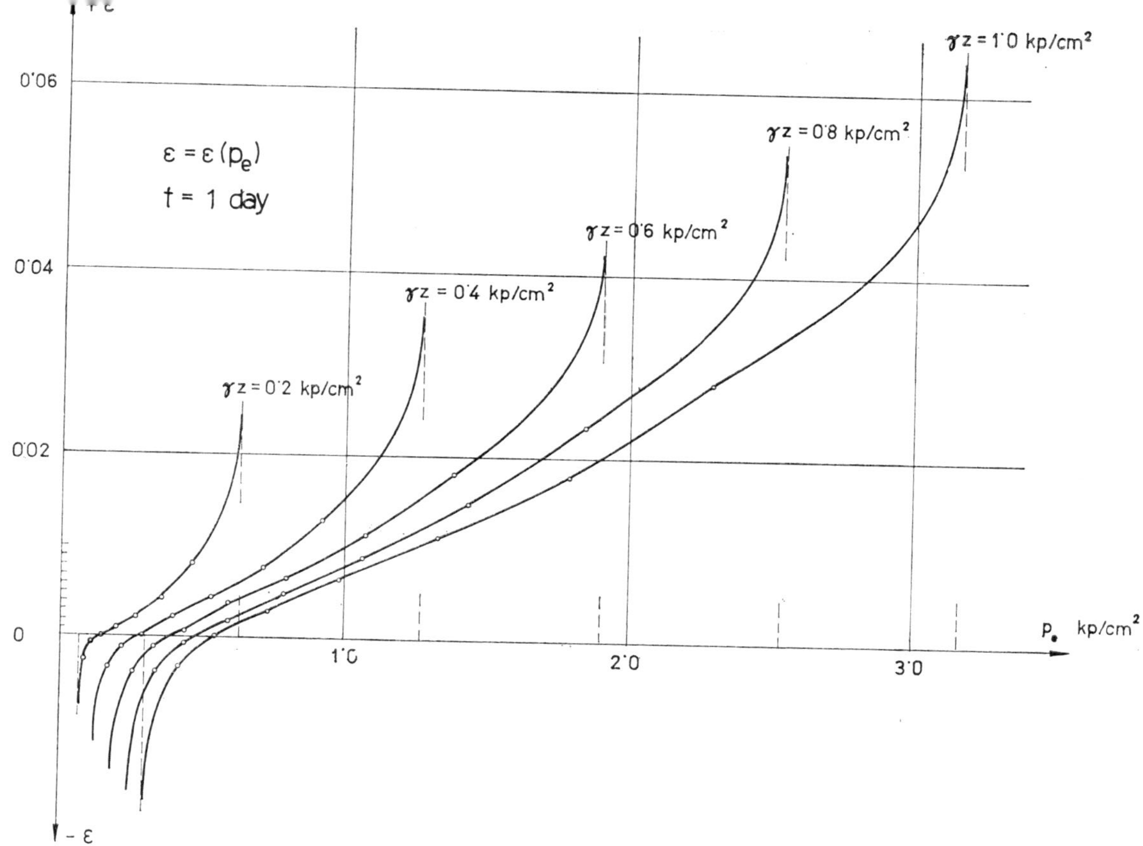

Fig. 22.3. Plots $\varepsilon_e = \{\varepsilon_e(p_e)\}_{\gamma z=\text{const},t=1\text{ day}}$ in Rankine's theoretical case with $\delta = 10^\circ$ according to stress–strain–time relationship presented in Fig. 17.2.

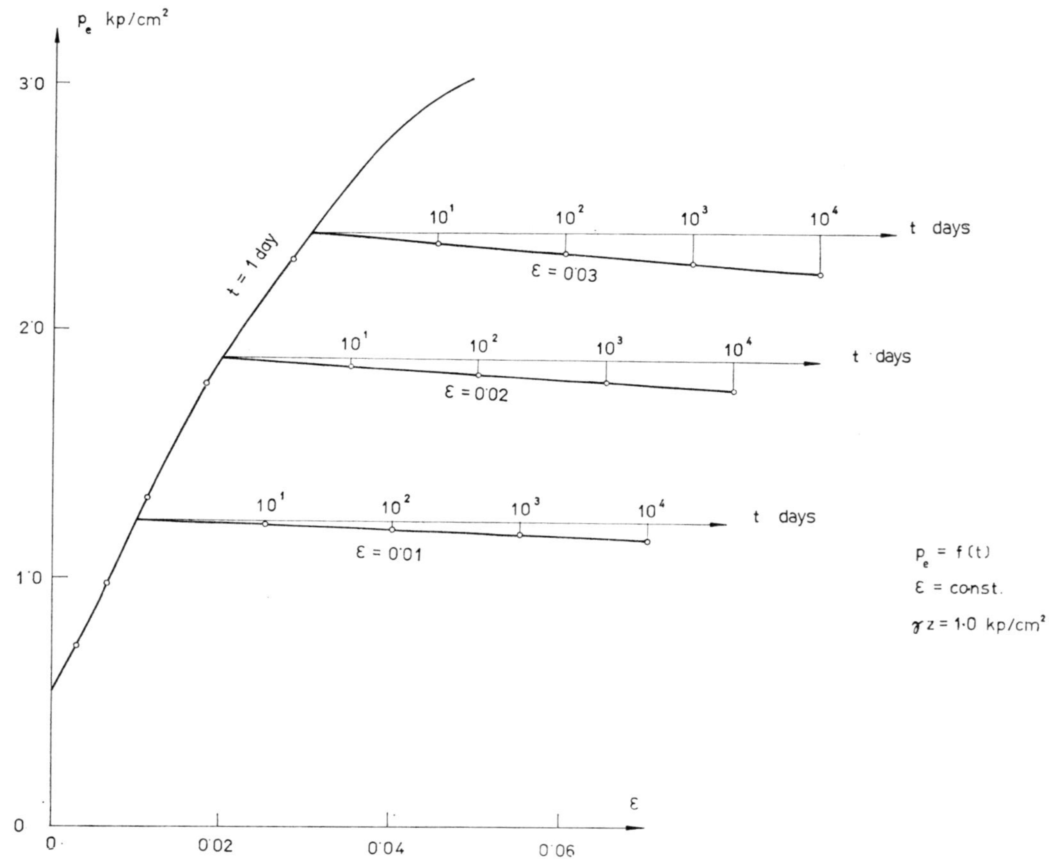

Fig. 22.4. Relaxation curves $p_e = \{p_e(t)\}_{\gamma z=\text{const},\, \varepsilon=\text{const}}$ in Rankine's theoretical case with $\delta = 10°$ according to stress–strain–time relationships presented in Fig. 17.2.

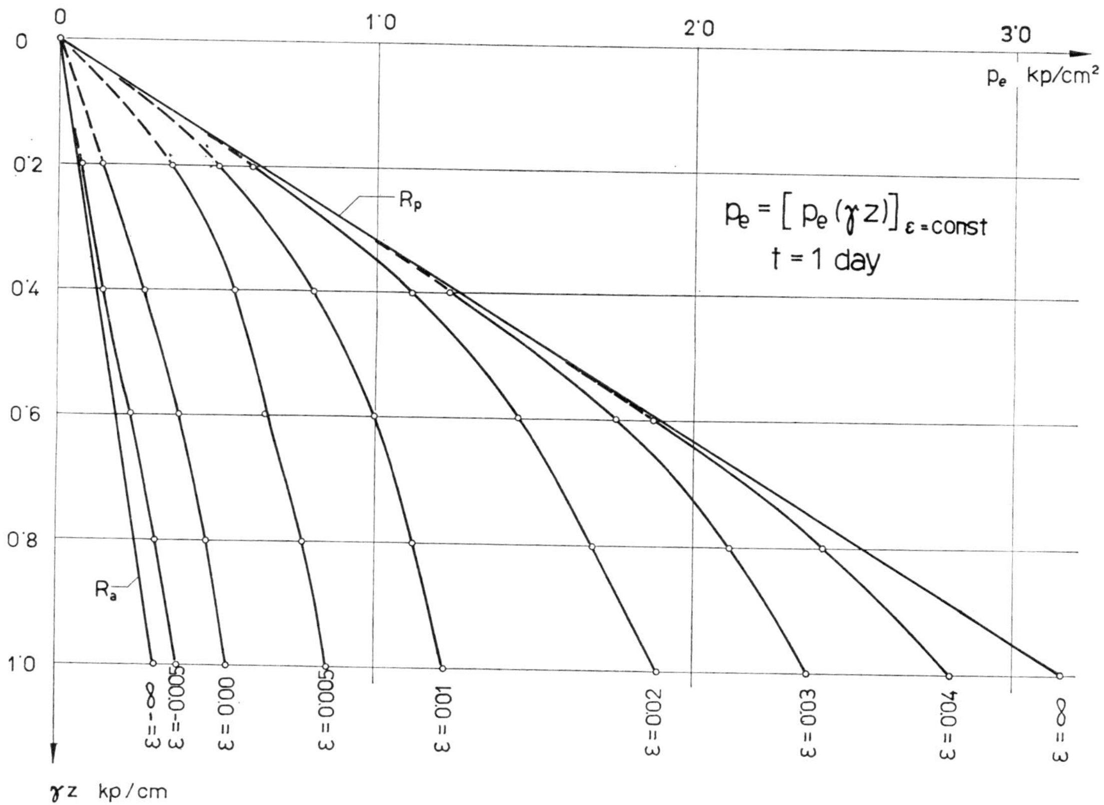

Fig. 22.5. Earth pressure distribution in Rankine's theoretical case ($\delta = 10°$) according to stress–strain–time relationships presented in Fig. 17.2

In the theoretical Rankine case a semi-infinite mass is assumed to be subjected to stretching or compressing. If this operation occurs at translatory displacement of vertical cross-sections, the strains are constant throughout the depth z. The earth-pressure diagrams corresponding to various ε_e values and to a given time t can be deduced from the $[\varepsilon_e = \varepsilon_e(p_e)]_{t=\text{const}}$ charts. The diagrams according to the plots of Fig. 22.3 are shown in Fig. 22.5. The boundary diagrams corresponding to unlimited outside ($\varepsilon_e \to -\infty$) and inside ($\varepsilon_e \to +\infty$) displacement, coincide with Rankine's limiting active (R_a) and passive (R_p) pressures.

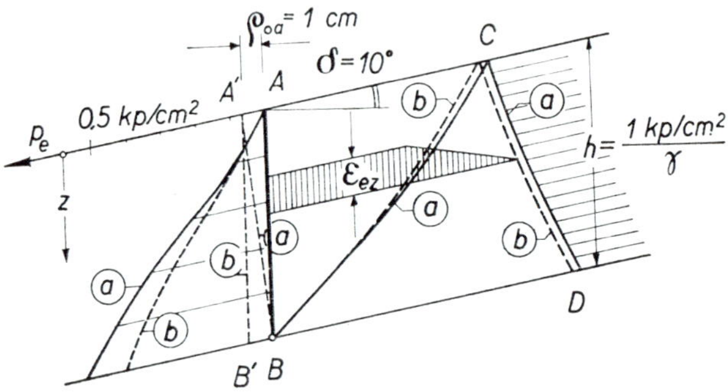

Fig. 22.6. Active earth-pressure distribution along a rigid wall corresponding to (a) rotation around its foot-line, (b) translation ρ_{oa} = 1 cm, t = 1 day, stress–strain–time relationships according to Fig. 17.2

In order to apply the results which were obtained for the theoretical Rankine case, to a rigid wall supporting a uniformly inclined slope of homogeneous soil, the following simplifying assumptions will be taken:

(1) Inside the wedge ABC (Figs. 22.6 and 22.7) the strains ε_{ez} in a given depth z below the slope surface are uniform. This wedge is surrounded by the retaining wall (AB), the slope surface (AC) and the line (BC) issuing from the foot-point of the wall and connecting the directions of the maximum mobilization of the strength of the soil.

(2) The strains ε_{ez} dissipate linearly in an adjacent wedge (BCD) limited by the above defined line BC, the base BD issuing at the slope δ from the foot-point of the wall, and the line CD connecting the directions of the stress characteristics of the at rest stress state ($\varepsilon_e = 0$).

The line BC can be found by iteration starting, e.g., with the direction of Rankine's rupture line. If we assume that a certain displacement

$$\rho_z = \rho_0 \left(1 - \frac{\gamma z}{\sigma_0}\right) \tag{22.8}$$

of the rigid wall occurs, the strains ε_{ez} corresponding to the dimensions of the wedges ABC and BCD can easily be obtained. The earth pressure p_e affecting the wall in the depth z can be then determined by using the charts of Fig. 22.3.

For the specifications $\rho_{oa} = 1$ cm. in the active case and $\rho_{op} = 15$ cm. in the passive case, and for

(*a*) $\sigma_0 = \gamma h$ (rotation about the foot-point),
(*b*) $\sigma_0 \rightarrow \infty$ (translation)

the resulting earth pressure diagrams are presented in Figs. 22.6 and 22.7*.

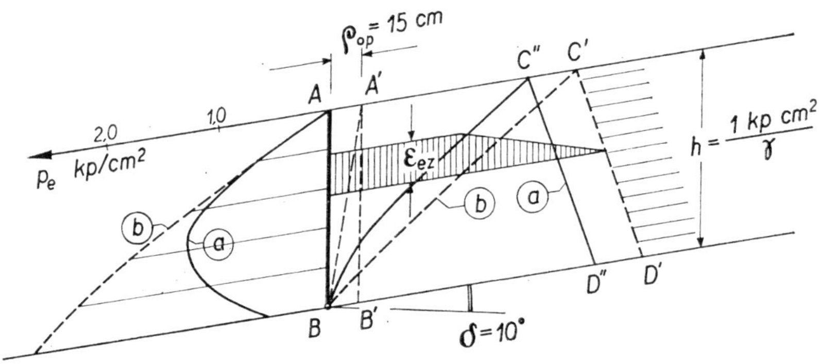

Fig. 22.7. Passive earth-pressure distribution along a rigid wall corresponding to (*a*) rotation around its foot-line, (*b*) translation $\rho_{op} = 15$ cm, $t = 1$ day, stress–strain–time relationships according to Fig. 17.2

22.1.2.2. *Some Special Solutions for a Rigid Vertical Wall with Horizontal Back-fill Surface*

For the simple Rankine case (smooth vertical rigid wall with horizontal back-fill surface) Jakobson (1958) was the first to deal with the problem in a way which, in principle, coincides with the concept explained in the foregoing section. Jakobson assumed a unique plastic field but did not insist on it being limited by the rupture surface passing through the foot of the wall. In the treated case with the above assumptions, the pressures in the vertical direction do not change after a small rotation of the wall. Thus the deformations of the plastic zone in the horizontal direction were computed for the linear state of additional stresses $\Delta\sigma_1 = (K_0 - K_a)\gamma z$,

* All diagrams presented in this Section are reproduced from a paper by Šuklje, Márczal and Podržajeva, 1968.

with γ denoting the unit weight of the backfill and K_0, K_a the at-rest and active earth pressure coefficients. Now, the deformation (elastic) moduli E in the horizontal direction are deduced from the compression moduli M_v of oedometer tests which were found to be governed by the relation:

$$M_v = M_1 \cdot p^a \tag{22.9}$$

where p is the normal stress and M_1, a constants to be found from oedometer test results. Thus

$$E = M_v\left(1 - \frac{2\nu^2}{1-\nu}\right) = M_1 p^a N \tag{22.10}$$

$$N = 1 - \frac{2\nu^2}{1-\nu} \tag{22.10a}$$

where ν is Poisson's ratio. The displacements compatible with the mobilization of a shear angle φ in the backfill and the corresponding

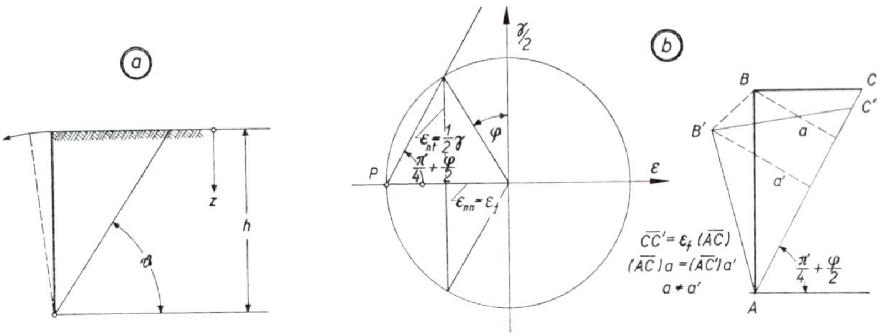

Fig. 22.8. (*a*) Scheme of stress in the backfill of a smooth vertical retaining wall rotating in the outside direction about its lower edge. (*b*) Displacement paths in zero volume change conditions; according to Brinch Hansen (1958)

earth pressures are given by the relation (see also Šuklje, 1958, and Jakobson, 1958-b):

$$\rho = C\zeta^{1-a}(1 - \zeta)h^{2-a} \tag{22.11}$$

where

$$\zeta = \frac{z}{h},\ C = \frac{(K_0 - K_a)\gamma^{1-a}}{\tan\vartheta M_1 N K_0{}^a} \tag{22.11-a,b}$$

In these equations, the preceding notation and that of Fig. 22.8 have been used. The angle of lines passing through the toe of the wall ($z = h$) and the point z are then given by the equation

$$\eta = \frac{\rho}{h - z} = Cz^{1-a} \tag{22.12}$$

and its maximum value ($z = h$)

$$\eta_{\max} = Ch^{1-a} \tag{22.12-a}$$

Now, by deducing the deformation moduli from oedometer tests under the assumption of a certain Poisson's ratio, the important influence of increasing distortional stresses on the increasing distortional strains cannot be taken into account. The consequence of neglecting this influence is that the displacements resulting from Jakobson's procedure are too small.

A different treatment of the same Rankine case (Fig. 22.8) was proposed by Ladanyi (1958). He used triaxial test results in order to express the mobilized shear angle φ_m as a function of the angular deformation γ. By accepting the Saint-Venant principle of the co-axiality of the stress and strain principal axes and of the mobilization of strength at constant volume, and using on the function $\varphi_m = \varphi_m(\gamma)$ as obtained from tests, Ladanyi has computed the displacement of the wall according to the degree of strength mobilization. The comparison of the active earth pressure coefficient plotted against the rotation angle η (Fig. 22.9), as obtained for the loose sand by Jakobson on one hand and Ladanyi on the other, proves that the distortional deformations increase with the degree of strength mobilization; and that these largely affect the movement of the wall which is necessary to enable the reduction of the earth pressure to the limiting active value. This effect seems to be even more important than demonstrated by Ladanyi because he did not consider the deformation outside the plastic zone and did not respect the condition that the Saint-Venant principle of zero volume strain does not permit the displacement path (Fig. 22.8) to be parallel to the rupture lines; the limiting rupture line AC must decrease in length corresponding to the linear strain ε_f (see Brinch Hansen, 1958). This shortcoming, however, cannot essentially affect the results presented in Fig. 22.9.

Although Ladanyi's solution is based on a special and limited treatment of the stress–strain relationship, it still remains within the scope of our general concept as explained in the foregoing sections. However, the solutions presented by Rowe (1954) and Kézdi (1958) are quite different. In principle, the method used by both is the same: (1) According to the results obtained by direct or triaxial shear tests they define the slip strain in the direction of the failure surface. (2) By applying this definition they compute the degree of mobilization of the friction in the backfill corresponding to a given rotation of the interior surface of the wall about its lower edge. The at-rest state is considered to be the initial stress state in

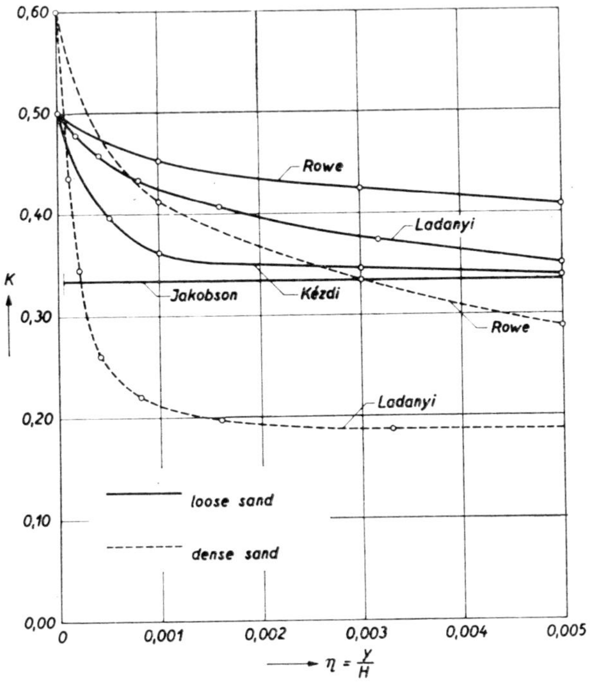

Fig. 22.9. Active earth-pressure coefficients versus rotation angle η around the foot of a smooth vertical retaining wall, according to procedures of Jakobson, Kézdi, Ladanyi and Rowe; after Šuklje, 1958

the plastic zone and the solutions obtained with the conditions of plastic equilibrium are assumed to be applicable also in cases of partial mobilization of the internal friction.

In Rowe's hypothesis the degree of mobilization of the internal friction depends on the degree of inter-locking of the grains which is a function of the movement of the grains per unit length of the slip surface.

Kézdi assumes that the displacement needed to mobilize a certain part of the friction does not depend on the length of the slip surface. This displacement can be determined by measuring the movements of specimens subjected to direct shear tests.

Both Rowe and Kézdi assume that the wall is rough. Nevertheless they neglect the influence of the friction along the wall on the direction of the rupture surfaces and assume that these surfaces are inclined towards the horizontal by

$$\alpha = \frac{\pi}{4} + \frac{\varphi}{2}$$

he fundamental hypothesis of both authors lead to the conclusion that the internal friction in the backfill cannot be entirely mobilized except

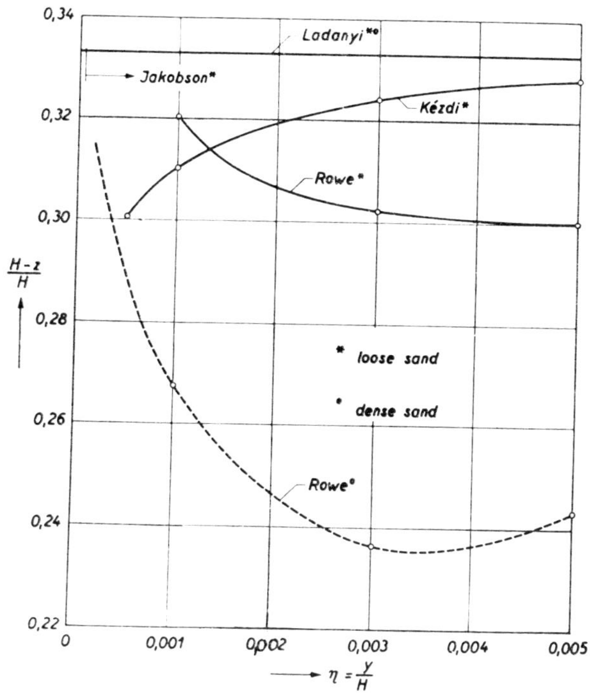

Fig. 22.10. Position of the point of application of the resultant earth pressure versus rotation angle η. (After Šuklje (1958))

above a limiting rupture surface. Below this surface, the active pressures on the wall decrease to the initial at-rest pressure which remains unchanged at the lower edge of the wall.

In Figs. 22.9, 22.10 and 22.11 the results obtained after Rowe's and Kézdi's suggestions are compared with those resulting, for the same wall and for the same backfill properties, from methods used by Jakobson and Ladanyi. The full lines are related to a backfill consisting of loose sand ($n_c = 46$ per cent, $\gamma = 1430$ kp/m³, $\varphi = 29° 40'$, $K_0 = 0{\cdot}50$) and to a height of wall $h = 315$ cm, and the dashed lines to a dense sand ($n_c = 39$ per cent, $\gamma = 1615$ kp/m³, $\varphi = 47°$, $K_0 = 0{\cdot}60$) and to a wall height of $h = 248$ cm. In order to eliminate the influence of different conditions other than those concerning the fundamentals of the methods, the wall was also assumed to be smooth in the case of both Rowe's and Kézdi's solutions. The divergences between the results obtained by the different methods are shown to be rather important.

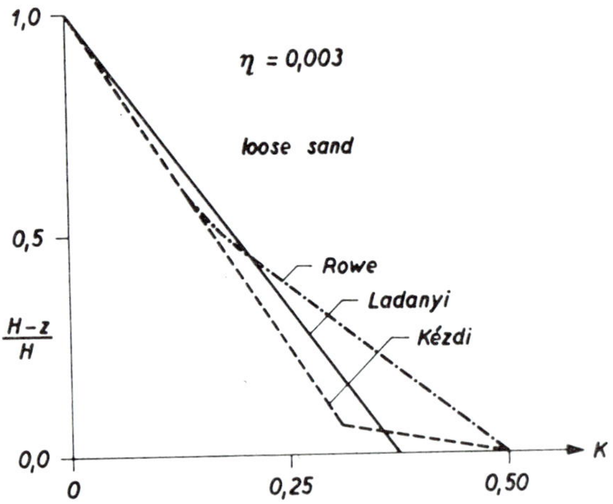

Fig. 22.11. Comparison of diagrams of the earth-pressure distribution along a smooth vertical retaining wall of 315 cm high for a rotation angle of $\eta = 0{\cdot}003$. (After Šuklje, 1958)

The fundamental definition of the slip strain in Rowe's procedure was based on the following facts: (*a*) Increase of interlocking with 'bunching' or change in formation in plan causes a fractional change of slip path (Fig. 22.12 (*a*)). (*b*) Any slip path through a soil mass may be considered as passing through a series of elements so that the summation of the deformation of the elements on the slip paths may be equated to the movement at the wall (Fig. 22.13 (*b*); Rowe, 1958). These facts seem to be correct, but their application in interpreting the results of direct shear tests has been subjected to criticism (Šuklje, 1958). Rowe does not seem to have insisted on the application of his hypothesis to earth pressure problems, it having been derived to demonstrate the order of displacements for walls pushed into soil and to calculate the foundation settlements on sand.

Rowe's concept certainly cannot be used for line ruptures occurring along surfaces of discontinuity. On the contrary Kézdi's hypothesis of independence of the slip displacements on the length of the failure surface seems to be suitable only to pure line ruptures (see Šuklje, 1958; Brinch Hansen, 1958; Ladanyi, 1958).

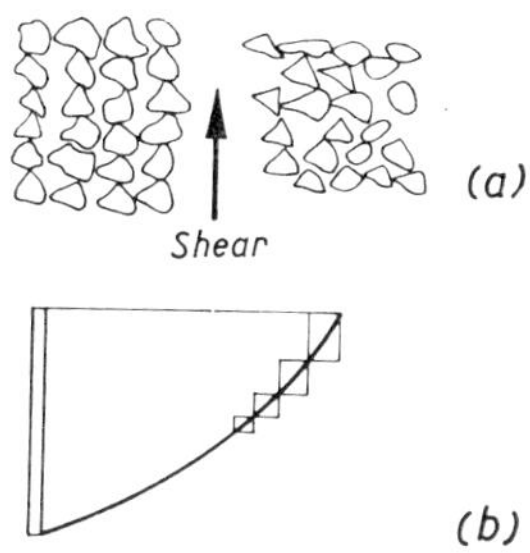

Fig. 22.12. Fractional change in length of the slip path before failure. (According to Rowe, 1958)

22.1.3 The Possibilities of Investigating Earth Pressure Problems by Model Tests

The difficulties occurring in connection with the investigation of the kinematic conditions of earth pressures by analytical methods lead to attempts to study them by model tests. The principles of the application of these experimental methods in soil mechanics have been thoroughly analysed and described by M. Rocha (1953). It is surprising that so little use has subsequently been made in soil investigations of these methods. It is probably due to the fact that it is not easy to choose appropriate test material satisfying the scale requirements. As in most earth pressure problems the state of stress arising from the dead weight cannot be neglected and the condition

$$\frac{1}{\alpha} = \frac{1}{\lambda\rho} \tag{22.13}$$

should hold if $1/\alpha$ represents the stress scale, $1/\lambda$ the length scale and $1/\rho$ the unit weight scale factors. Geometrically similar displacements ($1/\beta = 1$) can then be obtained with the model material having the following relations to the prototype material:

(*a*) The same shear angle, but the cohesion reduced by the scale factor ($1/\alpha = 1/\lambda\rho$);

(*b*) the effective stresses σ' in the stress–strain diagram $\sigma' = \sigma'(\varepsilon)$ reduced by the scale factor $1/\alpha$.

If the consolidation process should be investigated too, the further condition:

(*c*) $(1/\mu = \gamma/\lambda\rho\beta)$ (22.7) for the scale factor of the coefficient of permeability should be fulfilled, $1/\gamma$ representing the time scale and $1/\beta$ the strain scale factors.

22.2 GENERAL DISCUSSION OF CREEP EFFECTS ON THE MAGNITUDE OF THE EARTH PRESSURE

At the full mobilization of the shearing resistance of the soil the creep rate is initially either constant or even decreasing, but then it starts to increase until collapse occurs. For this reason the full shearing resistance cannot be mobilized in the backfill of a stable retaining wall.

Creep at constant speed appears usually at stress states which approach rupture. In order to reduce the active earth pressures to minimum values corresponding to a mobilization of the shearing resistance of the backfill of this order (e.g. to a 95 per cent mobilization), the retaining structure should be moved away at a speed corresponding to the creep speed of the soil. Further, in order to take advantage of the maximum value of the passive earth pressure the construction should be moved into the backfill at the corresponding speed.

If the degree of mobilization of the shearing resistance of the backfill is smaller, the creep speed decreases and if it is small enough—90 to 95 per cent in some available tests—the speed decreases approximately in inverse proportion with time, i.e. the creep develops as a logarithm function of time. Thus if the retaining construction still permits the initial creep after the hydrodynamic consolidation of the backfill, small further displacements will make it possible to sustain the mobilized stress states.

If, however, the movement of the wall is stopped, retention of the creep is expected to cause relaxation of the deviator stresses in the backfill and the corresponding increase of active or decrease of passive earth pressures on the wall. The smaller the creep speed at which the movement of the retaining structure is stopped, the smaller the relaxation effect is expected to be—and vice versa: the greater the preceding creep rate, the greater the relaxation effect.

The results of some model studies investigating these effects will be presented and discussed in the next section and their interpretation for retaining structures will be given.

22.3 RELAXATION EFFECTS ON THE EARTH PRESSURE

22.3.1 Experimental Data

The effects of the retention of soil creep have been investigated by S. Vidmar and the author (Šuklje and Vidmar, 1961-a; Vidmar, 1963) using long cylindrical consolidometers with roughened boundary surfaces and movable bases, as well as with model equipment. The movable base and retarding slab respectively are supported by dynamometers and the reaction forces are measured at successive phases of consolidation, motion of the base and slab respectively, as well as at the state of rest.

The results of tests effected in long consolidometers have been presented in the plots relating the translational displacements of the upper and lower piston, as well as the average values of the contact pressures, to the time. Two examples of such diagrams are reproduced (after Šuklje and Vidmar, 1961) in Fig. 22.13.

By dividing the difference of the upper (σ_a) and lower (σ_b) contact pressures by the boundary shear area of the cylinder, the average value of mobilized shear can be obtained:

$$\tau = \frac{(\sigma_a - \sigma_b)\dfrac{d^2\pi}{4}}{d\pi h} = \frac{d}{4h}(\sigma_a - \sigma_b) \tag{22.14}$$

d being the diameter of the piston and h the height of the specimen corresponding to the upper pressure σ_a and lower pressure σ_b. The ratio of the shear stresses τ after and before the stoppage of creep is assumed to be representative for the relaxation effect. In the cases presented in Fig. 22.13 this ratio is

$$\frac{\tau_0}{\tau_a} = \frac{0{\cdot}183\ \text{kp/cm}^2}{0{\cdot}445\ \text{kp/cm}^2} = 0{\cdot}41 \text{ in the first case } (a) \text{ and}$$

$$\frac{\tau_0}{\tau_a} = \frac{0{\cdot}341}{0{\cdot}480} = 0{\cdot}71 \text{ in the second case } (b);$$

yet the second case represents a test made in a series, with two preceding tests having been carried out on the same specimen. The ratio τ_0/τ_a with respect to the maximum shear strength mobilization over the whole series is:

$$\frac{\tau_0}{\tau_a} = \frac{0{\cdot}341}{0{\cdot}543} = 0{\cdot}63$$

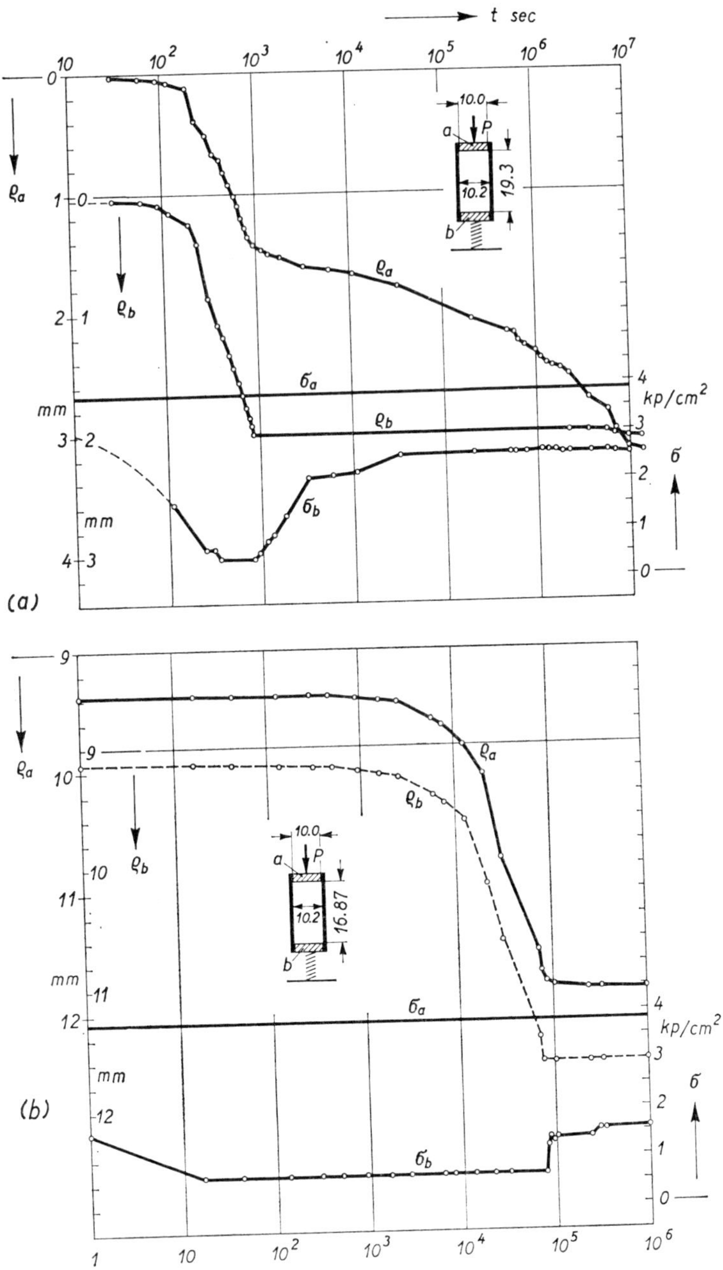

Fig. 22.13. Relaxation tests with clay in long consolidometers with rough walls and movable bases (after Šuklje and Vidmar, 1961-a): displacements (ρ) of the pistons and the corresponding contact pressures (σ) versus time (t)

The soil tested was a Tertiary clay with liquid limit $w_L = 40$ per cent and plasticity index $I_P = 24$ per cent; the initial consistency index was $I_c = 0{\cdot}63(w = 25$ per cent).

Fourteen tests were made in the model. The second group of these tests described in the paper by S. Vidmar (1963) contained 12 tests made with a Tertiary clay characterized by $w_L = 47$ per cent, $I_P = 32$ per cent. The initial water content of $w_0 = 32$ per cent decreased during tests to 25·5 per cent. The decrease was due to consolidation which was constrained by a drainage system of vertical sandy fill drains. The permeability coefficient corresponding to the consistency of the model clay was about

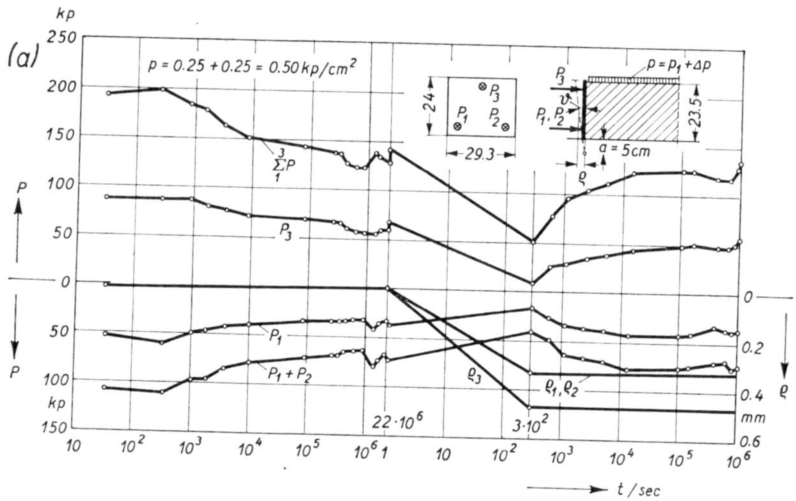

Fig. 22.14. Relaxation tests with clay by vertical retaining slabs (after Vidmar, 1963): displacements (ρ) of the slab and the corresponding horizontal components (σ) of Earth pressures versus logarithm of time (t):

(a) Stopped after consolidation and rotation of the wall in the external direction of $\vartheta = -2{\cdot}35 \,.\, 10^{-3}$;

$3{\cdot}10^{-9}$ cm/sec. The Mohr–Coulomb envelope of specimens consolidated from slurry at the liquid limit was found to be a straight line of slope $\varphi = 20°$ with $c' = 0{\cdot}01$ kp/cm².

Some examples of the results obtained are shown in Figs. 22.14 (a) to (e).

The horizontal components of the movements of the supporting plate (ρ), as well as the forces measured on each proving ring (P_1,P_2,P_3) and their sum $\sum_1^3 P_i$ are plotted against time t. In the cases when the supporting

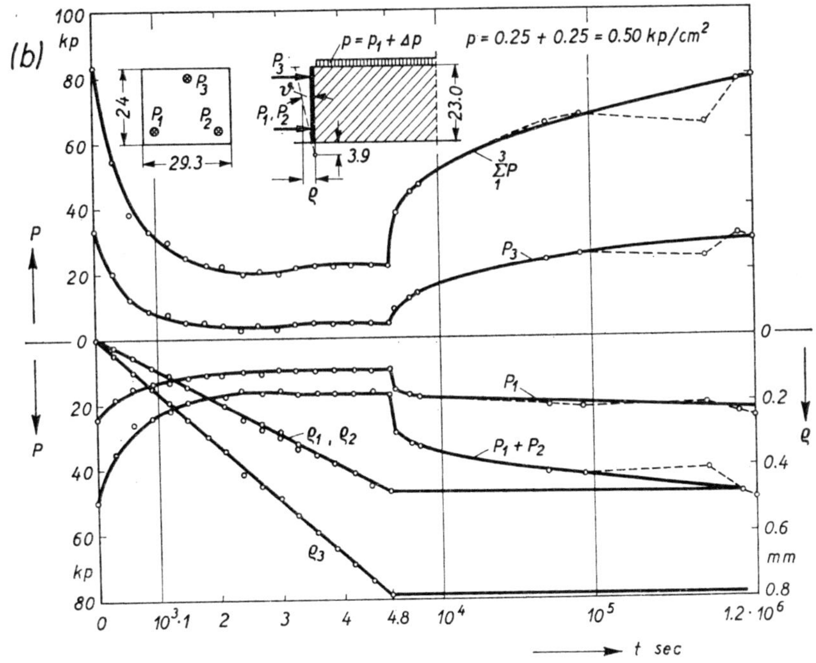

(*b*) Stopped after rotation of the wall in the external direction of $\vartheta = -3{\cdot}85 \,.\, 10^{-3}$;

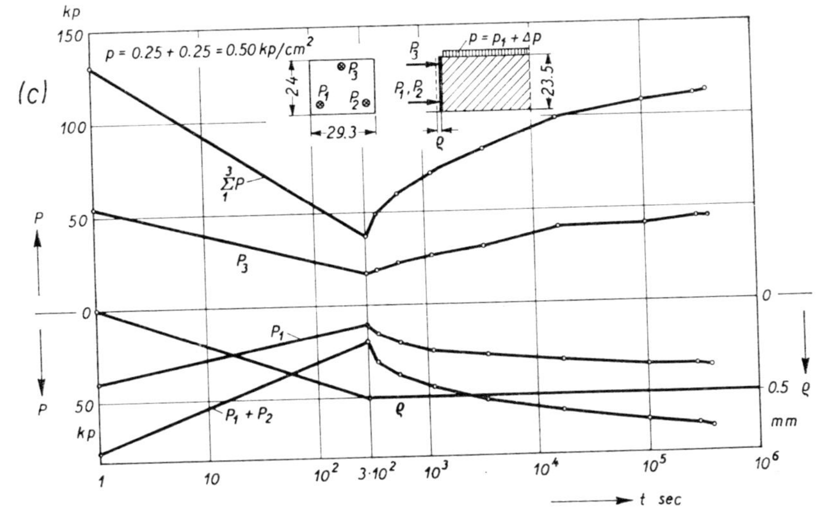

(*c*) Stopped after a translational movement of the wall in the external direction of $\rho = -0{\cdot}5$ mm;

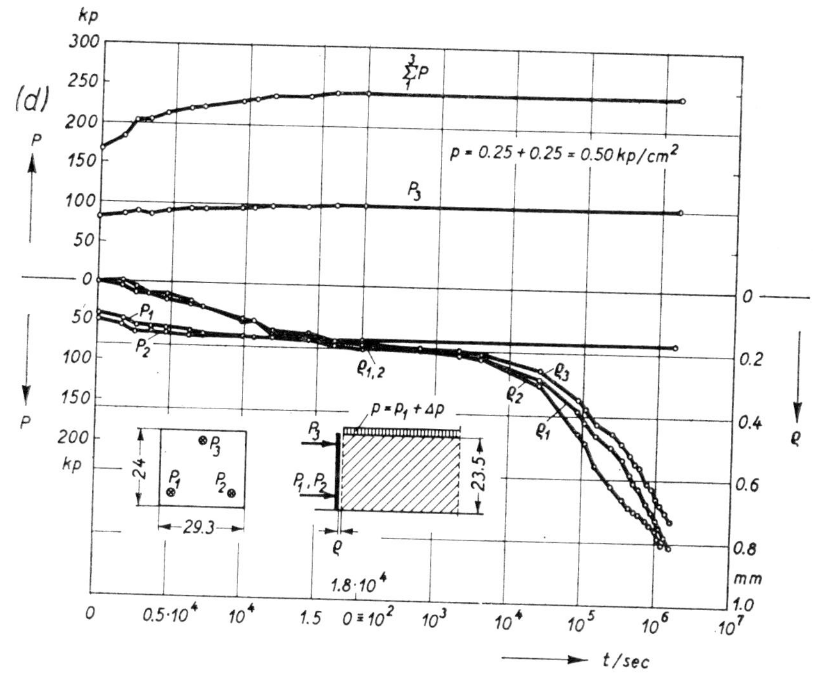

(*d*) Translational movement of the wall against the backfill, first at increasing and then at constant Earth pressure: $\rho <_{\max} = 0{\cdot}776$ mm;

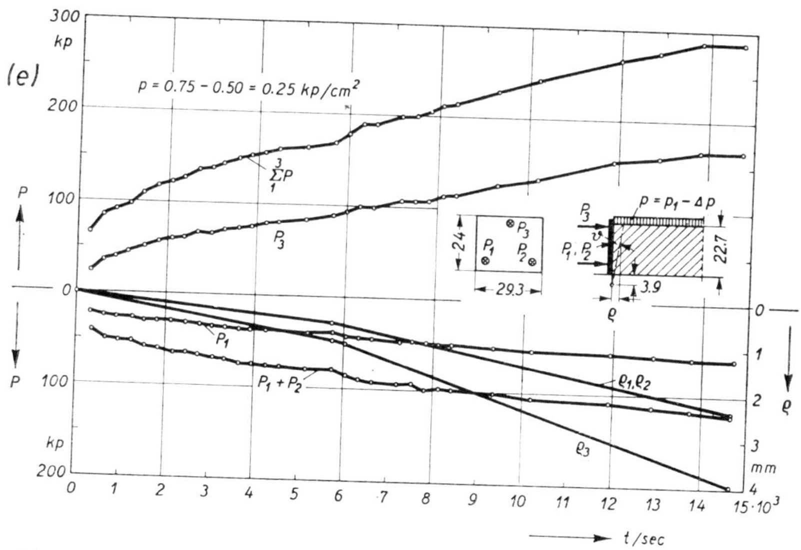

(*e*) Rotation of the wall against the backfill at increasing and finally at constant Earth pressure: $\vartheta_{\max} = 19{\cdot}10^{-3}$

plate rotated about a point below its lower edge, three displacements $\rho_1 = \rho_2$ and ρ_3 are given.

The shear stresses mobilized in subsequent phases of each test have been determined by failure plane analyses. The full strength of the soil has been assumed to have been mobilized in test No. 8 the results of which are presented in Fig. 22.14 (*b*). From the $\varphi = 0$ analysis the total average shearing resistance was found to be $\tau_f = 2{\cdot}14$ t/m². Related to this resistance the following degrees of strength mobilization have been established in the tests presented in Fig. 22.14 (*a*) to (*e*):

No. of test	Presented in Fig.	Phase of the test	Time t 10^3 sec	Degree of strength mobilization η
1	22.14 (a)	consolidation outside rotation stopped	2240·0 2240·3 3288·6	0·598 0·862 0·594
8	22.14 (b)	— outside rotation stopped	0 4·8 1184·4	0·753 1·000 0·770
2	22.14 (c)	— outside translation stopped	0 0·3 371·0	0·594 0·934 0·649
4	22.14 (d)	— inside translation inside translation	0 18·0 1558·8	0·455 0·250 0·250
12	22.14 (e)	— inside rotation inside rotation	0 5·7 14·7	0·417 0·050 −0·213*

* In the passive direction.

22.3.2 The Interpretation of Experimental Data for Actual Retaining Structures

Based on test results presented in the previous section and on similar results of other tests effected in models and in long consolidometers (Šuklje and Vidmar, 1961-a; Vidmar, 1963) the following immediate statements can be made (Vidmar, 1963):

The magnitude of the earth pressure acting on the retaining plate, and the corresponding degree of mobilization of the strength of the soil, due either to a translational or to a rotational movement of the support, depend: (*a*) on the magnitude of the translational movement and angle of rotation respectively, (*b*) on the speed of translation or rotation respectively, (*c*) on the preceding degree of mobilization of the strength of the soil, and (*d*) on the creep speed and on the preceding duration of the creep.

After stopping the retaining wall, the active earth pressure becomes greater and greater and approaches the value of the earth pressure at rest, acting on the wall before the movement. In the tests carried out so far the magnitude of the average earth pressure at rest corresponded to a 66 per cent mobilization of the shearing resistance of the soil. The final active earth pressures observed after stopping the wall corresponded on average to a 73 per cent mobilization of the shearing resistance of the backfill, showing in most cases a tendency towards further increase.

Regardless of either the translational or rotational character of the preceding wall movement, the relaxation coefficient has approximately equal values if the degree of the preceding mobilization of shearing resistance was approximately equal.

Considering the earth pressures produced by movements of the retaining wall against the backfill, the importance of the consolidation effects due to increasing horizontal pressures should be emphasized. Appreciable consolidation and distortional displacements are needed in order to mobilize the full passive earth pressure, and the kinematic conditions of the permissible movement of the retaining structure will rarely permit the full strength of the backfill to be mobilized. In comparison with the active stress state the importance of the relaxation phenomena increases accordingly.

Arising from the general statements above and the preceding general discussion of the model test conditions (see Section 22.1.3) some further interpretation of the tests that were discussed in the previous sections can be presented. These tests are not model tests in the true understanding of the term. They were carried out as part of basic general research onto relaxation effects and the scale relations do not correspond to any real case. Nevertheless, the following tentative conclusions can be made considering general model scale rules:

(1) Neglecting the influence of the dead weight, disregarding the drainage phenomena and taking the stress scale $1/\alpha = 1$, geometrically similar displacement conditions control the degree of shearing resistance mobilization in the backfill and consequently the earth pressure development.

(2) The influence of the dead load which was not important in Vidmar's tests but which cannot be ignored in common retaining constructions,

requires that the stress scale $1/\alpha = 1/\lambda\rho$ should be respected. Consequently, geometrically similar displacement conditions hold for a backfill material related to the model material as stated in Section 22.1.3. For the same material as used in the model it can be expected that greater relative displacements are needed in larger walls to mobilize the same degree of strength or to maintain it. Thus the important relaxation effects established by Vidmar's tests appear in actual retaining constructions too.*

In detailed model studies of these effects the role of the density change with stress and with creep should be considered as well as the decrease of the creep speed with increasing density and vice versa. Although such tests are rather complicated, a reasonable approach to real relations could give useful information.

* Recently, Vidmar (1968) presented results of similar model tests made with loose sand. The tests have proved surprisingly high relaxation effects also in non-cohesive soils.

23

Creep Effects on the Earth Pressures in The Bases of Foundations

23.1 TIME-DEPENDENCE OF THE BEARING CAPACITY OF SOILS. GENERAL DISCUSSION

The bearing capacity of a soil is defined as the load acting on the base of the foundation, or as the pressure acting in the contact surface between the foundation and the soil, at which limiting equilibrium is reached along a continuous rupture surface or within a continuous rupture zone. The limiting equilibrium is commonly considered to be defined by the Mohr-Coulomb failure envelope and analytically expressed by Kötter's differential equations (see Chapter 7). Thus various methods from plasticity theories, which have been dealt with in Chapter 7, can be applied in order to obtain the bearing capacity of a given soil subjected to a given foundation load.

Now, the stress–strain relationship does not appear in these computational methods and the kinematic conditions are respected only in a qualitative way, i.e. as far as they determine the shape of the rupture figures. Thus creep effects can only be considered by introducing the long-term strength parameters (c' and $\tan \varphi'$).

In order to obtain the admissible load or admissible pressure on the base of the foundation, appropriate safety factors have to be taken into account. As far as they should be governed by our lack of knowledge of the forces acting on the foundation and in the soil, they may be related separately to the forces appearing in the computation. All the other uncertainties of the calculation are mainly expressed as safety factors related to the strength parameters c' and $\tan \varphi'$. The required reduction of these parameters accounts for and is governed by: the perfection of our investigation and knowledge of soil properties, the degree of heterogeneity and anisotropy of the soil, the extent of the forces considered in the computation, the perfection of the applied method of computation, the importance, function (its purpose and use) and life of the structure to be founded.

The admissible load, however, depends also on the admissible deformation of the soil underlying the foundation. The methods of computing the displacements of the soil have been dealt with in Chapters 15, 16 and 17. These methods were either based on the fundamental assumptions of the theory of elasticity, or the stress state assumed to be known. Now, except for some special solutions applying to certain rheological schemes, the stress states can be established, at the present stage of investigation, either for linear stress–strain relationships by using the solutions given by the theory of elasticity, or for 'plastic fields' inside which all stress circles reach a certain linear envelope. By various methods of the theory of plasticity such 'plastic fields' can be obtained for strength parameters reduced by required safety factors or mobilized at the given foundation load. The deformations of these plastic fields can be computed by the methods presented in Chapters 13 and 17, though the procedure is rather cumbersome. In order to obtain the settlements of the foundation, the deformations of the soil outside the 'plastic fields' should also be considered. Yet this can be done only by making further assumptions. The combination of the displacements inside the 'plastic field' with the displacements outside the field, according to the elastic stress state, is one possible approximation but this, however, does not satisfy the compatibility conditions.

When dealing with the methods of computing settlements, the possibilities taking into account the secondary consolidation were discussed as well and methods have been presented enabling the computation of the entire development of both spherical and distortional parts of the total settlement. But as far as the failure criterion (and not the displacement criterion) is concerned, the initial, and not the final stages are decisive in the admissible loading of cohesive soils of low permeability. Thus the strength parameters obtained by quick tests and related to total stresses, have to be introduced into the computation.

The degree of consolidation and the corresponding mobilization of the strength of the soil at the end of construction can be computed and taken into account when analysing the safety of the structure against failure. Appropriate approximative methods of computation should be chosen according to the special conditions of the case to be analysed.

23.2 REVIEW OF BEARING CAPACITY SOLUTIONS FOR FLAT FOUNDATIONS

The limiting equilibrium conditions as expressed by Kötter's differential equations (Chapter 7) yield a closed solution when applied to a flat uniformly loaded strip foundation of infinite length with a perfectly smooth base, lying on the surface of the soil which is assumed to be weightless and subjected to a uniform lateral surface load q (see Fig. 23.1). The solution was first presented by Prandtl (1921), and independently, by Bonneau (1921).* The plastic field which appears consists of the active Rankine wedge AFC (Fig. 23.1), the passive Rankine wedge CGE and an intermediate wedge CGF in which the rupture lines are logarithmic

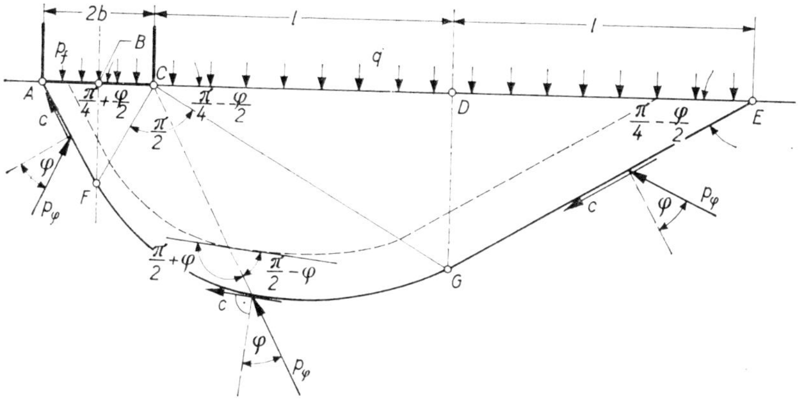

Fig. 23.1. Bearing capacity of a weightless soil subjected to a rigid, smooth infinite strip foundation (Prandtl's solution)

spirals with their centre at the edge C of the foundation and the pseudo-rupture lines are radial straight lines arising from the centre of the spiral. It can easily be proved that the limiting equilibrium conditions in all three parts of such a plastic field are satisfied if the foundation pressure has the following values:

$$p_f = q \tan^2\left(\frac{\pi}{4}+\frac{\varphi}{2}\right) \exp(\pi \tan\varphi) + c \tan\left(\frac{\pi}{4}+\frac{\varphi}{2}\right) \cdot$$

$$\cdot \left\{1 + \exp(\pi \tan\varphi) + \frac{\exp(\pi\tan\varphi) - 1}{\sin}\right\} \qquad (23.1)$$

For non-cohesive soils ($c = 0$) the above equation reduces to

$$(p_f)_{c=0} = q \tan^2\left(\frac{\pi}{4}+\frac{\varphi}{2}\right) \exp(\pi \tan\varphi) \qquad (23.2)$$

and for $\varphi = 0$:

$$(p_f)_{\varphi=0} = q + (2+\pi)c \qquad (23.3)$$

*See Caquot and Kerisel, 1949.

In order to consider the weight of the soil in the plastic zone, the methods of numerical integration of Kötter's differential equations explained in Chapter 7 should be applied. Caquot and Kerisel (1934, 1949, 1953) presented a general solution accounting for an arbitrary inclination of the earth pressure in the foundation planes. The influence of the self-weight of the soil and of the lateral pressure p are treated separately and the solutions are superimposed although different rupture lines result.

If the contact surface between the foundation and the soil is rough, there are no conditions for the limiting equilibrium state immediately below the base of the foundation. In a weightless soil the 'elastic wedge' appearing below the base is limited by a straight pseudo-rupture line (Fig. 23.2). According to Terzaghi (1943) it is inclined at the angle φ

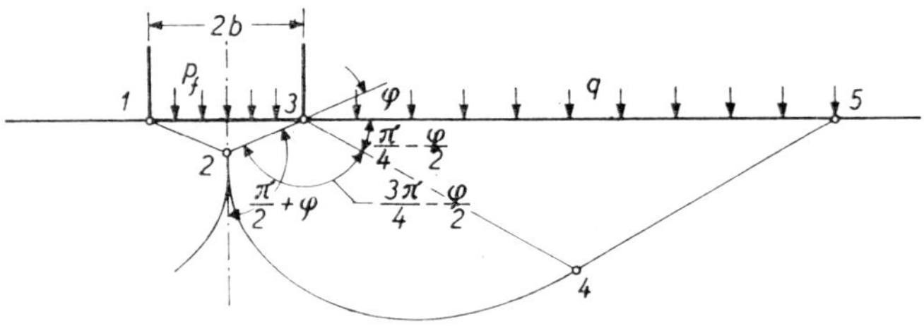

Fig. 23.2. Rupture figure below a rigid, rough infinite strip foundation according to Terzaghi (1943)

with the base. In this way the initial tangent of the rupture line coincides with the vertical axis of the foundation which is compatible with the translational vertical movement of the elastic wedge. The logarithmic rupture lines originating at the boundary (2–3) of the 'elastic wedge' pass into the straight rupture lines of the passive Rankine wedge (3–4–5). In Terzaghi's approximate solution the plastic field described has been assumed to hold also when considering the self-weight of the soil. The solution has been given in the form

$$p_f = \gamma b N_\gamma + q N_q + c N_c \tag{23.2}$$

where γ is the bulk density of the soil under the foundation base, c is the cohesion and N_γ, N_q, N_c are coefficients depending only on the inclination φ' of the limiting envelope (shear angle).

If the weight of the soil and the lateral loading in the foundation plane are considered simultaneously, the rupture surfaces do not follow Prandtl logarithmic spirals and Rankine straight lines. For certain strength parameters (c',φ') and certain foundation conditions (b,d) the bearing capacity has been computed by using Sokolovsky's method of numerical integration (Sokolovsky, 1960; Berezantzev, 1960). The resulting coefficients facilitating the determination of the bearing capacity for arbitrary soil and foundation conditions are tabulated (see also Isytovich, 1963).

For a rectangular load of finite dimensions Brinch Hansen (1961, 1966) has summarized the analytical and experimental data obtained by various authors: Terzaghi (1943), Skempton (1951), Meyerhof (1951, 1953, 1955), Schultze (1952), Lundgren and Mortensen (1953), and suggested the use of the following general formula:

$$p_a = \frac{P}{A} = \frac{\gamma'}{2} BN_\gamma s_\gamma i_\gamma + (c_m + q \tan \varphi_m) N_c s_c d_c i_c + q \tag{23.3}$$

where:

p_a = the admissible pressure related to the effective loading area ($B.L$) the centre of which is the point of application of the resulting force;

P = the total vertical admissible load;

B and L are the width and the length of the effective loading area as defined above;

γ' is the effective unit weight of the soil, i.e. the unit weight reduced by the uplift if present;

q is the minimum effective load in the plane of the foundation base outside the foundation;

φ_m is the admissible mobilized shear angle defined by the equation

$$\tan \varphi_m = \frac{\tan \varphi}{F}$$

if φ is the shear angle and F the corresponding safety factor;

c_m is the admissible mobilized cohesion defined by the equation

$$c_m = \frac{c}{F_c}$$

if c is the cohesion, i.e. the shear strength at zero effective pressure, and F_c the corresponding safety factor;

N_γ and N_c are bearing capacity factors related to a centrally and vertically loaded infinite load strip ($L \to \infty$, B = const) applied to the surface of the soil; they are functions of the mobilized shear angle φ_m;

s_γ and s_c are shape factors depending on the ratio B/L (Fig. 23.3)

$$s_\gamma = 1 - 0{\cdot}4\,\frac{B}{L}, \qquad s_c = 1 + 0{\cdot}2\,\frac{B}{L};$$

d_c is the depth factor depending on the ratio D/B between the depth D (Fig. 23.3) and the width of the effective load area, and on the mobilized shear angle φ_m; its approximate value for shallow foundations is:

$$d_c = 1 + 0{\cdot}35\,\frac{D}{B}$$

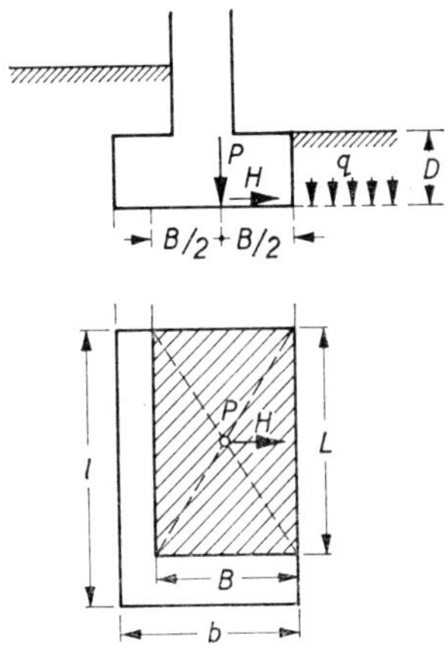

Fig. 23.3. Notation to Equation (23.3)

i_γ and i_c are inclination factors depending on the mobilized shear angle φ_m and on the ratio χ:

$$\chi = \frac{H}{Ac_m + V\tan\varphi_m}$$

where V and H are the vertical and horizontal components of the resultant force acting on the base of the foundation.

Brinch Hansen (1961) has presented diagrams where the coefficients N_γ, N_c, i_γ and i_c are plotted versus the mobilized shear angle φ_m and values of φ_m and χ respectively.

23.3 EARTH PRESSURE DISTRIBUTION IN THE BASES OF STIFF FOUNDATIONS

The formulae of the bearing capacity of the soil presented in the previous section are related to the safety of the foundation as a whole against sliding and no attention was paid to the distribution of the earth pressures in the foundation base. If the foundation base is not stiff, e.g. if the soil is loaded by an earth dam, the stresses in the base are mainly governed by the shape and unit weight of the dam material and by the surface and hydraulic loads transferred through the dam. The stability of the soil

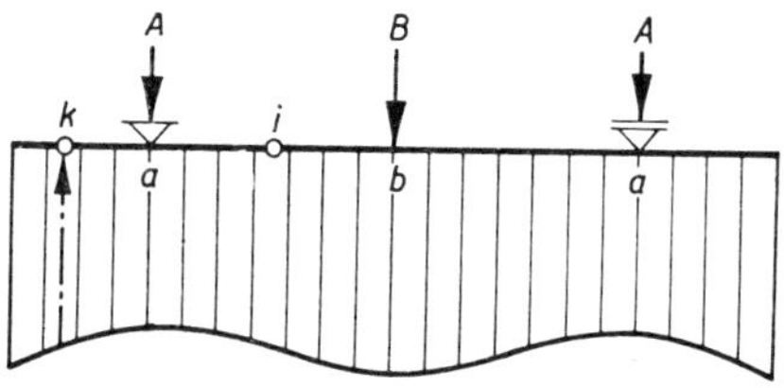

Fig. 23.4. To the computation of reaction stresses in the base of a symmetric continuous foundation beam subjected to three concentrated loads

subjected to such loads can be investigated by special analyses based either on a rigorous method of applying the theory of plasticity (see Chapter 7) or on the use of failure surfaces as explained in Chapter 19. For triangular, trapezoidal, parabolic and step-like infinite strip loads and for homogeneous cohesive subsoil layers of low permeability and of limited thickness, some approximate closed solutions have been presented by the author (Šuklje, 1952) and Habib (Habib–Šuklje, 1952).

The distribution of earth pressures in the base of a stiff foundation is governed by the stiffness of the structure on the one hand and the deformability of the soil and its degree of strength mobilization on the other. The condition that in the contact surface the deflections of the foundation must be in accord with the settlement of the soil should be satisfied. The realization of this principle may be demonstrated by the example presented in Fig. 23.4.

The foundation beam *a–b–a* is symmetrical and loads *A*, *B*, *A* acting at the top are assumed to be independent of the deformation conditions, i.e. the upper structure is assumed to be statically determinate. The distribution of contact pressures in the transverse direction is assumed to be uniform, the cross-section (the section modulus I) constant and the modulus of elasticity of the beam (E_c) constant also. The base of the foundation and the soil are assumed to remain in contact throughout the deformation process.

The diagram of the reaction pressures of the soil may be divided into n elements. The reaction force in an arbitrary element k may be denoted by X_k. The following further notation may be used:

a_{ik} = settlement of point i in the foundation base corresponding to the application of a force $X_{k1} = 1$ at point k*;
b_{ik} = deflection at point i of the beam relating to the straight line connecting a–a (Fig. 23.5) when the beam, as a freestanding static system supported at the points a, a, is exposed to the unit force $X_{k1} = 1$;
a_{ak} = settlement of point a corresponding to the application of the unit force $X_{k1} = 1$ at point k;
b_{ib} = deflection at point i when the freestanding beam a–a is subjected to the unit force $B_1 = 1$ acting at the point b.

The downward direction is taken to be positive.

The equality of deflections of the beam and settlement differences requires that for any point i the following equation should be satisfied**:

$$\sum_{k=1}^{n}(a_{ik}X_k) - \sum_{k=1}^{n}(a_{ak}X_k) = b_{ib}B - \sum_{k=1}^{n}(b_{ib}X_k)$$

yielding:

$$\sum_{k=1}^{n}(a_{ik} - a_{ak} + b_{ik})X_k - b_{ib}B = 0 \tag{23.4}$$

For symmetric loading with an odd number (n) of elements $(n - 1)/2$ non-homogeneous linear equations of the type (23.4) are obtained. Together with the equilibrium condition:

$$\sum_{k=1}^{n} X_k = 2A + B = \text{const} \tag{23.5}$$

there are $(n + 1)/2$ equations enabling the determination of $(n + 1)/2$ unknown reaction forces X_k.

* If the reaction forces X_k were substituted by concentrated forces, the settlement a_{ii} gets infinite value. Thus the settlement corresponding to a load $X_{i1} = 1$ distributed over the area of the element i should be taken.

** In this simplified procedure the compatibility conditions in the boundaries of single elements are disregarded.

If the forces A, B, A represent the reactions of a rigid structure the total weight of which is $(2A + B)$, the ratio A/B depends on the stiffness of the foundation and the compressibility of the soil and represents an additional unknown quantity (A/B). As in this case the deflection of point b is zero, there is also an additional condition:

$$b_{bb}B - \sum_{k=1}^{n} b_{bk}X_k = 0 \tag{23.6}$$

yielding:

$$B = \frac{1}{b_{bb}} \sum_{k=1}^{n} b_{bk}X_k \tag{23.6'}$$

Inserting it into Equation (23.4) we get:

$$\sum_{k=1}^{n} \left(a_{ik} - a_{ak} + b_{ik} - \frac{b_{ib}}{b_{bb}} b_{bk} \right) X_k = 0 \tag{23.7}$$

Together with condition (23.5) there are again $(n + 1)/2$ linear equations enabling the determination of $(n + 1)/2$ unknown reaction forces X_k.

Under the assumptions made (E_c = const, I = const) the deflections b_{ik}, b_{bk} can be expressed by the quotients:

$$b_{ik} = \frac{\beta_{ik}}{E_cI}, \quad b_{ib} = \frac{\beta_{ib}}{E_cI}, \quad b_{bb} = \frac{\beta_{bb}}{E_cI}, \quad b_{bk} = \frac{\beta_{bk}}{E_cI} \tag{23.8}$$

β_{ik}, β_{ib}, β_{bb}, β_{bk} being the deflections corresponding to the stiffness $E_cI = 1$.

If the subsoil is assumed to be a homogeneous isotropic elastic half-space whose deformations are governed by the modulus of deformation E_s and by the Poisson ratio ν, then the settlements a_{ik} and a_{ak} for a given value ν can be expressed by the quotients:

$$a_{ik} = \frac{\alpha_{ik}}{E_s}, \quad a_{ak} = \frac{\alpha_{ak}}{E_s} \tag{23.9}$$

α_{ik}, α_{ak} being the settlements corresponding to $E_s = 1$. In this way the equations of type (23.4) and (23.7) can be written as follows:

$$\sum_{k=1}^{n} \left(\alpha_{ik} - \alpha_{ak} + \frac{E_s}{E_cI} \beta_{ik} \right) X_k - \frac{E_s}{E_cI} \beta_{ib} B = 0 \tag{23.4'}$$

$$\sum_{k=1}^{n} \left[\alpha_{ik} - \alpha_{ak} + \frac{E_s}{E_cI} \left(\beta_{ik} - \frac{\beta_{ib}}{\beta_{bb}} \beta_{bk} \right) \right] X_k = 0 \tag{23.7'}$$

These equations prove that the distribution of contact pressures depends on the distribution of the active loads governing the coefficients β, on the

Poisson ratio ν influencing the coefficient α, and on the quotient E_s/E_cI.

Even in homogeneous deformation conditions, i.e. in the case of the deformation parameters being independent of octahedral normal and shear stresses, the Poisson ratio ν and the modulus of deformation E_s change during consolidation. Consequently the earth pressure distribution changes too and the most unfavourable distribution appearing during consolidation should be taken as the basis of the design of the foundation structure. Now, as the deformability of the soil varies with the magnitude of the spherical and distortional stress components, a rigorous application of the method of the determination of the stress distribution indicated above is not possible at all. In order to approach reality, the deformation parameters E_s and ν should be chosen as equivalent values (see Section 17.2) corresponding to a representative stress state expected below the foundation base.

With similar limitations, the settlements a_{ik}, a_{ak} appearing in equations (23.4) and (23.7) can be computed for the layered half-space so long as the corresponding fundamental solutions are available.

The resulting pressure distribution could be taken as the starting assumption in an iterative computation of the stress distribution in which the complex stress–strain relationships, as explained in Sections 13.3, 17.1, would have to be considered. An approximate pressure distribution can be reached by trial and error whereby the compatibility of the displacements of the structure and the soil in the foundation base must be respected. Such a computation is very cumbersome; even in this more perfect analysis not all the influential factors can be considered and thus the work entailed is hardly justified by the limited accuracy of the results obtained.

Appropriate model tests seem able to obtain clearer and more reliable information about the pressure distribution in the foundation base. It can be hoped that by systematic series of such tests appreciable data could be collected enabling the design of the earth pressure distribution for various foundation and subsoil conditions and in different phases of the consolidation process.

23.4 KINEMATIC CONDITIONS OF THE APPLICATION OF SHEAR FORCES IN THE FOUNDATION BASE

The normal and the tangential forces acting in the base of the foundation are transferred to the soil and, due to the reaction stresses, the soil undergoes deformations governed by the conditions of the initial, primary and secondary consolidation. The greater the contact stresses and the larger the contact surface, the greater the displacements. These increase further with increasing compressibility and with increasing mobilization of the strength of the soil. Finally, the smaller the lateral limitation on movement, the greater the tangential displacements.

Contact creep develops simultaneously with the consolidation of the ground beneath. It depends on the magnitude of the contact stresses, on the shearing resistance, density and water content of the immediate contact layer and, in cohesive soils, on the pore pressures in this layer; further it depends also on the roughness of the contact concrete surface. There is, however, no reason for the assumption that the magnitude of the shear displacements and the speed of contact creep should depend on the length of the rigid concrete base.

The stoppage of the shear displacement of the foundation requires the relaxation of the shear stresses in the contact surface. This relaxation is necessary to stop the creep in the contact surface as well as to stop the consolidation displacements of the ground beneath which, naturally, also causes displacement of the structure.

To allow for a certain degree of mobilization of contact shear strength, either the corresponding displacement must be allowed by the static character of the construction, or displacements of the foundation (supports) must be produced by preloading (by an amount exceeding the reactions of the structure to be supported), before the actual loading by the structure occurs.

In order to study the deformation conditions of the mobilization of the shear strength (friction and adhesion respectively) tests have been carried out by J. Brodnik and the author (Šuklje and Brodnik, 1963) using contact surfaces 20 cm in width and 60 cm in length and applying normal stresses up to 0·6 t/m². The concrete slab was made of fine-grained gravel and was in some tests smooth and in others rough. The soil consisted alternatively of fine and medium sand, fine gravel and clay. The limited extent of the investigations concerning the interval of stresses as well as the properties of the contact materials does not allow final conclusions to be made. The results obtained indicate the following tentative statements which need to be corroborated and completed by further tests: (*a*) The contact creep is not characterized by uniform movement unless the stress state approaches rupture; at smaller stresses the creep is a movement with negative acceleration. (*b*) The greater the preceding displacements, the smaller the relaxation of the stresses needed to stop the movement. (*c*) The tests with a cohesive base show more important relaxation effects than the tests with a non-cohesive base. (*d*) The friction between the smooth base and the coarse-grained soil is smaller than between the same base and the fine-grained soil. (*e*) The contact friction between a smooth concrete plate and the sandy or gravelly ground underneath is smaller than the internal friction of the sand and gravel respectively.

In the tests made the complete termination of creep required the reduction in mobilization of the full shearing strength of the contact between the concrete and sand or gravel of 37 to 11 per cent depending on the preceding degree of strength mobilization, on the preceding shear displacement and on the character of the contact materials; and by 64 to 47 per cent of the shearing strength of the concrete-clay contact. The absolute amount of the preceding shear displacement, however, was small—of the order of magnitude of 1 mm. In most cases the lateral displacements due to the consolidation of the subsoil are expected to be greater than the displacements needed to mobilize the shear stresses at the contact surface, and the admissible tangential force in the foundation base will mainly be governed by the deformation of the subsoil. Thus the interpretation of the laboratory tests has to be carried out carefully.

APPENDIX

Since the manuscript of this text was presented to the Editor in November 1965, several papers and some books have been published dealing with rheological behaviour of soils. The discussion of some of these publications was later inserted in the text; some others whose subject is closely related to the matters treated in the present book shall be summarized in this Appendix. The remainder are enclosed into the Bibliography recommended for further study.

APPENDIX

SOIL AS SATURATED LINEAR RHEOLOGICAL BODY

Verigin (1964, 1965) has presented solutions for the consolidation of the elastic half-space subjected to an external load normal to its boundary. Special solutions have been given for uniform loads acting along a straight line of finite length or along the perimeter of a rectangle and circle, respectively, as well as for a uniformly loaded circular area.

Schiffman, Ladd and Chen (1964) have based the solution of the consolidation process of a compressible layer loaded at the surface by a uniform load on linear stress–strain relationship corresponding to two Kelvin models and an elastic element in series.

Freudenthal and Spillers (1964) have given the solution for a viscous–elastic layer with deformation properties corresponding to Maxwell's body subjected to a partial surface load.

For the rheological model consisting of a Maxwell body connected in parallel with a Hookean spring in series with a Saint-Venant element (model M/V, see section 13.2.1) Kisiel and Lysik (1966) have developed three-dimensional relations between strains and stresses. In connection with Biot's equation of consolidation, general equations have been developed enabling the solution of the consolidation process provoked by loading the half-plate or half-space.

Examples of numerical solutions have been given for the development of volume deformations with time at plane state of stress (half-plate).

References of previous studies by Kisiel of the M/V model (from 1959 to 1964) are given in Kisiel's and Lysik's book *Rheology of Soils* (in Polish). This valuable text contains an excellent review of existing studies on stresses and strains in soils and on the development of displacements based on the assumption that the medium is either elastic or corresponds to linear rheological models.

De Josselin de Jong (1968) represents viscous grain skeleton by a generalized Kelvin's model. Instead of introducing a limited number of discrete parameters for the elements, the distribution of element parameters is continuous and stochastic. This means that the properties of the individual elements vary randomly and that by their abundance it is possible to represent the frequency of occurrence of their parameters by a continuous function.

The settlement time curves have been approximated by straight lines if plotted in a double logarithmic diagram.

The stochastic distribution of spring dashpot elements has been introduced in the consolidation theory by use of the frequency function. The result is a differential equation which resembles the original Terzaghi equation sufficiently to use the well known solutions.

As a second possibility suggested by K. Buisman (1940) the secular effect has been described by a model of cavities with various compressibility interconnected with channels of various conductivity. Again, the properties of the cavities and of the channels have been considered to be stochastic and their occurrence has been described by continuous frequency functions.

Mathematically the cavity channel elements behave very similarly to the spring dashpot elements. Their time-delaying effect on compressibility is in fact identical.

In the cavity channel model excess pore pressure is a stochastic variable. It is not possible to compute the excess pore pressure for the individual cavities since the equations do not describe these various individual pore pressures explicitly. Instead an average ambient excess pore pressure appears in the analysis, which is a weighted average of the pore pressures in the surrounding cavities. The weighted function turns out to be proportional to the conductivity of the connecting channels.

The differential equation governing the average ambient excess pore pressure has the same structure as the original Terzaghi consolidation equation and resembles the consolidation equation obtained for the spring dashpot assembly. This permits use of all the results obtained for spring dashpots for the cavity channel network as well. The difference is in permeability, which in consolidation equation appears now as a time dependent entity.

This theory gives for small values of time t the approximation of the form $t^{(1-\varepsilon)/2}$, where ε includes time dependence of both compressibility and permeability. The tests on spherical clay samples referred to by de Josselin de Jong and Verraijt, 1965, gave a positive value for ε of 0·1, and for the settlement, in the secondary period, power time laws of the form t^{α} with α equal to about 0·1.

SOIL AS SATURATED LINEAR RHEOLOGICAL BODY WITH COMPRESSIBLE PORE-LIQUID

An outstanding monography on rheological behaviour of soils has recently been published by Zaretsky (1967).

According to experimental data of Meschyan (1965) the soil skeleton is assumed to be governed by the law of linear hereditary creep theory of Boltzman–Volterra. The media governed by this theory have state epuation in the following integral form:

$$\varepsilon(t) = \frac{1}{E_0}[\sigma(t) + \int_0^t K(t,\tau)\sigma(\tau)\,\mathrm{d}\tau] \qquad \text{(A.1)}$$

where $K(t,\tau)$ is a function called the creep kernel; it is characteristic of the medium; σ is total stress, ε is strain and E_0 instantaneous linear modulus of deformation.

If the strain $\varepsilon(t)$ is known, equation (A.1) should be solved for the stress $\sigma(t)$. The integral equation of second degree of Volterra results having the form:

$$\sigma(t) = E_0[\varepsilon(t) - \int_0^t R(t,\tau)\varepsilon(\tau)\,\mathrm{d}\tau] \qquad \text{(A.2)}$$

where $R(t,\tau)$ is a function called the creep kernel resolvent.

The theory of creep characterized by the integral relation of the type (A.1) is called the theory of hereditary creep because it issues from the principle of the participation of the previous stress state and its influence onto the actual state.

The theory of hereditary creep includes all theories based on linear rheological models.

In order to clarify the mechanical meaning of the creep kernel $K(t,\tau)$ the stress in Equation (A.1) may be taken constant $\sigma(\tau) = \sigma_0 = \text{const}$ and both parts of equation differentiated by t:

$$\frac{1}{E_0}K(t) = \frac{1}{\sigma_0}\frac{\mathrm{d}\varepsilon}{\mathrm{d}t} \qquad \text{(A.3)}$$

Thus the function $K(t)$ denotes the speed of the creep at constant unit of stress. multiplied by E_0.

Special forms of the creep kernel used so far are:

(1) $$K(t-\tau) = \delta \exp[-\delta_1(t-\tau)] \qquad \text{(A.4)}$$

corresponding to Kelvin's rheological model,

(2) $$K(t-\tau) = \frac{\delta}{(t-\tau)^\gamma}\quad, 0 < \gamma < 1 \qquad \text{(A.5)}$$

including, with $\gamma = 1$, the logarithmic law of secondary compression,

(3) $$K(t-\tau) = \frac{\delta \exp[-\delta_1(t-\tau)]}{(t-\tau)^\gamma}, \; 0 < \gamma < 1 \tag{A.6}$$

representing the combination of previous forms.

Zaretsky's consolidation theory is not based on the classical concept of effective stresses. In his theory the entire volume deformation can be obtained by the superposition of two effects onto the soil skeleton: the effect of the total stress field governed by the compression operator of the soil skeleton, and the dilatation effect of the pore-pressure governed by the dilatation operator; the volume deformation is to be added to deviatoric deformation governed by the distortion operator of the soil skeleton. The compressibility of the pore-liquid due to the content of air bubbles is taken into account.

Denoting by α_w the compression modulus of the pore liquid, the differential equation of consolidation gets the form:

$$\frac{1}{\gamma_w}\frac{\partial e}{\partial t} + \frac{3e_{\text{med}}}{\alpha_w}\frac{\partial H}{\partial t} = \frac{(1+e_{\text{med}})k}{\gamma_w}\Delta H \tag{A.7}$$

$H = \dfrac{u}{\gamma_w}$ (u = excess pore pressure)

or

$$\frac{\partial e}{\partial t} = \frac{1+e}{\gamma_w}k\Delta u - \frac{3e}{\alpha_w}\frac{\partial u}{\partial t} \tag{A.7'}$$

The strain tensor ε_{ij} is composed of the component $\varepsilon_{ij}{}^{sk}$ expressing the effect of total stresses disregarding the effect of the pore liquid and $\varepsilon_{ij}{}^{p}$ due to the effect of the pore liquid:

$$\varepsilon_{ij} = \varepsilon_{ij}{}^{sk} + \varepsilon_{ij}{}^{p} = \varepsilon_{ij}{}^{sk} + \delta_{ij}\varepsilon^{p} \tag{A.8}$$

The volume strain ε_v equals:

$$\varepsilon_v = \varepsilon_v{}^{sk} + 3\varepsilon^p = \varepsilon_v{}^{sk} + \varepsilon_v{}^{p} \tag{A.9}$$

The dilatation $\varepsilon_v{}^p$ is related to the pore pressure $p(t)$ by the equation:

$$\varepsilon_v{}^p(t) = \frac{3}{\tilde{\alpha}_{vp}}\{p(t)\} = \frac{3}{\alpha_{vp}}\left[p(t) + \int_0^t K_{vp}(t,\tau)p(\tau)\,d\tau\right] \tag{A.10}$$

$K_{vp}(t,\tau)$ being the speed of the dilatation creep per unit of pore pressure and α_{vp} the dilatation modulus of the soil skeleton. If the dilatation effects are disregarded the stress tensor can be expressed in the form

$$\sigma_{ij}(t) = 2\tilde{G}\{\varepsilon_{ij}{}^{sk}(t)\} + \delta_{ij}\tilde{\lambda}\{\varepsilon_v{}^{sk}(t)\} \qquad (i,j = 1, 2, 3) \tag{A.11}$$

with the following equation for the first invariant of stresses:

$$I_1(t) = \sigma_{kk}(t) = \tilde{\alpha}_v\{\varepsilon_v{}^{sk}(t)\}; \tag{A.12}$$

the meaning of the above used operators is as follows:

$$\left.\begin{aligned} \tilde{G}\{y(t)\} &= G[y(t) - \int_0^t R_i(t,\tau)y(\tau)\,d\tau] \\ \tilde{\lambda}\{y(t)\} &= \tfrac{1}{3}[\tilde{\alpha}_v\{y(t)\} - 2\tilde{G}\{y(t)\}] \\ \tilde{\alpha}_v\{y(t)\} &= \alpha_v[y(t) - \int_0^t R_v(t,\tau)y(\tau)\,d\tau] \end{aligned}\right\} \tag{A.13}$$

α_v is the modulus of volume compressibility of the soil skeleton and G the shear modulus, $R_v(t,\tau)$ is the kernel resolvent of the volume creep and $R_i(t,\tau)$ the kernel resolvent of the distortional creep. Considering also the dilatation effects, the stress–strain relationship gets the form:

$$\sigma_{ij}(t) = 2\tilde{G}\}\varepsilon_{ij}(t)\} + \delta_{ij}[\tilde{\lambda}\{\varepsilon_v(t)\} - \frac{\tilde{\alpha}_v}{\tilde{\alpha}_{vp}}\{p(t)\}] \qquad (i, j = 1, 2, 3) \tag{A.14}$$

$$\sigma_{kk}(t) \equiv I_1(t) = \tilde{\alpha}_v\{\varepsilon_v(t)\} - 3\frac{\tilde{\alpha}_v}{\tilde{\alpha}_{vp}}\{p(t)\} \tag{A.15}$$

Denoting

$$\frac{1}{\tilde{\beta}}\{y(t)\} = \frac{\tilde{\alpha}_v}{\tilde{\alpha}_{vp}}\{y(t)\} = \frac{1}{\beta}[y(t) - \int_0^t R_\beta(t,\tau)y(\tau)\,d\tau] \tag{A.16}$$

$$\frac{1}{\beta} = \frac{\alpha_v}{\alpha_{vp}},\ \beta = \text{pore pressure coefficient} \tag{A.17}$$

and

$$P(t) = p(t) - \int_0^t R_\beta(t,\tau)p(\tau)\,d\tau \tag{A.18}$$

we obtain

$$\frac{1}{\tilde{\beta}}\{p(t)\} = \frac{1}{\beta}P(t) \tag{A.19}$$

and

$$\sigma_{ij}(t) = 2\tilde{G}\{\varepsilon_{ij}(t)\} + \delta_{ij}[\tilde{\lambda}\{\varepsilon_v(t)\} - \frac{1}{\beta}P(t)] \tag{A.20}$$

$$i, j = 1, 2, 3$$

$$\varepsilon_v(t) = \frac{1}{\tilde{\alpha}_v}\{\sigma_{kk}(t)\} + \frac{3}{\beta}\cdot\frac{1}{\tilde{\alpha}_v}\{P(t)\} \tag{A.21}$$

Expressing the strains by derivates of displacements

$$\varepsilon_{ij} = \frac{1}{2}\left(\frac{\partial u_i}{\partial x_j} + \frac{\partial u_j}{\partial x_i}\right) \tag{A.22}$$

and using the tensorial symbols

$$\varepsilon_{ij} = \tfrac{1}{2}(u_{i,j} + u_{j,i}),\ \varepsilon_v \equiv \varepsilon_{kk} = \mathrm{u}_{k,k} = \frac{\partial u_1}{\partial x_1} + \frac{\partial u_2}{\partial x_2} + \frac{\partial u_3}{\partial x_3} \tag{A.23}$$

Equation (A.20) becomes:

$$\sigma_{ij} = \tilde{G}\{u_{i,j} + u_{j,i}\} + \delta_{ij}[\tilde{\lambda}\{u_{k,k}\} - \frac{1}{\tilde{\beta}}\{p(t)\}] \tag{A.24}$$

$$i, j = 1, 2, 3$$

Finally, the expressions (A.24) have to be inserted into equilibrium equations giving:

$$\tilde{G}\{\Delta u_1\} + \frac{\tilde{G} + \tilde{\alpha}_v}{3}\{u_{k,k_1}\} = \frac{1}{\tilde{\beta}}\{p_{,1}\} = \frac{1}{\beta}P_{,1} \tag{A.25}$$

whereby the notation has been used:

$$u_{k,k_1} = \frac{\partial}{\partial x_1}(u_{1,1} + u_{2,2} + u_{3,3}) = \frac{\partial}{\partial x_1}\left(\frac{\partial u_1}{\partial x_1} + \frac{\partial u_2}{\partial x_2} + \frac{\partial u_3}{\partial x_3}\right) \tag{A.26}$$

$$p_{,1} = \frac{\partial p}{\partial x_1} \tag{A.27}$$

Three equations of the type (A.25) and the consolidation equation (A.7) are basic for the solution of the consolidation process.

Zaretsky (1967) has developed a general method for solving the above basic system of equations.

Concrete solution has been given for the consolidation process of a layer of finite thickness subjected to a uniformly distributed surface load, provided that the creep kernel is an exponential function of the type (A.4). Following assumptions were taken: $\alpha_w \to \infty$ (incompressible liquid), $R_i(t) = R_v(t) \equiv R(t)$.

The resulting pore pressures are presented in the form

$$p_w(z,t) = \beta_q H(z,t) \tag{A.27}$$

and the coefficients $H\left(\frac{h}{2}, t\right)$ as well as the degrees of consolidation $U_0 = \frac{s(t)}{s(\infty)}$ are plotted versus time t for certain values of δ_1, δ and $\frac{\pi^2 c}{h^2}$

c being the coefficient of consolidation:

$$c = \frac{k\beta}{\gamma_w \alpha_{sk}} \tag{A.28}$$

Solutions have been given also for the half-space subjected to a concentrated surface load. The incompressibility of the pore liquid ($\alpha_w \to \infty$) and the constancy of the Poisson ratio have been assumed again. General formulae have been presented for stresses and displacements in an arbitrary point inside the half-space. The solutions are given for an arbitrary form of creep kernel $K(t,\tau)$ which, however, does not appear in solutions for stresses. For non-viscous soils the settlement can be expressed in the form:

$$s(r,t) = \frac{P}{4\pi G_{sk}} \bar{s}(r,t) \tag{A.29}$$

Coefficients $\bar{s}(r,t)$ have been tabulated for various values ct, r and ν.

By using the principle of superposition, the results obtained for the concentrated load have been applied for a flexible uniformly distributed circular load. This solution can then further be used for settlements due to the effect of an arbitrary surface load.

For the uniformly loaded layer of limited thickness similar solutions have been developed by Ter-Martirosyan (1967).* There, too, the creep corresponding to Kelvin's model and the compressibility of the pore liquid have been simultaneously taken into account.

SOIL AS NON-SATURATED LINEAR RHEOLOGICAL BODY

For non-saturated soils (rocks respectively) the theory of linear hereditary creep was previously applied by Erzhanov (1964) with the creep kernel $L(t,\tau) = \delta(t - \tau)^{-\alpha}$. Fundamental equation and principles have

* See Isytovich and all (1967).

been developed enabling, in creep conditions, the determination of stress and strain fields in the underground structures and in the surrounding rocks. For several mining problems, closed solutions have been given facilitating the analysis of the time influence on the static behaviour of underground constructions and on the settlements of the ground surface due to underground excavations. Further solutions of mining problems based on the same memory function have been presented by Lapidus (1964), Aytaliev (1964) and Matveeva (1964).

REFERENCES

The following list gives the sources of material referred to in the text.

REFERENCES

W. E. A. Acum and L. Fox, Computation of Load Stresses in a Three-layer Elastic System, *Géotechnique*, **2**, 293–300 (1951).

P. Anagnosti, *Analiza standardnih metoda ispitivanja osobina zemljanih materijala sa stanovišta mehanike neprekidnih sredina* (The analysis of the standard methods of testing soil properties from the point of view of the mechanics of continua, in Serbo-Croatian), Thesis, Beograd (1962).

P. Anagnosti, Stresses, Deformations and Pore Pressure in Triaxial Tests Obtained by a Suitable Rheological Model, *Proc. Eur. Conf. Soil Mech. and Found. Engng., Wiesbaden*, **1**, 1–6 (1963).

S. Andrei, Discussion, *Proc. 5th Int. Conf. Soil Mech. Found. Engng, Paris*, **3**, 116 (1961).

Sh. M. Aytaliev, Issledovanie raboty obdelki napornogo tonnelya krugovoga ochertaniya pod dejstviem neustanovivshegosya gornogo davleniya (The behaviour of circular tunnels subjected to water and roof pressures, in Russian), *Rheologicheskie voprosy mekhaniki gornykh porod, Akad. Nauk Kazakhskoi SSR, Alma Ata*, 84–91 (1964).

L. Barden, Consolidation of Compacted and Unsaturated Clays, *Géotechnique*, **15**, 267–286 (1965-a).

L. Barden, Consolidation of Clay with Non-linear Viscosity, *Géotechnique*, **15**, 345–362 (1965-b).

E. de Beer, Shear Strength Characteristics of the "Boom Clay", *Proc. Geotech. Conf., Oslo, 1967*, **1**, 83–88 (1967).

V. G. Berezantsev, Raschet *prochnosti osnovanii sooruzhenii* (Computation of strength of the foundations of structures, in Russian), *Gosstroyizdat, Moskva* (1967).

M. A. Biot, Le problème de la consolidation des matières argileuses sous une charge, *Annls. Soc. scient. Brux.*, **B55**, 110–113 (1935).

M. A. Biot, General Theory of Three-dimensional Consolidation, *J. appl. Phys.*, **12**, 155–164 (1941).

A. W. Bishop, Discussion of Measurement of Shear Strength of Soils, *Géotechnique*, **2**, 113–116 (1950).

A. W. Bishop and G. Eldin, Undrained Triaxial Tests on Saturated Sands and Their Significance in the General Theory of Shear Strength, *Géotechnique*, **2**, 13–32 (1950).

A. W. Bishop and A. K. G. Eldin, The Effect of Stress History on the Relation Between and Porosity in Sand, *Proc. Third Int. Conf. Soil Mech. Found. Engng, Zurich*, **1**, 100–105 (1953).

A. W. Bishop, The Use of the Slip Circle in the Stability Analysis of Slopes, *Proc. Eur. Conf. Stabl. Earth Slopes, Stockholm*, **1**, 1–13 (1954), reprinted in Géotechnique, **5**, 7–17 (1955).

A. W. Bishop and D. J. Henkel, *The Measurement of Soil Properties in the Triaxial Test*, Arnold, London, First Edn., 1957, 190 pp.; Second Edn. (1962), 228 pp.

A. W. Bishop, The Principle of Effective Stress, *Teknisk Ukeblad* (*Oslo*), **106,** No. 39, 859–863 (1959). (Also published in Norwegian Geotechnical Institute, *Publ.* 32), Lecture delivered in Oslo in 1955.

A. W. Bishop, Test Requirements for Measuring the Coefficient of Earth Pressure at Rest, *Proceedings Brussels Conf.* 58 *on Earth Pressure Problems*, **1,** 2–14 (1958).

A. W. Bishop, Discussion, *Conf. Pore Pressure and Suction in Soils*, Butterworths, London, 63–66 (1960).

A. W. Bishop and J. B. Donald, The Experimental Study of Partly Saturated Soil in the Triaxial Apparatus, *Proc. 5th Int. Conf. Soil Mech. Found. Eng., Paris*, **1,** 13–22 (1961).

A. W. Bishop, The Strength of Soils as Engineering Materials (6th Rankine Lecture), *Géotechnique*, **16,** 91–128 (1966).

A. W. Bishop and A. L. Little, The Influence of the Size and Orientation of the Sample on the Apparent Strength of the London Clay at Maldon, Essex, *Proceedings of the Geotechnical Conference, Oslo, 1967*, **1,** 89–96 (1967).

L. Bendel, *Ingenieurgeologie*, 1. Hälfte, Springer, Wien, (1944), 832 pp.

L. Bjerrum and B. Kjaernsli, Analysis of the Stability of some Norwegian Natural Clay Slopes, *Géotechnique*, **7,** 1–6 (1957).

L. Bjerrum, N. Simons and I. Torblaa, The Effects of Time on the Shear Strength of a Soft Marine Clay, *Proc. Brussels Conf. 58 on Earth Pressure Problems*, **1,** 148–158 (1958).

L. Bjerrum, The Effective Shear Strength Parameters of Sensitive Clays, *Proc. 5th Int. Conf. Soil Mech. Found. Eng., Paris*, **1,** 23–28 (1961).

L. Bjerrum, S. Kringstad and C. Kummeneje (1961), The Shear Strength of a Fine Sand, *Proc. 5th Int. Conf. Soil Mech. Found. Eng., Paris*, **1,** 29–37 (1961).

L. Bjerrum, Engineering Geology of Normally Consolidated Marine Clays as Related to the Settlements of Buildings (7th Rankine Lecture), *Géotechnique*, **17,** 83–117 (1967).

M. Bonneau, Equilibre limite et rupture des milieux continus, *Ann. Ponts Chauss.*, **117,** 609–653, 769–801 (1947).

J. Boussinesq, *Applications des potentiels à l'étude de l'équilibre et du mouvement des solides élastiques*, Gauthier-Villars, Paris, (1885).

A. S. K. Buisman, Results of Long Duration Settlement Tests, *Proc. Int. Conf. Soil Mech. Found. Eng., Harvard University*, **1,** 103–106 (1936).

K. A. S. Buisman, *Grondmechanika* (Soil Mechanics, in Dutch), Waltman, Delft, (1941).

D. M. Burmister, Stress and Displacement Characteristics of a Two-layer Rigid Base Soil System: Influence Diagrams and Practical Applications, *National Academy of Sciences, Highway Research Board, Proc. 35th Annual Meeting, Washington*, 773–814 (1956).

A. Caquot, *Équilibre des massifs à frottement interne, Stabilité des terres pulvérulentes ou cohérentes*, Gauthier-Villars, Paris, (1943).

A. Caquot and J. Kérisel, *Traité de Mécanique des sols*, 2nd edn. (1949): 385 pp., 3rd edn. (1956): 260 pp., 4th edn. (1966): 506 pp.

A. Caquot and J. Kérisel, Sur le terme de surface dans le calcul des fondations en milieu pulvérulent, *Proc. 3rd Int. Conf. Soil Mech. Found. Eng., Zurich*, **1,** 336–337 (1953).

A. Casagrande and S. D. Wilson, Effect of Rate of Loading on the Strength of Clays and Shales at Constant Water Content, *Géotechnique*, **2,** 251–263 (1951).

A. Casagrande and S. D. Wilson, Testing Equipment, Techniques and Errors, Moderators' Report Session 2, *ASCE Research Conference on Shear Strength of Cohesive Soils, University of Colorado*, 1123–1130 (1960).

I. F. Christie, A Re-appraisal of Merchant's Contribution to the Theory of Consolidation, *Géotechnique*, **14,** 309–320 (1964).

Ch. A. Coulomb, Essai sur une application des règles de maximis et minimis à quelques problèmes de statique relatifs à l'architecture, *Mem. Div. Sav., Académie des Sciences, Paris*, **7** (1776).

J. R. Cox, The Alexander Dam, Soil Studies and Settlement Observations, *Proc. Int. Conf. Soil Mech. Found. Eng., Harvard University*, **2,** 296–298 (1936).

H. Darcy, *Les fontaines publiques de la ville de Dijon* (1856).

N. Ya. Denissov, *Stroitel'nie svoistva glinistykh porod i ikh ispolzovanie v gidrotekhnicheskom stroitel'stve* (Engineering properties of clayey soils and their role in hydraulic engineering, in Russian), *Gosenergoizdat, Moskva-Leningrad*, 287 pp. (1956).

N. Ya. Denissov, Pore Pressure and Strength of Underconsolidated Clay Soils, *Proc. 6th Int. Conf. Soil Mech. Found. Eng., Montreal*, **1,** 208–212 (1965).

J. Drnovšek and I. Šubic, *Duktilnost vezljivih zemljin, ugotovljena pri enoosnem napetostnem stanju* (Ductility of cohesive soils at uniaxial stress state, in Slovenian), *Gradbeni vestnik, Ljubljana*, 238–241 (1966).

D. C. Drucker, A Definition of Stable Inelastic Material, *J. Applied Mechanics*, **26,** 101–186 (1959).

Zh. S. Erzhanov, *Teoriya polzuchesti gornykh porod i ee prilozheniya* (The theory of the creep of rocks and its applications, in Russian), *Izd. Nauka, Alma Ata*, (1964) 175 pp.

W. Fellenius, *Erdstatische Berechnungen mit Reibung und Kohäsion und unter Annahme kreiszylindrischer Gleitflächen*, Ernst & Sohn, Berlin, (1927), 40 pp.

V. A. Florin, *K voprosu o gidrodinamicheskikh napryazheniyakh v gruntovoi masse* (Contribution to the problem of the hydrodynamic stresses in the soil, in Russian), *GONTI* (1938).

V. A. Florin, *Odnomernaya zadacha uplotneniya szhimaemoi poristoi polzuchei zemlyanoi sredy* (One-dimensional problem of the consolidation of a compressible porous viscous soil medium, in Russian), *Izvestiya AN SSSR*, OTN No. 5 (1953).

V. A. Florin, *Osnovy mekhaniki gruntov* (Fundamentals of soil mechanics, in Russian), **1,** *Gosstroyizdat, Leningrad-Moskva*, (1959), 356 pp.

V. A. Florin, *Osnovy mekhaniki gruntov* (Fundamentals of soil mechanics, in Russian), **2,** *Gosstroyizdat, Leningrad-Moskva*, (1961), 543 pp.

J. B. Folque, *Reologia de solos nao saturados* (Rheology of non-saturated soils, in Portuguese), *Laboratório Nacional de Engenharia Civil, Lisboa*, Mem. No. 176 (1961-a), 209 pp.

J. Folque, Rheological Properties of Compacted Unsaturated Soils, *Proc. 5th Int. Conf. Soil Mech. Found. Eng., Paris*, **1,** 113–116 (1961-b).

L. Fox, Computation of the Stresses in Two-layer System Resulting from Uniform Loading Symmetrically Applied to the Upper Surface, *Road Res. Tech. Pap. (L.)* No. 9 (1948).

A. M. Freundenthal and W. R. Spillers, On the Consolidating Viscoelastic Layer under Quasi-Static Loading, *IUTAM Symposium Grenoble 1964: Rheology and Soil Mechanics*, Springer-Verlag, Berlin, 1966, 196–202 (1964).

O. K. Fröhlich, *Druckverteilung im Baugrande mit besonderer Berücksichtigung der plastischen Erscheinungen*, Springer, Wien, (1934), 185 pp.

M. Fukuoka and T. Taniguchi, Research on Landslides, *Proc. 5th Int. Conf. Soil Mech. Found. Eng., Paris*, **2**, 603–605 (1961).

E. C. W. A. Geuze, Horizontal Earth Pressure against a Row of Piles, *Proc. Second Int. Conf. Soil Mech. Found. Eng., Rotterdam*, **4**, 135–140 (1948).

E. C. W. A. Geuze and Tan Tjong-Kie, The Mechanical Behaviour of Clays, *Proc. Second Int. Cong. Rheol.*, 247–259 (1953).

E. C. W. A. Geuze, Discussion, *Proc. 3rd Int. Conf. Soil Mech. Found. Eng., Zurich*, **3**, 119 (1953).

E. C. W. A. Geuze, The Effect of Time on the Shear Strength of Clays, *Am. Soc. Civ. Eng. Convention in New Orleans* (1960).

E. C. W. A. Geuze, Rheological and Mechanical Models of Saturated Clay Particle Systems, *IUTAM Symposium Grenoble 1964: Rheology and Soil Mechanics*, Springer-Verlag, Berlin, 90–98 (1966).

R. E. Gibson and I. Mc-Namee, The Consolidation Settlement of a Load Uniformly Distributed over a Rectangular Area, *Proc. 4th Int. Conf. Soil Mech. Found. Eng., London*, **1**, 297–299 (1957).

R. E. Gibson and K. Y. Lo, A Theory of Consolidation for Soils Exhibiting Secondary Compression, *Norwegian Geotechnical Institute—Publication No.* 41 (also *Acta Polytechnica Scandinavica*, 296/191, Ci 10) (1961).

I. I. Gol'denblat and N. A. Nikolaenko, *Teoriya polzuchesti stroitel'nykh materialov i ee prilozheniya* (Theory of creep of building materials and its applications, in Russian), *Stroyizdat* (1960).

M. Goldstein and G. Ter-Stepanian, The Long-term Strength of Clays, *Proc. 4th Int. Conf. Soil Mech. Found. Eng., London*, **2**, 311–314 (1957).

M. Goldstein, Discussion, *Proc. Brussels Conference 58 on Earth Pressure Problems*, **3**, 91–92 (1958).

P. Habib, La résistance au cisaillement des sols, *Annales de l'Institut Technique du Bâtiment et des Travaux Publics, Sixième Année, No. 61* (Sols et Fondations No. 12), Paris, 1–40 (1953-a).

P. Habib, Influence de la variation de la contrainte principal moyenne sur la résistance au cisaillement des sols, *Proc. 3rd Int. Conf. Soil Mech. Found. Eng., Zurich*, **1**, 131–136 (1953-b).

P. Habib and L. Šuklje, Étude de la stabilité des fondations sur une couche d'argile d'épaisseur limitée, *Annales de l'Institut Technique du Bâtiment et des Travaux Publics, Septième Année, No. 83* (Sols et Fondations No. 15), Paris, 1148–1166 (1954).

R. Haefeli, Mechanische Eigenschaften von Lockergesteinen, *Schweizerische Bauzeitung*, **3**, 299–303, 321–325 (1938).

R. Haefeli and Ch. Schaerer, Der Triaxialapparat, *Schweizerische Bauzeitung*, **128**, 51–53, 65–67, 81–84 (1946).

R. Haefeli, Creep Problems in Soils, Snow and Ice, *3rd Int. Conf. Soil Mech. Found. Eng., Zurich*, **3**, 238–251 (1953).

J. Brinch Hansen, *Earth Pressure Calculation*, The Danish Technical Press, Copenhagen, (1953).

J. Brinch Hansen, The Internal Forces in a Circle of Rupture, *Danish Geotechnical Institute, Bulletin No. 2* (1957).

J. Brinch Hansen, Discussion, *Proc. Brussels Conference 58 on Earth Pressure Problems*, **3,** 124–126 (1958).

J. Brinch Hansen and H. Lundgren, *Hauptprobleme der Bodenmechanik*, Springer, Berlin (1960), 282 pp.

J. Brinch Hansen, A Model Law for Simultaneous Primary and Secondary Consolidation, *Proc. 5th Int. Conf. Soil Mech. Found. Eng., Paris*, **1,** 133–136 (1961-a).

J. Brinch Hansen, A General Formula for Bearing Capacity, *Ingeniøren*, **5,** 38–46 (1961-b) (Danish Geotechnical Institute Copenhagen, *Bull.* 11; see also the "Note concerning GI Bulletin No. 11" in *GI Bulletin* No. 21, Copenhagen 1966).

J. Brinch Hansen, Some Stress-Strain Relationships for Soils, *Proc. 6th Int. Conf. Soil Mech. Found. Eng., Montreal*, **1,** 231–234 (1965).

H. Hasegawa and M. Ikeuti, On the Tensile Strength Test of Disturbed Soils, *IUTAM Symposium Grenoble 1964: Rheology and Soil Mechanics*, Springer-Verlag, Berlin, 405–412 (1966).

R. M. Haythornwaite, Mechanics of the Triaxial Test for Soils, *Journal of the Soil Mechanics and Foundations Division, Proc. ASCE*, **86,** No. SM 5, 35–61 (1960).

D. J. Henkel, *The Correlation between Deformation, Pore Water Pressure and Volume Change Characteristics of Saturated Clays*, Ph.D. Thesis, University of London (1958).

D. J. Henkel, The Relationships between the Strength, Pore Water Pressure and Volume Change Characteristics of Saturated Clays, *Géotechnique*, **9,** 119–135 (1959).

D. J. Henkel, The Relationship between the Effective Stresses and Water Content in Saturated Clays, *Géotechnique*, **10,** 41–54 (1960-a).

D. J. Henkel, The Shear Strength of Saturated Remoulded Clays, *ASCE Conference on Shear Strength of Cohesive Soils, University of Colorado*, 533–554 (1960-b).

O. Hoffman and G. Sachs, *Introduction to the Theory of Plasticity for Engineers*, McGraw-Hill, New York (1953), 276 pp.

S. Hultin, Grusfyllningar för Kajbyggnader (Gravel Fillings for Quay Constructions, in Swedish), *Teknisk Tidskrift*, **46,** 292 (1916).

M. J. Hvorslev, *Über die Festigkeitseigenschaften gestörter bindiger Böden.*, Thesis, 159 pp., published by Danmarks Naturvidenskabelige Samfund, *Ingeniørvidenskabelige Skrifter*, Series A, No. 45, Copenhagen (1937).

M. J. Hvorslev, Physical Components of the Shear Strength of Saturated Clays, *ASCE Research Conference on Shear Strength of Cohesive Soils, University of Colorado*, 169–273 (1960).

B. Jakobson, On the Influence of Wall Movement on Earth Pressure, *Proc. Brussels Conference 58 on Earth Pressure Problems*, **1,** 105–115 (1958).

N. Janbu, Earth Pressures and Bearing Capacity Calculations by Generalized Procedures of Slices, *Proc. 4th Int. Conf. Soil Mech. Found. Eng., London*, **2,** 207–212 (1957).

J. E. Jennings, A Revised Effective Stress Law for Use in the Prediction of the Behaviour of Unsaturated Soils, *Conference on Pore Pressure and Suction in Soils, London*, 26–30 (1960).

S. J. Johnson and R. V. Whitman, Shear Strength of Saturated, Remolded Clays, Moderators' report Session 3, *ASCE Research Conference on Shear Strength of Cohesive Soils, University of Colorado*, 1131–1135 (1960).

G. de Josselin de Jong and A. Verruijt, Primary and Secondary Consolidation of a Spherical Clay Sample, *Proc. 6th Int. Conf. Soil Mech. Found. Eng., Montreal*, **1**, 254–258 (1965).

G. de Josselin de Jong, Consolidation Models Consisting of an Assembly of Viscous Elements or a Cavity Channel Network, *Géotechnique*, **18**, 195–228 (1968).

T. C. Kenney, The Influence of Mineral Composition on the Residual Strength of Natural Soils, *Proceedings of the Geotechnical Conference Oslo 1967*, **1**, 123–130 (1967).

Á. Kézdi, Earth Pressure on Retaining Wall, Tilting about the Toe, *Proc. Brussels Conference 58 on Earth Pressure Problems*, **1**, 116–132 (1958).

Á. Kézdi, *Erddrucktheorien*, Springer-Verlag, Berlin, 1962.

W. M. Kirkpatrick, The Condition of Failure for Sands, *Proc. 4th Int. Conf. Soil Mech. Found. Eng., London*, **1**, 172–178 (1957).

I. Kisiel, B. Lysik, *Zarys Reologii Gruntów* (Outline of the rheology of soils, in Polish), *Arkady, Warszawa* (1966), 315 pp.

W. Kjellman, Do Slip Surfaces Exist?, *Proc. European Conference on Stability of Earth Slopes, Stockholm*, **1**, 14–20 (1954); reprinted in *Géotechnique*, **5**, 18–22 (1955).

Hon-Yim Ko and R. F. Scott, A New Soil Testing Apparatus, *Géotechnique*, **17**, 40–57 (1967).

H. Koning, Stress Distribution in a Homogeneous, Anisotropic, Elastic Semi-infinite Solid, *Proc. 4th Int. Conf. Soil Mech. Found. Eng., London*, **1**, 335–338 (1957).

V. G. Korotkin, *Zadacha uplotneniya pri prililozhenii k poverkhnosti grunta sosredotochennoi sily* (Consolidation of Soils whose surface is subjected to a concentrated load, in Russian), *Trudy LPI*, No. 2 (1951).

V. G. Korotkin, *Opredelenie naprazhenii v vodonasyshchenom osnovanii pri zagruzke ego sosredotochennoi siloi* (Determination of stresses in saturated soils submitted to a concentrated load, in Russian), *Trudy LPI*, No. 178 (1955).

F. Kötter, Die Bestimmung des Druckes an gekrümmten Gleitflächen, *Sitzungsber. Kgl. Preuss. Akad. der Wiss.*, Berlin (1903).

B. Ladanyi, The Mobilization of Shear Strength in the Active Rankine Case of Earth Pressure, *Proc. Brussels Conference 58 on Earth Pressure Problems*, **1**, 133–146 (1958).

T. W. Lambe, Methods of Estimating Settlements, *Journal of the Soil Mechanics and Foundations Division, Proc. ASCE*, **90**, No. SM 5, 43–67 (1964).

L. S. Lapidus, Vliyanie polzuchesti na process sdvizheniya gornykh porod (The influence of the creep on the development of displacements of rocky soils, in Russian), *Reologicheskie voprosy mekhaniki gornyh porod, Akad. nauk Kazakhskoi SSR, Alma Ata*, 72–79 (1964).

S. G. Lehnitski, *Nekotorye sluchai ploskoi zadachi teorii uprugosti anizotropnogo tela* (Some cases of the plane problem of the theory of elasticity of the anisotropic body, in Russian), Eksperimentalnye metody opredeleniya napryazhenii i deformatsii v uprugikh i plasticheskikh zonakh, *ONTI* (1935).

G. A. Leonards and J. Narain, Flexibility of Clay and Cracking of Earth Dams, *Journal of Soil Mechanics and Foundations Division, Proc. ASCE*, **89**, No. SM 2, 47–98 (1963).

K. Y. Lo, Stress-Strain Relationship and Pore Water Pressure Characteristics of a Normally Consolidated Clay, *Proc. 5th Int. Conf. Soil Mech. Found. Eng., Paris*, **1**, 219–224 (1961-a).

K. Y. Lo, Secondary Compression of Clays, *Journal of the Soil Mechanics and Foundations Division, Proc. ASCE*, **87**, No. SM 4, 61–87 (1961-b).

G. M. Lomize and A. L. Kryzhanovsky, On the Strength of Sand, *Proceedings of the Geotechnical Conference Oslo 1967*, **1**, 215–219 (1967).

H. Lundgren and K. Mortensen, Determination by the Theory of Plasticity of the Bearing Capacity of Continuous Footings on Sand, *Proc. 3rd Int. Conf. Soil Mech. Found. Eng., Zurich*, **1**, 409–412 (1953).

J. Mandel, Tassement d'une couche d'argile saturée d'eau sous l'effet d'une force concentrée à la surface du sol, *Proc. 3rd Int. Conf. Soil Mech. Found. Eng., Zurich*, **1**, 413–417 (1953).

J. Mandel, Consolidation des couches d'argiles, *Proc. 4th Int. Conf. Soil Mech. Found. Eng., London*, **1**, 360–367 (1957).

J. Mandel, Application du calcul symbolique à l'étude des corps viscoélastiques, *Introduction à l'étude de la rhéologie* by B. Persoz *et al.*, Dunod, Paris, 93–116 (1960).

J. Mandel, Tassements produits par la consolidation d'une couche d'argile de grande épaisseur, *Proc. 5th Int. Conf. Soil Mech. Found. Eng., Paris*, **1**, 733–736 (1961).

N. N. Maslov, Some Questions of the Stability and Deformation of Retaining Walls on Clay Soils, *Proc. Brussels Conference 58 on Earth Pressure Problems*, **3**, 199–203 (1958).

V. P. Matveeva, Raschet formirovaniya vo vremeni muldy osedaniya nad tonnelyami (The computation of the development with time of the depression due to the settlements above the tunnels, in Russian), *Reologicheskie voprosy mekhaniki gornykh porod, Akad. nauk Kazakhskoi SSR, Alma-Ata*, 80–83 (1964).

W. Merchant, *Some Theoretical Considerations on the One-dimensional Consolidation of Clay*, S.M. thesis, Massachusetts Inst. of Technology (1939).

S. R. Meschyan, *Nekotorye voprosy polzuchesti glinistykh gruntov* (Some creep problems of clayey soils, in Russian), *Izvestiya Akad. Nauk Armyanskoi SSR, Seriya fiz.-mat. nauk*. (1965).

G. G. Meyerhof, The Ultimate Bearing Capacity of Foundations, *Géotechnique*, **2**, 301–332 (1951).

G. G. Meyerhof, The Bearing Capacity of Foundations under Eccentric and Inclined Loads, *Proc. 3rd Int. Conf. Soil Mech. Found. Eng., Zurich*, **1**, 440–445 (1953).

G. G. Meyerhof, Influence of Roughness of Base and Ground-water Conditions on the Ultimate Bearing Capacity of Foundations, *Géotechnique*, **5**, 227–242 (1955).

A. Moos and R. Haefeli, Schweizerische Probleme auf dem Grenzgebiet von Bodenmechanik, Geologie und Glaziologie, *Mitteilungen aus dem Institut für Bodenmechanik Aachen*, No. 25, 237–264 (1962).

N. R. Morgenstern and V. E. Price, The Analysis of the Stability of General Slip Surfaces, *Géotechnique*, **15**, 79–93 (1965).

S. Murayama and T. Shibata, Rheological Properties of Clays, *Proc. 5th Int. Conf. Soil Mech. Found. Eng., Paris*, **1**, 269–273 (1961).

N. M. Newmark, Influence Charts for the Computation of Stresses in Elastic Foundations, *Bull.* 40, *No.* 12, *Univ. Illinois Eng., Expt. Sta. Bull. Ser.* 338 (1942).

E. Nonveiller and L. Šuklje, Landslide Zalesina, *Proc. European Conference on Stability of Earth Slopes, Stockholm*, **1**, 107–121 (1954); reprinted in *Géotechnique*, Vol. **5**, 143–153 (1955).

E. Nonveiller, *Klizište Zalesina* (Landslide Zalesina, in Serbo-Chroatian), *Geološki vjesnik*, **8–9**, 141–152 (1956).

E. Nonveiller, Settlement of a Grain Silo on Fine Sand, *Proc. European Conference Soil Mech. Found. Eng., Wiesbaden*, **1**, 285–294 (1963).

E. Nonveiller, The Stability Analysis of Slopes with a Slip Surface of General Shape, *Proc. 6th Int. Conf. Soil Mech. Found. Eng., Montreal*, **2**, 522–525 (1965).

N. N. Pavlovsky, *Teoriya dvizheniya gruntovykh vod pod hidrotekhnicheskimi sooruzheniyami* (Theory of the movement of the ground water under hydraulic structures, in Russian), 1922, Izd. Nauchno-melioratsionogo instituta, Sobranie sochinenii, t.II, *Izd. Akademii nauk*, 1956.

R. B. Peck and J. Lowe III, Shear Strength of Undisturbed Cohesive Soils, Moderators' report Session 4, *ASCE Research Conference on Shear Strength of Cohesive Soils, University of Colorado*, 1137–1140 (1960).

M. R. Peltier, Recherches expérimentales sur la courbe intrinseque de rupture des sols pulvérulents, *Proc. 4th Int. Conf. Soil Mech. Found. Eng., London*, **1**, 179–182 (1957).

K. E. Petterson, *Kajraset i Goteborg des 5te mars 1916* (Collapse of a quay wall at Gothenburg, 5th March, 1916, in Swedish), *Teknisk Tidskrift*, **46**, 289 (1916).

L. Prandtl, Eindringungsfestigkeit und Festigkeit von Schneiden, *Zeitschrift für angewandte Mathematik und Mechanik*, **1**, 1, 15–20 (1921).

W. J. N. Rankine, On the Stability of Loose Earth, *Phil. Trans. Roy. Soc., London*, **147**, 9–27 (1857).

R. M. Rappoport, *Zadacha Bussineska dlya sloistogo uprugogo poluprostranstva* (Boussinesq's problem for layered elastic halfspace, in Russian), *Trudy Leningradskogo politekhnicheskogo instituta*, No. 5 (1948).

J. Résal, *Poussée des terres, I: Stabilité des murs de soutènement, II: Théory des terres cohérentes*, Paris (1903, 1910).

M. Rocha, The Possibility of Solving Soil Mechanics Problems by the Use of Models, *Proc. 4th Int. Conf. Soil Mech. Found. Eng., London*, **1**, 183–188 (1957).

K. H. Roscoe, A. N. Schofield and E. P. Wroth, On the Yielding of Soils, *Géotechnique*, **8**, 22–53 (1958).

K. H. Roscoe and H. B. Poorooshasb, A Theoretical and Experimental Study of Strains in Triaxial Tests on Normally Consolidated Clays, *Géotechnique*, **13**, 12–38 (1963).

K. H. Roscoe and A. N. Schofield, Mechanical Behaviour of an Idealized "Wet-clay", *Proc. European Conference on Soil Mechanics and Foundation Engineering, Wiesbaden*, **1**, 47–54 (1963).

I. Th. Rosenquist, Physico-chemical Properties of Soils, Soil-water System, *Journal of Soil Mechanics and Foundations Division, Proc. ASCE*, **85**, No. SM 2, 31 (1959).

I. Th. Rosenquist, Physico-chemical Properties of Soils, Soil-water Systems, *Trans. ASCE*, **126,** Part 1, 745–765 (1961).

S. A. Roza, Osadki gidrotekhnicheskikh sooruzhenii na glinakh s maloi vlazhnostyu (Settlements of hydraulic structures on soils of small water content, in Russian), *Gidrotekhnicheskoe stroitel'stvo*, No. 9 (1950).

P. W. Rowe, A Stress–Strain Theory for Cohesionless Soil with Applications to Earth Pressures at Rest and Moving Walls, *Géotechnique*, **4,** 70–88 (1954).

P. W. Rowe, Discussion, *Proc. Brussels Conference 58 on Earth Pressure Problems*, **3,** 126–129 (1958).

P. C. Rutledge and F. C. Walker, Problems Associated with Practical Applications of Shear Strength Data, Moderators' report Session 6, *ASCE Research Conference on Shear Strength of Cohesive Soils, University of Colorado*, 1151–1155 (1960).

M. Saito and H. Uezawa, Failure of Soil Due to Creep, *Proc. 5th Int. Conf. Soil Mech. Found. Eng., Paris*, **1,** 315–318 (1961).

G. N. Savin, Napryazheniya v anizotropnom massive pri zadannoi nagruzke na poverkhnosti (*ploskaya zadacha*) (Stresses in anisotropic media subjected to surface loads—plane problem, in Russian), *Vestnik inzhen'erov i tekhnikov*, No. 3 (1940).

R. L. Schiffman and R. E. Gibson, The Consolidation of Non-homogeneous Clay Layers, *Journal of the Soil Mechanics and Foundations Division, Proc. ASCE*, **90,** No. SM 5, 1–30 (1964).

R. L. Schiffman, Ch. C. Ladd and A. T.-F. Chen, The Secondary Consolidation of Clay, *IUTAM Symposium Grenoble 1964: Rheology and Soil Mechanics*, Springer-Verlag, Berlin, 1966, 273–298 (1964).

E. Schultze, *Der Widerstand des Baugrundes gegen schräge Sohlpressungen*, Bautechnik, 1962.

R. F. Scott, *Principles of Soil Mechanics*, Addison-Wesley, Reading (Mass.) (1963), 550 pp.

H. B. Seed and R. C. Hirschfeld, Shear Strength of Compacted Cohesive Soils, Moderators' report Session 5, *ASCE Research Conference on Shear Strength of Cohesive Soils*, University of Colorado, 1141–1150 (1960).

A. W. Skempton, The Bearing Capacity of Clays, *Proc. Building Research Congress 1951, London, Division 1*, 180–189 (1951).

A. W. Skempton, The Pore-pressure Coefficients A and B, *Géotechnique*, **4,** 143–147 (1955).

A. W. Skempton and A. W. Bishop (1954), Building Materials, their Elasticity and Inelasticity, Edited by M. Reiner, North-Holland Pu. Co., Amsterdam, Pt D., Chap. 10, *Soils*, 417–482 (1954).

A W. Skempton and L. Bjerrum, A Contribution to the Settlement Analysis of Foundations on Clay, *Géotechnique*, **7,** 168–178 (1957).

A. W. Skempton, Address on Effective Stress in Soils, Concrete and Rocks, *Conference on Pore Pressure and Suction in Soils*, London, 4–16 (1960).

A. W. Skempton, Long term Stability of Clay Slopes (4th Rankine Lecture), *Géotechnique*, **14,** 77–101 (1964).

V. V. Sokolovski, *Teoriya plastichnosti* (Theory of plasticity, in Russian), *Gostekhteorizdat* (1950).

V. V. Sokolovski, *Statics of Soil Media* (Translated from the 2nd edition of

"Statika sypuchei sredy", in Russian, *Gostekhizdat* 1954), Butterworths Scientific Publications, London (1960), 237 pp.

I. Sovinc, Settlement Analysis of an Experimental Dyke on Normally Consolidated Moor Sediments, *University of Ljubljana, Acta Geotechnica*, **10–11**, No. 11, 1–7 (1964).

I. Sovinc, Stresses and Displacements in a Limited Layer of Uniform Thickness, Resting on a Rigid Base and Subjected to a Uniformly Distributed Load of Rectangular Shape, *Proc. 5th Int. Conf. Soil Mech. Found. Eng., Paris*, **1**, 823–828 (1961).

W. Steinbrenner, Taffeln zur Setzungsberechnung, *Strasse* (1934); *Schriftenreihe der Strasse*, Nr. 3, Volk und Reich Verlag (1936).

A. V. Stepanov, *Prichini osobennosti razrusheniya uprugo-anizotropnikh tel* (The reasons for the kind of rupture of elastic anisotropic bodies, in Russian), *Izvestiya AN SSSR, seriya fizicheskaya* T. 14, No. 1 (1950).

A. S. Stroganov, One-dimensional Deformation of Soil as Non-linear Viscoelastic Medium, *Proc. European Conference on Soil Mechanics and Foundation Engineering, Wiesbaden*, **1**, 55–60 (1963).

L. Šuklje, Discussion sur l'affaissement séculaire, *Proc. 3rd Int. Conf. Soil Mech. Found. Eng., Zurich*, **3**, 168–169 (1953-a).

L. Šuklje, Discussion sur les limites de l'analyse $\varphi = 0$, *Proc. 3rd Int. Conf. Soil Mech. Found. Eng., Zurich*, **3**, 211–212 (1953-b).

L. Šuklje, Capacité portante des couches cohérentes peu perméables et d'épaisseur limitée, *Proc. Yug. Soc. Soil Mech. Found. Eng., Ljubljana*, Nos 1–9 (No 3), 16–26 (1954).

L. Šuklje, Influence de l'épaisseur de la couche compressible sur le caractère et la vitesse de la consolidation, *Cinqième Congrès International des Grands Barrages, Paris*, **1**, 361–384 (1955).

L. Šuklje, The Analysis of the Consolidation Process by the Isotache Method, *Proc. 4th Int. Conf. Soil Mech. Found. Eng., London*, **1**, 200–206; **3**, 107–109 (1957).

L. Šuklje and I. Sovinc (1957), Landslides in Tertiary Clay in the Upper Sava Valley, *Proc. Yug. Soc. Soil Mech. Found. Eng.*, Nos 13–14 (No. 14), 17–25 (1957).

L. Šuklje, L'influence des déplacements cinématiques des dispositifs de soutènement et l'influence du facteur "Temps" sur l'amplitude de la pression des terres sur un écran de soutènement, Rapport Général, *Conférence de Bruxelles 58 sur les Problèmes de Pressions de Terres*, **3**, 99–116 (1958).

L. Šuklje, Modifikacija i proširenje newmarkove metode računa slijeganja tla pod proizvoljnim opterećenjem (Modification and extension of Nemark's method of settlement computation, in Serbo-Croatian), *Gradjevinar, Zagreb*, **12**, 393–398 (1960).

L. Šuklje, Facteurs gouvernant les gradients critiques dans la base d'excavation, *Association Internationale de Recherches Hydrauliques, Neuvième Assemblée Générale, Dubrovnik*, 428–436 (1961-a).

L. Šuklje, Discussion sur la résistance à la traction des sols, *Proc. 5th Int. Conf. Soil Mech. Found. Eng., Paris*, **3**, 355–356 (1961-b).

L. Šuklje and S. Vidmar, Essais sur les effets provoqués par la retenue du fluage des sols, *Proc. 5th Int. Conf. Soil Mech. Found. Eng., Paris*, **2**, 485–492 (1961-a).

L. Šuklje and S. Vidmar, A Landslide Due to Long Term Creep, *Proc. 5th Int. Conf. Soil Mech. Found. Eng., Paris*, **2**, 727–736 (1961-b).

L. Šuklje, Discussion on Pore-pressure Dissipation, *Proc. 5th Int. Conf. Soil Mech. Found. Eng., Paris*, **3**, 107–109 (1961-c).

L. Šuklje, *Raziskava hitrosti lezenja pobočij* (Investigation of the creep-speed of slopes, in Slovenian), Research report for the Boris Kidrič's Fond, available in Soil Mechanics Laboratory at the University of Ljubljana, 20 pp. + 34 figs. (1962).

L. Šuklje and I. Sovinc, An Applied Analysis of Distortional Displacements of Normally Consolidated Clays, *Proc. European Conference on Soil Mechanics and Foundation Engineering, Wiesbaden*, **1**, 193–198 (1963).

L. Šuklje, The Equivalent Elastic Constants of Saturated Soils Exhibiting Anisotropy and Creep Effects, *Géotechnique*, **13**, 291–309 (1963-a).

L. Šuklje, Discussion, *European Conference on Soil Mechanics and Foundation Engineering, Wiesbaden*, **2**, 45–46 (1963-b).

L. Šuklje and J. Brodnik, Deformation Conditions of the Mobilization of the Friction between Concrete and Soil, *University of Ljubljana, Acta Geotechnica* **4–6,** No. 4, 1–4 (1963).

L. Šuklje, Étude sur la consolidation sphérique des sols partiellement saturés soumis à la filtration linéaire, *IUTAM Symposium Grenoble 1964: Rheology and Soil Mechanics*, Springer-Verlag, Berlin, 1966, 305–315 (1964-a).

L. Šuklje, Kriticheskie fazy opolzaniya otkosov (Critical phases of the creep of slopes, in Russian), *Materialy soveshchaniya po voprosam izucheniya opolznei i mer bor'by s nimi*, Izdatel'stvo Kievskogo universiteta, 226–233 (1964-b).

L. Šuklje and J. Drnovšek, Investigation of the Tensile Deformability of Soils using Hollow Cylinders, *Proc. 6th Int. Conf. Soil Mech. Found. Eng., Montreal*, **1**, 368–372 (1965).

L. Šuklje, Étude supplémentaire sur la consolidation sphérique des sols partiellement saturés soumis à la filtration linéaire, *University of Ljubljana, Acta Geotechnica*, No. 14-b, 1–14 (1966).

L. Šuklje, Common Factors Controlling the Consolidation and the Failure of Soils, *Proceedings of the Geotechnical Conference Oslo*, 1967, 153–158 (1967).

L. Šuklje, G. Vogrinčič, G. Tratnikova and J. Jenko, Testing of the Flexibility of Clays using Photographic Registration of Displacements, *University of Ljubljana, Acta Geotechnica* **15–18,** No. 17, 1–4 (1967).

L. Šuklje, L. Márczal and J. Podržajeva, Dependence of Earth Pressures on Strains in Rankine's Case, *University of Ljubljana, Acta Geotechnica* **21–23**, No. 22, 1–21 (1968).

L. Šuklje and B. Kogovšek, *Isochrones of a Uniformly Loaded Layer of Viscous Soils*, III Sesja Naukowa Politechniki Wroclowskiej Wydzialu Rudovnictwa Ladowego 2–5 pazdzierniko 1968 r., Referaty **1**, 369–380 (1968).

Tan Tjong-Kie, *Onderzoekingen over de rheologische eigenschappen van klei* (Investigations on the rheological properties of clay, in Dutch with English summary), *Uitgeverij Excelsior, 's-Gravenhage* (1954), 152 pp.

Tan Tjong-Kie, Three Dimensional Theory on the Consolidation and Flow of the Clay-layers, *Scientia Sinica*, **6**, No. 1, 203–215 (1957).

Tan Tjong-Kie, Consolidation and Secondary Time Effect of Homogeneous, Anisotropic, Saturated Clay Strata, *Proc. 5th Int. Conf. Soil Mech. Found. Eng., Paris*, **1**, 367–373 (1961).

Tan Tjong-Kie, Determination of the Rheological Parameters and the Hardening Coefficients of Clays, *IUTAM Symposium Grenoble 1964: Rheology and Soil Mechanics*, Springer-Verlag, Berlin, 256–270 (1966).

D. W. Taylor and W. Merchant, A Theory of Clay Consolidation Accounting for Secondary Compression, *J. Maths and Physics*, **19**, 167–185 (1940).

D. W. Taylor, Research on Consolidation of Clays, *Dept. of Civil and Sanitary Eng., Massachusetts Inst. of Technology, Publ. Serial 82*, 1–147 (1942).

D. W. Taylor, *Fundamentals of Soil Mechanics*, Wiley, New York (1948), 700 pp.

G. P. Tchebotarioff, E. R. Ward and A. A. De Philippe, The Tensile Strength of Disturbed and Recompacted Soils, *Proc. 3rd Int. Conf. Soil Mech. Found. Eng., Zurich*, **1**, 207–210 (1953).

K. Terzaghi, Die Berechnung der Durchlässigkeitsziffer des Tones aus dem Verlauf der hydrodynamischen Spannungserscheinungen, *Sitzungsber. Akad. Wiss. Wien, mathem.-naturw. Kl., Abt. IIa*, 132.Bd, 3.u.4.H. (1923).

K. Terzaghi, *Erdbaumechanik auf bodenphysikalischer Grundlage*, Deuticke, Wien, (1925), 399 pp.

K. Terzaghi, Die Coulombsche Gleichung für den Scherwiderstand bindiger Böden, *Die Bautechnik*, **16**, 343–346 (1938).

K. Terzaghi and O. K. Fröhlich, *Theorie der Setzung von Tonschichten*, Deuticke, Leipzig (1936), 168 pp.

K. Terzaghi, *Theoretical Soil Mechanics*, Wiley, New York (1943), 510 pp.

B. Tiedeman, *Über die Schubfestigkeit bindiger Böden*, Bautechnik, **15**, 400–403 (1937).

N. A. Tsytovich, *Mekhanika gruntov* (Soil mechanics, in Russian), 4 *izd., Gosstroyizd., Moskva* (1963), 634 pp.

N. A. Tsytovich, Yu.K. Zaretsky, M. B. Malyshev, M.Yu. Abel'ev, Z. G. Ter-Martirosyan, *Prognoz skorosti osadok osnovanii sooruzhenii* (*konsolidatsiya i polzuchest mnogofaznikh gruntov*) (Prediction of the rate of settlements of foundations—Consolidation and creep of multiphase soils, in Russian), *Stroyizdat, Moskva* (1967), 240 pp.

N. N. Verigin, Soil Consolidation under the Influence of an External Load Normal to the Boundary of Half-space, *IUTAM Symposium Grenoble 1964: Rheology and Soil Mechanics*, Springer-Verlag, Berlin, 231–234 (1966).

N. N. Verigin, Consolidation of Saturated Soil upon Action of External Load Normal to Boundary of Half-space, *Proc. 6th Int. Conf. Soil Mech. Found. Eng., Montreal*, **1**, 398–401 (1965).

S. Vidmar, Vrste konsolidacijskih krivulj pri edometrskih preizkusih (The forms of consolidation curves of oedometer tests, in Slovenian), *Gradbeni vestnik, Ljubljana*, Nos. 37–38, 97–102 (1956).

S. Vidmar, Relaxation Effects on the Earth Pressure of Cohesive Soils, *Hungarian Academy of Sciences, Proceedings of the International Conference on Soil Mechanics and Foundation Engineering, Budapest*, 103–118 (1963).

S. S. Vyalov, *Polzuchest i dlitelnoe soprotivlenie merzlykh gruntov* (Creep and long-term strength of frozen soils, in Russian), *Dokl. AN SSSR*, 104, No. 6 (1955).

S. S. Vyalov and A. M. Skibitsky, Rheological Processes in Frozen Soils and Dense Clays, *Proc. 4th Int. Conf. Soil Mech. Found. Eng., London*, **1**, 120–124 (1957).

S. S. Vyalov, *Reologicheskie svoistva i nesushchaya sposobnost merzlykh gruntov* (Rheological properties and bearing capacity of frozen soils, in Russian), *Izd-vo AN SSSR* (1959).

S. S. Vyalov and A. M. Skibitsky, Problems of the Rheology of Soils, *Proc. 5th Int. Conf. Soil Mech. Found. Eng., Paris,* **1,** 387–391 (1961).

S. S. Vyalov, V. G. Gmoshinsky, S. E. Gorodetski, V. G. Grigoryeva, Yu.K. Zaretsky, N. K. Pekarskaya, E. P. Shusherina, *Prochnost i polzuchest merzlykh gruntov i raschety ledogruntovykh ograzhdenii* (Strength and creep of frozen soils, in Russian), *Izd. Akad. nauk SSSR, Moskva* (1962), 254 pp.

S. S. Vyalov, Reologiya merzlykh gruntov (Rheology of frozen soils, in Russian), *Prochnost i polzuchest merzlykh gruntov, Izd. Akad. nauk SSSR, Moskva,* 5–54 (1963).

H. M. Westergaard, A Problem of Elasticity Suggested by a Problem in Soil Mechanics: Soft Material Reinforced by Numerous Strong Horizontal Sheets, *Contributions to the Mechanics of Solids, Stephen Timoshenko 60th Anniversary Volume*, MacMillan, New York (1938).

R. V. Whitman, Some Considerations and Data Regarding the Shear Strength of Clays, *ASCE Research Conference on Shear Strength of Cohesive Soils, University of Colorado,* 581–614 (1960).

K. Wolf, Ausbreitung der Kraft in der Halbebene und im Halbraum bei anisotropen Materialen, *Zeitschrift für angewandte Mathematik und Mechanik*, H. 5, B. 15 (1935).

T. H. Wu, A. K. Loh and L. E. Malvern, Study of Failure Envelope of Soils, *Journal of the Soil Mechanics and Foundations Division, Proc. ASCE,* **89,** No. SM 1, 145–181 (1963).

Yu. K. Zaretsky, *Teoriya konsolidatsii gruntov* (Theory of consolidation of soils, in Russian), *Izd. Nauka, Moskva* (1967), 270 pp.

BIBLIOGRAPHY

List of publications recommended for further study.

BIBLIOGRAPHY

M. B. Abbot, One-dimensional Consolidation of Multilayered Soils, *Géotechnique*, **10**, 151–165 (1960).

H. Aboshi and H. Monden, Three-dimensional Consolidation of Saturated Clay, *Proc. 5th Int. Conf. Soil Mech. Found. Eng., Paris*, **1**, 559–562 (1961).

K. Akai and T. Adachi, Study on the One-dimensional Consolidation and the Shear Strength Characteristics of Fully Saturated Clay, in Terms of Effective Stress, *Proc. 6th Int. Conf. Soil Mech. Found. Eng., Montreal*, **1**, 146–150 (1965).

I. Alpan, The Dissipation Function for Unsaturated Soils, *Proc. 5th Int. Conf. Soil Mech. Found. Eng., Paris*, **1**, 3–6 (1961).

N. H. Arutyun'an, *Nekatorye voprosy teorii polzuchesti* (Some problems of the theory of creep, in Russian), Gostekhteorizdat (1952).

A. Balla, Stress Conditions in the Triaxial Compression Test, *Proc. 4th Int. Conf. Soil Mech. Found. Eng., London*, **1**, 140–143 (1957).

J. Biarez, J. M. Pierrard, K. Wiendieck and E. Absi, Rhéologie des sols, *Proc. 6th Int. Conf. Soil Mech. Found. Eng., Montreal*, **1**, 21–25 (1965).

J. Biarez, A. Belot, J. M. Pierrard and K. Wiendieck, Quelques aspects de la loi rhéologique des sols, *IUTAM Symposium Grenoble 1964: Rheology and Soil Mechanics*, Springer-Verlag, Berlin, 343–353 (1966).

L. Bjerrum, W. Jönson and C. Ostenfeld, The Settlement of a Bridge Abutment on Friction Piles, *Proc. 4th Int. Conf. Soil Mech. Found. Eng., London*, **2**, 14–18 (1957).

L. Bjerrum and B. Kjaernsli, Analysis of the Stability of some Norwegian Natural Clay Slopes, *Géotechnique*, **7**, 1–16 (1957).

L. Bjerrum, Secondary Settlements of Structures Subjected to Large Variations in Live Load, *IUTAM Symposium Grenoble 1964: Rheology and Soil Mechanics*, Springer-Verlag, Berlin, 460–471 (1966).

R. W. Christensen and T. H. Wu, Analysis of Clay Deformation as a Rate Process, *Journal of the Soil Mechanics and Foundations Division, Proc ASCE*, **90**, No. SM 6, 125–157 (1964)

I. F. Christie, Secondary Compression Effects during One-dimensional Consolidation Tests, *Proc. 6th Int. Conf. Soil Mech. Found. Eng., Montreal*, **1**, 198–202 (1965).

I. F. Christie, The Solution of Consolidation Problems by General Purpose Analogue Computer, *Géotechnique*, **16**, 131–148 (1966).

A. Croce, Secondary Time Effect in the Compression of Unconsolidated Sediments of Volcanic Origin, *Proc. 2nd Int. Conf. Soil Mech. Found. Eng., Rotterdam*, **1**, 166–169 (1948).

E. H. Davis and G. P. Raymond, A Non-linear Theory of Consolidation, *Géotechnique*, **15**, 161–173 (1965).

N. Ya. Denissov and B. F. Reltov, Elastic and Structural Deformations of Clayey Soils, *Proc. 4th Int. Conf. Soil Mech. Found. Eng., London*, **1**, 17–21 (1957).

D. C. Drucker and W. Prager, Soil Mechanics and Plastic Analysis or Limit Design, *Quart. Applied Mathematics*, **10**, 157–165 (1952).

D. C. Drucker, R. E. Gibson and D. J. Henkel, Soil Mechanics and Work Hardening Theories of Plasticity, *Proc. ASCE, Eng. Mech. Div.*, **81**, No. 798 (1955).

T. Edelman, The Consolidation of a Layer, which Modulus of Elasticity is Proportional to the Depth, *Proc. 3rd Int. Conf. Soil Mech. Found. Eng., Zurich*, **1**, 358–361 (1953).

M. Fukuoka, Landslides in Japan, *Proc. 3rd Int. Conf. Soil Mech. Found. Eng., Zurich*, **2**, 234–238 (1953).

N. M. Gersevanov and D. E. Polshin, *Teoreticheskie osnovy mekhaniki gruntov* (Theoretical fundamentals of soil mechanics, in Russian), *Stroyizdat, Moskva* (1948).

E. C. W. A. Geuze and C. M. A. de Bruyn, The First Law of Thermodynamics and the Consolidation Process, *Proc. 2nd Int. Conf. Soil Mech. Found. Eng.*, Rotterdam, **3**, 29–31 (1948).

E. C. W. A. Geuze and Tan Tjong-Kie, The Mechanical Behaviour of Clays, *Proc. 2nd Int. Cong. Rheol.*, 247–259 (1953).

R. E. Gibson, The Progress of Consolidation in a Clay Layer Increasing in Thickness with Time, *Géotechnique*, **3**, 171–182 (1958).

R. E. Gibson, An Analysis of System Flexibility and its Effects on Time-lag in Pore-water Pressure Measurements, *Géotechnique*, **13**, 1–11 (1963).

R. E. Gibson, Some Results Concerning Displacements and Stresses in a Non-homogeneous Elastic Half-space, *Géotechnique*, **18**, 58–67 (1967).

M. N. Goldstein, V. A. Misumsky and L. S. Lapidus, The Theory of Probability and Statistics in Relation to the Rheology of Soils, *Proc. 5th Int. Conf. Soil Mech. Found. Eng., Paris*, **1**, 123–126 (1961).

M. Goldstein, L. Lapidus and V. Misumsky, Rheological Investigation of Clays and Slope Stability, *Proc. 6th Int. Conf. Soil Mech. Found. Eng., Montreal*, **2**, 482–485 (1965).

S. E. Grechishchev, Issledovanie polzuchesti vyazko-sypuchei sredy pri slozhnom napryazhennom sostoyanii (The investigation of the creep of viscous soils subjected to complex stress state, in Russian), *Reologicheskie voprosy mekhaniki gornykh porod, Akad. nauk Kazakhskoi SSR, Alma-Ata*, 138–150 (1964).

P. Habib and A. Puyo, Evaluation des contraintes transversales pendant la consolidation, *Proc. 3rd Int. Conf. Soil Mech. Found. Eng., Zurich*, **1**, 32–34 (1953).

R. Haefeli and W. Schaad, Time Effect in Connection with Consolidation Tests, *Proc. 2nd Int. Conf. Soil Mech. Found. Eng., Rotterdam*, **3**, 23–28 (1948).

R. Haefeli, Investigation and Measurements of the Shear Strength of Saturated Cohesive Soils, *Géotechnique*, **2**, 186–208 (1951).

R. Haefeli, Ch. Schaerer and G. Amberg, The Behaviour under the Influence of Soil Creep Pressure of the Concrete Bridge Built at Klosters by the Rhaetien Railway Company, Switzerland, *Proc. 3rd Int. Conf. Soil Mech. Found. Eng., Zurich*, **2**, 175–179 (1953).

R. Haefeli, Creep and Progressive Failure in Snow, Soil, Rock, and Ice, *Proc. 6th Int. Conf. Soil Mech. Found. Eng., Montreal*, **3**, 134–148 (1965).

P. Homsen, Landslips in Norwegian Quick-clays, *Géotechnique*, **3**, 187–194 (1953).

K. Hoshino, A General Theory of Mechanics of Soils, *Proc. 4th Int. Conf. Soil Mech. Found. Eng., London,* **1,** 160–166 (1957).

K. Hoshino, An Analysis of the Volume Change, Distortional Deformation and Induced Pore Pressure of Soils under Triaxial Loading, *Proc. 5th Int. Conf. Soil Mech. Found. Eng., Paris,* **1,** 151–158 (1961).

N. Janbu, Soil Compressibility as Determined by Oedometer and Triaxial Tests, *European Conference on Soil Mechanics and Foundation Engineering, Wiesbaden,* **1,** 19–25; Discussion, **2,** 17–21 (1963).

N. Janbu, Consolidation of Clay Layers Based on Non-linear Stress-strain, *Proc. 6th Int. Conf. Soil Mech. Found. Eng., Montreal,* **1,** 249–253 (1965).

J. E. Jennings and J. B. Burland, Limitations to the Use of Effective Stresses in Partly Saturated Soils, *Géotechnique,* **12,** 125–144 (1962).

G. de Josselin de Jong, *Statics and Kinematics in the Failable Zone of a Granular Material,* Waltman, Delft (1959), 122 pp.

F. Kawakami and S. Ogawa, Strength and Deformation of Compacted Soil Subjected to Repeated Stress Applications, *Proc. 6th Int. Conf. Soil Mech. Found. Eng., Montreal,* **1,** 264–268 (1965).

A. W. Koppejan, A Formula Combining the Terzaghi Load–Compression Relationship and the Buisman Secular Time Effect, *Proc. 2nd Int. Conf. Soil Mech. Found. Eng., Rotterdam,* **3,** 32–37 (1948).

E. H. Leeuw, The Theory of Three-dimensional Consolidation Applied to Cylindrical Bodies, *Proc. 6th Int. Conf. Soil Mech. Found. Eng., Montreal,* **1,** 287–290 (1965).

G. A. Leonards and B. K. Ramiah, Time Effects in the Consolidation of Clay, *Symposium on Time Rate of Loading in Testing Soils, Am. Soc. Testing Materials, Special Technical Publication,* No. 254 (1959).

G. A. Leonards and P. Girault, A Study of the One-dimensional Consolidation Tests, *Proc. 5th Int. Conf. Soil Mech. Found. Eng., Paris,* **1,** 213–218 (1961).

G. M. Lomize *et al., Voprosy prochnosti i deformiruemosti gruntov* (Problems of strength and deformability of soils, in Russian), *Azerbaizdanskoe Gos. Izd., Baku* (1966), 108 pp.

J. Mandel, Consolidation des sols (Étude mathématique), *Géotechnique,* **3,** 287–299 (1953).

N. N. Maslov, *Prikladnaya mekhanika gruntov* (Applied soil mechanics, in Russian), *Mashstroyizdat, Moskva* (1952).

N. N. Maslov, *Usloviya ustoichivosti sklonov i otkosov* (Conditions of the stability of slopes and excavations, in Russian), *Gosenergoizdat, Moskva* (1955), 467 pp.

N. N. Maslov, Soprotivlyaemost glinistykh gruntov sdvigu v nekonsolidirovannom sostoyanii i voprosy polzuchesti (Shearing resistance of clayey soils in non-consolidated state and the creep problems, in Russian), *Sb. trudov Vsesoyuznoga soveshchaniya po stroitel'stvu na slabykh vodonasyshchennykh glinistykh gruntakh, Tallin* (1965).

J. B. Martins, Consolidation of a Clay Layer of Non-uniform Coefficient of Permeability, *Proc. 6th Int. Conf. Soil Mech. Found. Eng., Montreal,* **1,** 308–312 (1965).

J. Mc-Namee and R. E. Gibson, Displacement Functions and Linear Transforms Applied to Diffusion through Porous Elastic Media, *Quarterly Journal of Mechanics and Applied Mathematics,* 98–111 (1960-a).

J. Mc-Namee and R. E. Gibson, Plane Strain and Axially Symmetric Problems of the Consolidation of a Semi-infinite Clay Stratum, *Quarterly Journal of Mechanics and Applied Mathematics*, 210–227 (1960-b).

R. D. Mindlin, Force at Point in the Interior of a Semi-infinite Solid, *Physics*, **7**, 195–202 (1936).

S. Murayama and T. Shibata, On the Rheological Character of Clay, Part I, *Disaster Prevention Research Inst., Kyoto Univ., Bulletin No* 26 (1955).

S. Murayama and T. Shibata, On the Secondary Consolidation of Clay, *Proc. 2nd Japan Cong. on Testing Materials*, 178–181 (1959).

S. Murayama and T. Shibata, Flow and Stress Relaxation of Clays, *IUTAM Symposium Grenoble 1964: Rheology and Soil Mechanics*, Springer-Verlag, Berlin, 99–129 (1966).

P. L. Newland and B. H. Allely, Volume Changes in Drained Triaxial Tests on Granular Materials, *Géotechnique*, **7**, 17–34 (1957).

P. L. Newland and B. H. Allely, Volume Changes during Undrained Triaxial Tests on Saturated Dilatant Granular Materials, *Géotechnique*, **9**, 174–182 (1959).

P. L. Newland and B. H. Allely, A Study of the Consolidation Characteristics of a Clay, *Géotechnique*, **10**, 62–74 (1960).

A. D. Northey and R. F. Thomas, Consolidation Test Pore Pressures, *Proc. 6th Int. Conf. Soil Mech. Found. Eng., Montreal*, **1**, 323–327 (1965).

J. Ohde, Zur Theorie des Erddruckes unter besonderer Berücksichtigung der Erddruckverteilung, *Bautechnik*, **16**, 150, 176, 241, 331, 480, 570, 753 (1938).

W. Olszak and P. Perzyna, On Elastic Visco-plastic Soils, *IUTAM Symposium Grenoble 1964: Rheology and Soil Mechanics*, Springer-Verlag, Berlin, 47–57 (1966).

W. H. Perloff, K. Nair and J. G. Smith, Effect of Measuring System on Pore Water Pressures in the Consolidation Test, *Proc. 6th Int. Conf. Soil Mech. Found. Eng., Montreal*, **1**, 338–341 (1965).

B. Persoz, *et al.*, *Introduction à l'etude de la rhéologie*, Dunod, Paris (1960), 251 pp.

K. E. Petterson, The Early History of Circular Sliding Surfaces, *Géotechnique*, **5**, 275–296 (1955).

M. Reiner, *Twelve Lectures on Theoretical Rheology*, North Holland Publishing Company, Amsterdam, 1949.

K. H. Roscoe and A. Thurairajah, On the Uniqueness of Yield Surfaces for Wet Clays, *IUTAM Symposium Grenoble* (1964): *Rheology and Soil Mechanics*, Springer-Verlag, Berlin, 364–384, (1964).

P. W. Rowe, Stress-dilatancy, Earth Pressures and Slopes, *Journal of the Soil Mechanics and Foundations Division, Proc. ASCE*, **89**, No. SM 3, 37–61 (1963).

M. Saito, Forecasting the Time of Occurrence of a Slope Failure, *Proc. 6th Int. Conf. Soil Mech. Found. Eng., Montreal*, **2**, 537–541 (1965).

J. A. J. Salas and J. M. Serratosa, Compressibility of Clays, *Proc. 3rd Int. Conf. Soil Mech. Found. Eng., Zürich*, **1**, 192–198 (1953).

R. L. Schiffman and A. A. Fungaroli, Consolidation Due to Tangential Loads, *Proc. 6th Int. Conf. Mech. Found. Eng., Montreal*, **2**, 188–192 (1965).

R. F. Scott, New Method of Consolidation-Coefficient Evaluation, *Journal of the Soil Mechanics and Foundations Division, Proc. ASCE*, **87**, No. SM 1, 29–42 (1961).

R. A. Sevaldson, The Slide in Lodalen, October 6th, 1954, *Géotechnique*, **6**, 167–182 (1956).

A. W. Skempton and R. D. Northey, The Sensitivity of Clays, *Gótechnique*, **3**, 30–53 (1952).

Z. Sobotka, Rhéologie des problèmes de déformation plane des milieux continus, *IUTAM Symposium Grenoble 1964: Rheology and Soil Mechanics*, Springer-Verlag, Berlin, 218–230 (1966).

A. S. Stroganov, Visco-plastic Flow of Soils, *Proc. 5th Int. Conf. Soil Mech. Found. Eng. Paris*, **2**, 721–726 (1961).

A. S. Stroganov, Sur certains problèmes rhéologiques de la mécanique des sols, *IUTAM Symposium Grenoble 1964: Rheology and Soil Mechanics*, Springer-Verlag, Berlin, 237–250 (1966).

A. Thurairajah and K. H. Roscoe, The Correlation of Triaxial Compression Test Data on Cohessionless Granular Media, *Proc. 6th Int. Conf. Soil Mech. Found. Eng., Montreal*, **1**, 377–381 (1965).

S. S. Vyalov, Plasticity and Creep of Cohesive Medium, *IUTAM Symposium Grenoble 1964: Rheology and Soil Mechanics*, Springer-Verlag, Berlin, 79–81 (1966).

S. S. Vyalov, Plasticity and Creep of Cohesive Medium, *Proc. 6th Int. Conf. Soil Mech. Found. Eng., Montreal*, **1**, 402–406 (1965).

H. E. Wahls, Analysis of Primary and Secondary Consolidation, *Journal of the Soil Mechanics and Foundations Division, Proc. ASCE*, **88**, No. SM 6, 207–231 (1962).

H. Yamaguchi, Strain Increments and Volume Change in the Plastic Flow of a Granular Material, *Proc. 5th Int. Conf. Soil Mech. Found. Eng., Paris*, **1**, 413–418 (1961).

Y. Yoshimi and J. O. Osterberg, Compression of Partially Saturated Cohesive soils, *Journal of the Soil Mechanics and Foundation Division, Proc. ASCE*, **89**, No. SM 4, 1–24 (1963).

Yu. K. Zaretsky, Ob obobshchenii teorem N. H. Arutyunyana dlya sredpodchinyayushchikhsya nelineinoi polzuchesti (On the generalization of the theory of N. H. Arutyunjan for the media subjected to non-linear creep, in Russian), *Reologicheskie voprosy mekhaniki gornykh porod, Akad. nauk Kazakhskoj SSR, Alma-Ata*, 120–131 (1964).

NOTATION

This list contains the majority of symbols used in this book. The symbols listed are frequently used with special or additional subscripts designating special meaning or components of the corresponding quantities. Some items of minor importance or infrequent appearance are not included below. They are explained in the section in which they are used.

In a few instances a given symbol is used herein to represent two or more entirely different quantities. In general these different quantities do not appear in the same chapter and, when are used at the same time, special subscripts are introduced in order to avoid confusion.

The nomenclature and the symbols agreed by the International Society of Soil Mechanics and Foundation Engineering at the 5th International Conference in Paris, 1961 (*Proceedings*, Vol. III, 69–70) have been strictly respected. The dimensions of physical quantities have been expressed in cm (length), sec (time) and kp (force) measuring system; dimensionless coefficients are signified by 1, and angles are measured in radians (1) or degrees (1°). The dimensions are given before the meaning. Numbers appearing after the meaning refer either to the section in which the item is described or firstly used, or to the equation (given in parentheses) by which the item is defined, or to the figure (Fig.) explaining the meaning of the symbol. Different meanings of the same symbol are separated by semicolons. If after the semicolon the dimension is not given, the preceding dimension is to be kept.

a	cm, radius, Fig. 5.3; longer side of a rectangle, 5.4.2; kp cm^{-2}, adhesion, 7.3.2; kp^{-1} cm^2, (linear) compressibility, (9.44), (10.35), (10.50)
a, b	cm^{-1}, parameters, (7.11)
a_0, a_1	kp^{-1} cm^2, compressibility, 10.1.2.2
a_1	1, length ratio, (5.40)
b	cm, shorter side of a rectangle, 5.4.2; kp^{-1} cm^2, linear compressibility, (10.35); kp cm^{-2} $sec^{1/n}$, coefficient of nonlinear viscosity, (11.66)
c	cm sec^{-1}, constant velocity, 4.2.2; kp cm^{-2}, cohesion intercept (in terms of total stress), 6.1; cm^2 sec^{-1}, coefficient of consolidation, (15.30′); cm^4 sec^{-1} kp^{-1}, (modified) coefficient of consolidation, (11.64)
c_e	kp cm^{-2}, effective cohesion component, Fig. 18.6
c_m	kp cm^{-2}, mobilized cohesion, 19.3.1
c_o	kp cm^{-2}, cohesion intercept of normally consolidated soils, 6.1
c_p	cm^2 sec^{-1}, coefficient of consolidation, (10.56)
c_s	cm^2 sec^{-1}, coefficient of consolidation, (11.40)
c_u	kp cm^{-2}, ultimate cohesion component, Fig. 18.6
c_v	cm^2 sec^{-1}, coefficient of consolidation, (9.45); kp cm^{-2}, rheological cohesion component, Fig. 18.6

c'	kp cm^{-2}, effective cohesion component (in terms of effective stress), 6.1
c_r'	kp cm^{-2}, effective cohesion component of the "shear strength line", 18.1; residual effective cohesion intercept, 18.5
c_v'	cm^2 sec^{-1}, coefficient of consolidation, (14 g)
d_c	1, depth factor, 23.3
e	= 2·7183, base of natural logarithms, 4.4.2; 1, void ratio, 8.1
e_{ii}, e_{jj}, e_{kk}	1, principal strains, 3.1.1
e_{oct}	1, octahedral linear strain, spheric(al) strain component, (3.21)
e_{rs}	1, strain tensor, (3.3)
e_{rs}^d	1, distortional (deviatorial) component of the strain tensor, (3.20)
e_{rs}^0	1, spheric(al) (isotropic) component of the strain tensor, (3.19)
e_v	1, volume strain, 3.1.1
$e_{\alpha\beta}$	($\alpha = x, y, z$, $\beta = x, y, z$) 1, strain components of the strain tensor e_{rs}, (3.4)
e_1	1, void ratio at the beginning of the observed stress or time interval, 8.1
e_1, e_2, e_3	1, principal strains, 3.1.2
$\bar{e}$	1, average void ratio, 11.1.2
$\bar{e}_v$	1, finite volume strain, 3.3
$\bar{e}_{\alpha\alpha}$	($\alpha = x, y, z$) 1, finite linear strains, 3.3
$\dot{e}_{rs}$	sec^{-1}, tensor of strain velocities, (3.12)
$\dot{e}_{rs}^d$	sec^{-1}, distortional (deviatorial) component of the tensor of strain velocities, 4.1.5
$\dot{e}_{rs}^0$	sec^{-1}, spheric(al) (isotropic) component of the tensor of strain velocities, 4.1.5
$\dot{e}_{\alpha\beta}$	($\alpha = x, y, z$; $\beta = x, y, z$) sec^{-1}, strain velocity components of the tensor of strain velocities, (3.13)
exp (α)	$\equiv e^{\alpha}$ (e = 2·7183)
f	1, = tan $\alpha = \tau/\sigma$, (7.93)
$f(t)$	–, creep function, 4.4.1
$f^*(p)$	–, image of the creep function, 4.4.2
$f_1(T), f_2(T)$	–, functions, (13.116)
g	1, coefficient, (10.85)
h	cm, thickness, height, 5.4.1; length, 7.3.2
h_s	cm, height of the voidless solid phase of the soil, Fig. 11.2
h_1	cm, thickness of the sample or layer at the beginning of the observed stress interval, 8.1
i	–, arbitrary subscript, 4.4.1; 1, hydraulic gradient, 9.2.1;
i_c, i_γ	1, inclination factors of the bearing capacity, 23.2
i_0	1, limiting hydraulic gradient, 9.2.1
i, j, k	–, principal directions, 2.1.1; –, subscripts indicating the position of a point in the network, Fig. 16.1
i	–, arbitrary subscript, 4.4.1
k	cm, length of the chord of a circular rupture line, 7.3.1; 1, stress coefficient, Fig. 7.13; cm sec^{-1}, coefficient of permeability, 9.2.1; kp^{-1} cm^2, deformation parameter, (21.9)

k_x, k_y, k_z	cm $\sec^{-1}$, coefficients of permeability in the x, y and z directions, 14.1.1
k_1, k_2	1, modulus ratios, (5.40)
l	cm, length, 3.3; $kp^{-1}\ cm^2$, deformation parameter, (21.10)
m	–, final value of the subscript i, 4.4.1; 1, any natural number, or zero, 9.3.2; number of circular rings, 5.3.3; $= 2k - 1$, (7.65); exponent, (11.18), (13.122); secondary consolidation ratio, (10.46)
m_v	$kp^{-1}\ cm^2$, coefficient of volume change, (8.24)
m_{vp}	$kp^{-1}\ cm^2$, coefficient of primary compressibility, (10.2)
m_{vs}	$kp^{-1}\ cm^2$, coefficient of secondary compressibility, (10.3)
n	–, normal, 2.1.1; final value of the subscript j, 4.4.1; 1, number of elements, 5.3.2, 5.3.3; exponent, (11.1), (13.123); porosity, 9.1.1
n_a	1, porosity related only to voids filled with air, 12.2
n_w	1, porosity related only to voids filled with water, 12.2
$\vec{n}$	1, unit vector of the normal n, 2.1.1
(n,x), (n,y), (n,z)	1, direction cosines of the normal (components of the unit vector $\vec{n}$), 2.1.1
	$\sec^{-1}$, parameter of the transformed function (image), 4.4.2; kp cm^{-2}, pressure, stress, 4.1.3; uniformly distributed load, 5.4.2; mean normal effective stress, octahedral effective pressure, (13.65)
p_a	kp cm^{-2}, active earth pressure (stress), 7.4.2; admissible pressure, 23.2
p_b	kp cm^{-2}, total stress acting, in Rankine's case, on planes parallel to the slope surface, Fig. 7.11; "bond" = plastic resistance at the end of primary consolidation, 10.1.2.1
p_c	kp cm^{-2}, $= c \cot \varphi$ or $c' \cot \varphi'$, Fig. 14.7
p_e	kp cm^{-2}, earth pressure (stress) at rest in Rankine's case, 7.4.1
p_g	kp cm^{-2}, static resistance, 10.1.2.1
p_{ii}, p_{jj}, p_{kk}	kp cm^{-2}, principal stresses σ_1, σ_2, σ_3, 2.1.1
p_n	kp cm^{-2}, stress acting on the elementary plane of the normal n, 2.1.1
p_{oct}	kp cm^{-2}, octahedral total stress, (2.20)
p_p	kp cm^{-2}, passive earth pressure (stress), 7.4.3
p_{rs}	kp cm^{-2}, (total) stress tensor, (2.1)
$p_{rs}{}^d$	kp cm^{-2}, distortional (deviatorial) component of the stress tensor, (2.26)
$p_{rs}{}^0$	kp cm^{-2}, spheric(al), (isotropic) component of the stress tensor, (2.24)
p_v	kp cm^{-2}, viscous structural resistance, 10.1.2.1
p_w	kp cm^{-2}, pore pressure, (A.27)
$p_{\alpha\beta}$	($\alpha = x, y, z$ or ξ, η, ζ or i, j, k or n, t, $\beta = x, y, z$ or ξ, η, ζ or i, j, k or n, t) kp cm^{-2}, stress components of a stress tensor, α = direction of the normal to the stressed plane element, β = stress direction, 2.1.1, 2.2
p'	kp cm^{-2}, $= \frac{1}{2}(\sigma_1' + \sigma_3')$, 17.3.3

q	kp cm^{-2}, uniformly distributed load, 5.3.1; principal effective stress difference $= \sigma_1' - \sigma_3'$, (13.66)
q'	kp cm^{-2}, $= \frac{1}{2}(\sigma_1' - \sigma_3')$, 17.3.3
r	cm, length along a pseudo-rupture line, 7.3.1
r, φ, z	cm, 1(1°), cm, cylindrical coordinates, 5.2.1
$r(t)$	–, relaxation function, 4.4.1
$r^*(p)$	–, image of the relaxation function, 4.4.2
s	cm, length along a rupture line, 7.3.1; kp cm^{-2}, sum of pore pressures in adjacent points of the network, 16.2.2
$s = s(\zeta)$	1, stress coefficient, 7.2.2
s_c, s_γ	1, shape factors of the bearing capacity, 23.2
t	–, tangential direction, 2.2; sec, time, 3.1.1
t_0	sec, time at the beginning of a given time interval, 4.2.2
t_r	sec, time to failure, 20.3.2
t_s	sec, consolidation parameter, 11.2.1
u	kp cm^{-2}, pore pressure, pore pressure excess, 6.1, 9.2.1
u_p	kp cm^{-2}, pore-air pressure, 12.1
u_0	kp cm^{-2}, maximum pore pressure in a parabolic isochrone, 11.1.2; pore pressure at time $t = 0$, 14.2.1
u_p	kp cm^{-2}, preconsolidation pore pressure (excess), 6.1
u_w	kp cm^{-2}, pore-water pressure, 12.1
$u_{z,t}$	kp cm^{-2}, pore pressure in the depth z, at the time t, $= u(z,t)$, 9.3.3
u^*	kp cm^{-2}, total pore pressure, 9.2.1
u'	1, pore pressure ratio $= u/\Delta p$, 11.4
$\bar{u}_x$	kp cm^{-2}, equivalent pore pressure defined by the equation $\dfrac{\partial \bar{u}_x}{\partial x} = \dfrac{\partial u}{\partial x} - \gamma_w i_{0x}$, 9.2.1
$\vec{u}(u_x, u_y, u_z)$	cm, displacement vector, 3.1.1
v	1 (1°), angle between the rupture line and horizon, 7.3.1; "stress angle", (13.157); sec^{-1}, consolidation speed $= \dfrac{\partial e}{\partial t}$, 8.3; cm sec^{-1}, velocity of flow, (apparent) seepage speed (related to the gross cross-section area of the soil), 9.2.1
v_e	cm sec^{-1}, average value of the true seepage speed $= v/n$, (9.13)
$\mathring{v}$	cm sec^{-1}, apparent velocity of grains related to the gross cross-section area of the soil, 9.2.2
$\mathring{v}_e$	cm sec^{-1}, average value of the true velocity of grains, (9.24)
$\vec{v}(v_x, v_y, v_z)$	cm sec^{-1}, vector of displacement velocity, 3.1.1
w	cm, displacement component in the direction z, 5.2.1; cm sec^{-1}, (apparent) velocity of the air flow (related to the gross cross-section area of the soil), 12.2
w_e	cm sec^{-1}, average value of the true velocity of the air flow $= \dfrac{w}{n_a}$, 12.2
w_0	cm, vertical displacement of a point lying in the axis of a circular load area, 5.3.1
w_s	cm, displacement in the axis of the circular area caused by a sectorial load, 5.3.2

w_L	1, liquid limit, 9.4.2
w_P	1, plastic limit, 9.4.2
w_∞	cm, vertical displacement (settlement) of the surface of the half space, 5.4.1.2
x	1, $=\frac{z}{H}$, 11.4
x, y, z	cm, orthogonal coordinates, 2.1.1
$\dot{x}, \dot{y}, \dot{z}$	cm sec^{-1}, components of the displacement velocity, 4.1.2
x'', y'', z''	cm sec^{-2}, acceleration components, 2.1.1
z	cm, ordinate, depth, 5.2.1
z_p	cm, $= \overline{2\text{-}1}$, Fig. 7.6
A	cm^2, area, 5.3.1; kp cm^{-2}, modulus of linear compression of the spring "a", 10.1.1.2; kp^{-1} cm^2 sec^{-1}, consolidation parameter, (13.25); 1, pore-pressure coefficient, 14.3.1; kp cm^{-3}, one of the components (X, Y, Z) of body forces per unit of volume, 5.1
A, A_m	kp cm^{-2}, coefficients, 9.3.2
A_2	kp cm^{-2} sec, viscosity parameter, 13.2.2
B	–, rheological symbol for the Bingham body, 4.2.1; kp cm^{-2}, modulus of linear compression of the spring "b", 10.1.1.2
B_2	1, dimensionless viscosity parameter, 13.2.2
C	(kp cm^{-2}), constant, 2.1.2; integration constant, 7.3.1; 1, $= 2{\cdot}3026/C_c$, (8.22)
C_c	1, compression index, (8.18); kp^{-1} cm^2, compressibility of the soil skeleton, 9.1.1; compressibility, 13.1.3
C_s	kp^{-1} cm^2, compressibility of the solid, 9.1.1
C_w	kp^{-1} cm^2, compressibility of the fluid, 9.1.1
C_1, C_2, C_3	kp cm^{-2}, stress constants, 2.1.2
D	cm, polar distance, Fig. 15.3
E	kp cm^{-2}, modulus of elasticity, modulus of linear deformation, deformation modulus, 4.1.4; kp, earth pressure, 7.3.2
E_i	kp cm^{-2}, deformation modulus in the direction "i", 13.1.2
F, F_s	1, safety factor, 17.1.3
F_c	1, factor of safety related to the cohesion part of the shearing resistance, 19.3.2
F_φ	1, factor of safety related to the friction part of the shearing resistance, 19.3.2
G	kp cm^{-2}, modulus of shear deformation, shear modulus, distortional modulus, 4.1.4; kp, weight of an earth (backfill) wedge, 7.3.2
H	–, rheological symbol for the Hookean spring, 4.2.1; 1, thickness ratio, (5.40); Henry's coefficient of the solubility of the air in the water, 12.2; cm, hydraulic (piezometric) head (height), 9.2.1; kp, horizontal load (force), Fig. 23.4
I_c	1, consistency index, 9.4.2
I_P	1, plasticity index, 9.4.2
I_1, I_2, I_3	kp cm^{-2}, kp^2 cm^{-4}, kp^3 cm^{-6}, invariants of the stress tensor, (2.7)–(2.9)

J	1, ratio of time factors, (10.65)
J_1, J_2, J_3	1, invariants of the strain tensor, (3.8)–(3.10)
$J_0(\,), J_1(\,)$	1, Bessel's progression of the first degree, 14.2.1
K	–, rheological symbol for Kelvin's body, 4.2.1; kp cm^{-2}, modulus of compressibility, compression modulus, (4.8); 1, $= \dfrac{C_c}{2{\cdot}3026(1+e)}$, (8.23); principal effective stress ratio, (17.30)
K_a	1, coefficient of the active earth pressure, 7.4.2
K_c	kp cm^{-2}, modulus of compressibility of the soil skeleton, (9.10)
K_0	1, coefficient of the earth pressure at rest, 14.4.3
K_p	1, coefficient of the passive earth pressure
K_s	kp cm^{-2}, modulus of compressibility of the solid, 9.1.1; modulus of secondary consolidation, 11.2.1
$K(t,\tau)$	sec^{-1}, creep kernel, App.
$\bar{K}$	1, apparent earth pressure coefficient, (16.54)
M	–, rheological symbol for the Maxwell body, 4.2.1; point, 5.3.1; 1, shear modulus ratio, 5.4.1.2; sec^{-1}, consolidation factor, 9.3.2.; 1, tangent of the slope of the critical state line = η_{cr}, 13.3.1.1; kp cm^{-2}, deformation modulus, 13.5; kp cm, moment, bending moment, 21.4.1.2
M_v	kp cm^{-2}, modulus of linear compressibility, (18.6)
M^x, M^y, M^z	1, coefficients depending on angles α and φ, 7.3.1
M_R	kp cm, resultant moment, 7.3.1
N	–, rheological symbol for the Newton liquid, 4.2.1; 1, number of sectorial ring elements, 5.3.3; 1, coefficient, 11.1.2; kp, (resultant) normal force, 7.3.1
N^x, N^y, N^z	1, coefficients depending on angles α and φ, 7.3.1
N_γ, N_q, N_c	1, bearing capacity factors, 23.2
0	–, point, origin of the coordinate system, 2.2
P	–, point, 2.1.2; kp, concentrated (resultant, total) load, force, 5.2.2
P_b	kp, total force acting, in Rankine's case, on planes parallel to the slope surface, Fig. 7.10; resultant water pressure acting on a slip-surface element, Fig. 19.1
P_e	kp, total earth pressure at rest in Rankine's case, Fig. 7.10
P_w	kp, resultant water pressure acting on the vertical side of a slice, Fig. 19.1
Q	kp, total load, 5.4.1.2
$Q(t)$	–, creep function, 13.4.2
R	cm, polar distance, Fig. 5.2; 1, consolidation parameter, 11.4; residual factor, (18.28)
$R(t)$	–, relaxation function, 13.4.2
$R_i(t,\tau)$	sec^{-1}, kernel resolvent of the distortional creep, (A.13)
$R_v(t,\tau)$	sec^{-1}, kernel resolvent of the volume creep, (A.13)
S_d	1, structural parameter, 13.1.3
S_r	1, degree of saturation, (12.1)
Sch Sc B	–, rheological symbol for body of Schofield and Scot Blair, 4.3
Schw	–, rheological symbol for the Schwedoff body, 4.3

St V	–, rheological symbol for Saint-Venant's body, 4.2.1
T	kp, (resultant) shear (tangential) force, 7.3.1; kp cm^{-2}, intensity of shear stress, (13.113); 1, time factor $\equiv T_v$, (9.64)
T_k	1, pore-phase time factor, (10.64)
T_v	1, time factor, (9.64)
T_λ	1, time factor of the secondary consolidation, (10.47)
T^x, T^y, T^z	1, coefficients depending on angles α and φ, 7.3.1
U	1, degree of consolidation, consolidation ratio, 9.3.2, 10.1.2.1
U_c	1, degree of consolidation, 15.3
U_T	consolidation degree according to Terzaghi's theory, 10.2.1.2
U_z	1, consolidation ratio, (10.59)
U_z'	1, dissipation ratio, (10.66)
V	cm^3, volume before deformation, 3.3; kp, vertical load (force), 23.2
V_a	cm^3, volume of the pore-air, 12.4.2
V_s	cm^3, volume of the voidless solid, 12.4.2
V'	cm^3, volume after deformation, 3.3
W	kp, gravity force, 7.4.1; kp cm^{-2}, deformation energy per unit of volume
W_d	kp cm^{-2}, distortional energy per unit of volume, 6.2.2
X, Y, Z	kp cm^{-2}, components of body forces per unit of volume, 2.1.1
Y	kp, resultant tangential force in the vertical side of a slip slice, 19.2
α	variable, orthogonal coordinate $= x, y, z$ or ξ, η, ζ, the corresponding subscript, 2.1.1, 3.1.1; 1(1°), angle between the normals of an arbitrary plane and the principal plane, 2.2; boundary value of the angle ϑ, Fig. 5.3; central half-angle of a circular rupture line, Fig. 7.8; angle of the stress obliquity, 7.5.1; $= \frac{\pi}{4} + \frac{\varphi'}{2}$, Fig. 18.5; slope angle of the failure surface, Fig. 19.1; cm, settlement increase in time intervals $\frac{t}{t_1} = 10$, (8.11); 1, coefficient of the pore-pressure dissipation, (16.28); coefficient, (17.26); total strain increase in time intervals $\frac{t}{t_1} = 10$, (17.6)
α_e	1, void ratio decrease in time intervals $\Delta t = t - t_1$, $\frac{t}{t_1} = 10$, (8.12);
α_i	($i = 1, 2, 3$) 1(1°), angle between the octahedral normal and the principal axes, 2.1.2
α_m	1(1°), slope angle of the failure surface, Fig. 18.5
α_p, α_s	kp^{-1} cm^2, constant of the secondary consolidation, (9.70)
α_v	kp cm^{-2}, modulus of volume compressibility of the soil skeleton, (A.13)
α_w	kp cm^{-2}, compression modulus of the pore liquid, (A.7)
α^d	1, deviatoric strain increase during a time interval $\Delta t = 10^{n+1} - 10^n$, $n = \pm 0, 1, 2, \ldots$, 17.1.1

α^0	1, spheric strain increase during a time interval $\Delta t = 10^{n+1} - 10^n$, $n = \pm 0, 1, 2, \ldots$, 17.1.1
$\bar{\alpha}_e$	1, average value of α_e, 11.1.2
$\frac{1}{\alpha}$	1, stress scale, 22.1.3
β	variable, orthogonal coordinate = x, y, z, the corresponding subscript, 2.1.1; 1(1°), slope of the Coulomb failure envelope in the plane of symmetry (σ_1', σ_d')—for compression tests, Fig. 6.4; slope angle, Fig. 7.6; angle between an arbitrary ray and the vertical, Fig. 7.13; $kp^{-1}\,cm^2$, consolidation parameter of partial saturation (12.24); 1, pore pressure coefficient, (A.17)
β_a	$kp^{-1}\,cm^2$, consolidation parameter of partial saturation, (12.19)
β_0	1(1°), slope of the octahedral normal in the plane (σ_1', σ_d'), Fig. 6.4
β^*	1(1°), slope of the Coulomb failure envelope in the plane of symmetry (σ_1', σ_d')—for extension tests, Fig. 6.4
$\bar{\beta}_e$	1, average void-ratio decrease in time intervals $t_j - t_i$ corresponding to the condition $\sqrt{t_j} - \sqrt{t_i} = 1$, Fig. 11.3
$\frac{1}{\beta}$	1, strain scale, 22.1.3
γ	1, shear strain, 3.2; $kp\,cm^{-3}$, unit (volumetric) weight of soil (bulk density), 7.2.1
γ_{oct}	1, octahedral shear strain, (3.22)
γ_w	$kp\,cm^{-3}$, unit weight (bulk density) of the (pore) water, 9.2.1
γ_{xy}	1, shear strain in the (x,y) plane, (3.2)
γ_1, γ_2	$kp^{-1}\,cm^2$, Hookean spring constants, 13.2.1
γ'	$kp\,cm^{-3}$, effective unit weight of the soil (the unit weight reduced by the uplift when it appears), 23.2
$\dot{\gamma}_1$	sec^{-1}, strain velocity of Newton's dashpot, 10.1.2.2
$\dot{\gamma}_{oct}$	sec^{-1}, octahedral shear-strain velocity, 4.1.6
$\frac{1}{\gamma}$	1, time scale, 22.1.3
δ	1(1°) slope angle, Fig. 7.10; angle of wall friction, Fig. 7.8
δ_{rs}	1, Kronecker's delta, (2.25)
$\delta(t)$	–, Dirac's function, 4.4.1
ε	1, linear strain, 3.2
$\varepsilon_x = \varepsilon_{xx}$	1, linear strain in the direction x, 3.1.1
ε_v	1, volume strain, 3.1.1
ε_1, ε_2	1, principal strains, 3.2
$\varepsilon_{\alpha\beta}$	($\alpha = x, y, z$ or r, φ, z; $\beta = x, y, z$ or r, φ, z) 1, $\equiv e_{\alpha\beta}$, 5.2.1
$(\varepsilon_1)_{v0}$	1, undrained axial strain, 13.3.1.1
$(\varepsilon_1)_1$	1, total axial strain ε_1 at the time $t = 1$, 17.1.1
$(\varepsilon_1{}^d)_1$	1, deviatoric axial strain $\varepsilon_1{}^d$ at the time $t = 1$, 17.1.1
$(\varepsilon_1{}^0)_1$	1, spheric axial strain $\varepsilon_1{}^0$ at the time $t = 1$, 17.1.1
$\dot{\varepsilon}$	sec^{-1}, linear strain velocity, 4.1.7
ζ	cm, displacement (settlement) in the direction z, 15.3

ζ_0	1, earth pressure coefficient at rest, 14.1.2
η	1, $= \chi - \psi$, (7.8); kp cm^{-2} sec, coefficient of viscosity, 4.1.5; 1, $= \frac{q}{p}$, Fig. 13.13; 1(1°), rotation angle, 22.1.2
η^d	kp cm^{-2} sec, coefficient of distortional viscosity, 4.1.5
η°	kp cm^{-2} sec coefficient of spheric(al) (isotropic) viscosity, 4.1.5
$\bar{\eta}$	kp cm^{-2} sec, coefficient of (structural) viscosity, (10.51)
ϑ	1(1°), direction angle, polar coordinate, Fig. 5.2; dilatancy angle (angle of the surface energy), 18.5
ϑ_0, ϑ_1	1(1°), boundary slopes of a wedge-shaped plastic domain, 7.2.2
$\bar{\vartheta}_e$	1, average value of the void-ratio decrease, either $\bar{\alpha}_e$ or $\bar{\beta}_e$, 11.1.2
κ	1, dimensionless cohesion intercept, 6.1; void ratio parameter, (13.92)
λ	kp cm^{-2}, the first Lamé coefficient, 4.1.4; kp cm^{-2} sec, coefficient of proportionality in Saint Venant's rheological equation, 4.1.6; kp^{-1} cm^{2} sec^{-1}, viscosity $= \frac{1}{\eta}$, 10.1.1.1; $1 = 1 - \mu$ (μ = degree of consolidation), 11.4; cm^{-1}, variable, (14.38)
λ_e	1, slope of the consolidation line, 13.3.1.1
$\frac{1}{\lambda}$	1, length scale, 22.1.3
μ	kp cm^{-2}, the second Lamé coefficient, 4.1.4; 1(1°), $= \frac{\pi}{4} - \frac{\varphi}{2}$, 7.2.1; sec^{-1}, viscosity factor, (10.8); 1, coefficient of friction on octahedral area = tan ψ, 11.3; 1, degree of consolidation, 11.4; 1, coefficient of reduction, 17.3.2
μ_t	1, factor (12.45)
$\frac{1}{\mu}$	1, permeability scale, 22.1.3
ν	1, Poisson's ratio, 4.1.4
ν_{ik}	1, Poisson's ratio governing the influence of the stress σ_i on the strain ε_k, 13.1.2
ξ	1, $= \chi + \psi$, (7.8); $= \frac{p_0}{p}$, (13.69); kp cm^{-2}, variable (stress), 13.4.1
ξ, η, ζ	cm, rectangular (Cartesian) coordinates, 2.1.1; displacements in the x, y, z directions, 14.1.2
ξ', η', ζ'	cm, Cartesian coordinates after deformation, 3.3
π	= 3·141592
ρ	cm, displacement in radial direction, 5.2.1; settlement (displacement in vertical direction), 8.1; radial distance, 15.1.2; kp cm^{-4} sec^{2}, mass per unit of volume, 2.1.1; density of the air, 12.2
ρ_a, ρ_1, ρ_m	1, coefficients of rheological decrease in strength, (18.17), (18.19);
ρ_t	cm, settlement at the time t, 9.3.2

ρ_∞	cm, final settlement, 5.4.1.2
$\frac{1}{\rho}$	1, unit weight scale, 22.1.3
σ	kp cm^{-2}, normal stress, 2.2; apparent mean normal stress, (7.20)
σ_e	kp cm^{-2}, equivalent consolidation pressure, 6.1
σ_l	kp cm^{-2}, lateral pressure, 10.2.1.2
σ_m	kp cm^{-2}, mean (isotropic, octahedral) normal stress, 13.5
σ_0	kp cm^{-2}, constant value of the stress, 4.4.1
σ_{oct}	kp cm^{-2}, octahedral normal stress, spherical stress, (2.19)
σ_p	kp cm^{-2}, preconsolidation pressure, 6.1
σ_r, $\sigma_t \equiv \sigma_\varphi$, σ_z	kp cm^{-2}, normal stresses in the cylindrical system of co-ordinates, 5.2.1
σ_1, σ_2, σ_3	kp cm^{-2}, principal stresses, 2.1.2
σ^0	kp cm^{-2}, $\equiv \sigma_{oct}$, 16.2.3
σ'	kp cm^{-2}, normal effective stress, effective pressure, 6.1
σ_d'	kp cm^{-2}, $= \sigma_3'\sqrt{2}$, 6.2.1
σ_p'	kp cm^{-2}, effective part of the preconsolidation pressure, 6.1
$\sigma_{10}'(\sigma_{10}'^*)$	kp cm^{-2}, intercept on the axis σ_1' of the Coulomb failure envelope in the plane of symmetry (σ_1', σ_d'), for compression (extension) tests, Fig. 6.4
$\bar{\delta}'$	kp cm^{-2}, average value of the effective normal stress, 11.1.2
τ	kp cm^{-2}, shear (tangential) stress, 2.2; viscous resistance of the clay skeleton, 11.4; sec, time preceding the time t, 4.2.2, 9.2.3
τ_d	kp cm^{-2}, dilatancy component of the shear strength, 18.2
τ_f	kp cm^{-2}, shear strength, shear stress at failure in the failure plane according to Coulomb-Mohr's failure theory, 6.1; peak (value of the shear) strength, 18.5
τ_{oct}	kp cm^{-2}, octahedral shear (tangential) stress, (2.21)
τ_r	kp cm^{-2}, residual (value of the shear) strength, 18.5
τ_φ	kp cm^{-2}, effective friction component of the shear strength, 18.2
τ^x, τ^y, τ^z	1, coefficient depending on angles α and φ, 7.3.1
φ	1 (1°), angle, polar coordinate, 5.3.1; apparent angle of shearing resistance (in terms of total stress), apparent shear angle, 22.1.2
φ_c	1 (1°), slope of the straight line connecting the cohesion intercepts as function of the preconsolidation pressures, 6.1
φ_f	1 (1°), internal friction angle, 18.5
φ_m	1 (1°), mobilized shear angle (angle of the mobilized shearing resistance), 14.4.3; slope of the tangent to the Mohr circle, (18.23)
φ_r	1 (1°), friction angle, shear angle of preconsolidated soils, 6.2; shear angle in Hvorslev's law of shearing resistance, (6.6)
φ'	1 (1°), effective angle of shearing resistance (in terms of effective stress), effective shear angle, 6.1
φ_e'	1 (1°), effective friction angle, (18.11)
φ_r'	1 (1°), slope of the effective "shear strength line", Fig. 18.5; residual effective shear angle, 18.5

χ	1, stress dependent function, (7.7); pore-pressure parameter of partly saturated soils, 12.11; bearing capacity parameter, 23.2
ψ	1 (1°), angle between the direction x or ϑ and the direction of the principal stress σ_1, Fig. 7.1, 7.3; slope angle of the $T = T(\sigma_{oct})$ straight line for Γ = const, Fig. 13.21
ω	1 (1°), angle of the stress obliquity, 2.2; slope angle of the chord of a circular rupture line, Fig. 7.8; slope angle, Fig. 7.13; $= \arccos \frac{h}{D}$, Fig. 15.3; 1, consolidation parameter, (12.26);
ω_b, ω_e, ω_r	1 (1°) angles of stress obliquity, Fig. 7.10
Γ	1, void ratio constant, 13.3.2.1; intensity of shear strains, (13.114)
Δ	Laplace's operator
Θ	$cm^2\ sec^{-1}$, coefficient of consolidation, (10.30)
Π, Π'	–, planes, Fig. 2.3
Φ	1, settlement function $= \Phi(\omega, M, \nu_1, \nu_2)$, 5.4.1.2; numerical value of $\Phi_\nu(\alpha)$, 5.3.2
$\Phi_\nu(\alpha)$	1, function dependent on ν and α, (5.27)
$\Psi(\alpha)$	1, $= 1 - \cos^3 \alpha$, (5.39)
$\Psi(H)$	1, $= \frac{\sigma_z}{q}$, (5.41)
—	–, symbol, for the connection in series of two rheological elements, 4.2.1
‖	–, symbol for the parallel connection of two rheological elements, 4.2.1

Author Index*

*The References and Bibliography are not taken into account in this Index.

Subject Index